SEALS and SEALING HANDBOOK

3rd Edition

Melvin W. Brown

ISBN 0 946395 82 9

Other books in this series include:
Hydraulic Handbook
Pumping Manual
Handbook of Hose, Pipes, Couplings and Fittings
Handbook of Power Cylinders, Valves and Controls
Pneumatic Handbook
Filters and Filtration Handbook
Pump Users' Handbook
Submersible Pumps and their Applications
Centrifugal Pumps
Handbook of Valves, Piping and Pipelines
Handbook of Fluid Flowmetering
Handbook of Noise and Vibration Control
Handbook of Mechanical Power Drives
Industrial Fasteners Handbook
Handbook of Industrial Materials

Published by
ELSEVIER SCIENCE PUBLISHERS LIMITED
Mayfield House, 256 Banbury Road, Oxford OX2 7DH
England.

Printed in Great Britain by Image Plus Limited, East Molesey, Surrey, England.

PREFACE

One of the most vital engineering components in use today, is a seal. To prevent leakage of a fluid between two moving parts, where there is friction and consequent wear, by means of a form of seal, is a standard recurring problem for designers, engineers, managers, plant operators and maintenance professionals.

To prevent leakage of a viscous, corrosive, inflammable or incompatible fluid under high pressure or temperature on continuous running machinery in a sterile, explosive or other critical environment, can be an engineer's recurring nightmare. Whether in automotive, chemical and process, manufacturing industry, oil and gas, plant engineering, hydraulics, pneumatics, marine or power generation; equipment end users demand more power, higher temperatures, increased speed, greater reliability and extended running life. All of which in turn present a mandate for new materials, innovative design, improved quality control and better methods of sealing.

It is for these reasons that the SEALS AND SEALING HANDBOOK has been compiled. This remarkable reference manual contains in one volume, all necessary and essential plus much useful information and data in ready reference format on seals and sealing technology. It is a vital contribution to modern industry and an indispensable part of the engineer's bookshelf.

The Publishers

ACKNOWLEDGEMENTS

AES Engineering Ltd
American Petroleum Institute
Asberit Ltda
Freudenberg Angus Ltd
S. Backmann
L. E. Bayliss
R. F. Battilana
J. Bowler
British Standards Institute
John Crane International
R. Couto
A. W. Chesterton Co
Dowty Seals Ltd
Du Pont de Nemours International SA
Durametallic Corporation
Ferrofluidics Corp
Flexibox International
Fluid Seal Manufacturers Association
A. C. Gregory
G. W. Halliday
B. Halligan
Hall & Hall Ltd
International Standards Organisation
K. Irving
Richard Klinger Ltd
Latty International
Marston Lubricants
D. P. MacDonald
R. Maloney
J. Plumridge
Pioneer Weston
Sandvik AB
Sealmaster Corporation
E.G.&G. Sealol
Dr L. P. Smith
L. Sterling
Shambau Europa (UK) Ltd
G. O. Thompson
Tom-Pac Ltd
James Walker & Co Ltd
N. M. Wallace
R. H. Weeks
Weir Polypac Ltd
Wellworthy
H. J. Whitgreave
Wynn's Precision Canada

CONTENTS

$\Delta p = 1$ bar

ACKNOWLEDGEMENTS - ILLUSTRATIONS AND TABLES

Page Number	Company
4	Du Pont de Nemours International SA
7	Dowty Seals Ltd
18/19/20	Weir Polypac
31	Flexibox International
33/4	Durametallic Corp.
46	Du Pont de Nemours International SA
48/50	Du Pont de Nemours International SA
55/58, Table 4	Du Pont de Nemours International SA
62	Du Pont de Nemours International SA
65, top	Dowty Seals Ltd
80, table	United Kingdom Atomic Energy Authority
91/2	Richard Klinger Ltd
103, Fig 8a	Henry Crossley (Packings) Ltd
103, Fig 8b	James Walker & Co Ltd
104, Fig 8d	James Walker & Co Ltd
105	Richard Klinger Ltd
107	Flexitallic Engineering Ltd
108	Dowty Seals Ltd
109, Fig 12	James Walker & Co Ltd
bottom	Latty International
110, top	Parker Hannifin Group
111	A.W. Chesterton
113/15	Latty International
116	James Walker & Co Ltd
118/9	Dowty Seals Ltd
127/8	Richard Klinger Ltd
133	Loctite UK Ltd
135	Loctite UK Ltd
136	A.W. Chesterton Ltd
137	Bintcliffe Turner Ltd
138, top right	Bintcliffe Turner Ltd
centre	Bintcliffe Turner Ltd
bottom	The British Screw Company
141	Advanced Products NV
142, Fig 2	Wood Bros & Co
143, Figs 4, 5	Dowty Seals Ltd
144, Fig 6	Dowty Seals Ltd
145, top	Parker- Prädifa

Page Number	Company
146	James Walker & Co Ltd
147	Latty International
173, table	Parker Hannifin Corp
187, Figs 15, 16	Freudenberg Angus Ltd
189, Fig 20	Latty International
192, Table 1	Fothergill & Harvey Ltd
198, Fig 3	Dowty Seals Ltd
199	Dowty Seals Ltd
204, Fig 14	Advanced Products NV
208, left	Garlock GmbH
213, top	Parker Seals
bottom	Hallite Seals International Ltd
215, Fig 4	Dowty Seals Ltd
Fig 5	Freudenberg Angus Ltd
216, bottom	Martin Merkel
219, top	James Walker & Co Ltd
bottom	Latty International
223, top	Freudenberg Angus Ltd
centre	James Walker & Co Ltd
224, Fig 17	Microdot Products
225, centre	Parker-Prädifa
bottom	Microdot Products
226, Fig 20	Microdot Products
227, centre	Shamban
230, bottom left	James Walker & Co Ltd
bottom right	Shamban
231, top left	Martin Merkel
top right	Shamban
centre	James Walker & Co Ltd
bottom	Caterpillar Industrial Products Inc
232, top left	James Walker & Co Ltd
centre left	James Walker & Co Ltd
bottom	Martin Merkel
bottom right	BTR Seals
233/4	Weir Polypac
236, Figs 5, 6	Hallite Seals International Ltd
237, Figs 7a, 7b	Microdot Products
238, centre	Parker-Prädifa
Fig 9	Polypac B.A.L.
239, Fig 10	Martin Merkel
bottom	Dowty Seals Ltd
240, top	Latty International
Figs 11, 12	Hallite Seals International Ltd
241, Fig 14	James Walker & Co Ltd

Page	Source
242/3	Tetrafluor Inc
246	Polypac B.A.L.
248, bottom right	Microdot Products
249, Fig 5	Dowty Seals Ltd
250, top	Latty International
centre	Parker-Prädifa
251, Fig 8	Hallite Seals International Ltd
252, Fig 9	BTR Seals
bottom	Dowty Seals Ltd
253, top	Hallite Seals International Ltd
255	Dowty Seals Ltd
264	Latty International
266, centre	A.W. Chesterton Co
bottom	W.L. Gore
273/4	Latty International
275	Tom-Pac Ltd
278	John Crane International
288	Tom-Pac Ltd
296, top	Flexibox International
298, bottom	Flexibox International
303/4	John Crane International
306	Flexibox International
308/9	Flexibox International
310/11	Latty International
313/5	Flexibox International
316, top	Flexibox International
table	John Crane International
317, top	Flexibox International
centre	EG&G Sealol
323, top	John Crane International
324/7	John Crane International
341, centre	Flexibox International
right	John Crane International
342, top	John Crane International
Fig 5	EG&G Sealol
343	Borg Warner
344	EG&G Sealol
345, Fig 7	John Crane International
Fig 8	EG&G Sealol
352/5	AES Engineering Ltd
356/7	Flexibox International
359, top	Deep Sea Seals Ltd
361	EG&G Sealol
362	Deep Sea Seals Ltd
364, top	Deep Sea Seals Ltd
365	Deep Sea Seals Ltd
366/68/69	Deep Sea Seals Ltd
378	John Crane International
381, top	John Crane International
384	Borg Warner
385	Freudenberg Angus Ltd
388	Freudenberg Angus Ltd
389, bottom	Freudenberg Angus Ltd
392, top	James Walker & Co Ltd
Fig 8	Freudenberg Angus Ltd
393, top	Martin Merkel
394	Freudenberg Angus Ltd
396, Fig 12	Dowty Seals Ltd
397, top	Garlock
right	RHP
399, Fig 15	Freudenberg Angus Ltd
centre	John Crane International
bottom	Dowty Seals Ltd
400	Freudenberg Angus Ltd
401, bottom	INA
402/3	Freudenberg Angus Ltd
408, top	Aeroquip UK Ltd
bottom	James Walker & Co Ltd
409, top	Hammes of Steinbach
centre	Freudenberg Angus Ltd
bottom	Martin Merkel
410	Du Pont de Nemours International SA
411, Fig 20	Freudenberg Angus Ltd
412, Table 8	Eriks Allied Polymer Ltd
418, centre	Freudenberg Angus Ltd
table	Martin Merkel
423, left	F.T.L. Corp
Fig 6	Dowty Seals Ltd
424, Fig 8	Freudenberg Angus Ltd
425, top	Weir Polypac
Fig 10	Martin Merkel
427, top	F.T.L. Corp
428	Martin Merkel
433, centre	Latty International
435, bottom	F.T.L. Corp
437, Fig 3	Freudenberg Angus Ltd
438, top	F.T.L. Corp
Figs 5, 6	Freudenberg Angus Ltd
448, Fig 2	Freudenberg Angus Ltd
450	Freudenberg Angus Ltd
452/3	Freudenberg Angus Ltd
461	Henry Crossley (Packings) Ltd
470	B.F. Goodrich
471	EG&G Sealol
474	Ferrofluidic Corp USA
476	Ferrolfuidic Corp USA
478	Ferrofluidic Corp USA
488, Figs 1, 2	F.T.L. Corp
490, Figs 3, 4, 5	GMN-Leidenfrost

SECTION 1

Basic Principles

DEFINITIONS
MECHANICS OF SEALING
SEAL FRICTION
WEAR AND SEAL LIFE
SURFACE TEXTURE
SAFETY AND ENVIRONMENTAL HEALTH

Definitions

A SEAL is basically a device for closing (sealing) a gap or making a joint fluid-tight (the fluid in this case being either a gas or liquid). Seals fall broadly into two categories:

(i) *Static Seals* where sealing takes place between surfaces which do not move relative to one another.

(ii) *Dynamic Seals* where sealing takes place between surfaces which have relative movement, *eg* rotary movement of a shaft relative to a housing, or reciprocating movement of a rod or piston in a cylinder.

A distinction can be drawn between a *seal* and a *joint* in that the former application normally involves the sealing of an annular relative movement – a dynamic seal. A *joint* is used to seal between two permanently or relatively static components or surfaces – a static seal. The seal material so employed may be specifically referred to as a joint, jointing or gasket. A number of dynamic type seals may, however, equally well be employed for static type seals, for which application they may be referred to as seals rather than joints or gaskets.

Another name for a dynamic seal is a *packing* or *gland packing*. This arises from the original method of providing a dynamic seal for a rod, shaft or piston *via* a housing or gland packed with a resilient or semi-resilient material providing a localized contact area offering a physical barrier to leakage. This method is still widely used, and remains virtually the standard for specific applications, although both the variety and performance of the various packing materials available have increased enormously.

There are also types of seals which do not conform exactly to the basic definitions of static seals and dynamic seals. Thus some static seals (by function) may also be designed to accommodate limited movement of the surfaces being sealed, *eg* to accommodate swivelling motions in flexible couplings for pipes. These are sometimes called *semi-static* seals, *flexible (static)* seals, or *pseudo-static* seals. More usually they are simply described by purpose or application.

Similarly there are types of seals designed specifically to prevent access of dirt, dust or other harmful contaminants into a system. They may take the form of bellows or gaiters or similar geometric forms, which are basically semi-static seals although directly capable of accommodating large relative movements. Alternatively they may be forms of ring seals, or basically dynamic seals designed to provide a wiping or scraping action, or merely provide a dirt barrier, rather than work as a seal. Both of these categories may be described as *exclusion* seals, Figure 1.

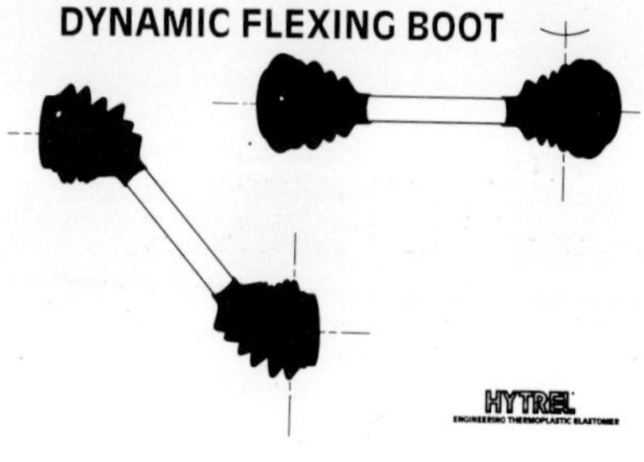

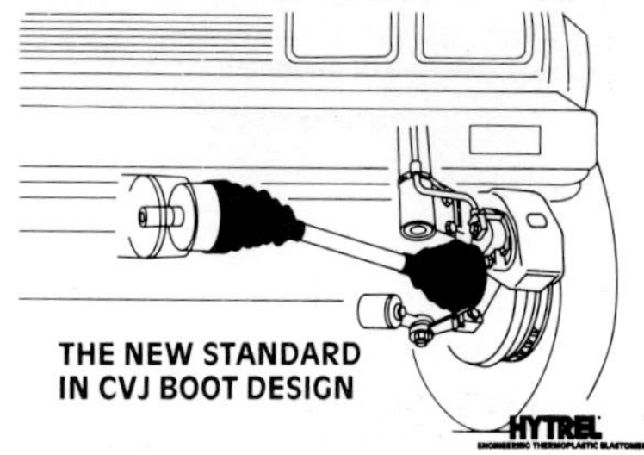

Figure 1
A typical exclusion seal – rack and pinion bellows in Du Pont Hytrel.
Hytrel was selected for resistance to brake fluids and fuel, flex fatigue and impact, flexibility at low temperature (−40°C), hot strength (including tear) cost reduction *vs* other materials, thinner walls and lower weight, rapid cycle, very low scrap.

Seal categories and types

Static seals are normally described by type, *eg* gaskets, *ring* seals, *etc.* Gaskets may be further categorized by material as: non-metallic (fibrous), elastomeric, metallic, semi-metallic, *etc;* or by construction: flat gaskets, spiral-wound gaskets, *etc.* Sealants are regarded as a separate category, also separately described as liquid gaskets, liquid jointing, thread sealants, *etc.*

Dynamic seals fall into two main categories:

(i) *Contact* seals where the seal bears against its mating surface under positive pressure.

(ii) *Clearance* seals which operate with positive clearance (*ie* no rubbing contact).

The majority of dynamic seal types in general use come in the former category, operating with rubbing contact separated only (and lubricated) by a thin oil film. They fall into two distinct classifications – *compression* seals and *pressure-energized* seals. Compression seals generate radial pressure for sealing by 'squeeze' imparted to a soft gland material or packing ring(s) when the gland is tightened, expanding the material radially – Figure 2. The packing material is usually in the form of a cut length of suitable section inserted into the gland, forming in effect a series of cut rings which are then compressed by tightening the gland nuts to close the effective gland length and apply radial pressure between the face of the packing and the surface over which it slides. Compression seals of this type are true packings and may be employed for both dynamic and static applications, both as reciprocating and rotary seals.

They are also suitable for sealing in either direction of (reciprocating) motion and are thus double-acting seals. Their particular application is for heavy duty (reciprocating) rod seals, or rotary shaft seals.

There are other types of compression seals which employ rigid ring sections which generate their contact pressure by being sprung into place (*eg* piston rings); or by applied spring pressure (*eg* metallic split-ring seals).

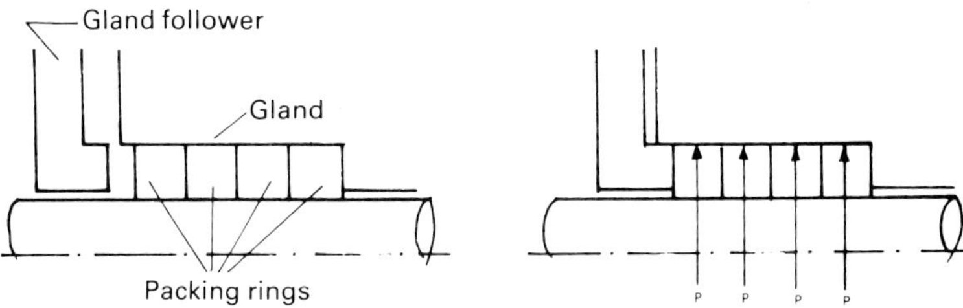

Figure 2

Pressure-energized dynamic seals fall into two categories. The first comprises *solid elastomeric* rings (such as *O-rings* and *rectangular rings*) which are assembled in grooves with an interference fit. This imparts a 'squeeze' or preload pressure, providing sealing in the static condition (they may also be used as purely static seals). Under fluid pressure acting on one side of the seal through the clearance gap the elastomeric section is deformed, increasing the interface pressure by an amount equal to the fluid pressure. Thus, if the preload pressure is p and the fluid pressure is P, the effective interface pressure under 'working' conditions is P + p (Figure 3). Because this is greater than the actual fluid pressure (P), sealing is maintained. This holds good as long as the seal does not extrude into the clearance gap: if the clearance gap is too great, a close clearance flat spacer washer should be fitted. Seals of this type may be *double-acting, ie* capable of sealing in either direction with reciprocating motions.

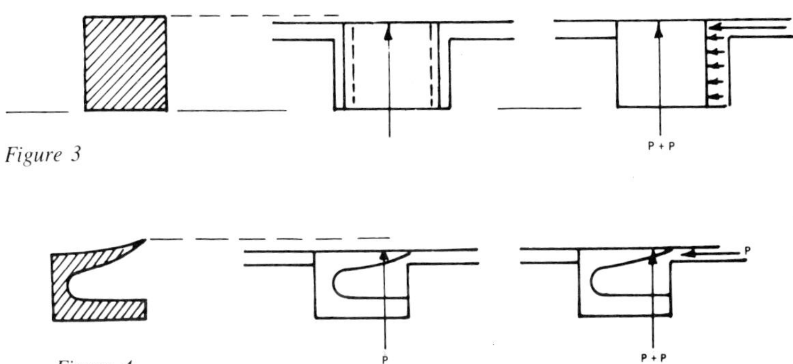

Figure 3

Figure 4

The other category of pressure-energized seal employs a 'hollow' section with a flexible lip or lips (Figure 4). Again it is assembled with an interference fit giving a preload pressure p. Fluid pressure (P) acting on the section then further increases the interface pressure to P + p. Typical seal sections in this category are the *U-ring, V-ring* and its variants; and sections with a single flexible lip, such as the *C-ring* and its derivatives. Seal rings of this type are *single-acting, ie* sealing by pressure energization is effective in one direction only.

There are further types of ring seals which combine both modes of working, *ie* incorporate a flexible lip section fitted with a solid elastomeric ring. Some of these may also incorporate back-to-back configuration to produce double-acting seals.

Ring seals may be asymmetric in the sense that they have only one 'working' side, *eg* they will work only as a *rod* seal or a *piston* seal, and may be referred to as such. These are sometimes referred to as *internal* or *external* seals, respectively (Figure 5).

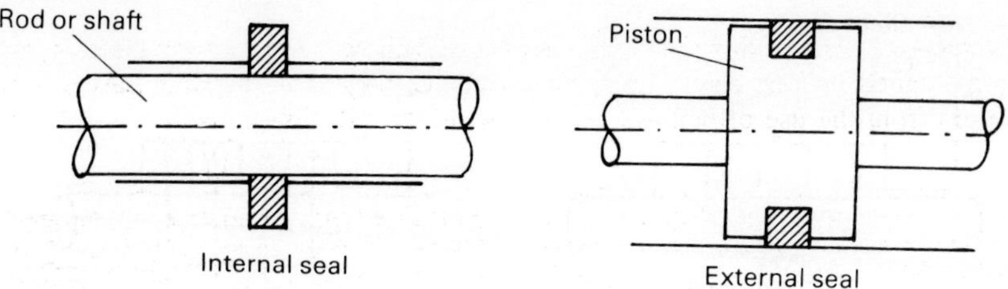

Figure 5

This particular description is best avoided as the same description may be applied to physical positioning of a seal relative to a particular component, *ie* it may be positioned internally or externally. Note also that some forms of rod seals will work equally well as piston seals; and *vice versa.* In other words, they are symmetric in this sense.

Yet another type of dynamic (rotary) seal is the *narrow lip* seal encircled by a garter coil spring. Again it is assembled with an interference fit, preload pressure being produced in this case by stretching the spring. This type of seal has low friction (because of the narrow lip size) and relatively limited pressure sealing capability (increasing the preload pressure *via* a stronger garter spring would cause excessive wear of the narrow lip). It is used exclusively as an *oil* seal for rotating shafts, which is the best description for the type. It is, however, commonly called a *radial lip* seal, which is not a clear-cut definition because many other types of ring seals with lips can work as radial shaft seals.

Face seals

In its simplest form a *face* seal comprises an elastomeric diaphragm section supported by a metal casing or metal inserts rubbing against a shoulder or flange on the shaft, pressure at the interface being maintained by a spring or series of springs (Figure 6). The sealing surface thus offered is substantially greater in area than that of a lip seal, resulting in a positive seal and also one which is less likely to be affected by the presence of dirt because the sealing surface cannot lift and tends to exclude dirt. Being spring loaded the sealing face is also automatically compensated for wear, whilst wear itself is directly proportional to the face loading and dependent on the rubbing materials involved. With a suitable choice of materials, such as a hard wearing sealing ring bonded to the diaphragm in contact with a rubbing face of cast iron, phosphor bronze or stainless steel, wear can be reduced to negligible proportions.

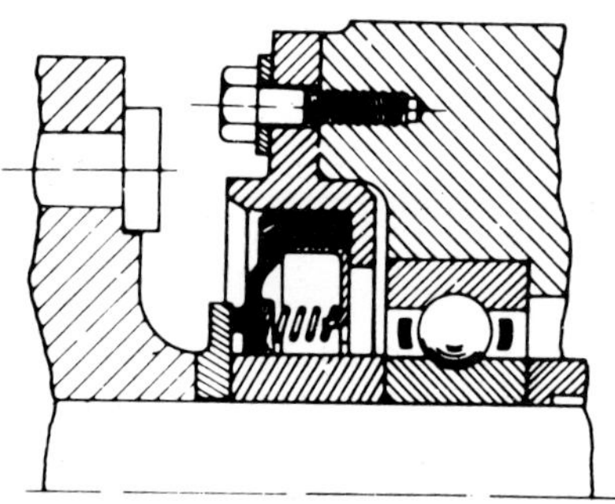

Figure 6
Simple face seals.

Simple face seals of this type may incorporate bellows as the spring element, when they are called *bellows* seals. Note, however, that these are dynamic seals and quite different from the use of bellows as static seals or exclusion devices.

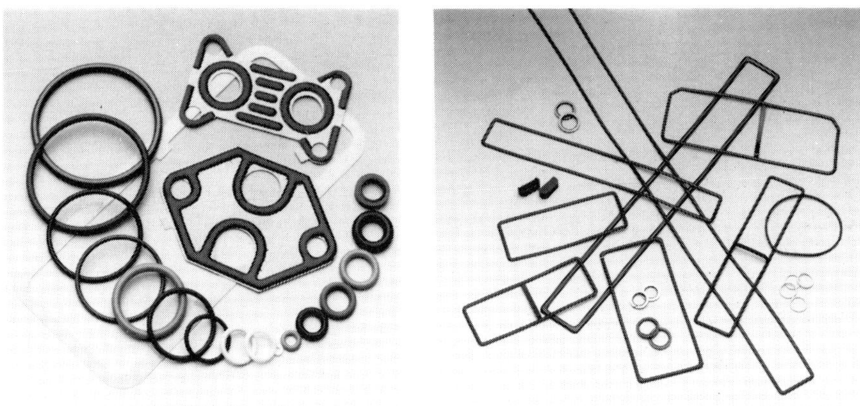

Mechanical face seals

Rather more complicated forms of spring-loaded face seals used for more arduous duties are called *mechanical seals.* These are, again, face seals but employ two rigid mating rings, one static and one rotating, held in close contact by spring pressure. These rings are of low friction materials (or a low friction combination) and thus can, if necessary, operate at high rotational speeds and rubbing velocities without excessive friction, wear or overheating. They can also be flooded with coolant, if necessary, when handling high temperature fluids.

See also *Glossary,* Section 9.

Mechanics of Sealing

STATIC SEALS aim at providing a complete physical barrier in a potential leakage path to which they are applied. They are *zero-leakage* seals. To achieve this the seal must be resilient enough to flow into and fill any irregularities in the surfaces being sealed and at the same time remain rigid enough to resist extrusion into the clearance gap between the surfaces under the full system pressure being sealed. Both requirements must be long term. Resilient flow is produced by closure loading, stressing the seal in compression. Contact pressure is then maintained by stored elastic strain energy in the complete seal system. Performance will be degraded by any stress relaxation which may occur in the system. This may be caused by stress relaxation in the seal material itself (which may also be associated with creep into the clearance space); differential thermal expansion; or in the case of gaskets, flange deflection or bolt stretch.

All elastomeric ring seals require the seal material to have an interference fit with one of the mating parts on assembly. Thus in the case of a solid rubber ring, such as an O-ring or rectangular ring, the material can be in direct compression or tension, or part compression and part tension. Similarly, a *flexible lip* seal can be in compression or tension, depending whether the section is sealing on the o.d. or the i.d.

Thus, irrespective of the type of seal, a load is generated between the seal contact point and the mating area. This interface load depends on the amount of interference or 'squeeze' produced when the seal is assembled, together with the modulus of elasticity of the material. On *lip type* seals the interface load can vary with lip radial thickness, length of the flexible leg(s) and the modulus of elasticity of the material. The distribution of the interface load depends on the section geometry, *eg* Figure 1. Such load diagrams give a general indication of the friction and leakage characteristics. For dynamic seals the load generated by the seal must be multiplied by the coefficient of friction to obtain the dynamic load to move the seal assembly, which is effectively a power loss.

In theory at least, the greater the stored elastic strain energy of the seal and thus the greater the margin available to resist any relaxation effects in service. This is largely true up to the point where the seal material itself is damaged by excessive compression and suffers a permanent loss of properties. This is more likely to occur with gaskets than other types of static seals, although there are again specific limits to which elastomeric materials may be prestressed since they are themselves incompressible (*ie* they are capable of accommodating distortion, but not volume reduction).

The requirements of a dynamic seal are conflicting and call for a compromise. In the case of contact seals, good sealing calls for a satisfactorily high contact pressure on the

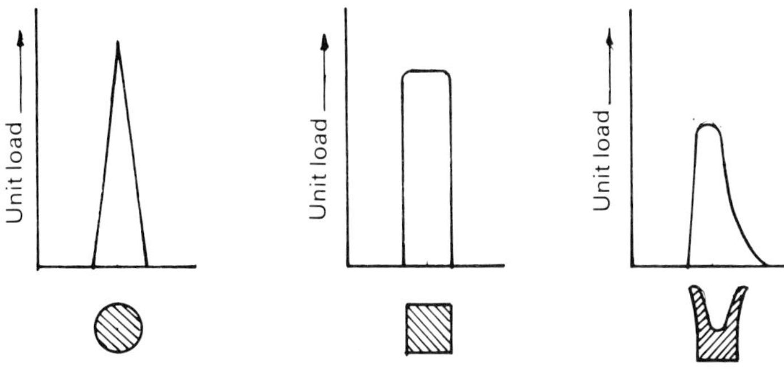

Figure 1

surface being sealed (depending on the pressure to be sealed); and at the same time the seal needs to have minimum possible friction and rubbing wear.

With a compression type seal a high preload, and thus high friction, are inevitable. The latter can be relieved to a certain extent by the choice of packing or seal material, and also adjusted by tightening the gland in the case of a packing to achieve minimum compression consistent with an acceptable degree of sealing, at the same time accepting a certain amount of leakage. Periodic readjustment will then be required to take up wear which may occur on the face of the packing. This is a normal requirement for most types of packed glands.

With a compression seal, seal friction is relatively high, regardless of the actual pressure. Thus if a gland is adjusted to seal at a particular high pressure, the compression (and thus friction) will be higher than necessary when operating against lower pressures. Static friction, and thus breakout force, will also be higher than running friction, although this difference may be negligible with PTFE impregnated or coated packings. The other main disadvantage of a compression seal for dynamic applications is that it is usually much bulkier than a pressure-energized seal.

There are, however, distinct applications where one or the other is a preferred type. Thus for rotary motions, particularly of a heavy duty nature, the packed gland generally offers greater durability and may be the only type in which suitable materials can be employed to resist the high temperatures or severe service conditions involved. Flexible (pressure-energized) seals or seal sets, on the other hand, can generally offer a superior performance for reciprocating duties, even under arduous conditions (within the range of the maximum service temperature of the materials employed), with appreciably less friction and wear. All flexible seal sections, however, may be subject to definite limitations as to the pressure they can withstand without extrusion or excessive deformation and, due to the immense variety of different designs available, need individual study as regards application and installation. Figure 2 is a general guide to the pressure and speed ranges for which various types of dynamic seals are suited.

With almost all types of dynamic seals – compression or pressure-energized – lubrication of the seal face plays an important part in determining seal performance and seal life. The possible exception is where the slip surface is PTFE, as this has an extremely low coefficient of friction when sliding on most other surfaces and can rub with negligible wear without lubrication. In 'wet' applications the fluid being sealed may itself provide effective lubrication. For sealing dry gases or aqueous solutions or steam, where the fluid itself is not a lubricant, it may be necessary to provide the seal face with a source

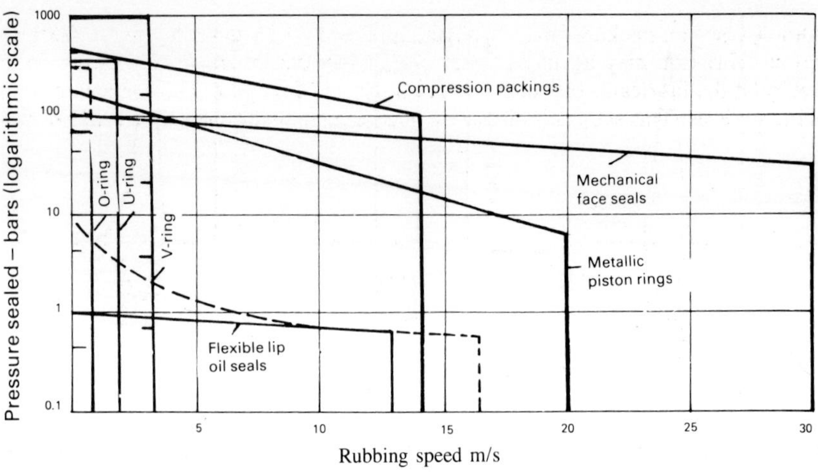

Figure 2

of lubricant or impregnate the seal material with a lubricant. This may even call for a special design of seal, or materially reduce the choice of a suitable type of seal or packing.

Compression packings may be designed to operate on dry surfaces, in which case the lubricant is contained within the packing material itself. Simple compression sealing rings and pressure-energized seals or seal sets are normally designed to operate in a fully lubricated condition. In either case the seal itself rides on a thin film of lubricant which provides the final sealing barrier, retained in position by the surface tension of the film (Figure 3).

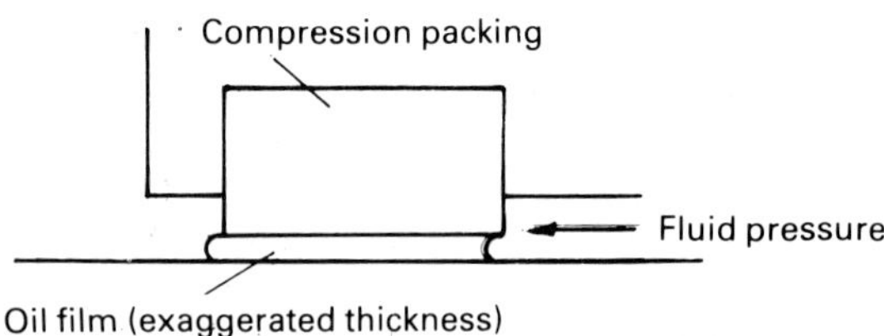

Figure 3

The thickness of this film is critical. If too small it can be bridged by surface irregularities, producing high friction and rapid seal wear. If too thick, the meniscus will break down, producing a high leakage rate. In practice, the seal will not be perfect under dynamic conditions (*ie* it will not be a zero-leakage seal without excessive pre-load pressure), and its performance will also depend on load, speed and fluid viscosity.'

When at rest the lubricant film will tend to become squeezed out to a thickness of less than 0.25 μmm (0.0001 in) under pre-load pressure, leading to relatively high 'break-out' friction because only boundary lubrication is present. As speed increases, full hydrodynamic lubrication is established, with minimum friction (Figure 4). Under these conditions the film thickness may be anything between 0.25 and 2.5 μmm (0.0001 and 0.001 in). Friction may again rise with further increase in speed, depending on the viscosity of the lubricant, the form and detail design of the seal, and the nature of the rubbing surfaces. This subject is discussed in more detail in the chapter on *Seal Friction.*

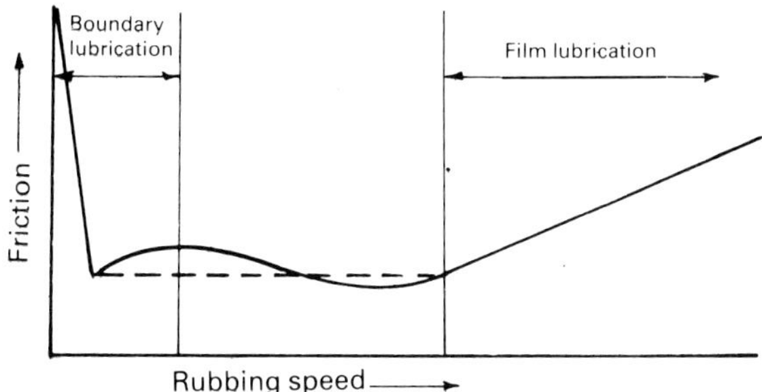

Figure 4

Leakage

It is possible with most compression seals to reduce leakage to near zero or even zero (depending on both the viscosity of the fluid and the pressure being sealed), simply by increasing the pre-load (compression) pressure – *ie* virtually turning them into static seals with rubbing contact. With, at best, only boundary lubrication then being present, they are not generally suitable for working as practical dynamic seals. Equally, with pressure-energized seals or face seals, which normally leak anyway under dynamic working, leakage can be reduced by increasing their contact pressure, but only at the expense of higher friction and shorter and less reliable seal life. A successful design (or suitable choice) of seal is therefore one that provides the required degree of pressure sealing at acceptable friction and wear levels, and with acceptable leakage rates.

In the case of *reciprocating* seals a further loss can be caused by fluid transport. Thus, in the case of a *reciprocating rod* seal, the emergent rod will carry with it an oil film, which is not necessarily an indication of leakage but rather a positive sign that the seal face is receiving lubrication. This film will normally be withdrawn past the seal again on the in-stroke of the rod, unless the seal has a pronounced 'wiping' action, as would be the case when using a compression type seal.

The true leakage in such cases is the amount of oil or fluid which actually falls off the rod (or progressively accumulates on the 'dry' side of the seal). In the case of a compression seal this may need to be a continuous positive leakage to ensure adequate and continued lubrication of the seal face, unless the packing is self-lubricating when compression can be increased to stop the leak. In the case of a flexible seal, the true leakage represents the difference between the emergent film and the re-entry film as actually deposited, and may well be negligible. If apparently excessive, this may be due to a leaking seal, or it could be caused by an excessive wiping action on the re-entry stroke, such as, for example, by fitting a wiper ring which is wiping or scraping too closely. It is possible, in fact, to have a wiper ring in combination with a main seal ring act as a pump, the wiper removing the bulk of the quite normal emergent film on the rod on each in-going stroke, although the wiping action has to be prettty severe for this to be achieved.(Figure 5).

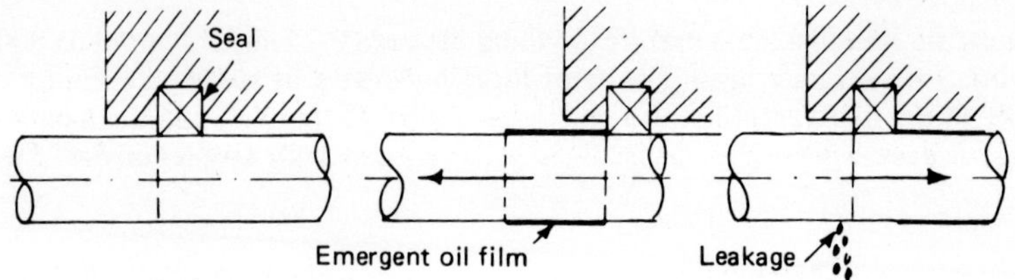

Figure 5

The main points of practical interest to emerge from this are: (i) that there is a distinction between apparent leakage and true leakage, particularly in the case of lip-type seals; (ii) the presence of oil on the surface on the 'dry' side of a seal is normal and desirable (as indicating adequate lubrication); and (iii) perfectly dry seal operation may well indicate lack of lubrication and the likelihood of early seal wear or failure. The presence of an oil film on an emergent rod as in (ii) is, of course, a reason why wipers may also be required and is desirable in an atmosphere contaminated with dust or dirt. A wiper will remove such particles as may cling to the oil film on the re-entry stroke and prevent them being carried back into the seal with the re-entry film.

On typical hydraulic systems fluid loss by oil transport is insignificant. Normal transport of oil, depending on the design of the seal, should lie between 0.001 and 0.004 cm^3 per square metre of seal surface area (0.0001 in^3 per square yard of seal surface area); seal surface area being the stroke length multiplied by the sealing circumference (or π times the seal diameter).

Here it should be noted that with pressure-energized seal types operating in an *unpressurized* state, fluid transport loss can be very much higher than with the same seals operating in a pressurized state. Also double-acting seals (*eg* piston seals) can promote higher oil transport losses than single-acting seals.

Seal Friction

THE FRICTION between a dynamic seal and the sealing surface depends on a number of factors such as seal design and material, fluid and fluid pressure, temperature, rubbing speed and surface finish. The frictional load resulting may not be significant in many applications (except in pneumatic cylinders and valves where minimum friction is desirable for optimum performance), but friction itself can be harmful in generating heat which can cause degradation to the seal material and lubricating film, and/or increase leakage by lowering fluid viscosity. Degradation is more significant since this can yield abrasive products further contributing to friction and wear.

Seal performance in this respect is difficult to analyze in general terms since a number of empirical factors are involved, specific to the design of seal. However, as a basis, friction is obviously proportional to a function of the actual contact pressure, although the actual coefficient of friction involved may vary with speed, time, *etc,* as well as material and surface finish.

In the absence of wedging, seal friction is proportional to the effective contact pressure, *ie:*

Seal friction $= \mu \ P_e \times a \ b$

where

μ = coefficient of friction

P_e = effective contact pressure

a = seal face contact width (inches)

b = seal face contact length (inches)

= πd for a circular seal, where d is the contacting diameter

The coefficient of friction will be characteristic of the materials involved in rubbing contact, and also the lubricant present. It will, however, also vary with speed and so a more complete general formula is:

Seal friction $= \mu V \ P_e \times a \ b$

where = rubbing speed

The use of such a formula is strictly limited, particularly as the assembly or 'interference' pressure is normally unknown and thus the effective pressure (P_e) largely indeterminate. (P_e is equal to the 'interference' pressure plus the fluid pressure in the case of pressure-energized seals). The following semi-empirical formula is possibly more useful, although it ignores such factors as rubbing speed and surface finish.

Seal friction $= K\ \mu P_e \times D$
where K is an empirical factor specific to the design of seal installed and working under design conditions and D is the seal diameter

In the case of a true compression seal the section is substantially rigid when compressed, wedging is unlikely and friction will normally be largely independent of internal pressure. The effective pressure (P_e) can thus be taken as the actual pressure of compression, although this will be difficult or impossible to determine. The coefficient of friction (μ) may also be unknown, and variable, although it may be possible to estimate this with a reasonable degree of accuracy; this is discussed later. Specific solutions are difficult to obtain, therefore, unless evaluated on empirical lines, or on the basis of comparative data. Thus formulas can only be used directly to investigate possible differences in performance and friction on compression seals of the same type and material, but different size.

Friction coefficients

The dry friction of typical seal materials rubbing on smooth, dry seal surfaces may be anything from $\mu=0.4$ to 1.0. For lubricated surfaces the range is very much lower, for example, $\mu=0.02$ to 0.10. This is particularly true in the case of elastomers. Fabric materials and impregnated fabrics show similar values of μ, but usually with rather less variation, for example, $\mu=0.04$ to 0.08 for lubricated conditions. Lubricated leather normally has very low friction, but values of μ for leather seals are often higher than those for elastomers, commonly because leather rings are used in conjunction with rougher rubbing surfaces.

In general, the harder the material the higher the friction, and the softer the material the lower the friction, although this only holds good at low pressures. It is more applicable to elastomers, where hardness can vary, than to comparisons with other seal materials, where the frictional coefficient does not necessarily correspond with the hardness of the material.

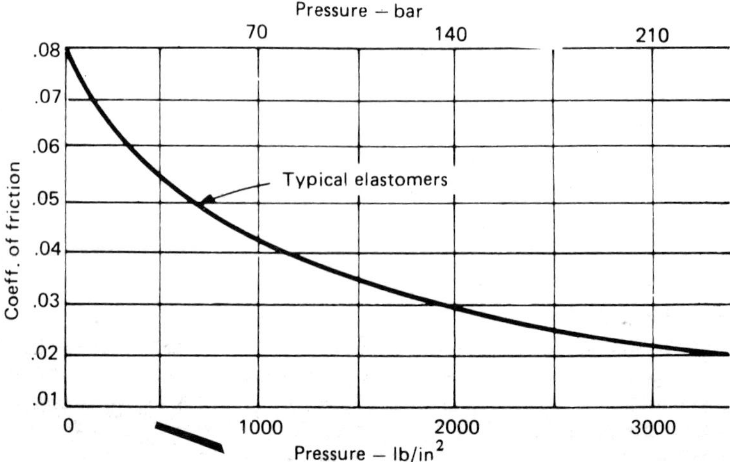

Figure 1

The coefficient of friction is also a function of pressure, although the actual relationship is not clearly established. Basically friction will be highest at low pressures, with a minimum value achieved at some high pressure (Figure 1). This can vary with different forms of seal sections as it is dependent on section deformation.

The variation of friction with pressure is also dependent on the surface finish, and particularly the production method in the case of cylinders and piston seals. Figure 2 shows typical differences in friction with three different cylinder finishes. The more rapid increase in friction with increasing working pressure is marked with the rougher surface and texture of a cold hammered finish, compared with honed or burnished tubes. Conventional cylinder finishes are produced by honing, resulting in a precisely-controlled surface, having a roughness average of between 0.25 μmm and 0.625μmm (0.0001 in to 0.00025 in). The biggest problem for the seal designer, however, has been caused by the recent tendency to use hydraulic cylinders produced directly from 'as-drawn' tube, without subsequent finishing treatment.

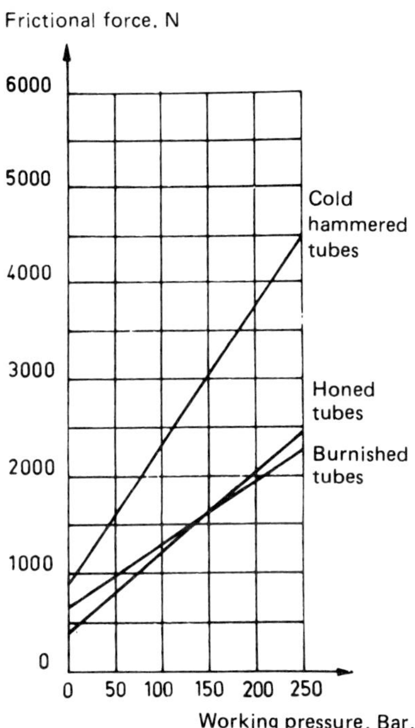

Figure 2

Friction and speed

The variation of friction with rubbing speed is more clearly defined and follows three stages (Figure 3). Static friction is normally high, but once 'break-out' has been initiated the frictional coefficient falls to a low value at low speeds and increases with increasing speed up to a first peak. From then on, with further increase in speed the frictional coefficient falls to a minimum value and then rises again as the speed continues to increase. This is a general presentation of the friction/speed relationship and can be modified by other conditions and by different seal materials and construction. The friction of an O-ring seal, for example, is very high at low rubbing speeds; whilst seals with PTFE rubbing faces have low 'stiction' and a more substantially constant coefficient of friction.

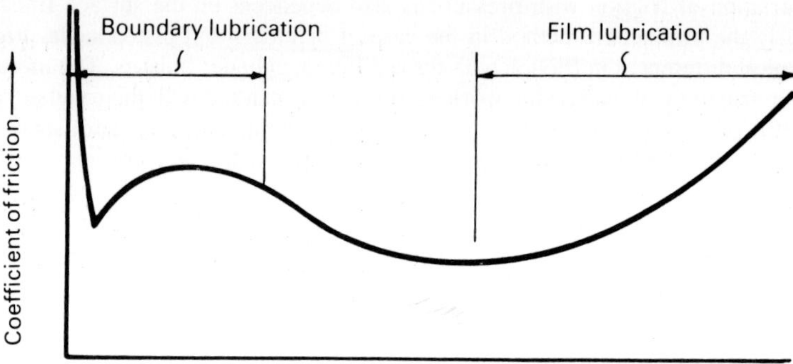

Figure 3

Effect of idle time

The effect of standing time on friction is shown diagrammatically in Figure 4. The increase of frictional coefficient with standing time is quite rapid, contributing to 'stiction'. The effect is accelerated by conditions favourable to the drying out of lubricant on the seal face and under such conditions the friction/time curve is asymptotic to the dry friction value. The dry friction coefficient can be up to ten times that of the same seal operating under lubricated conditions (except where the rubbing surface is PTFE).

Favourable factors to minimize the adverse effects of idle time are, primarily, to ensure that the seal cannot dry out (particularly in the case of elastomeric seals); and also to have a good finish on the rubbing surface to inhibit any tendency for the deformable seal material to adhere to the raw metal surface.

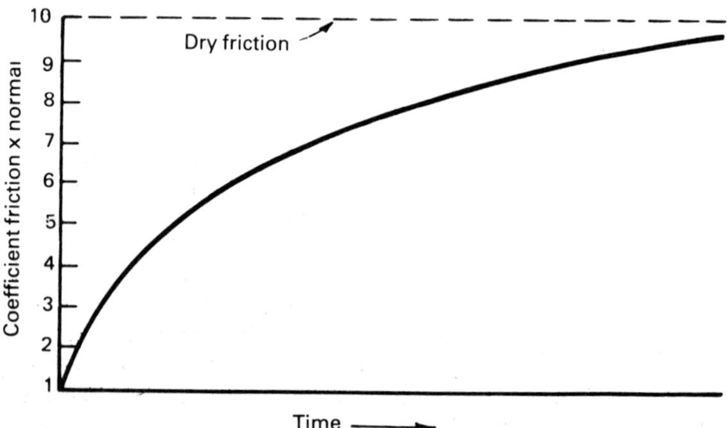

Figure 4

At high operating pressures the starting friction can be critical, *eg* it can limit the reserve force available in a working hydraulic or pneumatic cylinder. Logically, therefore, seals should be chosen which give the smallest possible friction under these circumstances.

Pressure-energized seals

In the case of a flexible seal of the pressure-energized type the effective pressure is equal to the sum of the preload pressure and the fluid pressure. The higher the fluid pressure the less significant the value of the preload pressure, and it may be ignored in many cases for high pressure evaluation. If required for low pressure calculation, the value of preload pressure is best determined by direct test.

Pressure-energized seals are prone to extrusion and wedging at higher pressures, particularly if the clearance gap is generous. Should wedging occur, friction will be increased considerably and is substantially proportional to the square of the effective pressure. The basic friction formula then becomes:

$$\text{Seal friction} = \mu V\ (p\ 2P)^2\ \pi\ a\ b$$
$$\text{or} = K\mu V\ (p + P)^2 \times D$$

where p = preload interference pressure bar (lb/in^2)
P = fluid pressure bar (lb/in^2)
V = rubbing velocity m/sec (ft/min)
K = a constant depending on seal type

There are no specific rules as to when wedging may occur as this will depend on the form of the seal as well as the internal pressure and working conditions. Thus the tendency for the seal to wedge will be enhanced by a rough surface, lack of lubrication and high reciprocating speeds. In general, wedging is unlikely under any circumstance within the normal operating range of the seal if the clearance gap is small enough – not more than 0.055 to 0.127 mm (0.002 to 0.005 in). For clearance gaps greater than 0.25 mm (0.010 in) there is always the possibility of wedging. Thus if a preliminary calculation based on friction being proportional to pressure yields a low result and the clearance gap is generous, the second formula is probably more appropriate to the operating conditions.

Friction of reciprocating seals

The frictional coefficient of reciprocating seals will vary primarily with seal design, and also with pressure. The frictional loss will vary as will the friction coefficient and rubbing speed. Both have their significance, *ie* frictional coefficient in selecting a type of seal, and friction loss in deriving mechanical performance loss under operating conditions. Reliable data in either case can only be determined empirically.

Frictional coefficients for typical U-rings, V-ring sets and three proprietary seal designs are given in Figure 5, plotted against pressure. This can be compared with the friction loss expressed as a percentage of the theoretical pulling power when used as rod seals (Figure 6a) and piston seals (Figure 6b) in a cylinder.

Friction increases with increasing operating pressure, but the ratio of friction loss to the theoretical pulling power of the cylinder is reduced. This means that the friction coefficient (μ) is smaller, because the liquid pressure in the lubricant film increases, and the roughness of the sealing surface is therefore filled more effectively with lubricant.

Friction generates heat, and heat affects the sealing material at temperatures above +50°C. At such temperatures the pressurizing medium, usually oil, may cause the seal material to swell or lose its resistance or hardness. The friction is increased and the possibility of permanent damage to the seal is greater.

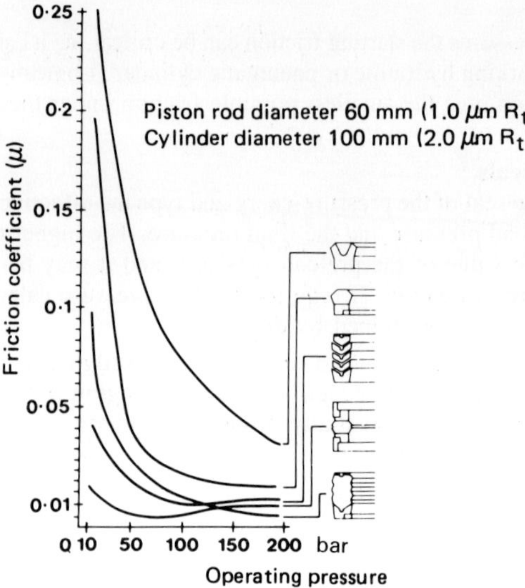

Figure 5
Friction coefficient (μ) for various seal designs.

At high temperatures, oil 'ages'. Oxidation occurs, with a reduction in lubricating properties. The result is further damage to the seal material.

Friction loss of some seals plotted against speed are shown in Figures 7a and 7b. With all reciprocating seals friction rises sharply when the speed falls below 0.05 to 0.15 m/s

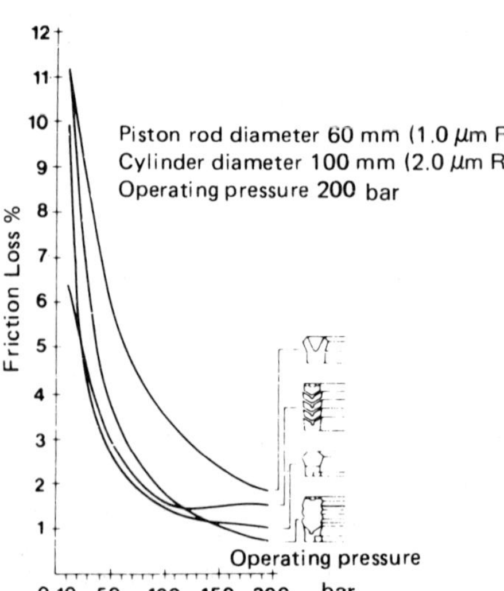

Figure 6a
Friction loss for piston rod seals as a % of the theoretical pulling power of the cylinder.

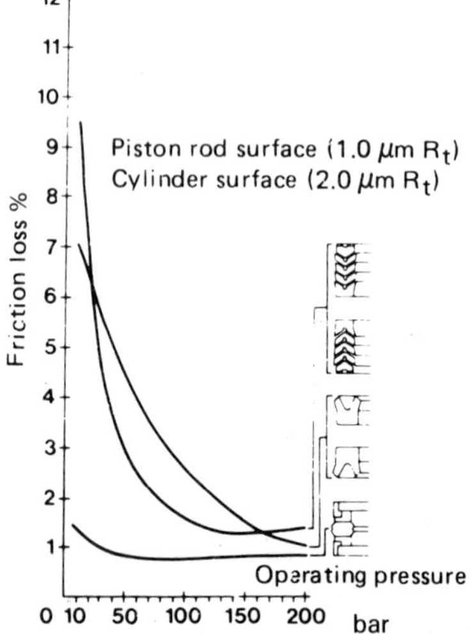

Figure 6b
Friction loss with piston seals as a % of the theoretical pulling power of the cylinder.

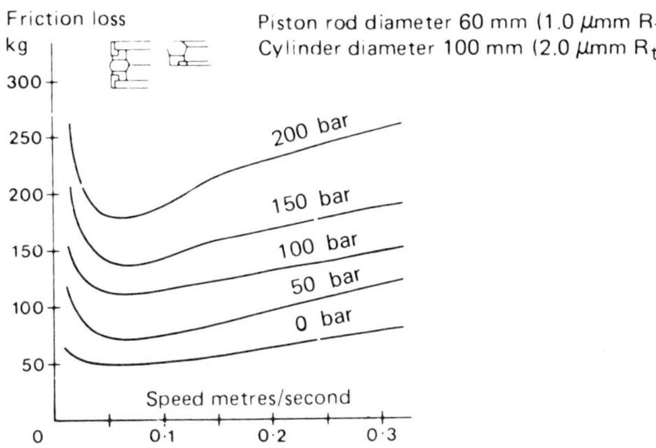

Figure 7a
Frictional loss of Polypac Balsele piston seal type D11W and piston rod seal type S11E, at various operating pressures.

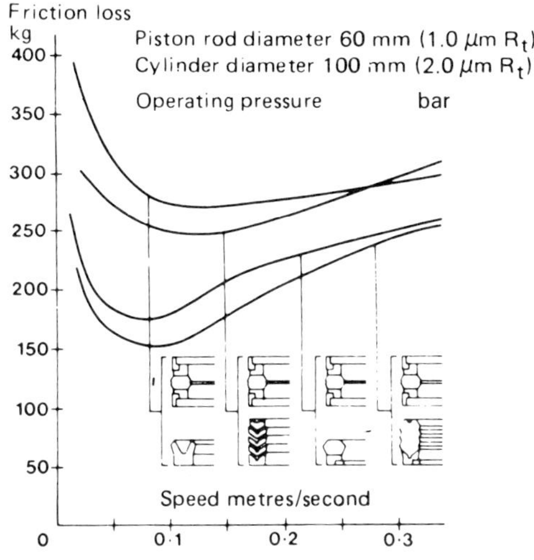

Figure 7b
Friction loss of a cylinder with Balsele piston seal type D11W and piston rod seal type S11E or Selemaster, Veepac and U-rings from polyurethane.

(10 to 30 ft/min); but it also increases at high speeds when the oil film cannot build up to lubricate the sealing surfaces. This is particularly so where a seal surface comes into contact with a dry surface, for example when a piston rod enters the cylinder. However, high speeds normally occur only during movement without load, when the operating pressure, and therefore the pressure at the sealing surface, is small. Typical frictional values dependent on cylinder tube production finish are shown in Figure 8.

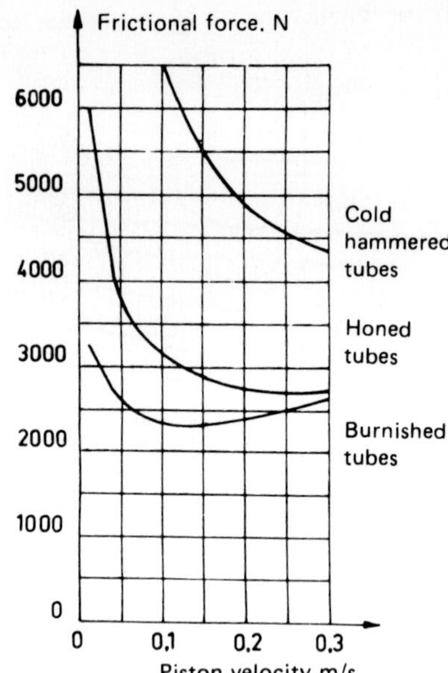

Figure 8
Frictional force against piston velocity for different types of surface at a working pressure of 250 bar.

It is difficult to evaluate the maximum speed of an *hydraulic* seal as it is dependent upon many factors. The most important are: direction of movement, against a dry or wet surface, operating pressure, viscosity of the pressurizing liquid and the wear resistance of the seal material.

The operating speeds of hydraulic cylinders are usually of the order 0.1 to 0.3 m/sec (0.3 to 0.9 ft/s). With pneumatic cylinders, operating speeds are usually much higher.

Friction of O-ring seals

The coefficient of friction of O-rings used as dynamic seals is relatively insignificant compared with other factors which are adjustable to control friction, *eg* see Table 1.

TABLE 1 – VARIABLE FACTORS AFFECTING O-RING FRICTION

To Decrease Friction	Factor	To Increase Friction
Decrease squeeze	**Unit load (squeeze)**	Increase squeeze
Decrease R_a	**Surface finish (metal)**	Increase R_a
Decrease hardness	**Hardness**	Increase hardness
Increase speed	**Speed of motion**	Decrease speed
Smaller section	**Cross section of O-ring**	Larger section
Decrease pressure	**Pressure**	Increase pressure
Lubricate	**Lubrication**	Omit lubrication
Increase temperature	**Temperature**	Lower temperature
Increase width	**Groove width**	Decrease width
Decrease diameter	**Diameter of bore or rod**	Increase diameter
Increase smoothness	**Smoothness (O-ring)**	Reduce smoothness
Compress O-ring	**Joule effect***	Stretch O-ring

*Applies to rotary seals

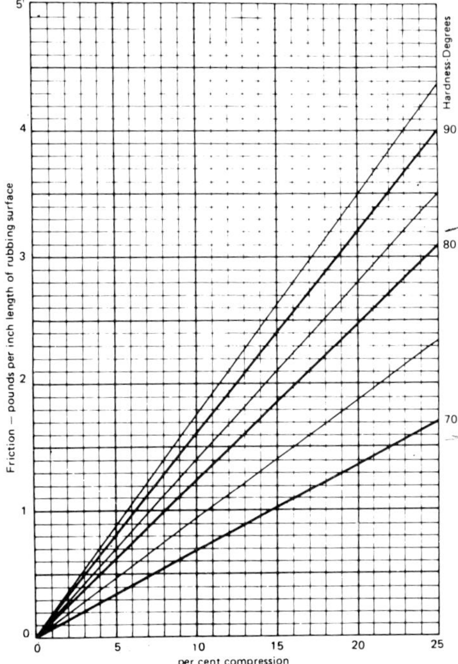

Figure 9
O-ring friction due to compression.

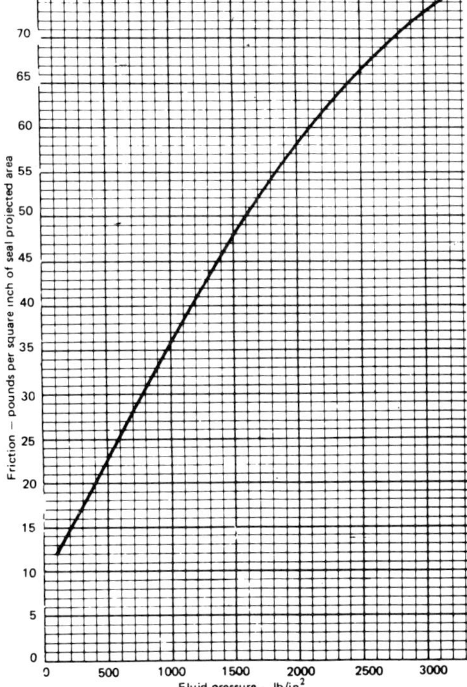

Figure 10
O-ring friction due to fluid pressure.

TABLE 2 – PROJECTED AREA AND RUBBING LENGTHS OF INCH SIZE O-RINGS

O-ring Size o.d. W*	Projected Area sq in	Rubbing Length (L)	
		Piston Groove	Rod Groove
1/4 x 1/16	0.03	0.79	0.39
9/32	0.04	0.89	0.49
5/16	0.05	0.98	0.58
11/32	0.05	1.08	0.68
3/8	0.06	1.18	0.78
7/16	0.07	1.38	0.98
1/2	0.08	1.57	1.17
9/16	0.09	1.77	1.37
5/8	0.10	1.97	1.57
11/16	0.11	2.16	1.76
3/4	0.12	2.36	1.96
13/16	0.14	2.56	2.16
7/8	0.15	2.75	2.35
15/16	0.16	2.95	2.55
1	0.17	3.14	2.75
1.1/16	0.18	3.34	2.94
1.1/8	0.19	3.54	3.14
1.3/16	0.20	3.73	3.33
1.1/4	0.21	3.93	3.53
1.5/16	0.22	4.13	3.73
1.3/8 x 1/16	0.24	4.32	3.92
1.7/16	0.26	4.52	4.12
1.1/2	0.25	4.72	4.32
1.1/2 x 3/32	0.40	4.72	4.12
1.9/16	0.42	4.91	4.32
1.5/8	0.43	5.11	4.51
1.11/16	0.45	5.30	4.71
1.3/4	0.47	5.50	4.90
1.13/16	0.49	5.70	5.10
1.7/8	0.50	5.89	5.30
1.15/16	0.52	6.09	5.49
2	0.54	6.29	5.69
2.1/16	0.56	6.48	5.89
2.1/8	0.58	6.68	6.08
2.3/16	0.59	6.88	6.28
2.1/4	0.61	7.07	6.47
2.5/16	0.63	7.27	6.67
2.3/8	0.65	7.46	6.87
2.7/16	0.66	7.66	7.07
2.1/2	0.68	7.86	7.26

*W = width of section

One of the most significant characteristics is the friction/time relationship, particularly as such seals are normally employed with a fairly high degree of 'squeeze'. This would appear to generate definite cold flow of the O-ring into the surface irregularities of the mating part, irrespective of whether or not lubricant is present. As a general rule the resulting increase of friction or break-out friction on standing can be taken as three times the running friction for a typical rubber hardness (70 degrees) and a typical fine surface of 0.2 to 0.25 μm (8 to 10 μin). This break-out friction value is modified in more or less direct proportion to differences in hardness and surface finish, but more particularly it can be reduced by employing a softer rubber.

A basic formula for calculating O-ring friction is:

$$\text{Friction} = (f_c + L) + (f_h \times A) \text{ in consistent units}$$

where L = length of seal rubbing surface (that is, circumference of outer diameter with a piston seal; or circumference of inner diameter with a rod type seal)

A = projected area of seal, (the same in either application)

f_c = friction due to O-ring compression

f_h = friction due to fluid pressure

Typical values of f_c and f_h can be calculated from Figures 9 and 10 respectively. Table 2 gives values of projected area and lengths of rubbing surfaces for a limited range of standard O-ring sizes.

Wear and Seal Life

BECAUSE OF their differing designs, and because they are produced from different materials, sealing systems have varying behaviour patterns at increasing operating pressures.

When a hard material is used the danger of damage by compression is reduced. On the other hand, a hard material does not have such good sealing characteristics as a soft material, particularly at low operating pressures.

For the best sealing system, effective at high and low operating pressures, a seal constructed from several types of material with different properties is needed. The ideal would be a solid seal made from several materials having an increasing hardness and reaching a maximum hardness in the rear section of the seal space, where a gap occurs. However, it is not practicable to achieve this fully, although a number of proprietary

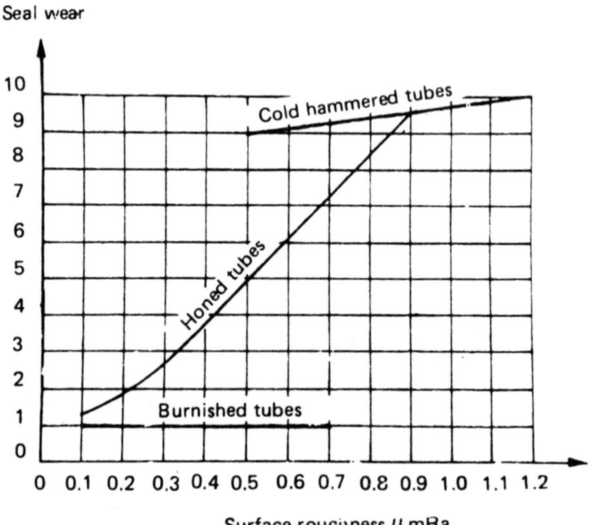

Figure 1

designs of seals are constructed on a multi-stage principle in an attempt to approach the ideal.

Seals lose their ability to function because of normal wear of the seal material, which is greatest at the instant of starting and at low speeds and through erosion of the seal material which occurs when the pressurizing fluid flows over the sealing surface and impinges on an area of deterioration.

The first indication is seen at low pressures when, because of wear, the seal is no longer capable of maintaining the required contact with the sealing surface. At high pressure, because the deformation is greater, sealing may continue to be adequate so long as the pressure is maintained.

The life of a seal cannot be predicted in exact terms because it depends on so many

TROUBLE SHOOTING – RECIPROCATING SEALS

Fault	Cause	Action
High friction	(i) Improper assembly.	(i) Check against recommended assembly for that type of seal, reduce interference fit or pressure if necessary.
	(ii) Wrong size of seal.	(ii) Check geometric specification
	(iii) Poor surface finish.	(iii) Improve surface finish or use a seal material capable of rubbing on a rougher surface.
	(iv) Excessive rubbing speed.	(iv) Different seal type may be required.
	(v) Excessive pressure being sealed.	(v) Replace seal with a different type, or with different type of elastomer.
Slip-stick action	(i) Seal allowed to dry out.	(i)
	(ii) Poor surface finish.	(ii) As (iii) above.
	(iii) Inadequate lubricating film.	(iii) Amend operating conditions or change seal type (*eg* use PTFE composite seal).
Excessive leakage	(i) Seal fitted wrong way round..	(i) Check —or use double-acting seal when required.
	(ii) Insufficient preload.	(ii) Check geometry and pre-load specification.
	(iii) Seal shrinkage.	(iii) Check that seal material is compatible with fluid; if not replace.
	(iv) Seal wear.	(iv) Replace seal; if life is low, consider an alternative seal type. Check for cause if seal is damaged.
Seal damage	(i) Incorrect initial assembly.	(i) Replace seal following manufacturer's assembly instructions.
	(ii) Spiral failure (applicable to O-rings).	(ii) Check geometry; also suitability of O-ring for the application.
	(iii) Extrusion damage.	(iii) Reduce extrusion gap in seal assembly; or incorporate back-up ring(s).
	(iv) Rubbing (O-rings U-rings).	(iv) Check geometry; reduce extrusion gap if necessary.

factors – starting with a suitable choice of seal for the job and correct installation. Wear can then be aggravated by lack of lubrication, shaft irregularities, excessive frictional heat, a seal compound which is too soft, *etc.* The normal life expectancy of seals will also vary considerably from one application to another as acceptable conditions and even the type of seal recommended differ widely. Thus 4000 hours can be considered a normal life for an *hydraulic cylinder* seal, properly designed for its purpose and functionally efficient. The normal life of a lip-type seal, on the other hand, may be only 1000 hours.

If the life of a seal is significantly less than average for a particular application, then it is probable that an unsuitable seal was chosen in the first place so that the operating conditions have turned out to be more severe than was expected when the seal was chosen.

Seal wear is heavily dependent on the finish of the surface against which the seal rubs, this in turn being determined to a large extent by the production method. Figure 1 illustrates this for typical hydraulic cylinders with three different finishes. Seal wear here is graded visually from 0 for no apparent wear to 10 for a worn out seal. These particular figures were taken after 100000 cycles of cylinder operation at a working pressure of 250 bar.

A significant fact relative to the aforementioned is that with burnished tubes seal wear was largely unaffected by surface finish throughout the range 0.08 μm to 0.7 μm; but rather more so in the case of cold hammered tubes with surface finish ranging from 0.4 μm to 1.25 μm.

See also chapters on *Seal Friction* and *Surface Texture.*

TROUBLE SHOOTING – ROTARY SEALS

Fault	Cause	Action
Excessive leakage	(i) Incorrect installation.	Seal wrong way round or improperly installed.
	(ii) Insufficient lip pressure (oil seals).	Check that seal size is correct and garter spring is functioning properly.
	(iii) Scarred or rough shaft.	Improve shaft finish, or use softer seal or different seal material.
	(iv) Cracked seal lip.	Check for cause; may be rough shaft, wrong seal material or too high a speed or temperature.
	(v) Seal damaged or worn.	See below.
High friction	(i) Inadequate lubrication.	May be necessary to change to a PTFE composite seal.
	(ii) Excessive preload	
	(iii) Excessive pressure.	May be necessary to change to a different type of seal.
Seal damaged	(i) Improper or careless assembly.	Replace seal using proper technique and tools.
	(ii) Rough shaft.	Observe surface finish and any chamfer requirements.
	(iii) Abrasive present (will also show as shaft wear).	Fit exclusion type seal in dirty atmospheres.
Excessive seal wear	(i) Wrong type of seal or seal material.	Fit more suitable type of seal for operating conditions.
	(ii) Excessive pressure.	Fit more suitable type of seal for operating conditions.
	(iii) Excessive temperature.	Fit more suitable type of seal for operating conditions.

Surface Texture

THE TEXTURE of a surface against which a seal rubs has a significant effect on friction, wear and seal life. Texture in this context refers both to surface roughness (or surface irregularities) and to the pattern of these irregularities. The former is capable of sampling measurement. Pattern can only be described empirically, *eg* in terms of lay or the direction of the predominant pattern; and waviness.

The standard method of measuring roughness is by an average value of the profile variation from a centre line over a reference (sampling) length L (Figure 1). This is known as Centre Line Average (CLA), now commonly expressed as Ra in British Standards and also adopted as ISO standard. In the United States it is designated AA (Arithmetical Average), and in Holland Ru.

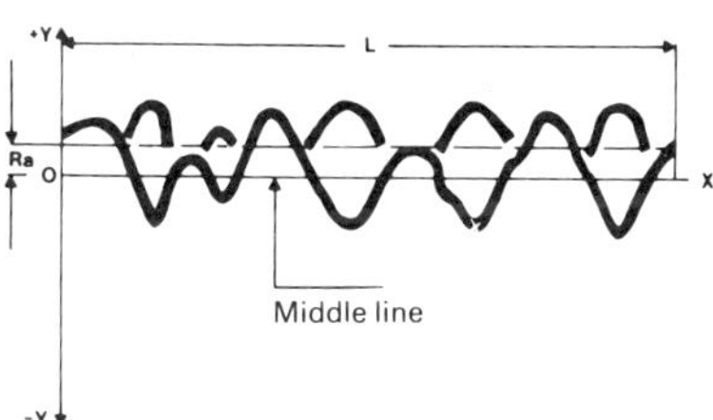

Figure 1
Diagrammatic representation of average roughness value (Ra).

Most surface measuring instruments give a direct reading of the Ra value, either in micrometers (μm) or micro inches (μin). The ISO standard is μm (1 μm = 40 μin and 1 μin = 0.025 μm).

ISO R1302 relates nominal Ra values to equivalent roughness grade numbers, which are useful for avoiding misinterpretation of numerical values where different units may be quoted for Ra values.

Surface roughness may also be expressed in terms of maximum roughness, depth or

Nominal Ra values		ISO roughness grade number
μm	μin	
50	2000	N12
25	1000	N11
12.5	500	N10
6.3	250	N9
3.2	125	N8
1.6	64	N7
0.8	32	N6
0.4	16	N5
0.2	8	N4
0.1	4	N3
0.05	2	N2
0.025	1	N1

the distance between the peak and the base line measurement over the sampling length (Figure 2). Maximum roughness is designated Rt and is measured in the same units as Ra. Both values can be significant in determining the optimum surface finish required for use with seals.

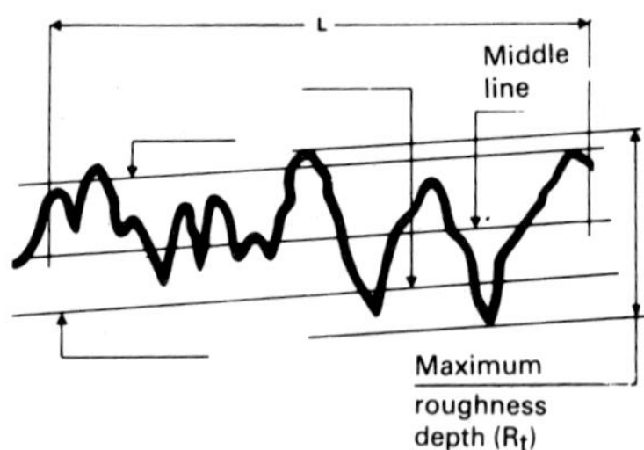

Figure 2
Diagrammatic representation of roughness depth (Rt).

Recommended surface finishes

The aim of all types of surface finishing is to provide a surface which causes the least wear to the seal.

Rod seals, which seal against moving surfaces, can be damaged by fine abrasive particles which may adhere to a rough surface. Rods therefore should have a low surface roughness value, and a surface similar to the best hard chrome and a resistance to corrosion. The ideal surface roughness value lies between 0.16 and 0.40 μm (Ra) or 1.0 and 2.5 μm (Rt).

Piston seals, which seal against the inner surface of a cylinder, are not subjected to the same extent to the action of abrasive dirt particles entering from the atmosphere, and can therefore tolerate a rougher surface. The ideal surface roughness value lies between 0.25 and 0.63 μm (Ra) or 1.6 and 4.0 μm (Rt).

The operating pressure must also be taken into account when evaluating surface properties. At higher operating pressures, the oil film between the seal and the sealing surface is thinner, and the friction is greater. Under such conditions of operation a surface quality which approaches the lower values given should be chosen.

The surface in the seal housing, where the seal is static, should have a surface quality of about 1.6 μm (Ra) or 10 μm (Rt).

Process	Surface finish	
	μm	μin
Planing	1.5–12.5	60–500
Shaping	1.5–12.5	60–500
Milling	0.9–6.25	35–250
Broaching	0.9–3.00	35–120
Reaming	0.9–3.00	35–120
Boring	0.5–6.25	20–250
Turning	0.5–6.25	20–250
Diamong bored and turned	0.25–0.5	10–20
Grinding	0.125–1.75	5–70
Honing	0.125–1.625	5–65
Buffing, burnishing	0.125–0.5	5–20
Lapping	0.05–0.5	2–20
Polishing	0.05–0.5	2–20
Super finishing	0.025–0.25	1–10

Finishes from machining processes

Both the roughness and pattern of the surface finish produced can vary widely with different machining processes. The following gives typical values likely to be achieved with different processes (but can also be variable depending on the quality of the machine tool and the material being processed).

As an example, optimum seal life with a hydraulic piston rod seal is given when the rod is circular ground (or roller burnished) to a surface finish of better than 0.2 μm (Ra). (It should then also be chrome plated and polished).

See also chapter on *Wear and Seal Life*.

Safety and Environmental Health

ENVIRONMENTAL HEALTH, as far as seals and sealing is concerned, works both ways. The environment itself has a considerable influence on seal selection. At the same time one purpose of the seal is to contain a fluid and so protect the immediate environment from contamination. The meaning of contamination in this context may range simply from avoiding 'messiness' (although avoiding product loss is probably far more significant in such cases) to the elimination of contamination by nauseous, toxic or hazardous fluids by fully containing them within the system.

Safety has a two-fold meaning. Unless the environment is 'safe' for the seal, its useful life will be short and contamination and product loss problems will be increased. Equally, a seal which permits external leakage of a toxic or otherwise dangerous product is an environmental health hazard.

From either aspect, the basic problem involved is the same, and starts with the selection of the most suitable seal for the job. This must take into account all the environmental factors involved. If necessary, this can then be subjected to environmental control(s), depending on the particular environmental problems remaining.

Environmental factors involved embrace both the ambient surroundings (external factors) and internal factors. The most obvious external ambient factors are *temperature* and *pressure*, although these are not usually significant except in specialized applications. The likely maximum range of ambient temperatures is from −50°C (arctic) to +50°C (tropic), which is within the working temperature range of most elastomeric materials used for seals. Special problems here are only likely to arise in the case of refrigeration equipment, systems for handling liquefied gases, *etc*, involving much lower ambient temperatures; and the handling of heated products, combustion equipment, rocket motors, *etc*, involving much higher ambient temperatures.

Most seals, too, operate under atmospheric conditions as far as external pressure is concerned. Exceptions here are seals on high flying aircraft when the ambient pressure can be very much lower than normal (sea level); and in deep tunnelling equipment where pressures may be of the order of two or three atmospheres, or submarine equipment where very much higher external (fluid) pressures may be involved.

External environmental problems are far more likely to arise where the atmosphere is corrosive, dirty or otherwise incompatible with simple materials. In chemical plant the air is commonly contaminated to a degree which, while not being injurious to health, can cause metallic corrosion, as can plant or equipment which operates in a salt-laden atmosphere. Other operating environments are contaminated with abrasives ranging from mild (dry dust atmospheres) through moderate (dry abrasive grits) to severe (*eg* wet abrasives and slurries). Seal design will often be influenced by the severity of such conditions.

Non-electrical seal condition monitor gives audible and visual alarms.

Atmospheres containing ozone can also be difficult. Ozone-rich atmospheres occur in the presence of electrical equipment such as switchgear and motors, and some types of elastomers are dramatically degraded by even moderate concentrations of ozone. These can affect not only operational seals, but seals held in storage.

Radiation is another external factor which may have to be considered in the field of nuclear engineering. Radiation can affect both elastomers used for seals, and lubricants in the system.

Internal factors

Internal factors are primarily concerned with the sealed fluid, the operating parameters and the installation design. Some, or all, of these factors may need consideration:

(i) Fluid pressure
(ii) Fluid temperature
these determine the seal type, seal material and design.
(iii) Fluid viscosity, which can influence wear on mechanical seals.
(iv) Fluid nature – gas seals may require special attention because of absence of lubrication.
(v) Cleanliness – presence of abrasive particles will accelerate seal wear.
(vi) Chemical activity of product, which affects choice of seal material and/or type of seal.
(vii) Toxicity, which may require special designs of seals.
(viii) Vapour pressure, which may dictate degree of temperature control required.
(ix) Entrained air, which can cause accelerated wear at faces of mechanical seals.

Design and operating parameters

(i) Available space, which may restrict the type of seal available (or require redesign of equipment).

(ii) Motion, reciprocating or rotary. Significant factors are maximum rubbing speed and (for reciprocating seals) number of strokes per minute. If shafts are stationary for long periods, special design of seals may be required.

(iii) Quality control – both in surface finishes on equipment and seal installation procedure.

(iv) Eccentricity, which can seriously influence seal performance and lift, particularly in the case of lip seals.

(v) Vibration, which can be significant in both static and dynamic seals with elastomeric elements.

Vapour emission

A small amount of leakage is commonplace with most types of dynamic seals. Such liquid leakage may vaporize after escaping from the seal, contaminating the surrounding atmosphere and possibly presenting a health hazard. Direct vapour emission is, however, also possible from so-called 'zero-leakage' or 'dry' seals, which can be a particular and, possibly, unsuspected danger where the contained product is toxic or hazardous, especially as the vapour emission may be invisible.

Legislation concerned with the UK Health and Safety at Work Act has emphasized the need for data on hazardous vapour emissions, with Guidance Note EH 15/77 issued in 1977 tabulating both time-weighted threshold limit values (TLV-TWA) and short-term exposure limits (TLV-STEL) for toxic substances. Particular research on this subject, and especially on suitable vapour detection instrumentation, has been undertaken by BHRA Fluid Engineering (amongst others).

Field work undertaken included site investigations made with the co-operation of three petro-chemical companies on both running and stand-by pumps employing mechanical face seals. Particular findings were that single seals on both running and stationary stand-by pumps commonly emit vapour even when there is no visible liquid or vapour leakage. Also that, in some cases, vapour leakage can increase when a pump stops. Specifically, however, factors most likely to lead to increased vapour emission are:

(i) Increase in fluid pressure being sealed.

(ii) Higher fluid volatility.

(iii) Pump stop-start or intermittent duty.

Vapour emission from seals is, in fact, a subject on which much further work is being carried out, particularly in the design of seals for use in critical conditions.

Environmental controls

There are three basic types of environmental controls:

(i) Temperature controls.

(ii) Controls for dirty or incompatible environments.

(iii) Safety controls.

The temperature at which a seal works is a vital operating parameter. It governs seal wear and life, and also its resistance to chemical attack. It also affects the behaviour of the fluid being contained. Options available for temperature control are:

(a) Jacket cooling (or heating).

(b) Bypass flush.

(c) Cooling *via* circulating rings.

d) Heat exchanger.
(e) Air cooler.

The most direct method of keeping contaminants out of the system is by the use of *exclusion* type seals. These also normally offer the simplest solution. Where the protection afforded is not adequate, internal or external flushing systems may have to be employed, or double seals (with or without intermediate flushing).

Safety controls
The prime purpose of any seal is to contain the fluid to an acceptable degree. The type and design of seal is selected accordingly. The system is then as 'safe' as the integrity of the seal. In the event of seal failure (partial or complete), there is the further safety factor of whether the leakage imposes a threat to equipment, personnel or the environment. If it does, then provision must be made for controlling the leakage.

There are two basic approaches to controlling leakage in the event of seal failure. One is the use of *tandem* seals which will prevent leakage of the process fluid to atmosphere in the event of failure of the *inner* or *primary* seal (*eg* Figure 1). The other is to use a single seal but make provision for restraining and collecting leakage should seal failure occur. (Note: additional options are available in the case of packed glands – see chapter on *Compression Packings,* Section 4.)

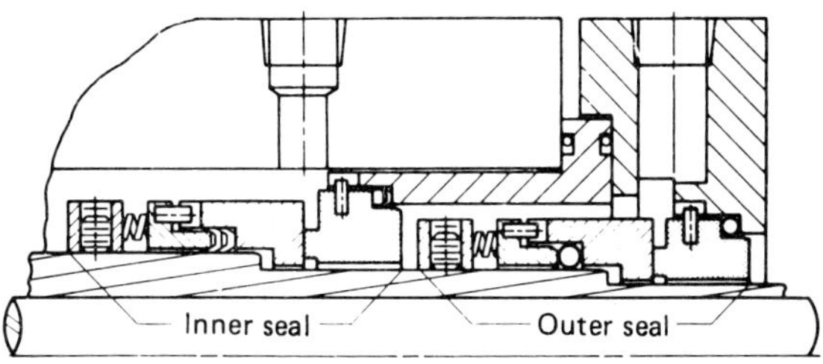

Figure 1
Double tandem seal.

In the case of *mechanical face* seals, double tandem seals with a barrier liquid between the seals can be particularly effective.

If the inner seal fails, the resulting pressure rise in the area between the seals is sensed at the reservoir, where it can either be registered on a gauge or activate an alarm. In any event, a failure of the inner seal can be detected while the outer seal assumes the responsibility of sealing the shaft until repairs can be made.

The use of tandem seals requires the pressure between the two seals to be lower than the product pressure. If it were higher, it would open the inner seal faces, and the liquid between the seals (commonly referred to as the sealant) would mix with the product. Over a period of time, the tank containing the sealant would be completely drained.

If the product pressure is high enough to necessitate the use of a balanced seal, then both the inner and outer seals should be balanced.

When selecting a sealing liquid system for double tandem seals, one must not only consider the amount of cooling that is required to dispose of the heat generated by the outer or secondary seal but also how to dispose of the primary seal weepage that occurs during normal operation. In a double tandem seal, the primary seal weepage is contained in the sealing liquid system for the secondary seal. Unless this collection of normal primary seal weepage is properly drained or vented, pressures will build up in the *secondary* seal system.

Figure 2 shows a recommended method of piping double tandem seals using thermal convection to cool the secondary seal. Should thermal convection prove to be inadequate to dissipate the heat generated by the secondary seal, a circulating ring can be added to induce circulation through the reservoir.

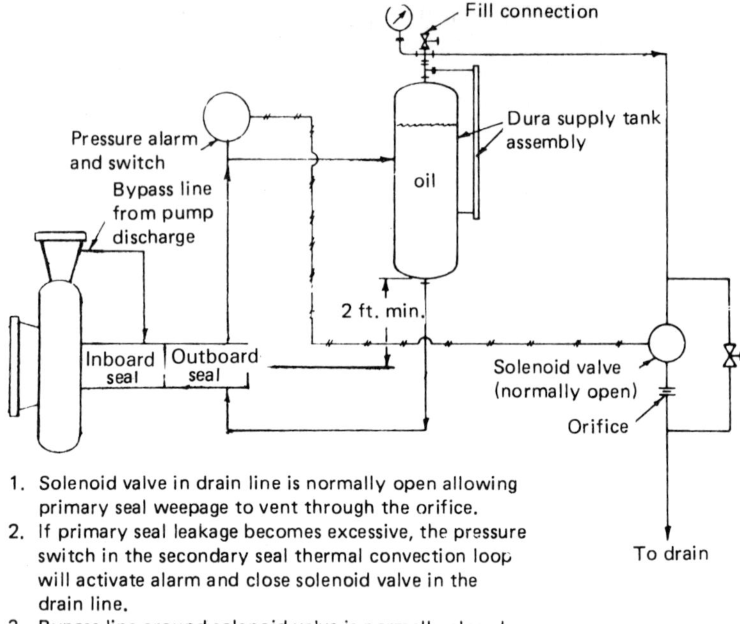

Figure 2
Recommended piping diagram for double tandem seal.

Ring vent and drain

A vent and drain can be applied to the auxilliary packing rings of internal mechanical seals. The purpose of a vent and drain is as the term implies. As an example, when handling liquefied petroleum gases, any leakage past the seal will be a highly explosive gas which can be 'vented' to safe external disposal, normally the flare stack, where it will be burnt off in the atmosphere. The drain connection at the bottom of the gland, 180°C from the vent, permits hazardous liquid leakage to be carried away to a point where it can be safely collected.

The vent and drain is provided in the gland ring by machining two drilled and tapped openings behind the stationary inserts as illustrated in Figure 3. Leakage is discouraged from travelling along the shaft to the atmosphere by a restriction, such as a *throttle bushing*, incorporated into the rear of the gland ring. With the vent and drain connections backed up by a throttle bushing, leakage takes the path of least resistance to the point of collection or disposal. Note that the cross-section area of the drain hole should be greater than that of the difference between the shaft diameter and the bore of the *throttle bush*.

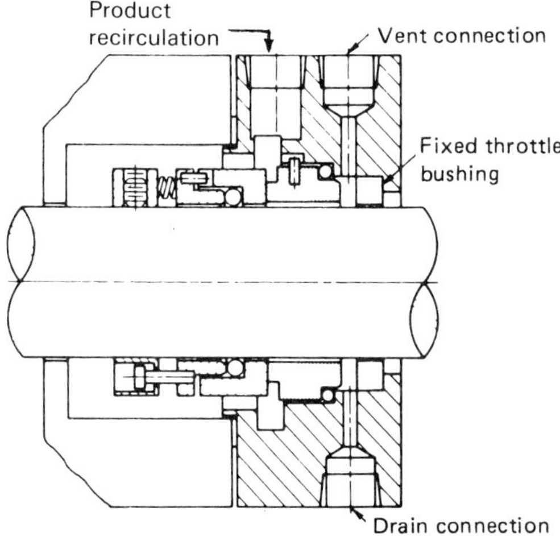

Figure 3
Vent and drain with fixed throttle bushing.

(1) Throttle bushings
Most throttle bushings used in conjunction with vent and drain connections are of the *fixed type,* being pressed into a bore at the rear of the gland ring and adequately backed up by the back of the gland ring itself. Throttle bushings are made from carbon, bronze, graphite or glass-loaded PTFE, all of which are non-sparking materials.

Figure 3 shows a typical fixed throttle bushing application. The throttle bushing is not free to float and centre itself on the shaft centre, so 0.63 mm (0.025 in) diameter clearance is provided between the i.d. of the bushing and o.d. of the shaft. It has been found that this is the optimum clearance to adequately accommodate slight vertical misalignment of the gland ring and/or shaft runout.

An alternative *gland throttle* bushing arrangement is the floating variety. When studying Figure 4, it becomes apparent that this type of bushing is free to centre itself on the shaft, thereby allowing the application of a closer radial clearance of 0.05 mm (0.002 in) between the bushing and the shaft. The additional restricting efficiency provided by the floating throttle bushing is especially popular for applications where it is desirable to achieve maximum restriction by the use of a throttle bushing.

(2) Auxiliary stuffing box
The application of an *auxiliary stuffing box* behind a mechanical seal is illustrated in Figure 5. In this case, an *auxiliary packing follower flange* compresses two rings of suitable packing against the shaft. This arrangement will provide a nearly positive *auxiliary shaft* seal in the event of primary mechanical seal failure. However, it is necesssary that the packing be properly lubricated and cooled to avoid excessive wear on the shaft and

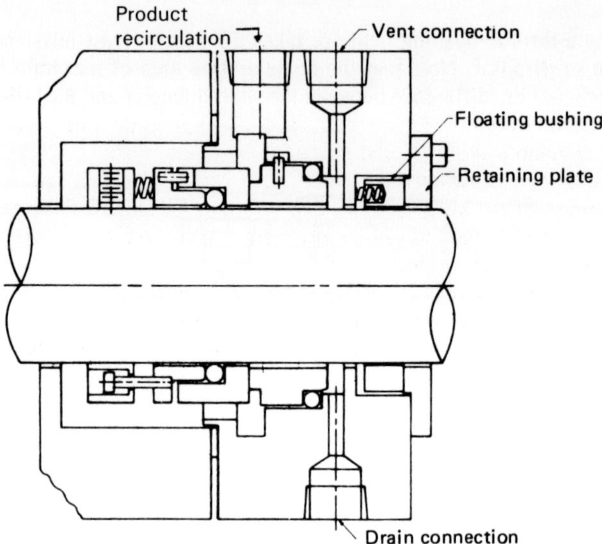

Figure 4
Vent and drain with floating throttle bushing.

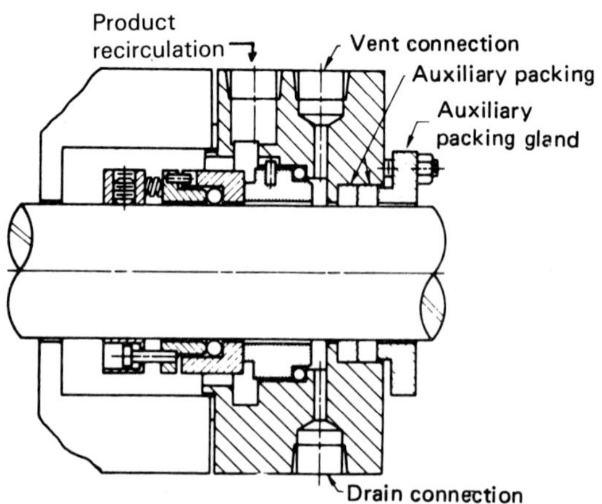

Figure 5
Vent and drain with auxiliary stuffing box.

packing. This entails circulating a liquid, usually water, through the vent and drain connections. The shaft immediately under the packing should preferably be hard coated to avoid excessive wear.

Although an auxiliary stuffing box provides the user with a tight back-up seal, the necessity of furnishing lubrication to the packing may prove objectionable. Therefore, one may prefer to use a throttle bushing design that does not require the circulation of any liquid through the vent and drain connections, other than when it is necessary to quench or dilute any leakage because of its potentially hazardous nature.

The application of double inside seals does not impose any problems, other than the selection of a suitable sealing liquid system to maintain pressure between the seals and dissipate the heat load and the space to contain them. There may be objections to the potential leakage of the sealing liquid into the pump product in the event of an inner seal failure. Double tandem seals as described previously, where the barrier liquid between the seals is at low pressure, offer a solution. The purpose of this seal is not to create an artificial environment, but to provide a back-up seal in the event of inner seal failure. The inner seal functions in a manner identical to a conventional single inside seal. The cavity between the inner and outer seal is flooded from a closed reservoir. The liquid in the reservoir provides lubrication to the outer seal. Because the space between the seals is only flooded and not under pressure, the product, not the liquid in the reservoir, lubricates the faces of the inner seal.

SECTION 2

Materials

PROPERTIES OF ELASTOMERS
ELASTOMERIC MATERIALS
PLASTIC POLYMERS
CEMENTED CARBIDES
MISCELLANEOUS SEALS MATERIALS
COMPATIBILITY OF SEAL MATERIALS

If an O-ring can cost you more than

If you're looking to reduce maintenance and downtime significantly, it's nice to know you can rely on O-rings made of Du Pont KALREZ.*

Only KALREZ O-rings are capable of handling more than 1,500 chemical reagents at up to 316 °C.

Unlike other materials, KALREZ O-rings have proven their performance handling harsh, corrosive and hazardous fluids under extreme temperatures in actual plant operations for more than 10 years.

KALREZ O-rings are resilient and non-rigid, offer excellent compression set and creep resistance characteristics, and are serviceable over a broad range of pressures with very low outgassing.

And KALREZ O-rings are available from a network of authorized distributors who will assist you in selecting the right compound

Properties of Elastomers

ALL RUBBER compounds deteriorate over a period at high temperatures, especially in air: they become brittle or resinous and lose their elastic properties. This effect is time-dependent, irreversible and varies considerably from one compound to another.

At temperatures below 0°C rubber compounds progressively stiffen and lose their resilience as the temperature falls, until brittleness ultimately occurs, and cracks appear if the rubber is flexed. In service, this adverse effect on resilience and compression set may seriously impair the dynamic performance of the rubber component. This brittleness and stiffening is reversible when the temperature rises again. Specific low temperature materials, such as silicone rubber, are readily available, or, within an elastomeric group such as Nitrile rubbers, special compounding techniques may be employed, to provide the material with a greater resistance to stiffening at low temperatures. However, in doing so, the resistance to oil or fuel may be reduced.

Any decision based on the working temperature range must be made with due consideration given to the environment in which it is expected to function. It is known that when immersed in certain fluids the tolerance of a rubber compound to significantly higher temperatures can be much lower compared to its tolerance on operating in dry heated air. The figures for working temperature range given under the individual elastomer headings in the chapter on *Elastomeric Materials*, may therefore be modified by actual service conditions.

Ageing covers the progressive and permanent deterioration of the seal material with age. Initially this is inherently a material characteristic which can be controlled to some extent by compounding but it is also controlled by the life history of the seal from the time of manufacture. Thus a material particularly prone to oxidation could be aged rapidly in storage before actual use by being exposed to a hot atmosphere, sunlight, or other unfavourable surroundings. On the other hand, a seal stored and applied in service under favourable conditions could remain virtually free from ageing effects for years.

Accelerated ageing tests can indicate the general ageing characteristics of specific elastomers, but such data are comparative only. Service experience is the most reliable information of this type, although age tests may indicate a promising alternative material. The actual service life of a seal depends on so many factors that it cannot be estimated on simple data basis, particularly as the main criteria are usually how the seal is stored and handled during assembly and the actual working conditions in service.

In general, ageing has the effect on tensile strength, hardness and reduction in elongation (increasing brittleness) as shown in Figure 1. Wide variations in behaviour time occur

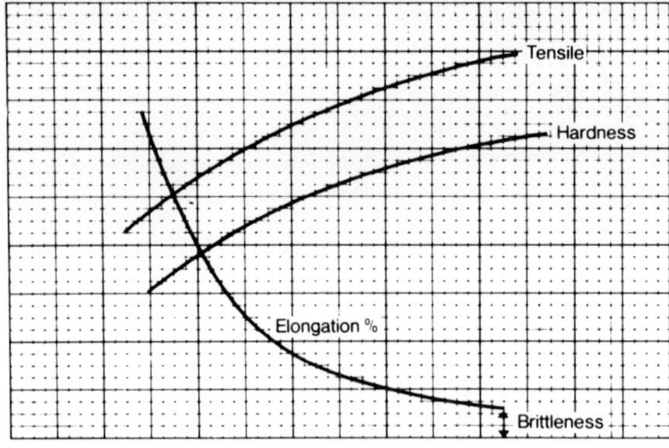

Hours exposure high temperature

Figure 1

with different rubbers, however. This can be expressed as a *heat ageing index* or time to embrittlement at specific temperatures. Comparative performance of various elastomers is given in Table 1.

TABLE – HEAT AGEING INDEX – TIME TO EMBRITTLEMENT

Materials	**125**°C	150°C	200°C	300°C
Perfluoroelastomer*	>1000	>1000	>1000	>1000
Fluorocarbon	>1000	>1000	>1000	
Fluorosilicone	>1000	>1000	>1000	–
Silicone	>1000	>1000	>1000	–
Polyacrylic	>1000	>1000	>60	–
Chlorohydrin	>1000	>600	–	–
Ethylene propylene	>1000	400	–	–
Hypalon	850	70	–	–
Nitrile	500	100	–	–
SBR	500	45	–	–
Chloroprene	350	30	–	–
Polysulphide	60	3	–	–

*Perfluoroelastomer never embrittles

Compatibility

An elastomer in contact with, or immersed in a fluid will normally tend to absorb a certain amount of fluid, and the material will swell or increase in volume. This will tend to modify the properties of the elastomer, notably hardness, strength, resilience and abrasion resistance, such changes being more or less proportional to the amount of swell or percentage increase in volume. Such conditions are acceptable – and usually inevitable – provided the necessary seal properties are retained and there is no positive physical or chemical attack. However, excessive swell may indicate lack of compatibility (for example, excessive modification of properties) as well as producing excessive dimensional change. In general the lower the swell the greater the likely compatibility; measurement of swell being a basic test of compatibility, *eg* see Figure 2. On the other

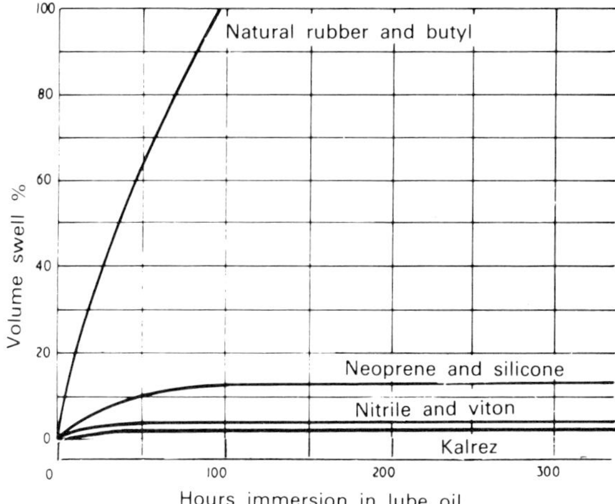

Figure 2

hand, excessive swell may be accompanied by definite chemical attack, such as solution of, or interaction with some of the material constituents, or embrittlement of the surface leading to cracking. In such cases the fluid and material are incompatible.

While there are no specific limits to swell (provided all the other features are compatible), a volume increase of as much as 100% may be acceptable for gasket type seals; 50% for static seals (O-rings); and about 15-20% as a maximum for dynamic seals. The *actual* change in volume may be appreciably less than that indicated by swell alone, the linear dimension involved (approximately one third of the volumetric change) may well be offset by compression set.

If continuously immersed in fluid, swell will increase with time of immersion up to a point where no more fluid will be absorbed and the volumetric expansion remains at a constant figure. This will also depend on temperature – *eg* see Figure 3. The higher

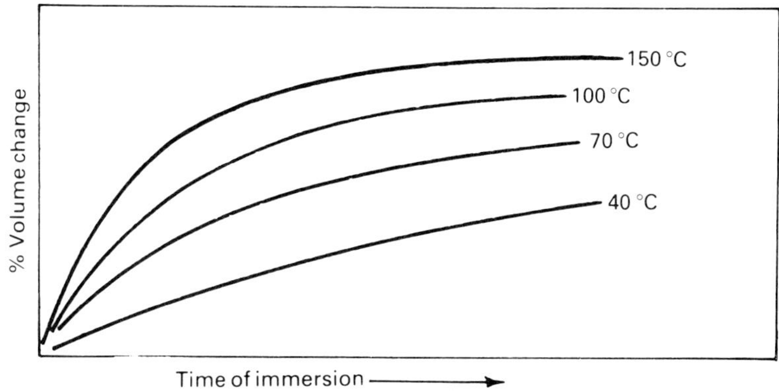

Figure 3
Variation of swelling with temperature.

the temperature, as a general rule, the greater the amount of fluid absorbed by a given material, *ie* the more rapidly it will absorb that fluid and the greater will be the final (constant) volumetric expansion.

If the fluid concerned is non volatile, the swell will remain permanent. If the fluid is volatile, and immersion is not continuous or contact with fluid is only partial, the amount of absorbed liquid, and thus the swell, can vary – the seal can swell and shrink. It will also shrink if removed or otherwise allowed to dry out and this may or may not be serious. The seal may not reseat properly on re-swelling, or it may accommodate swelling cycles quite satisfactorily. Any shrinkage in service, however, is very likely to lead to leakage if it exceeds about 3 to 4% in the case of dynamic seals.

The usual cause of shrinkage is drying out and this seldom arises where a seal is continuously in contact with the fluid. However, in some cases a fluid may dissolve out a proportion of the plasticizer in the elastomer which can modify the composition, or even cause the seal to shrink rather than swell under working conditions. This can result in a loose seal and leakage.

Swelling (due to fluid absorption) generally has the same effect as adding plasticizer, by making the seal more flexible with decrease in hardness. Shrinkage will normally produce an increase in hardness and a reduction in flexibility. In both cases the change can be estimated on the basis of volumetric change. Swell will also tend to decrease the tensile strength of the elastomer – see Figure 4. However, it is the actual conditions pertaining at the seal under working conditions which are significant. Simple tests based on immersion of the material in the fluid are only a guide. Swell can also vary with seal geometry. Thus Figure 5 shows typical variation in swell with seal material thickness.

Immersion tests can be virtually valueless unless taken over a reasonably long duration in order to ensure that a stabilized condition is reached. To be realistic they are best performed on a typical seal rather than a material specimen as the section area can govern the rate of swell, although this does not affect the issue if the test is immersed long enough to ensure complete stabilization. The main practical variation then likely to be involved is the amount of fluid, bearing in mind, of course, that immersion exposes all surfaces

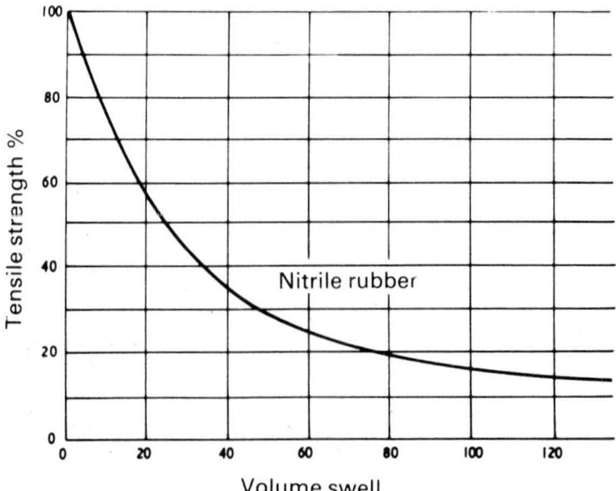

Figure 4
Heat and oil resistance of elastomers.

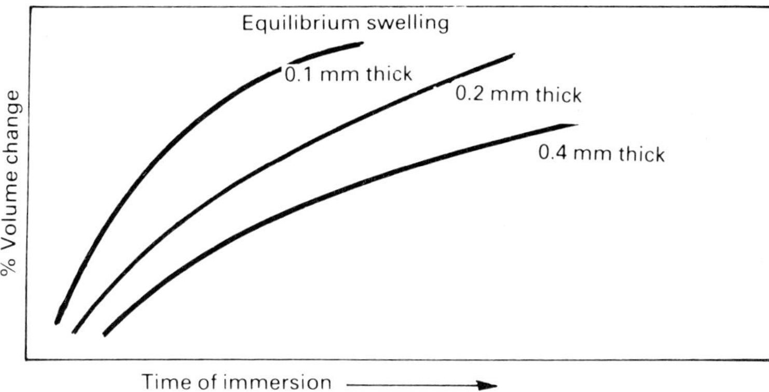

Figure 5
Variation of swelling with thickness.

to the fluid. In a practical seal installation only one side of the seal is likely to be continually in contact with fluid, and even on that side fluid contact may not be continuous.

In the case of fluids, material selection may be based on the aniline point of the oil and the known performance of seal materials with regard to standard test fluids. The aniline point is basically a measure of the aromatic content of the fuel, which is the most significant single factor affecting the compatibility of an oil and elastomer. Figure 6 illustrates this effect on different rubber groups.

Oils with the same aniline point will normally have a similar effect on elastomers, the higher the aromatic content the greater the tendency to produce swelling. Compatibility of elastomers with oils can thus be evaluated against a number of test fluids of known and different aromatic content. Any other oil with the same aromatic content (same or specified aniline point) can then be expected to behave in a similar manner to the corresponding test fluid. The particular advantage of this is that the aniline point of an oil can be determined easily and quickly.

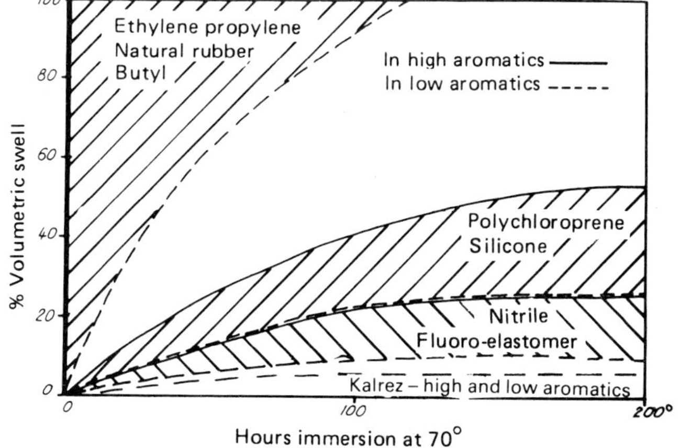

Figure 6

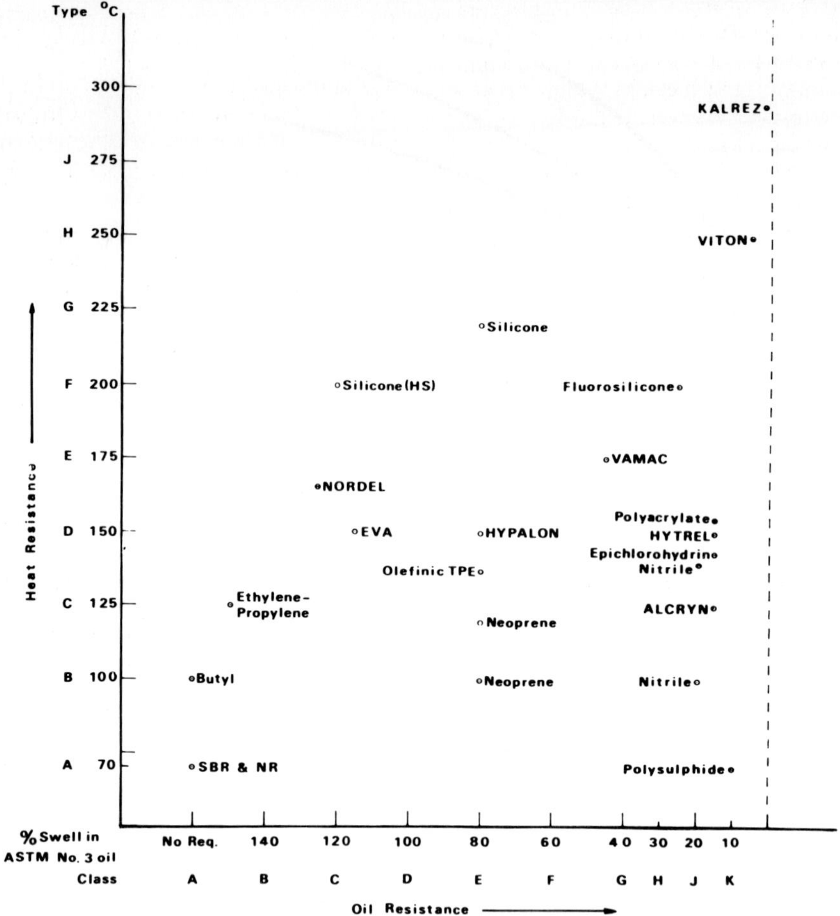

Note:

(i) The SAE J200/ASTM D200 classification only covers crosslinked elastomers. HYTREL, ALCRYN and the other thermoplastics are included for comparison.

(ii) The temperature of testing in ASTM No.3 oil varies with the heat resistance classification (Type) as follows:-

Type	A	B	C	D-J
Test temperature, °C.	70	100	125	150

(iii) Specific points are shown for clarity. In practice, there will be a "spread" depending upon compounding and grades of polymers used.

Heat and oil resistance of elastomers.

Elastic recovery

Elastic recovery or resilience is a measure of the ability of an elastomer to return to its original shape when a compression load is removed. Ideally a seal should have good resilience, which can be largely controlled by compounding. It is a peculiarity of many elastomers, however, that resilience can vary widely with temperature, with a sharply defined minimum value which commonly occurs in the range −20 to +20°C. Some elastomers, notably silicones, retain substantially constant resilience over a wide range of temperatures (Figure 7).

A low rating is desirable for dynamic seals, where recovery is important – this may refer to *compression set* or *tension set* and is a measure of the permanent change in original dimension after being compressed, or stretched, under compressive or tensile loading, respectively.

Compression set is significant in that because most seals are loaded in compression some permanent reduction in dimension or shrinkage will occur, by the amount of set characteristic for that material. This may, however, be offset by other factors, such as swelling of the seal in contact with the fluid or can be allowed for in seal design. The compression set figure is not necessarily of particular importance.

Excessive tension set may result in a seal ring being a loose fit after being stretched in place over a rod because the seal does not recover to its original i.d. to which the groove was matched. This would probably be offset by compression set on completion of the assembly and so tension set is normally ignored. The latter may be significant in plastomers which have low elongation and slow recovery, particularly if over-stretched. Both elastomers and plastomers, however, if loaded in tension, or with residual tensile stresses, will tend to contract with an increase in temperature.

Elastic modulus

Elongation is a reciprocal indication of rubber stiffness. It is defined as the increase in length expressed as a percentage of the original length, which for the ultimate case is the elongation at the point of breaking. Elongation is also used to define an elastomer – that is, a material capable of 100% elongation. Permissible elongation – the percentage 'stretch' which can be applied without permanent damage or *permanent set* – determines the amount by which a ring seal can be stretched to fit in place.

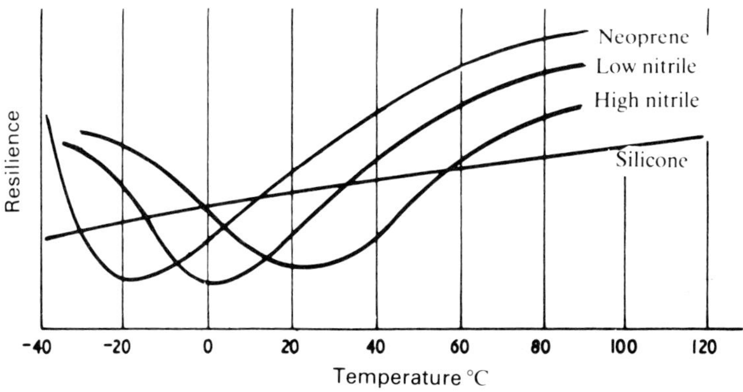

Figure 7
Resilience of typical elastomers.

The term *modulus* is also used in connection with elongation, and is generally taken to refer to the modulus (in tension) or the stress produced in the material at a predetermined elongation, for example, 100% elongation. This can be used as a measure of quality control. Modulus can also refer to stress per specified distortion in shear and compression (modulus in shear and modulus in compression, respectively). A change of modulus of a material indicates a change in material characteristics, a loss of modulus, for example, indicates a degradation of the product.

As a general safety rule O-rings and similar elastomer seals should not be *permanently* stretched more than about 5% as otherwise the resulting residual stresses can cause early deterioration, further accelerated by any rise in temperature. This applies particularly to the more generally used elastomers such as Nitrile, SBR and natural rubbers. Some elastomers, notably EP, can accommodate a relatively high amount of permanent stretch with no adverse effects.

Joule effect

Joule effect is a peculiar property of rubber for which there is no standard method of measurement.

Rubber band stretched 200-300%

Rubber band contracts when heated by match —Joule effects" can also make rubber springs stiffer in warm weather, and can make a rubber seal ring seize a rotating shaft when heated by friction.

When a stretched rubber band is warmed by a match flame, almost everyone expects it to elongate slightly because of thermal expansion. But it does not elongate; it contracts and lifts the attached weight about ¼ in (6.35 mm). This is a demonstration of the 'Joule effect', a phenomenon of great practical importance, but one that many machine designers are not acquainted with. Stated in general terms: when rubber is stretched, then heated, it tries to contract. Its modulus of elasticity, or stiffness, or ability to carry load, increases with rise in temperature. If the rubber is under constant load it will contract; if under constant strain it will exert greater stress.

(Another related thermo-elastic effect: rubber evolves heat when rapidly stretched, absorbs heat when allowed to contract rapidly. A rubber band held to the lips feels warm when rapidly stretched, cool when rapidly relaxed.) If rubber is under no strain it does expand when heated, in accordance with its coefficient of expansion. Joule effect occurs only when the rubber is strained first, then heated.

A contradiction of the Joule effect appears to lie in stress-strain curves similar to one shown in Figure 8. The literature abounds with such curves, obtained by running a standard tension test on a composition at room temperature, and at some elevated temperature. The curves seem to say that rubber's modulus of elasticity is lower at the elevated temperature. According to the Joule effect its modulus should be higher at the

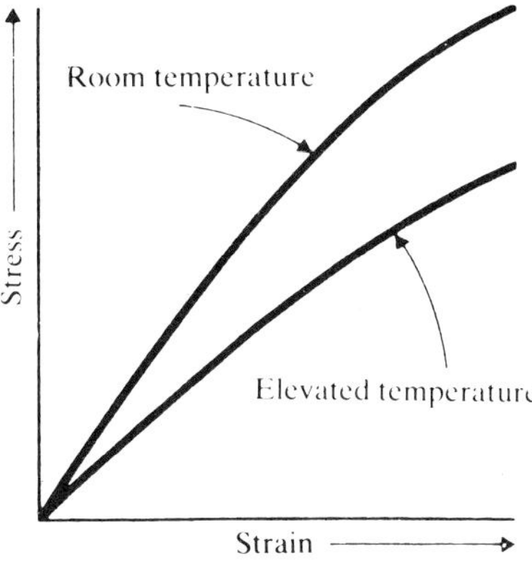

Figure 8

elevated temperature. Again, the contradiction is not real. Ira Williams, formerly with Du Pont's rubber laboratory, reasoned that at the higher temperature another effect, plastic flow, was overriding the Joule effect. If plastic flow were eliminated, said Williams, the stress-strain curves would come out as shown in Figure 9 with higher modulus at the higher temperature. He confirmed his idea by running tension tests at high speed so plastic flow would not have time to occur. Other experimenters have eliminated plastic flow by 'conditioning' the rubber. They get the same result – higher modulus at higher temperature. (To 'condition' the rubber band in the drawing , they would bounce the weight up and down or vibrate the rubber band like a harp string. When the weight came to rest they would measure the stress-strain relationship. In the standard tension test, great care is taken not to stretch the specimen before it is put in the testing machine.)

So Joule effect is real – and the counter evidence of standard tension tests is only evidence that the standard method stretches the rubber so slowly that plastic flow predominates.

What is the practical importance of Joule effect? It is the chief stumbling-block to torsion springs in automobiles. It is almost always the cause when rubber O-rings fail as rotary shaft seals.

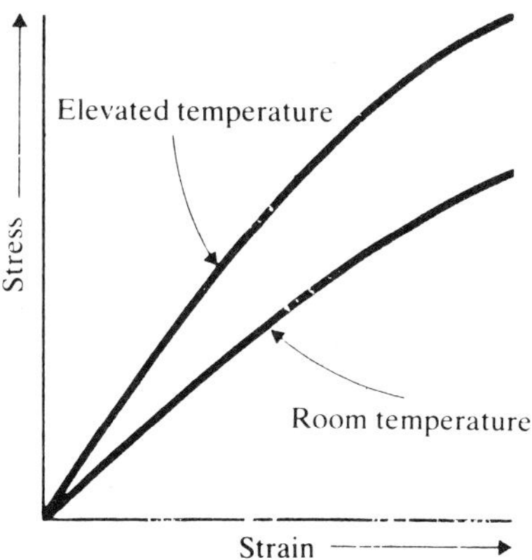

Figure 9

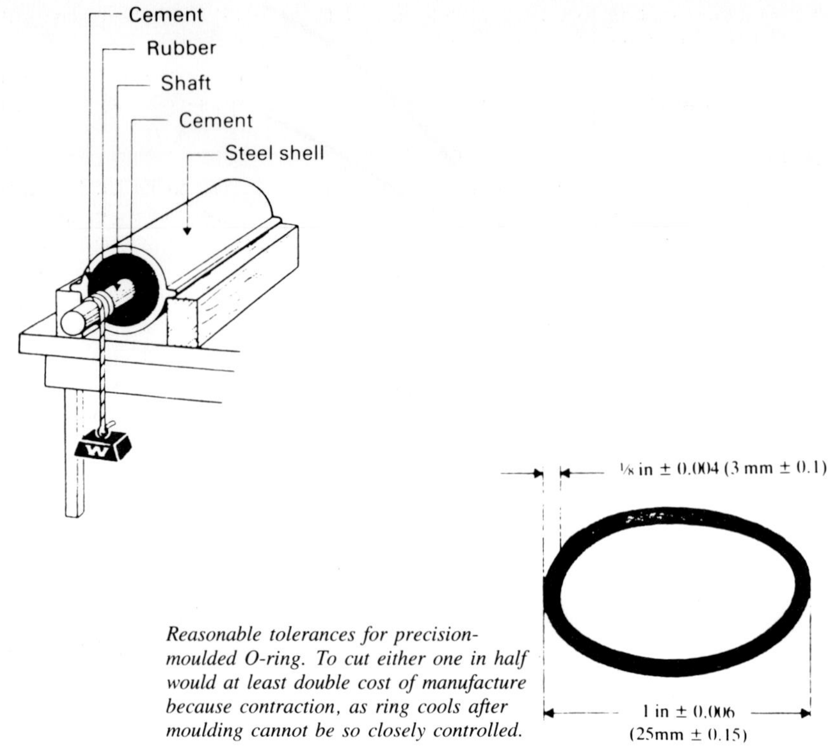

Reasonable tolerances for precision-moulded O-ring. To cut either one in half would at least double cost of manufacture because contraction, as ring cools after moulding cannot be so closely controlled.

Joule effect on O-rings

When an O-ring is used to seal a rotating shaft, there is always a temptation to make the ring a little smaller than the shaft so it will snap on and make a better seal. That is a fatal mistake. Friction heats the O-ring. The ring is under tension so joule effect comes into play. The ring tries to contract – more friction, more heat. The rubber chars, cracks, leaks. On a high-speed shaft, failure can occur in minutes.

The solution is to make the ring oversize, not undersize, and to obtain a seal by compressing the ring against the shaft. The i.d. of the ring should be about 5% more than the shaft diameter, according to experiments carried out by Precision Rubber Products Corp.

In order to offset Joule effect and the dimensional changes resulting from it, it is necessary to use an oversized ring in relation to the shaft. In other words, apply the seal under peripheral compression. O-ring seals do not have the continual tendency to build up heat. Instead, the heat curve levels off after a moderate temperature rise. Whereas a slightly stretched seal causes the temperature to continue to rise until failure occurs.

This conclusion is based on experiments like this. A shaft, driven by a 3.450 r/min motor, was enclosed in a cylinder into which oil was pumped under pressure. One end of the cylinder was sealed with a diaphragm (not under test) and the other end with the O-ring under study. Shaft diameter was 25.35 mm (0.998 in). An undersized ring, with the i.d. of 24.99 mm (0.984 in), failed in four minutes; in that time oil temperatures rose from 16 to 146°C (60 to 295°F) and was climbing rapidly. An oversized ring, with the i.d. of 26.52 mm (1.044 in), was run for 100 hr and the test discontinued with no leakage or other malfunction; oil temperature rose to 67°C (152°F) and leveled off. Under similar conditions seals have run for 4,000 hr without trouble.

There is much more to proper design than i.d. of the ring, but the concept of peripheral compression, instead of tension, is the critical idea that brings success within reach.

Flexibility

Initial flexibility is largely dependent on the basic elastomer concerned and the hardness. In service flexure performance may be substantially modified by the working temperature of the seal, the presence of fluid in contact with the seal, and other factors. True flexure performance, therefore, can only be evaluated by experience or measurement under actual service conditions. This applies also to failure by flex cracking, although here the temperature of operation may be the main factor with specific elastomers, and lead to crystallization at low temperatures or embrittlement at high temperatures.

In general terms with decreasing temperature below about −10°C to 50°C rubber characteristics change first from 'rubbery' to 'leathery' and, at still lower temperatures, lose all flexibility and become brittle (Figure 10). The critical low temperature at which marked loss of flexibility occurs varies widely with elastomer type and composition – *eg* see Figure 11.

Composites of rubber and fabric are widely used in the manufacture of seals to provide strength and rigidity in much the same way as reinforcement in concrete. They restrain the stretch of the rubber but essential flexibility is not lost. Controlled extensibility can be built into a fabric by careful choice of the yarn and weave construction, to give a specific degree of stretch to the rubber fabric composite. Rubberized fabrics are extensively used in seals to improve the modulus and stiffness and to increase the strength, wear and tear resistance. Many types of fabrics are used, depending on the temperature and service conditions. Up to about 100°C there is no significant degradation of either natural or synthetic fabrics. Although many fabrics may be affected by exposure to oxygen or water, the fact that they are embedded in a rubber compound protects them from degradation. Above 100°C, however, it is necessary to be more selective with rubber compound as well as the fabric used, because both (with the exception of glass or asbestos fabric) are increasingly affected by a rise in temperature.

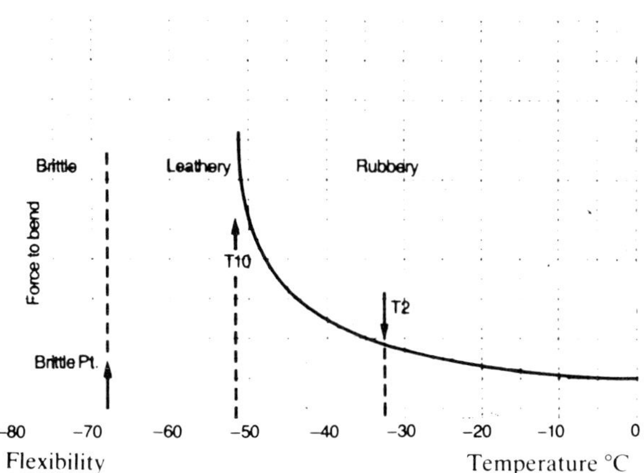

Figure 10
Low temperature behaviour of rubber

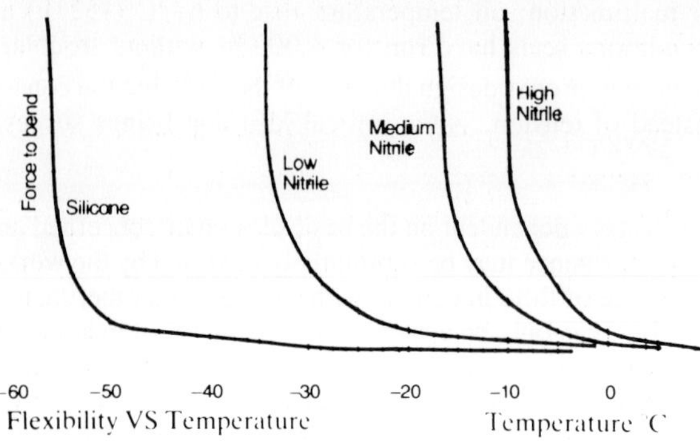

Figure 11
Low temperature flexibility of elastomers.

Hardness

In general terms rubbers with low hardness (*ie* softer materials) are more flexible and thus seal more readily on rougher surfaces or have better conformity, although they are more susceptible to wear, abrasion and extrusion.

A *decrease* in hardness can also be expected to reduce breakout friction with dynamic seals but running friction is reduced with *increasing* hardness (provided the unit surface load is reduced by reducing the squeeze). Basically, therefore, a higher hardness figure should give lower rubbing friction in any dynamic seal, although this depends specifically on adjusting the squeeze to compensate. Retaining the same squeeze and increasing the hardness of the elastomer can increase both the breakout friction and the running friction.

The hardness of any basic rubber type can be modified by compounding. Hardness values used for seal materials may range from as soft as 40 to 50° up to very hard compounds of 90 to 95° hardness – see Tables 2 and 3. Hardness is thus a control factor in the formulation of an elastomer for specific applications. For the majority of applications a hardness figure of 70 to 80° is normally selected.

TABLE 2 – AIR PERMEABILITY OF ELASTOMERS

	cc/sec/cm²/cm x 10⁻⁸			
Elastomer	**24 °C (75 °F)**	**80 °C (175 °F)**	**120 °C (250 °F)**	**175 °C (350 °F)**
Butyl	0.2	3.2	13	–
Low nitrile	1.3	8.0	22	–
High nitrile	negligible	4.1	15	–
SBR (Buna S)	2.5	29	47	–
Natural rubber	4.9	44	71	–
Neoprene	1.0	9.8	26	–
Polyacrylate	1.9	18	48	94
Polyurethane	0.5	9.7	31	–
Fluoro silicone	–	128	–	–
Silicone	115	350	–	690
Viton A	–	8.8	36	146

TABLE 3 – GAS PERMEABILITY OF ELASTOMERS

			cc/sec/cm^2/cm x 10^{-8} at		
Elastomer	**Oxygen**	**Nitrogen**	**Carbon Dioxide**	**Hydrogen**	**Methane**
Butyl	0.3 to 1.0	0.11 to 0.35	0.03 to 0.10	0.03 to 0.10	0.01 to 0.33
SBR	13.0 to 34.5	4.8 to 14.5	94.0 to 195.0	30.5 to 74.0	16.0 to 43.0
Neoprene	3.0 to 10.0	0.9 to 3.55	19.5 to 56.5	10.3 to 28.5	2.5 to 9.8
Nitrile	3.2 to 10.5	0.9 to 3.7	23.0 to 66.0	11.5 to 31.5	2.4 to 10.1
Viton	1.1 to 6.7	0.3 to 2.6	5.9 to 29.8	–	–
Natural rubber	18.0 to 19.5	6.6 to 22.5	102.0 to 220.0	39.0 to 97.0	22.0 to 64.0

Note: Figures refer to permeability at 25 °C and 50 °C, respectively.
Viton measured at 30 °C and 60 °C.

Standards for measurement of rubber hardness are: IRHD (International Ssu-1TRubber Hardness Degrees), BS (British Standard) degrees and Shore Durometer A. The first two are identical. However, Shore Hardness (Shore A) may be measured as an instantaneous reading or as a 30 second reading. In the former case the value obtained is about five degrees higher than IRHD. Instantaneous Shore A Hardness figures are used in many American specifications. Shore Durometer or hardness degrees (now an obsolete measurement) are virtually identical except for rubbers softer than 30°C degrees. BS Hardness BD 903 Part A7:1957 has been accepted as the ISO standard and is in agreement with USA ASTM/D/1415-56T and the German DIN 53519 standard.

In practice hardness values can be measured to plus or minus one degree, although it is more usual to quote the hardness figure to the nearest five degrees. Shore Durometer hardness measurement is normally subject to plus or minus five degrees tolerance and thus Shore hardness figures are normally quoted to the nearest ten degrees. Thus a difference in hardness of a few degrees on any of these scales has virtually no significance.

A hardness figure alone is not a reliable, or even an accurate criterion, due largely to the difficulty of measuring the *actual* hardness of a seal specimen accurately.

Hardness (and therefore friction) is also affected by swelling. Swelling tends to reduce hardness and at the same time will increase the squeeze. Similarly, hardness is affected by temperature, decreasing with increasing temperature, and *vice versa.* At the same time the squeeze is modified by the expansion or contraction of the seal, (remembering that this expansion or contraction will be greater than that of the metal surrounds).

Tensile strength and linear resistance

Tensile strength is a measure of the mechanical strength of material and in the case of rubbers can be taken as a general indication of resistance to deterioration under stress, *ie* wear and cracking. However, there is no direct relationship between tensile strength and wear resistance. Other factors, such as roughness of the surface against which the seal rubs, and service temperature, can be more significant in practice.

The tensile strength of elastomers is generally low to moderate, polyurethane rubbers being the exception – see Figure 12. Tensile strength also degenerates with increasing temperature. The actual value of tensile strength is not normally of critical importance in seals except that materials with a tensile strength below 70 kg/cm^2 (1000 lb/in^2) may not be suitable for dynamic seals.

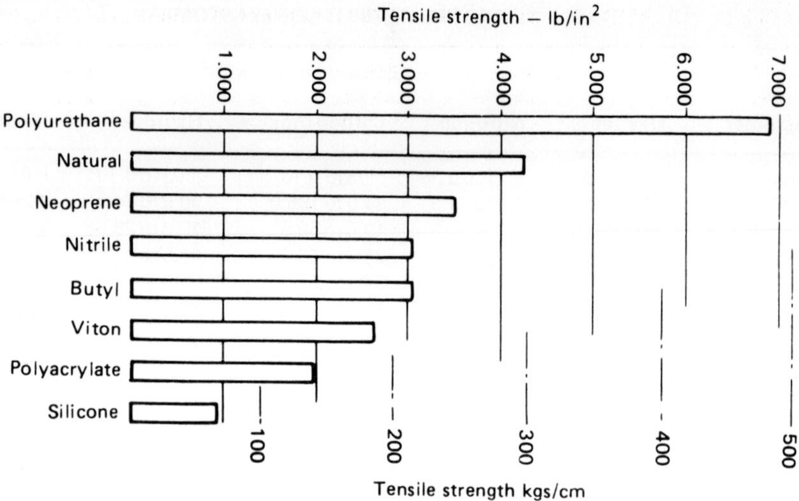

Figure 12
Tensile strength of rubbers at 20°C.

Abrasion resistance

This is an important parameter as far as dynamic seals are concerned, but difficult to assess other than on purely empirical lines. Experience shows that certain materials, such as natural rubber, polyurethane and leathers, have outstanding resistance to abrasion, while others, such as silicone rubbers, may have poor characteristics in this respect. In general, as far as elastomers are concerned, abrasion resistance generally improves with increasing hardness for a particular basic elastomer and may be further enhanced by compounding. Good abrasion resistance is also often allied to high tear resistance, and *vice versa.* See also Table 4.

TABLE OF ASTM MATERIAL DESIGNATIONS

ASTM D-2000, SAE J200 Material Designation (Type and Class)	Type of Polymer Normally Used	ASTM D-2000, SAE J200 Material Designation (Type and Class)	Type of Polymer Normally Used
AA	Natural rubber, reclaimed rubber, SBR, butyl, ethylene propylene, polybutadiene, polyisoprene	CE	Chlorosulphonated polyethylene (Hypalon®)
		CH	NBR polymers
AK	Polysulphides	DF	Polyacrylic (butyl-acrylate type)
BA	Ethylene propylene, high temperatures SBR and butyl compounds	DH	Polyacrylic polymers
		FC	Silicones (high strength)
BC	Chloroprene polymers (Neoprene)	FE	Silicones
BE	Chloroprene polymers (Neoprene)	FFKM	Perfluoroelastomers
BF	NBR polymers	FK	Fluorinated silicones
BG	NBR polymers, urethanes	GE	Silicones
BK	Polysulphides, NBR	HK	Fluorinated elastomers (Viton®, Fluorel®, *etc.*)
CA	Ethylene propylene		

Example of how to read an ASTM designation: ASTM AA 625

AA = Natural Rubber
6 = 60 ± 5 Durometer
25 = 2500 Minimum Tensile Strength

TABLE 4 – COMPARATIVE PROPERTIES OF TYPICAL COMMERCIAL ELASTOMERS

PROPERTIES Chemical Name	**Natural Rubber** Polyisoprene	**SBR or GR-S or Buna S** Butadiene styrene	**Butyl** Isobutylene isoprene	**Nitrile** Butadiene acrylonitrile
Material Designation (ASTM D-2000, SAE J200 Classification)	AA	AA	AA	BF, BG, BK, CH
Tensile Strength, MPa (Psi)				
Pure Gum	Over 20.7 (3000)	Below 6.9 (1000)	Over 10.3 (1500)	Below 6.9 (1000)
Black loaded stocks	Over 20.7 (3000)	Over 13.8 (2000)	Over 13.8 (2000)	Over 13.8 (2000)
Hardness Range (Durom. A.)	30 to 90	40 to 90	40 to 75	40 to 95
Specific Gravity (Base Material)	0.93	0.94	0.92	1.00
Adhesion to Metals	E	E	G	E
Adhesion to Fabrics	E	G	G	G
Tear Resistance	G to VG	F	G	F
Abrasion Resistance	E	G to E	G	G
Compression Set	G	G	F	G
Rebound				
Cold	E	G	P	G
Hot	E	G	VG	G
Dielectric Strength	E	E	E	P
Electrical Insulation	G to E	G to E	G to E	P
Permeability to Gases	FL	FL	VL	L
Acid Resistance				
Dilute	F to G	F to G	E	G
Concentrated	F to G	F to G	G	G
Solvent Resistance				
Aliphatic hydrocarbons	P	P	P	E
Aromatic hydrocarbons	P	P	P	G
Oxygenated (Keytones, *etc.*)	F to G	G	G	P
Lacquer solvents	P	P	F to G	F
Resistance to:				
Swelling in lubricating oil	P	P	P	VG
Oil and Gasoline	P	P	P	E
Animal and Vegetable Oils	P to G	P to G	VG	VG
Water absorption	VG	G to VG	VG	G
Oxidation	G	F	E	G
Ozone	P to F	P	E	F
Sunlight ageing	P	P	VG	P
Heat ageing (upper limit cont. service)	85°C (185°F)	90°C (194°F)	120°C (248°F)	115°C (239°F)
Flame[3]	P	P	P	P
Heat	G	F to G	VG	G
Cold	E	VG	G	F to G

Key: O – Outstanding, E – Excellent, VG – Very Good, G – Good, F – Fair, P – Poor, FL – Fairly Low, L – Low, VL – Very Low.

*** Du Pont's registered trademark.**

[1] Available only as finished parts only from Du Pont.

[2] With additives

[3] These evaluators are qualitative and comparative only. They should not be construed as recommendations. Specific compounding is required to optimize performance. Elastomer choice should be based upon a practical consideration of the potential fire hazards involved in each individual case and, if applicable, the results of appropriate flame tests.

See footnote:-
Measuring Performance Characteristics
on page 429.

cont...

TABLE 4 – COMPARATIVE PROPERTIES

PROPERTIES Chemical Name	**Neoprene** Polychloroprene	**Hypalon** Chlorosulfonated polyethylene	**Nordel* (EPDM Rubber)** Ethylene propylene polymer	**Epichloro-hydrin Rubber**
Material Designation	BC, BE	CE	CA	CH
(ASTM D-2000, SAE J200 Classification)				
Tensile Strength, MPa (Psi)				
Pure Gum	Over 20.7 (3000)	Over 17.2 (2500)	Below 6.9 (1000)	Below 6.9 (1000)
Black loaded stocks	Over 20.7 (3000)	Over 20.7 (3000)	Over 20.7 (3000)	17.2 (2500)
Hardness Range (Durom. A.)	40 to 95	40 to 95	40 to 90	40 to 90
Specific Gravity (Base Material)	1.23	1.12 to 1.28	0.86	1.36 to 1.27
Adhesion to Metals	E	E	G to E	F to G
Adhesion to Fabrics	E	G	G	F to G
Tear Resistance	G	F	F to G	F to G
Abrasion Resistance	E	E	G to E	F to G
Compression Set	F to G	F	G	P
Rebound	VG	F to G	VG	G
Cold				
	VG	G	VG	G
Hot				
	G	VG to E	E	G
Dielectric Strength	F to G	G	E	G
Electrical Insulation	L	VL	FL	L to FL
Permeability to Gases				
Acid Resistance	E	E	E	F to G
Dilute	G	VG	E	F
Concentrated				
Solvent Resistance	F to G	G	P	E
Aliphatic hydrocarbons	F	G	P	G
Aromatic hydrocarbons	P	F	G to VG	P
Oxygenated (Keytones, *etc.*)	P	P	P to F	F
Lacquer solvents				
Resistance to:	G	G to E	P	E
Swelling in lubricating oil	G	G	P	E
Oil and Gasoline	G	G	G	E
Animal and Vegetable Oils	G	VG	VG to E	G
Water absorption				
Oxidation	VG to E	E	E	G
Ozone	VG to E	O	O	E
Sunlight ageing	VG	O	O	G
Heat ageing (upper limit cont. service)	95°C (203°F)	135°C (275°F)	145°C (293°F)	135°C (275°F)
Flame[3]	G	G	P	P to F
Heat	VG	E	E	VG
Cold	G	G	E	G to VG

Key: O – Outstanding, E – Excellent, VG – Very Good, G – Good, F – Fair, P – Poor, FL – Fairly Low, L – Low, VL – Very Low.

*** Du Pont's registered trademark.**
[1] Available only as finished parts only from Du Pont.
[2] With additives
[3] These evaluators are qualitative and comparative only. They should not be construed as recommendations. Specific compounding is required to optimize performance. Elastomer choice should be based upon a practical consideration of the potential fire hazards involved in each individual case and, if applicable, the results of appropriate flame tests.

OF TYPICAL COMMERCIAL ELASTOMERS (cont'd.)

Vamac* Ethylene/acrylic elastomer	**Silicone** Polysiloxane polymer	**Viton*** Fluoroelastomer	**Kalrez**** Perfluoro-elastomer[1]	**Adiprene** Urethane rubber
AEM	GE	HK	FFKM	BG
–	Below 10.3 (1500)	Over 12.4 (1800)	Not applicable	Over 27.6 (4000)
Over 17.2 (2500)	Over 10.3 (1500)	Over 13.8 (2000)	Over 13.8 (2000)	–
40 to 95A	40 to 85	55 to 95	65 to 95A	60 to 99 +A (up to 80D)
1.08 to 1.12	1.14 to 2.05	1.85	2.01	1.06
VG to E	E	F to G	F	E
G	E	G to E	G	VG to E
G	P	F	F	E
G	P	G	G	O
G	F	F to G (Exc. at high temp.)	F	F
P	E	F to G	Not available	Good (poor at very low temp.)
F	E	G (Exc. at high temp.)	Not available	G at R.T.
G	G	G	E	E
F to G	E	F to G	E	F to G
VL	FL	VL	F	FL
G	E	E	E	F
P	F	E	E	P
G	P	E	E	G to E
F	P	E	E	F to G
P	F	P	E	P
P	P	P to F	E	P
G	F	E	E	G to E
G	F	E	E	G to E
G	G to E	E	E	G to E
VG up to 100°C (212°F)	E	VG	VG	G at R.T. P at 100°C (212°F)
O	E	O	O	E
O	E	O	O	E
O	E	O	O	G
165°C (329°F)	205°C (401°F)	205°C (401°F)	290°C (554°F)	85°C (185°F)
P	F to G	E	E	F (will melt)
E	O	O	VO	G
G	O	G	G	E

Measuring Performance Characteristics
Test methods have been established by the American Society for Testing Materials (ASTM) and the Society of Automative Engineers (SAE) for measuring physical and chemical properties of products made from elastomers. Properly used and interpreted, these tests provide an excellent basis for specifying to the rubber supplier the performance characteristics you require in a rubber product. A complete description of testing procedures is contained in the publication, 'ASTM Standards on Rubber Products', issued by the ASTM, 1916 Race Street, Philadelphia, Pennsylvania, 19103.

cont...

TABLE 4 – COMPARATIVE PROPERTIES OF TYPICAL COMMERCIAL ELASTOMERS (cont'd.)

PROPERTIES			Hytrel*		Alcryn
Chemical Name			Polyester elastomer		
Material Designation			YBPO		Not assigned
(ASTM D-2000, SAE J200 Classification)					
Tensile Strength, MPa (Psi)					
Pure Gum	25.5 (3700)	35.8 (5200)	38.0 (5500)	38.0 (5500)	–
Black loaded stocks	–	–	–	–	–
Hardness Range (Durom. A.)	92A	55D	63D	72D	60 to 80A
Specific Gravity (Base Material)	1.17	1.20	1.22	1.25	1.21
Adhesion to Metals	G	G	G	G	–
Adhesion to Fabrics	G	G	G	G	–
Tear Resistance	E	O	O	O	G
Abrasion Resistance	O	VO	VO	VO	E
Compression Set	F	F	P	P	F to G
Rebound	VG	G	F	F	P
Cold					
	E	VG	G	G	P
Hot					
	F to G	F to G	F to G	G	–
Dielectric Strength	F to G	F to G	F to G	G	F
Electrical Insulation	F	F	F	G	–
Permeability to Gases					
Acid Resistance	F	F	F	F	E
Dilute	P	P	P	P	–
Concentrated					
Solvent Resistance	E	E	E	E	G
Aliphatic hydrocarbons	G	G	G	G	G
Aromatic hydrocarbons	F	G	G	G	F
Oxygenated (Keytones,*etc.*)	F	F to G	G	G	–
Lacquer solvents					
Resistance to:	G	E	E	E	E
Swelling in lubricating oil	VG	E	E	E	E
Oil and Gasoline	VG	E	E	E	E
Animal and Vegetable Oils	VG up to	VG up to	VG up to	VG up to	–
Water absorption	100°C (212°F)	100°C (212°F)	100°C (212°F)	100°C (212°F)	G
Oxidation	E	E	E	E	E
Ozone	E	E	E	E	O
Sunlight ageing	VG[2]	VG[2]	VG[2]	VG[2]	O
Heat ageing (upper limit cont. service)	100°C (212°F)	110°C (230°F)	110°C (230°F)	110°C (230°F)	–
Flame[3]			G[2] (will melt)		–
Heat	VG	E	E	E	VG
Cold	E	E	E	E	VG

Key: O – Outstanding, E – Excellent, VG – Very Good, G – Good, F – Fair, P – Poor,FL – Fairly Low, L – Low, VL – Very Low.

*** Du Pont's registered trademark.**

[1] Available only as finished parts only from Du Pont.

[2] With additives

[3] These evaluators are qualitative and comparative only. They should not be construed as recommendations. Specific compounding is required to optimize performance. Elastomer choice should be based upon a practical consideration of the potential fire hazards involved in each individual case and, if applicable, the results of appropriate flame tests.

See footnote:-
Measuring Performance
Characteristics
on page 429.

Tear resistance

In general tear resistance tends to be moderate to low with elastomers and high with fabricated materials or leather. The higher the tear resistance, the less the likelihood of a seal failing should it be accidently scratched or nicked, as may happen during fitting. Such materials with low tear resistance need particular care in handling and fitting to avoid all possibility of such damage occurring.

Permeability

Permeability is not usually a significant factor with liquid seals but can be important when sealing gases and critical for vacuum work. Permeability varies considerably with different elastomers – Table 5 – with the hardness of individual elastomers, and also with temperature.

In the case of vacuum services, further troubles may be experienced through outgassing of adsorbed or absorbed gases, and evaporation or sublimation of volatile constituents of the elastomer at very low pressures. The problem is one which requires close co-operation with seal manufacturers at an early stage of equipment design.

Operating Temperature

The operating temperature of a seal is a vital factor because any substantial difference between this and normal ambient temperature will normally modify the material characteristics, particularly in the case of elastomers. The changes which occur at low temperatures are quite different from those produced by elevated temperatures.

With decreasing temperature the tendency is for all elastomers to become progressively harder with loss of flexibility and slower recovery from deformation. Hardness/temperature curves in themselves do not give any particularly useful information as hardness may reach a nominal or actual maximum value while the material still retains good flexibility. Direct measurement of flexibility or torsional stiffness is much more significant, and if this is plotted against temperature it will show a curve with a characteristic bend, *eg* T2 point onwards in Figure 10. From this can be determined

TABLE 5 – HARDNESS RANGE AND SERVICE TEMPERATURE OF ELASTOMERS

Elastomer	Hardness Range* Degrees	SERVICE TEMPERATURE RANGE Minimum °C	Minimum °F	Maximum† °C	Maximum† °F
Natural rubber	30 to 90	–50	–58	70	160
Butyl	45 to 85	–40	–40	80	175
Chloroprene (Neoprene)	40 to 90	–40	–40	80	175
Nitrile (low/medium)	40 to 90	–60	–76	135	275
Nitrile (medium/high)	50 to 95	–40	–40	120	250
Polyurethane	60 to 85	–40	–40	100	212
Fluorinated rubbers	65 to 80	–20	– 4	250	480
Fluoro-silicone	60 to 80	–60	–76	200	392
Silicone	40 to 80	–20	–94	250	480
Ethylene propylene	45 to 85	–50	–58	150	300
Perfluoroelastomer	65 to 95	–50	–58	200	554

* A hardness of 100° is equivalent to that of glass.

† Conservative rating for continuous exposure.

the *freeze point* where a marked loss of flexibility starts. Beyond the freeze point stiffness increases very rapidly with further decrease in temperature until the *brittle point* is reached – that is, the material becomes brittle and will break if flexed.

For design purposes the freeze point can be determined as the temperature at which the original (or 20°C) stiffness is doubled (×2 freeze point). The freeze point equivalent to an increase in stiffness to ten times the original may also be given (×10 freeze point) as a close indication of the temperature at which the material becomes quite unusable for flexing and is rapidly approaching the brittle condition. The ×2 freeze point represents a safe minimum temperature limit for working.

With certain elastomers a decrease in temperature may promote definite crystallization of the material in addition to normal stiffening. This may build up slowly, or even be localized to give a 'flat spot' on the seal. The material may well be still usable under such conditions, due to the fact that it is nowhere near its brittle point, when in such cases the necessary resilience can be provided by spring loading if there is no immediate or economic choice of alternative elastomer.

For any basic elastomer, low temperature characteristics can be modified to some extent by compounding. Thus an increase in hardness will usually lower the brittle point but make the material less flexible generally, whilst improvements in chemical resistance will often raise the brittle point.

It should also be emphasized that laboratory tests on the material alone at low temperatures will not necessarily be characteristic of the material performance in service as a seal. This is largely because the fluid in contact with the seal can affect the degree of plasticization – it can be absorbed, for instance, and increase the effective degree of plasticization, or leach out a proportion of the original plasticizer. Control of these effects is largely a matter of compounding, although compatability with the fluid may be a prior requirement, in which case it may be necessary to sacrifice some low temperature performance.

At elevated temperatures, all elastomers lose strength and thus tend to become softer and more flexible. Normally recovery is complete on reduction of temperature, but if the temperature is high enough some changes may be permanent. Also ageing characteristics are accelerated by heat, normally taking the form of a progressive increase in hardness and modulus with loss of elastomeric properties.

A further important effect which may have to be considered when the operating temperature of the seal differs substantially from normal room temperature is the relative thermal expansion or contraction of the seal and its surrounds. The thermal coefficient of expansion is very much higher that that of metals (roughly ten times that of steel) – see Table 6.

This is normally most significant at elevated temperatures where thermal expansion of the seal is substantially greater than that of its surrounds and actual volumetric expansion may be further increased by swelling in contact with the fluid.

Compression (squeeze)

Many simple seals, such as O-rings, rely specifically on squeeze to close the seal, this being provided by a matching dimension for groove depth and clearance gap. Excessive squeeze is to be avoided as this can overstress the material and produce premature ageing, as well as leading to unnecessarily high friction in the case of dynamic seals. It may, however, be employed deliberately on static seal closures to increase the pressure rating.

Maximum squeeze which can be tolerated without deterioration of seal properties depends very much on the elastomer concerned. In the case of Nitrile rubbers a figure

TABLE 6 – COEFFICIENT OF EXPANSION OF ELASTOMERS

Elastomer	Coefficient of Expansion cm/cm/°C	Coefficient of Expansion (Varying slightly with compositon)* in/in/°F
Natural rubber	14.4 to 15.3 x 10^{-5}	80 to 85 x 10^{-6}
Butyl	15.0 x 10^{-5}	83 x 10^{-6}
Chloroprene	13.7 x 10^{-5}	76 x 10^{-6}
Nitrile	11.2 x 10^{-5}	62 x 10^{-6}
Fluorocarbon	16.2 x 10^{-5}	90 x 10^{-6}
Silicone	18.5 x 10^{-5}	103 x 10^{-6}

*For comparison –		
	Steel	11 x 10^{-6}
	Stainless steel	10.4 x 10^{-6}
	Aluminium	13 x 10^{-6}
	Perfluoroelastomer	23 x 10^{-5}
	Fluoroelastomer	16 x 10^{-5}

SUMMARY OF ELASTOMER PROPERTIES

	Resistance to non-polar fluids	Resistance to polar fluids	Heat	Low temperature	Comp. Set	Mech. str.	Resilience	Ozone resistance	Gaseous permeability	Comparative polymer cost (volume)
Perfluoroelastomer	E	E	E	M	VG	G	M	E	VG	(Parts)
Fluoroelastomer	E	P	E	M	E	G	M	VG	VG	3547
Polysulphide	E	G	M	G	P	P	G	E	E	290
High nitrile	E	P	G	M	G	G	M	M	VG	110
Chlorohydrin	E	P	G	G	VG	G	VG	E	G	250
Fluorosilicone	VG	P	E	E	E	P	G	E	P	6600
Medium nitrile	VG	P	G	G	G	G	G	M	G	100
Polyacrylic	VG	P	VG	M	G	G	M	VG	M	300
Low nitrile	G	P	G	VG	VG	G	G	M	M	90
Urethanes	G	P	P	VG	M	E	VG	VG	VG	230
Silicone	G	G	E	E	E	M	VG	E	P	690
Chloroprene	M	M	G	G	G	G	VG	VG	G	120
Chlorosulphonated polyethylene	M	VG	VG	G	M	G	M	E	VG	130
Polybutadiene	P	VG	G	VG	G	VG	E	M	G	40
SBR	P	VG	G	VG	G	G	G	M	M	40
Butyl	P	E	G	VG	G	G	P	VG	E	60
Ethylene-propylene	P	E	VG	VG	VG	G	G	E	G	60
Polyester	G	G	E	E	M	G	G	E	P	–

Key:

E – Excellent M – Moderate
G – Good P – Poor
VG – Very good

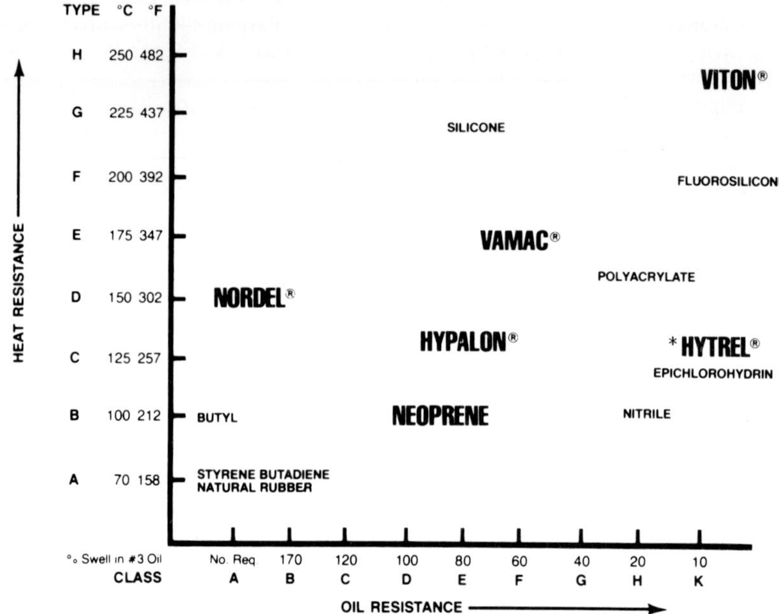

Example: The first two steps in any elastomer specification are to designate the *Type* (heat resistance) with a single letter, such as 'B' for 100 °C (212 °F) and to designate *Class* (oil resistance) with a second letter, such as 'C' for 120% (maximum percent volume swell in ASTM No.3 oil at 100 °C).

Thus, an elastomeric application with these requirements would be designated as 'BC'.

**HYTREL ® is a thermoplastic elastomer and, as such, is not included in the thermoset elastomer SAE J200 specification system. It has been positioned in this chart to demonstrate its resistance characteristics to oil and heat.*

of 30% compression is generally recommended. For elastomers such as silicones, fluoroelastomers and ethylene-propylene it can be higher.

Moulding shrinkage

All elastomers used for seals shrink by a certain amount during moulding. This is not normally significant because the mould dimensions are usually adjusted to compensate for this in the production of standard sizes of seals, for example O-rings. Where the same standard moulds are used for other elastomers the resulting mouldings may shrink more or less than the mould allowance. Thus the shrinkage with silicone rubber will be measurably greater than Nitrile or butyl, for example, in the same mould. To hold close tolerances in such cases may call for the use of separate moulds.

ASTM D2000 Line Call Out

The ASTM D2000 Line Call Out is a method of specifying the principal material and engineering parameters for a rubber compound according to the following code:

Grade number showing the complexity of property requirements.
B Specification of operating temperature.
G Specification of acceptable volume swell.
Shore A hardness in 10's (*eg* 9 would specify 90°).
Tensile strength in 100 lb/in^2 unit (*eg* 15 would specify 1500 lb/in^2).
A Heat resistance.
Compression set (22 h at 100°C to ASTM D395).
Fluid Resistance (ASTM Oil No 3 – 70h at 100°C – to ASTM D471).
Water Resistance (70h at 100°C to ASTM C471).
Low Temp Resistance (to ASTM D2137).
As an example, a complete Line Call Out could read: 7 BG 915 A14 B14 E14 E34 L14 F16.

See also Seal Selection Guides (Section 8).

Elastomeric Materials

THE MAJORITY of seals utilize synthetic rubber compounds, the prime reason for using rubber being its inherent elasticity due to the shape of the rubber molecule. Also rubber compounds offer an extremely wide choice of properties through compounding. Thus rubber compounds may be designed by the seal manufacturer specifically to meet a particular application, whilst most plastic compounds used in seals are normally selected from a range of standard materials offered by the large chemical product manufacturers.

Rubber compounds are based on the raw polymer, into which are incorporated a wide range of fillers and chemical agents which provide the processing, vulcanization and physical properties required. This has led to the development of specific general purpose rubber types for seals (*eg* Nitrile rubber which is the most popular synthetic rubber polymer for seal manufacture); and also special purpose rubbers for specific applications. Thus most reputable rubber seal manufacturers have developed a standard range of well tried compounds to meet most rubber applications, but in addition produce special compounds designed to satisfy particular conditions. Examples of the chain structure of different rubbers are shown in Figure 1.

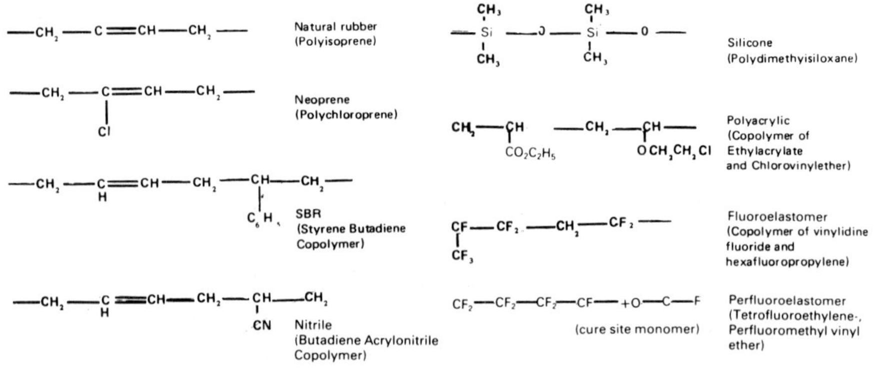

$—CH_2—C{=}CH—CH_2—$ Natural rubber (Polyisoprene)

$—CH_2—C(Cl){=}CH—CH_2—$ Neoprene (Polychloroprene)

$—CH_2—C(H){=}CH—CH_2—CH(C_6H_5)—CH_2—$ SBR (Styrene Butadiene Copolymer)

$—CH_2—C(H){=}CH—CH_2—CH(CN)—CH_2$ Nitrile (Butadiene Acrylonitrile Copolymer)

$—Si(CH_3)_2—O—Si(CH_3)_2—O—$ Silicone (Polydimethylsiloxane)

$CH_2—CH(CO_2C_2H_5)\quad—CH_2—CH(OCH_2CH_2Cl)—$ Polyacrylic (Copolymer of Ethylacrylate and Chlorovinylether)

$CF(CF_3)—CF_2—CH_2—CF_2—$ Fluoroelastomer (Copolymer of vinylidine fluoride and hexafluoropropylene)

$CF_2—CF_2—CF_2—CF—+O—C—F$ (cure site monomer) Perfluoroelastomer (Tetrofluoroethylene-, Perfluoromethyl vinyl ether)

Figure 1
Chain structure of different rubbers.

Elastomeric shielding products

Natural rubber
Initials: NR
Working Temperature Range: $-50°C$ to $+°C$

Physical Characteristics
Strength: high 2500/3500 lb/in^2
Resilience: high 60/80%
Resistance to abrasion: good 80 mm^3
Resistance to ageing/weathering: poor/medium.

Fluid/Chemical Resistance
Resistant to non-mineral, automotive brake fluids.
Not resistant to mineral based oils and greases, or hydrocarbon fuels.

Natural rubber has a basic formula C_5H_8. Unvulcanized rubber has limited uses in industrial and commercial applications, as it is adversely affected by environmental conditions, it is soft and sticky in hot weather, and stiff and rigid in cold weather. (Figure 2).

The vulcanization of rubber, which is the process of incorporating sulphur into the rubber and applying heat, has the effect of tying the polymer chains together at many of the double (C+H) bond sites. (Figure 3)

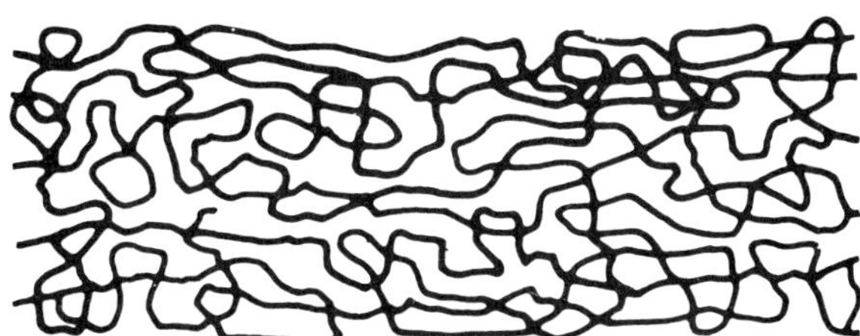

Figure 2
Unvulcanized rubber.

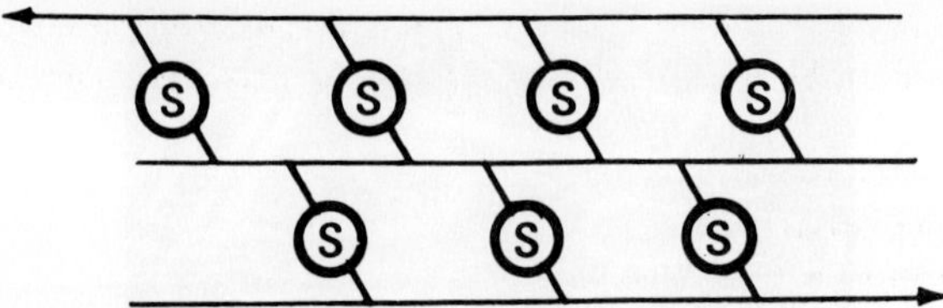

Figure 3
Two-dimensional diagram illustrating the nucleus of vulcanization where the rubber chainis are 'welded' with sulphur bridges (S).

Vulcanized rubber, illustrated in Figure 4, can be easily stretched (as can unvulcanized rubber), but when the stretching force is removed it will return, almost to its original shape.

A typical rubber mix would be as follows:

Rubber	100
Sulphur	2
Accelerator	1.5
Zinc Oxide	5
Stearic Acid	50
Carbon Black	50
Plasticizer	10
Antioxidant	1
Cure 10/150°C	

The use of natural rubber for seals is limited by its poor resistance to mineral-based oils. The generally excellent low temperature characteristics of natural rubber were originally a factor in favour of using a castor-based hydraulic fluid for systems having to operate at low temperatures, when natural rubber seals could be employed, although some of the synthetic elastomers now offer comparable low temperature performance. The cost of natural rubber is also favourable being appreciably less than all of the synthetics with the exception of butyl and *Buna S.*

Natural rubbers can be compounded to meet a wide variety of conditions, coupled with high resilience and strength, and are generally well worth considering for applications

Figure 4
Vulcanized rubber.

where mineral oil fluids are not involved. Flexibility is maintained down to about −55°C, and only the considerably more expensive and weaker silicone rubbers compare in this respect.

Styrene-butadiene rubber

Common Name: Buna S

Initials: SBR (originally GRS)

Working Temperature Range: −50°C to° + 100°C

Physical Characteristics

Strength: medium 1500/2500 lb/in^2

Resilience: medium 40/60%

Resistance to abrasion: medium 110 mm^3

Resistance to ageing/weathering: medium.

Fluid/Chemical Resistance

Resistant to non-mineral, automative brake fluids.

Not resistant to mineral based oils and greases, or hydrocarbon fuels.

Styrene-butadiene rubber was produced during the second World War as a replacement or substitute for natural rubber but has generally inferior properties, although its abrasion resistance may be comparable. It can be considered as a direct alternative to natural rubber, as an example for use with castor-based hydraulic fluids, and this is about its only application for seals.

Butadiene

Polybutadiene or butadiene rubber, generally known as *BR* is again a direct substitute for natural rubber with similar, but generally inferior properties. Low temperature performance is somewhat better than SBR rubber, but its application to seals is strictly limited.

Acrylonitrile-butadiene rubber

Common Name: Nitrile, Buna N (USA)

Initials: NBR

Working Temperature Range: −40°C to° +120°C (typical composition)
−60°C+135°C (specially compounded rubbers).

Physical Characteristics

Strength: medium 1500/2500 lb/in^2

Resilience: medium 40/60%

Resistance to abrasion: medium

Resistance to ageing/weathering: medium.

Fluid/Chemical Resistance

Resistant to mineral oil and greases, water and many other chemicals, hydrocarbon fuels, etc.

Not resistant to non-mineral automotive brake fluids.

Note: Increase in acrylonitrile content improves resistance to mineral oil but adversely affects low temperature resistance.

Low Nitrile = 18 to 24% acrylonitrile
Medium Nitrile = 28 to 33% acrylonitrile
High Nitrile = 38 to 40% acrylonitrile

Nitrile rubbers form the most important group of elastomers for general sealing use. Chemically Nitrile is a copolymer of butadiene and acrylonitrile content, typically varying between about 18% and 48%. Nominal designations are 'low', 'medium' and 'high' Nitrile. Resistance to petroleum based oils and hydrocarbons increases with increasing Nitrile content, but at the same time low temperature flexibility decreases with increasing Nitrile content. In order to obtain good low temperature performance with Nitrile rubbers it is usually necessary to sacrifice some high temperature fuel and oil resistance.

Nitrile rubbers have good physical characteristics and are superior to most other rubbers as regards cold flow, tear, and abrasion resistance. They are not particularly resistant to ozone, weathering and sunlight, but their properties in this respect can be improved by compounding. Due to their susceptibility to ozone attack Nitrile rubber seals should not be stored near any possible source of ozone, *eg*, near an electric motor or electrical equipment, or in direct sunlight.

Butyl

Butyl rubber is a copolymer of isobutylene with isoprene, yielding a group of rubbers which are extremely resistant to water (better than chloroprene and Nitrile) and certain other fluids. A further favourable characteristic is that butyl is very resistant to gas permeation, making it attractive in seals for vacuum systems.

Butyl is also resistant to vegetable oils, but is not suitable for use with mineral based oils and aromatic solvents. Until comparatively recently it was a first choice seal material for use with phosphate ester type hydraulic fluids, although it is now rivalled and largely superseded in this application by ethylene-propylene rubbers.

Carboxylated nitrile

Initials: XNBR

Working Temperature Range: −30°C to +120°C.

Physical Characteristics

Strength: high 2500/2800 lb/in^2

Resilience: medium 40/60%

Resistance to abrasion: high 90 mm^3

Resistance to ageing/weathering: medium.

Fluid/Chemical Resistance

Resistant to mineral oil and grease, water and many chemicals.

Not resistant to non-mineral, automative brake fluids.

Chlorosulphonated polyethylene

Trade Mark Name: Hypalon

Also known as CSM (chlorosulphonated monomer). The low temperature properties of this elastomer tend to be low for sealing applications. However, the material has good resistance to acids and heat and may be used in such applications. It is also completely resistant to ozone.

Working Temperature Range: −30°C to +120°C.

Ethylene-propylene
Initials: EP, EPM, EPDM*
Working Temperature Range: −50°C to +150°C.

Physical Characteristics
Strength: medium 15000/2500 lb/in^2
Resilience: medium 40/60%
Resistance to abrasion: high 110 mm^3
Resistance to ageing/weathering: good.

Fluid/Chemical Resistance
Resistant to non-mineral, automative brake fluids.
Resistant to phosphate ester fluid.
Resistant to water/steam and many chemicals.
Not resistant to mineral oils and grease or hydrocarbon fuels.

*The general class is EP rubber, also known as EPM in America. EPDM is a further polymer with enhanced properties for certain services.

Polychloroprene
Common Name: Neoprene
Initials: CR
Working Temperature Range: −40°C to +80°C.

Physical Characteristics
Strength: medium 1500/2500 lb/in^2
Resilience: medium 40/60%
Resistance to abrasion: medium 110 mm^3
Resistance to ageing/weathering: very good.

Fluid/Chemical Resistance
Moderate resistance to mineral oils and greases.
Not resistant to non-mineral automative brake fluids.

Polychloroprene is one of the best general-purpose synthetic rubbers, although its use in seal applications is somewhat limited in present times. Its main advantage is its excellent resistance to weather-ageing. It is also superior to natural rubber in performance at higher temperatures, but tends to harden or stiffen at low temperatures and may also crystallize at low temperatures if under stress. This tendency can be reduced by the correct choice of polymer type and by compounding.

Polyacrylic
Polyacrylic (ACM) rubbers are polymerized products of acrylic acid esters and form, in effect, a group midway in properties between Nitrile and fluorocarbon rubbers. One of the most attractive features from the sealing point of view is their excellent resistance to mineral oils, hypoid oils and greases up to temperatures of 175°C. Excellent resistance to ageing and flex-cracking also favours use for rotary shaft seals. Low temperature characteristics are not outstanding and mechanical strength and resistance to water are generally inferior.

Polysulphide
This rubber is noted for its excellent resistance to fuels and solvents, oxygen, ozone and ageing. Its mechanical properties and heat resistance are, however, relatively poor and so this elastomer is not widely used as a seal material unless no suitable alternative can be found for the chemical resistance offered.

Polyurethane
Initials: PU

Working Temperature Range: −40°C to +100°C.

Physical Characteristics

Strength: very high plus 3500 lb/in^2

Resilience: good 60/80%

Resistance to abrasion: very high 40 mm^3

Resistance to ageing/weathering: medium.

Fluid/Chemical Resistance

Resistant to mineral oils and greases.

Not resistant to non-mineral, automative brake fluids, water, or acids.

Polyurethane is one of the more recently developed elastomers with exceptional strength, tear, and abrasion resistance (better than all other rubbers) and retaining excellent flexibility at low temperatures. Resistance is good to petroleum products, hydrocarbons, ozone and weathering. Performance is generally unsatisfactory in contact with aqueous solutions of an acid or alkaline nature, chlorinated hydrocarbons, ketones, hot water, steam or glycol. Compression and permanent set characteristics also tend to degrade rapidly with increasing temperature.

Polyurethane rubbers are, therefore, most attractive from the point of view of their mechanical strength rather than 'chemical' or 'temperature' properties. They may be used to advantage, if compatible, under abrasive conditions and are particularly favoured for wipers.

Fluoroelastomer
Initials: FKM

Trademark Name: *Viton, Technoflon, Fluorel*

Working Temperature Range: −40°C to +250°C.

Physical Characteristics

Strength: medium 1500/2500 lb/in^2

Resilience: poor 20/40%

Resistance to abrasion: poor 150 mm^3

Resistance to ageing/weathering: excellent.

Fluid/Chemical Resistance

Excellent resistance to mineral oils and hydrocarbon fuels.

Resistant to many chemicals except ketones, alcohols and acids.

Silicone
Initials: MQ

Working Temperature Range: −70°C to +250°C.

Physical Characteristics

Strength: poor 500/1000 lb/in^2

Resilience: medium 40/60%

Resistance to abrasion: poor 200 mm^3

Resistance to ageing/weathering: excellent.

Fluid/Chemical Resistance

Resistant to mineral oils and greases.

Not resistant to water, acids and non-mineral automative brake fluids.

Basically, silicones have poor strength and tear and abrasion resistance, although mechanical performance can be enhanced by special compounding. Resistance is generally good to alkalis; the chemical properties can be enhanced by special by compounding to provide better resistance to oils and fuels for instance. In general, however, silicone rubbers are not recommended for use with hydrocarbons such as petrols and paraffin, the lighter mineral oils, or steam above a pressure of about 50 lb/in^2 as otherwise considerable swelling and softening of the elastomer can result.

The chief advantage of this type of elastomer is that it retains its flexibility down to very low temperatures, and can also withstand continuous heating at high temperatures without hardening, making it suitable for both high temperature and low temperature seals over a broader range than that covered by the other elastomers. A further application is for rotary seals where the operating temperatures may be higher than that permissible with conventional elastomers due to the friction developed, where again a silicone rubber may provide an answer. The cost of silicone rubber is, however, substantially higher than that of most other elastomers.

Fluorosilicone

Initials: FMQ or FVMQ

Working Temperature Range: −60°C to +200°C.

Physical Characteristics

Strength: poor 500/1000 lb/in^2

Resilience: medium 40/60%

Resistance to abrasion: poor 200 mm^3

Resistance to ageing/weather: excellent

Typical working values depend on compound formulation.

Fluid/Chemical Resistance

Good resistance to mineral oils and greases and hydrocarbon fuels.

Working characteristics of fluorosilicone rubbers are generally similar to those of ordinary silicones, but with a more restricted service temperature range. The main advantage offered is that fluorosilicone rubbers can have an oil resistance comparable with or closely approaching that of Nitrile rubbers. They can thus be used for service temperature limits of Nitrile rubbers and where ordinary silicone elastomers do not have the necessary compatibility with the fluid.

Perfluoroelastomer

Initials: FFKM

Trademark Name: Kalrez

Working Temperature Range: −50°C to +316°C.

Physical Characteristic

Strength:

Kalrez 4079 – 2450 psi (16.9 MPa). BLACK

Kalrez 1050 – 2300 psi (15.9 MPa). BLACK

Kalrez 1045 – 2200 psi (15.2 MPa). WHITE

Fluid Resistence

Excellent all-round resistance to acids, alkalis, ketones, esters, amines, oils, hot water/steam, solvents, *etc*.
(similar to Teflon* PTFE).

Applications

Suitable for general chemical processing industry, refinery, pharmaceutical, biochemical, semi-conductor, instrument, paint/adhesive and oil/gas production industry.

*Du Pont's registered trademark.

Perfluoroelastomer

Trademark Name: Chemraz

Working Temperature Range: −30°C to +230°C.

Physical Characteristics

Strength:

Chemraz 504 1125 lb/in^2

Chemraz 505 1700 lb/in^2

Chemraz 510 2100 lb/in^2

Chemraz 515 1450lb/in^2

Fluid/Chemical Resistance

Excellent all round resistance to acids, alkalis, ketones, esters, amines, solvents, etc.

Suitable for oil and gas production, pharmaceuticals, semi-conductor, liquid chromotography and biomedical uses.

Polyester elastomer

Initials: YBPO

Trade Mark Name: Hytrel

Working Temperature Range: −40°C to +135°C.

Physical Characteristics

Strength: very high 3600/6380 lb/in^2

Resistance to abrasion: outstanding

Resistance to ageing/weathering: excellent.

Fluid/Chemical Resistance

Outstanding resistance to solvents, hydrocarbon fluids including petrol and lubricating oils.

Halogenated polyolefin

Trademark Name: Alcryn

Working Temperature Range: −40°C to +120°C.

Physical Characteristics
Strength: medium 1885 lb/in^2
Resistance to ageing/weathering: excellent.

Fluid/Chemical Resistance
Outstanding oil resistance.

See also *Seal Selection Guides* (Section 8).

Plastic Polymers

NON-ELASTIC POLYMERIC materials may be used as support members with rubber seals and sometimes as seal members in their own right. PTFE is an outstanding example, noted for its almost complete resistance to chemical attack, outstanding service temperature range and a co-efficient of friction when rubbing on steel of the same order as ice sliding on ice. Its mechanical strength is, however, on the low side without filling or reinforcement.

These properties approach the ideal for many sealing applications, provided the basic limitations involved can be overcome. Thus PTFE may be used as a low friction impregnant for fabricated seal or packing sections, the fabricated section itself providing the necessary mechanical strength. In other applications it may be the only material, or by far the most suitable material, for resisting attack by the fluid being handled. In this case a standard form of seal ring may be moulded in the material, possibly with some modification of design to accommodate the lower 'elastic' properties, or the basic material itself may be slightly modified in order to produce an acceptable degree of elasticity. The latter is strictly limited in extent and a maximum 'stretch' which can usually be expected from moulded PTFE rings is of the order of 10%. This can present problems in fitting, such as stretching the rings sufficiently to locate in their grooves. Recovery from stretch will also be slow, so that assembled rings will have to be left for some considerable period before they recover their original size, although this process of recovery can be accelerated by gentle heating. For these reasons PTFE rings are commonly of split form.

Apart from composite seal constructions (and lock-up rings), the most extensive application of PTFE to moulded seals is for O-rings which are produced in a wide range of standard sizes. The PTFE is normally compounded in a form giving a permissible stretch of not more than 10% of the original, i.d. which may demand the use of split assemblies for the smaller diameter sizes.

PTFE O-rings can normally be fitted in the same groove sizes as those recommended for elastomer O-rings, although in particular applications it may be desirable, or necessary, to reduce the amount of squeeze, *ie*, to increase the groove depth slightly. Cost is higher than that of elastomeric O-rings, but a particular advantage which may be offered where a number of different services is involved is that stock sizes of PTFE rings will provide compatibility in all cases instead of having to stock rings in a variety of different elastomers. In the more general case, PTFE O-rings would only be chosen on a compatibility basis, where, for instance, no elastomer offers the necessary chemical resistance to enable an O-ring to be employed or where extremely low friction characteristics and lack of stiction are desirable with an O-ring assembly.

The application of PTFE to moulded rings of other standard seal sections is more limited, both as regards scope and the shapes which are practical to mould in the material.

A very useful extension of PTFE to seals, however, is in composite ring construction, where a machined or moulded PTFE section may be employed as a low friction rubbing surface backed by an elastomeric section or as back-up washers for O-rings, *etc.*

In the case of back-up washers, again the limited stretch of PTFE presents a problem in fitting and rings in this material are commonly split or spiral form to facilitate assembly. While this has the disadvantage of presenting an interrupted surface against which the O-ring bears under pressure this is largely offset by the other advantages offered.

Two similar plastomers to PTFE are PTFCE (polytrifluoro-chloroethylene) and FEP (fluorinated-propylene). PTFCE is not so good and its working temperature range is somewhat reduced. It has the main advantage that it is more amenable to moulding than PTFE and can be injection moulded, if necessary. Again general properties are similar to PTFE but its maximum service temperature is reduced to the order of 200°C.

Other materials which may be used for back-up rings include leather, rubberized fabric, thermoset laminates such as *Tufnol,* and nylon. None of these materials, with the exception of leather and rubberized fabric, is normally used for seals as such.

Cemented Carbides

CEMENTED CARBIDE is a group of powder metallurgical composite materials in which the uniform microstructure is dominated by one or more hard phases, usually tungsten carbide (WC) while the remaining part is a metallic binder phase. In most cases this binder phase is based on cobalt (Co), which has very high wetting properties towards the WC grains during the liquid sintering process, but other metals in the iron group can also be used, *eg* nickel (Ni).

Certain cemented carbide grades contain other types of hard principles like titanium carbide (TiC), niobium carbide (NbC) and tantalum carbide (TaC). These grades are used especially in tools for the turning and milling of metals, *ie* operations in which the tool is subjected to extremely high surface temperatures. Such high temperatures are not obtained in the surface of a mechanical seal and therefore straight WC-Co grades are mostly used in this application. These grades are characterized by high wear resistance, high thermal conductivity and a rather unusual combination of high strength and toughness. As a consequence of this they also exhibit a high resistance to mechanical and thermal stress fatigue, which is important for maintaining the sealing function during periods of more or less dry running. The microstructure of a typical straight WC-Co grade used in seal rings is shown in Figure 1.

Design

After pressing and pre-sintering, the material can be machined by conventional methods. When finally sintered it is completely dense and can only be worked by methods such

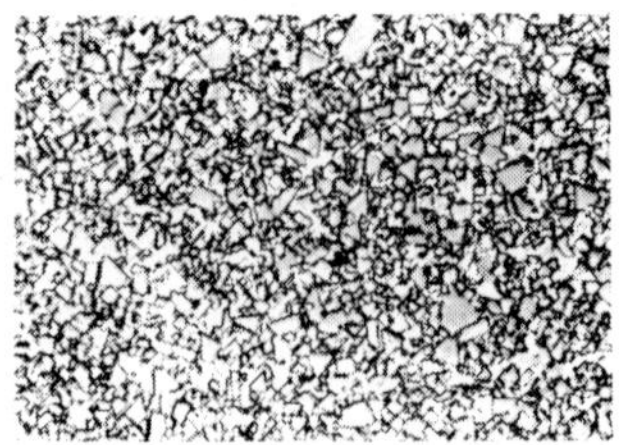

Figure 1
The microstructure of a typical straight WC-Co grade used in seal rings.

as grindling, lapping, honing, polishing and spark erosion. Normally cemented carbide seal rings are lapped to a surface flatness of two light bands (monochromatic sodium light) and a surface smoothness around 0.1 μm Ra.

The largest rings that can be manufactured in one piece have a diameter of about 450 mm. However, for larger dimensions a divided ring construction can be considered. Such constructions function very satisfactorily in applications such as larger dredging pumps.

Seal rings in cemented carbide can be made either wholly of cemented carbide or the carbide ring can be attached to a backing ring. In the latter case methods like brazing, cementing, vulcanizing and mechanical fastening can be employed.

Properties

The commercial straight WC-Co grades cover a large composition range, about 5 to 40% by volume in binder metal content. Furthermore, the average WC grain size may also be varied, normally in the range of 1 to 5 μm. As a consequence of these variation possibilities, materials with widely differing properties can be produced.

Wear resistance

The influence of the Co content and the WC grain on the abrasive wear resistance is shown in Figure 2. The diagram is made from results obtained in a test on DIN 50330 in which the flat end face of a pin was pressed against a rotating dry carborundum paper. The relative wear resistance value is proportional to the inverse of the volume loss during the test.

Similar results are obtained in other abrasive wear resistance tests when, *eg*, water is present in the contact surface.

Hardness

The hardness is influenced by the Co content and the WC grain size in the same way as the abrasive wear resistance. The range starts from the hardest tool steels and reaches up to about 2000 HV.

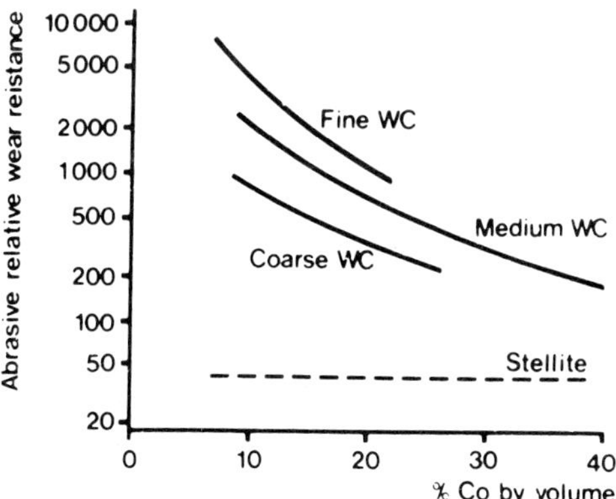

Figure 2
The influence of the Co content and the WC grain size on the abrasive wear resistance.

Density

The straight WC-Co grades cover a range of 12.6 to 15.2 g/cm^3 due to the high density of WC (15.7 g/cm^3). The values for TiC based grades are much lower as the density of TiC is 4.9 g/cm^3.

Mechanical properties

The compressive strength is very high for cemented carbides. It increases with decreasing Co content and decreasing WC grain size. For extreme grades the compressive strength is of the order of 7000 N/mm^2.

The transverse rupture strength (TRS) as a function of the Co content and the WC grain size is shown in Figure 3. The values are obtained in a three-point bending test with a ground prismatic specimen placed upon two cylindrical rests.

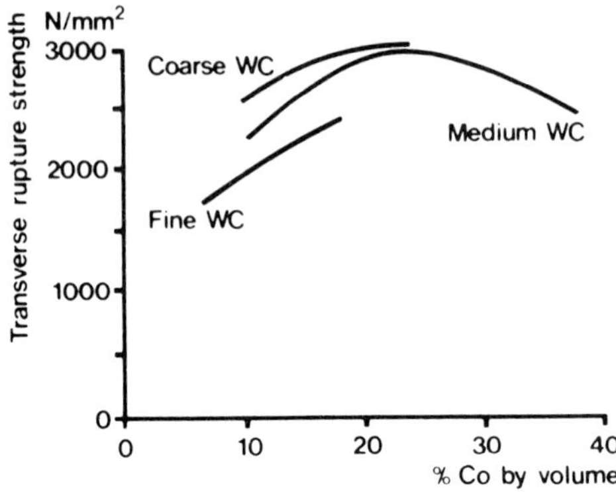

Figure 3
The influence of the Co content and the WC grain size on the transverse rupture strength.

The tensile strength geometry can be evaluated from the TRS value using the *Weibull* theory. Due to the TRS specimen geometry it is 60% of the TRS value and accordingly the tensile strength range is 1000 to 1800 N/mm^2. This has been verified in conventional uniaxial tension tests.

The modulus of elasticity (Young's modulus) is high, about two to three times that of steel. For WC-6% Co (by weight) E is 630000 N/mm^2 and, as it decreases with increasing Co content, WC-11% Co (by Weight) has an E-value of 575000 N/mm^2.

The fracture toughness expressed as the critical stress intensity factor in tension (K_{IC}) increases with increasing Co content and increasing WC grain size. This is shown in Figure 4. The values for the toughest cemented carbide grades, about 25 MN/m$^{3/2}$, are well within the range of the high speed steels.

Thermal properties

The thermal conductivity of straight WC-Co cemented carbides is approximately twice that of unalloyed steels and one third of that of copper. The influence of the Co content is shown in Figure 5. The WC grain size has no noticeable effect but the presence of cubic carbides (TiC, NbC and TaC) largely decrease the thermal conductivity. Resistance

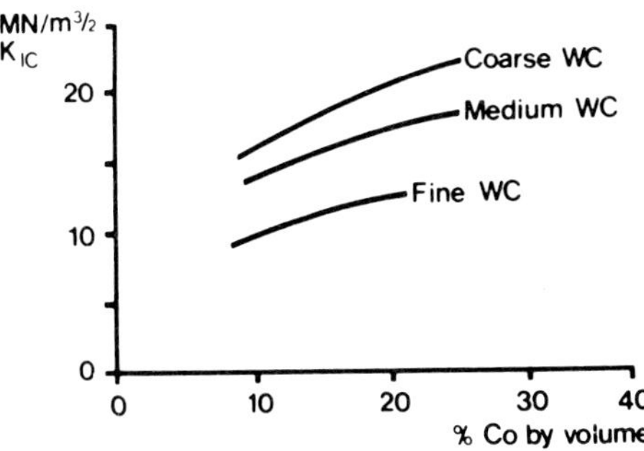

Figure 4
The influence of the Co content and the WC grain size on the critical stress intensity factor in tension (K_{IC}).

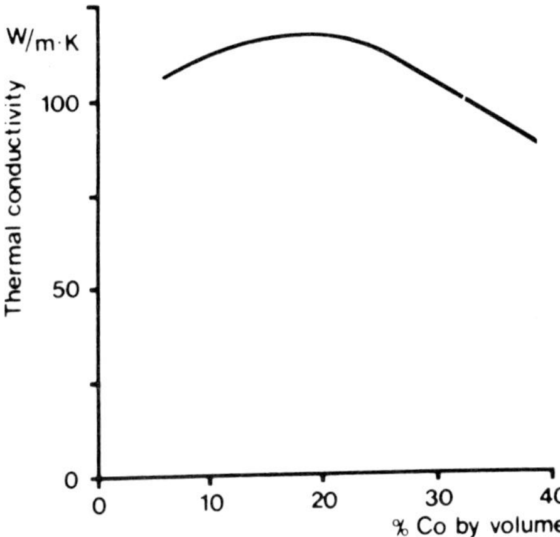

Figure 5
The thermal conductivity as a function of the Co content.

to thermal fatigue is high, due to the beneficial, mechanical properties and the high thermal conductivity. Under identical thermal cycle conditions the propagation rate of thermal fatigue cracks decreases with increasing Co content and increasing WC grain size.

At room temperature the specific heat capacity for straight WC-Co grades is about 200 J/(kg × K), which is nearly half the value of unalloyed steels.

As WC has a low linear expansion coefficient, straight WC-Co grades have values which are about a third of that of austenitic steels. The -value increases with increasing Co content (Figure 6).

In practice WC-based cemented carbides are resistant to oxidation in air up to about 650 to 700 S0008TgC. The constituent affected is the WC which will be transformed to WO_3. TiC is more resistant to oxidation and in air TiC based cemented carbides start to oxidize rapidly when about 900°C has been reached.

COMPARATIVE PROPERTIES OF 'REFEL' SILICON CARBIDE AND OTHER MATERIALS

Material	Density g/cc	Hardness kg/mm^2	Rupture Modulus ' ' M/N/m^2 (x 10^3 lb/in^2)	Young's Modulus 'E' GN/m^2 (x 10^6 lb/in^2)	Poissons Ratio	Thermal Expansion Coefficient ' ' x 10^{-8}/°C 0 to 1000 °C	Thermal Conductivity 'K' W/m °C (cal/cm sec) °C 500 °C	1200 °C	Thermal Shock Parameter Kα/Eα at 500 °C (Cal/cm sec)
REFEL Silicon Carbide	3.10	2500 to 3500	525 (76)	413 (60)	0.24	4.3	83.6 (0.2)	38.9 (0.093)	59
Hot Pressed Silicon Nitride	3.20	2500 to 3500	689 (100)	310 (35)	0.27	3.2	17.5 (0.042)	14 (0.033)	29
Reaction Bonded Silicon Nitride	2.60	900 to 1000	241 (35)	220 (32)	0.27	3.2	15 (0.036)	14.2 (0.034)	13
Hot Pressed Beryllia	3.03	–	207 (30)	400 (58)	0.34	8.5	62.7 (0.15)	16.7 (0.04)	9
Hot Pressed Alumina	3.90	2500	480 (70)	365 (53)	0.27	9.0	8.4 (0.02)	5.0 (0.012)	3
Tungsten Carbide (6% Co)	15.0	1500	1412 (205)	606 (88)	0.26	4.9	86 (0.02)	–	–
Nimonic 105	8.0	350 to 400	–	220 (32)	–	18.8	–	–	–

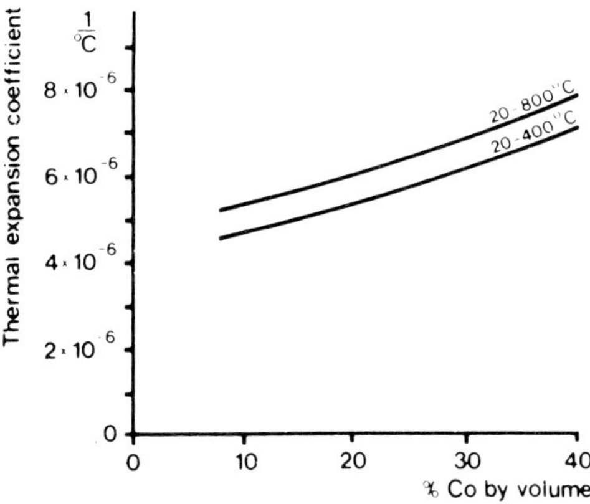

Figure 6
The linear expansion coefficient as a function of the Co content.

Friction

The dynamic friction coefficient in both lubricated and unlubricated sliding is comparable to that of ceramics. The value of the dynamic friction coefficient during steady state of unlubricated sliding decreases somewhat with increasing Co content.

Corrosion resistance

Corrosion in cemented carbides leads generally to a surface depletion of the binder phase. Normally straight WC-Co grades are resistant in solutions down to pH 7. This is also valid for WC-Co grades containing TiC, NbC and TaC. Alloyed TiC-Ni grades are resistant down to pH 1 but, compared to the straight WC-Co grades, they are brittle and have inferior thermal conductivity. They also have the disadvantages of being difficult to grind and braze.

In most corrosion-wear situations the optimum choice is specially alloyed WC-Ni grades which are resistant down to pH 2 to 3 as a general limit. However, they have also been shown to be resistant to corrosion in certain solutions having pH value less than 2. As WC is maintained as the hard principle and Ni and Co are similar metals in most respects their mechanical and thermal properties are comparable to those of the straight WC-Co grades.

The pH limits are only given as general guidelines as the corrosivity of a medium is also influenced by other factors such as temperature, ion concentration and galvanic effects.

Miscellaneous Seal Materials

LEATHER IS composed of high strength interlocking fibres giving the inherent characteristics of flexibility, toughness and resistance to abrasion. It is also a relatively soft material itself, so that its abrading effect on other materials with which it is in rubbing contact is small. Equally it has limitations notably the extent to which it can be moulded (which limits the form of seal sections possible with leather) and lack of rigidity and stretch. Modern processing and treatment methods have, however, considerably enhanced the mechanical properties available with leathers.

For sealing duties, the two chief advantages of leathers are the manner in which it can seal, rubbing against relatively rough surfaces, and its ability to absorb and retain lubricant. The former means that a leather seal can often be used on surfaces which are too rough to be sealed effectively by an elastomer (or will produce excessive wear and friction with an elastomeric seal). A leather seal may also often be employed as a replacement seal for a worn shaft or a worn bore as an alternative to reconditioning the shaft or bore to accommodate an elastomeric seal satisfactorily. Its lubricant retention properties means that a leather seal may provide a better solution than an elastomer where lubrication is restricted or the seal has to continue operating on initial lubricant alone.

Most modern industrial leathers are of the chrome retan type with a tensile strength in excess of 6000 lb/in^2. Maximum service temperature may be as high as 150 to 160°C, depending on the treatment and also, to some extent, on the medium involved. Mechanical performance may be further improved or modified by coating or impregnating the material. Thus silicone rubber-impregnation can be used for maintaining flexibility down to very low temperatures (−60 to −90°C), although this will appreciably increase the cost of the material. Triglyceride and Stearine stuffing improves resistance to water and produces a flexible greasy leather for seals and packings, although in this case the impregnant is soluble in oils. Paraffin wax impregnant produces a stiffer leather which is exceptionally water-resistant and again is used for water seals and packings. Acrylic emulsion treatments produce a tough, very flexible grain surface which is oil-resistant, but permeable to water. Polyurethane impregnant yields good resistance to oils and water, coupled with good flexibility.

The choice is wide, for leather characteristics can be modified in the same way as elastomers, although not to the same extent. The main variable – the quality of the original hide – is a matter of selection and production quality control. Ox hides are normally preferred for leather seals.

The limited mouldability of leathers generally restricts the application of the material to cup, flange and hat sections, U packings, and V packings. Flanged sections are also widely used in oil seals.

Compatibility is rather more limited than that offered by the range of elastomers. In general, leather seals are suitable for use with all oils but have limited compatibility with aqueous solutions, and are not recommended for use with acids and alkalis, gases and organic chemicals. Working temperature range is −40°C to +90°C.

Felt

Felt is a resilient, absorbent material which offers considerable possibilities for certain light-duty sealing applications, a felt seal requires very little contact pressure to work effectively. Felt has capillary properties to absorb and retain oil up to about 80% of its volume and can trap and retain abrasive particles without harm to the rubbing surface on which the seal operates. A lubricated felt seal whether 'clean' or 'dirty' has a polishing action.

Operating temperature for wool felts is of the order of −50°C to +110°C, but higher working temperatures are possible with felts made from synthetic fibres. Rubbing speeds up to 10 m/sec (2000 ft/in) are generally acceptable with lubricated seals and up to twice this figure if special attention is given to continuous lubrication.

Untreated wool felt is resistant to oils, greases and most solvents. It has limited resistance to mild acids and is degraded by alkalis. Synthetic fibre felts which are resistant to strong acids and alkalis are also available.

Plain felt seals are cut from standard grades of felt. For specific applications they can be impregnated with parafin, petroleum or colloidal graphite. All are normally saturated with oil or grease of slightly higher viscosity than that of the lubricant present in the system before installation.

Laminated felt seals are produced by combining one or more layers of felt with layers of elastomer. Different grades of felt may be used in different layers for specific purposes, *eg* filtration and retention. The presence of the impervious elastomeric layer(s) stops leakage of any fluid through the seal.

Fabrics

Sealing systems often contain fabric reinforcement for added mechanical strength. Cotton, nylon, asbestos and *Terylene* are all used in the appropriate applications, impregnated and coated with unvulcanized rubber, then fabricated and moulded in very much the same way as rubber seals.

Cotton is the most widely used fabric, obtained by separation from the seeds of various species of Gossypium, spun into yarn and woven into cloth. A typical cotton fabric would have a weight of eight ounces to the square yard, 0.02 in thick with a breaking strength of more than 60 lbf/in to the in width.

When exceptionally high strength is required, nylon or Terylene fabrics are specified. These are both long-chain polymers, similar in many ways to the rubber polymers which they reinforce.

Nylon refers to a family of materials – polymeric amides with protein-like chemical structure. Amides are organic compounds containing carbon, oxygen, nitrogen and hydrogen. Nylon has more elasticity than cotton, is stronger and has better chemical resistance.

Terylene (polyethylene terephthalate) is strong, abrasion-resistant and heat-resistant, and has high flex life and good chemical resistance. It is also used in pump diaphragms and similar hard working applications.

Asbestos is derived from mineral ores made up largely of magnesium silicate ($3MgSiO_3 5H_2O$). It has an exceptional resistance to heat, even to flames. It is used where seals and mouldings are required to withstand temperatures above the range of those catered for by cotton, nylon and Terylene.

Nomex aramid fibres are used when exceptional resistance to high temperatures is required. Aramid fibres do not melt and retain good physical properties for long periods at temperatures of up to 200°C. Above 370°C this fibre will degrade rapidly.

Compatibility of Seal Materials

COMPATIBILITY OF seal materials can be assessed by prolonged immersion of a seal (or seal material specimen) in the fluid concerned, followed by measurement of the volumetric change. This will indicate whether or not the swell is within acceptable limits. Flexing of the specimen will further indicate whether or not the surface has been embrittled (*ie* cracks under flexing).

In the case of oil fluids the aniline point of the oil is often used as a guide to compatibility. Oils with the same aniline point will normally have a similar effect on elastomers, the higher the aromatic content (*ie* the higher the aniline point), the greater the tendency to produce swelling of elastomers. If the performance of the elastomer is known for a number of test fluids, any other oil of the same aniline point can be expected to behave in a similar manner to that of the corresponding test fluid.

Seal Compatibility Index

A quantitative method of establishing the compatibility of an elastomer with a fluid is given in BS4892:1972. Specifically this involves the use of standard Nitrile rubber test rings for assessment of swell when immersed in different (oil) fluids, *ie* it is used for determining the compatibility of fluids rather than elastomers, in terms of a *Seal Compatibility Index* (SCI). The same principle can, however, be extended to deriving SCI values for different elastomers.

The test involves immersion of a test ring in fluid maintained at 100°C for 24 hours, with measurement of the i.d. of the ring before and after immersion (after cooling). From these measurements the percentage diametral swell (SD) can be calculated as:

$$SD = 100 \frac{D2 - D1}{D1}$$

where D1 is the diameter before immersion
D2 is the diameter after immersion.

The Seal Compatibility Index (SCI) is then taken as the volumetric swell (SD) expressed to the nearest whole number.

The percentage volumetric swell (SV) can also be calculated on a similar basis, *viz*:

$$SV = 100\left[\left(\frac{D2}{D1}\right)^3 - 1\right]$$

Alternatively, Table 1 can be used to convert diametral swell (SD) directly to volumetric swell (SV).

TABLE I – CONVERSION OF DIAMETRAL SWELL TO VOLUMETRIC SWELL

SD	SV	SD	SV	SD	SV	SD	SV	SD	SV	SD	SV
%	%	%	%	%	%	%	%	%	%	%	%
0.1	0.30	2.6	8.00	5.1	16.09	7.6	24.58	10.1	33.46	12.6	42.76
0.2	0.60	2.7	8.32	5.2	16.43	7.7	24.92	10.2	33.83	12.7	43.14
0.3	0.90	2.8	8.64	5.3	16.76	7.8	25.27	10.3	34.19	12.8	43.52
0.4	1.20	2.9	8.95	5.4	17.09	7.9	25.62	10.4	34.56	12.9	43.91
0.5	1.51	3.0	9.27	5.5	17.42	8.0	25.97	10.5	34.92	13.0	44.29
0.6	1.81	3.1	9.59	5.6	17.76	8.1	26.32	10.6	35.29	13.1	44.67
0.7	2.11	3.2	9.91	5.7	18.09	8.2	26.67	10.7	35.66	13.2	45.06
0.8	2.42	3.3	10.23	5.8	18.43	8.3	27.02	10.8	36.03	13.3	45.44
0.9	2.72	3.4	10.55	5.9	18.76	8.4	27.38	10.9	36.39	13.4	45.83
1.0	3.03	3.5	10.87	6.0	19.10	8.5	27.73	11.0	36.76	13.5	46.21
1.1	3.34	3.6	11.19	6.1	19.44	8.6	28.08	11.1	37.13	13.6	46.60
1.2	3.64	3.7	11.52	6.2	19.78	8.7	28.44	11.2	37.50	13.7	46.99
1.3	3.95	3.8	11.84	6.3	20.12	8.8	28.79	11.3	37.87	13.8	47.38
1.4	4.26	3.9	12.16	6.4	20.46	8.9	29.15	11.4	38.25	13.9	47.76
1.5	4.57	4.0	12.49	6.5	20.79	9.0	29.50	11.5	38.62	14.0	48.15
1.6	4.88	4.1	12.81	6.6	21.14	9.1	29.86	11.6	38.99	14.1	48.54
1.7	5.19	4.2	13.14	6.7	21.48	9.2	30.22	11.7	39.37	14.2	48.94
1.8	5.49	4.3	13.46	6.8	21.82	9.3	30.58	11.8	39.74	14.3	49.33
1.9	5.81	4.4	13.79	6.9	22.16	9.4	30.93	11.9	40.12	14.4	49.72
2.0	6.12	4.5	14.12	7.0	22.51	9.5	31.29	12.0	40.49	14.5	50.11
2.1	6.43	4.6	14.44	7.1	22.85	9.6	31.65	12.1	40.87	14.6	50.51
2.2	6.75	4.7	14.77	7.2	23.19	9.7	32.01	12.2	41.25	14.7	50.90
2.3	7.06	4.8	15.10	7.3	23.54	9.8	32.38	12.3	41.62	14.8	51.30
2.4	7.37	4.9	15.43	7.4	23.88	9.9	32.74	12.4	42.00	14.9	51.69
2.5	7.69	5.0	15.76	7.5	24.23	10.0	33.10	12.5	42.38	15.0	52.09

Note: The Seal Compatibility Index (SCI) is the SV expressed to the nearest whole number

SECTION 3

Static Seals

GASKETS
LIQUID SEALANTS
SELF-SEALING FASTENERS
MISCELLANEOUS STATIC SEALS

SECTION 3

Static Seals

GASKETS
LIQUID SEALANTS
SELF-SEALING FASTENERS
MISCELLANEOUS STATIC SEALS

Gaskets

THEORETICALLY AT least, the finer the surface finish of mating flange or machined surfaces the better they should seal when closed tight. However, even wrung surfaces clamped together would not necessarily produce a perfect, continuous seal. There is also the fact that although it would be a static seal in the sense that the seal itself is not subject to movement, the operating conditions of static seals may well involve substantial variations in pressure loading, temperature and relative physical movement. Under certain conditions this may be met satisfactorily by the use of a jointing compound applied to the surfaces, but again this has distinct limitations. Thus highly finished surfaces (such as those produced by lapping) sealed with a compound may well fail in service due to breakdown or extrusion of the compound. A definite jointing material offers more positive sealing and long term performance, at the same time substantially reducing the requirements, and thus the cost, of a suitable surface finish.

The terms joint, jointing and gasket are often used synonymously to describe static seals cut or fabricated from compressible flat sheet material, in preference to the term 'seal'. Academic definition would restrict the term 'jointing' to describe the sheet material, when the actual seal cut from it is then either a joint or gasket (synonymous). Where the final shape incorporates a more complex construction, however, such as compression rims, metal facings, or other reinforcement, it is more generally described as a gasket rather than a joint.

Basic gasket materials range from paper to solid metal. Primarily, choice must be based on compatibility with the working media involved and the performance required, that is, specifically the pressure to be sealed but also taking into account temperature, variations of pressure and temperature, and the type of joint involved. The latter may place restrictions on the size of gasket which can be accommodated and also the compression pressure by virtue of the maximum permissible bolt or flange stress.

Permissible, or available, bolt load also affects the choice of material thickness. Basically the material must be thick enough to deform sufficiently to accommodate any irregularities or inequalities in the flange faces under the available bolt load. The lower this load the greater the thickness which may be required, and *vice versa*, although this will also depend on the compressibility of the material, *eg* a soft material will obviously compress more readily than a hard one but at the same time will be more restricted in compressed pressure.

Selection of gasket thickness can, therefore, be somewhat arbitrary. Standard material thicknesses typically range from 0.8 or 1.6 mm up to 7.6 mm $^1/_{32}$ or $^1/_{16}$ in up to 0.3

in) in the case of cork materials; and from about 0.2 to 3.2 mm (0.008 in) for asbestos materials. In the latter case a thickness of 1.59 mm ($^1/_{16}$ in) is a normal initial choice for use with assembly loads up to about 700 bar (10 000 lb/in^2). Thickness for final choice can then be adjusted to meet any individual factors involved.

How gaskets seal

When closed, a gasket seal is subject to a compressive stress produced by assembly. Under working conditions this load may be relieved by hydrostatic end thrust (Figure 1). The gasket itself, however, is also subject to a side load due to internal pressure tending to extrude it through the flange clearance space. To resist extrusion the effective compressive pressure (that is, assembly load less hydrostatic end thrust) should be greater than the internal pressure, and remain so. A factor of at least two is usually recommended to allow for relaxation of gasket compression stress which is normally inevitable. This in turn will depend on the material. A material with a low relaxation is preferable as it can be employed with lower initial compression pressure, or maintain a higher factor of safety at the same pressure.

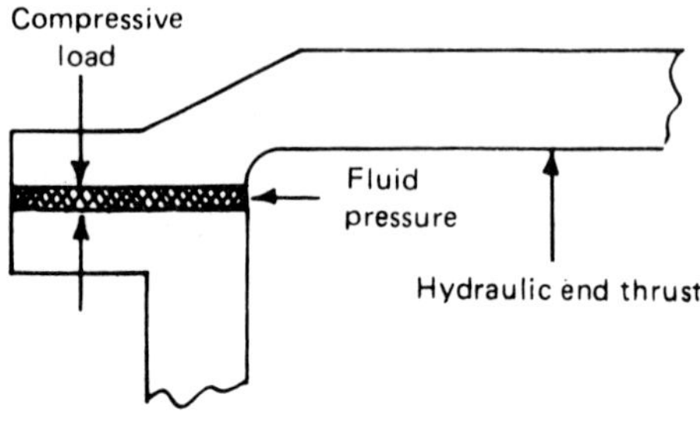

Figure 1

Assembly load

In *closed vessels* the internal pressure exerts a thrust on the cover; similarly, in *closed pipe* systems the pressure tends to force flanges apart. This force, referred to as the bursting thrust or hydrostatic end thrust is represented by the working pressure multiplied by the area of the aperture in the gasket. It is equal to the load which must be applied by the bolts to prevent the components blowing apart. It does not take into account the additional bolting load required to actually maintain compression on the gasket when the system is under pressure.

Residual gasket stress is stress remaining on the gasket at operating conditions and is determined as a function of the internal pressure and a factor known as the gasket factor. This gasket factor is an empirical figure derived from a consideration of the fluid involved and the operating conditions. The *residual gasket load* is then the actual load required to produce the residual stress. The sum of these two loads becomes the *assembly load, ie:*

Assembly load = Bursting thrust + Residual gasket load

Assembly stress

Once the assembly load has been calculated the assembly stress is determined from the equation:

$$\text{seating stress} = \frac{\text{assembly load}}{\text{gasket area}}$$

This assembly stress is obtained by the load exerted by the bolts on assembly and is a function of:

(a) Number of bolts.

(b) Bolt material, *ie* carbon steel, high tensile steel, *etc.*

(c) Bolt diameter.

(d) Permissible stress each bolt can carry.

Tables covering the properties of bolts should be consulted in order to arrive at the most suitable configuration within the boundaries set by the design and dimensions of the plant and equipment involved.

The aim should be to distribute the load evenly over the whole area of the gasket rather than have a few points of high loading with reduced stress at mid-point between the bolts. A more satisfactory arrangement is achieved by employing a larger number of smaller diameter bolts rather than fewer bolts of larger diameter. The use of a torque wrench is essential.

Sealing and surface stress

Even at very low operating pressures a gasket must be pressed against the flanges with a definite minimum surface stress to ensure the material flows into the surface irregularities. This deformation stress depends on the nature and compressibility (or deformity) of the gasket material.

Gasket design also takes into account the total stress applied and effect of temperature on the materials used. The fact that pressures in excess of normal operating conditions are encountered from time to time should not be overlooked. For example during pressure testing of plant the pressure involved can be double the value of the working pressure. Similarly, thermal expansion of pipe lines generates forces which could crush the gasket unless compensated by suitable expansion devices.

All materials suffer, to a varying degree, as a result of over-stressing; they lose strength and the ability to recover on removal of the applied stress, as noted previously. It is important that the gasket designer is acquainted with these limitations to avoid over-stressing gaskets on assembly or when operating conditions impose stresses beyond their capabilities.

Maximum permissible stress levels for high quality compressed asbestos fibre are given in Table 1. The performance of other materials can be assessed on a comparative basis (but see also Table 7).

TABLE 1 – MAXIMUM PERMISSIBLE SURFACE STRESS FOR HIGH QUALITY COMPRESSED ASBESTOS FIBRE

Thickness		Cold		300 °C	
mm	in	kp/cm^2	lb/in^2	kp/cm^2	lb/in^2
0.5	0.020	2000	28000	1400	20000
0.75	1/32	1600	23000	1100	15000
1.0	0.040	1400	20000	950	13000
1.5	1/16	1100	15500	750	10500
2.0		900	13000	600	8500
3.0	1/8	600	8500	400	5500

Surface finish

Surface finish is important only in that it governs the thickness and compressibility necessary in the gasket material to complete a physical barrier in the clearance gap between the flanges. However, this can govern the type of material which can be employed, and with it the ultimate performance of the gasket. Thus a resilient material which could provide good closure with comparatively rough surfaces may extrude at the working pressure required, or not be compatible with the fluid involved or the temperature of the service, in which case a finer surface finish may have to be employed in order to accommodate a harder material with a high closure pressure.

Thinner materials are capable of carrying higher compressive stresses than thicker materials of the same type, and this applies generally to the majority of materials used for gaskets. Again, with thinner materials it becomes necessary to provide a better quality surface finish on the metal faces.

Table 2 gives recommended minimum gasket thickness for (Klinger) standard sheets for various surface stresses (at working pressure) and degree of surface finish.

Turned flanges generally have peaked grooves which in effect reduce the area of gasket carrying the flange load. This gives an increased surface stress and increased gasket compression. In such cases the gasket thickness may be reduced to the values shown at the right of each column, (provided the flanges are rigid and undistorted).

Although a high degree of surface finish would appear desirable in all cases to relieve the gasket of having to compensate for surface irregularities this can be unnecessarily expensive. Surface finish requirements should be related to the working conditions involved and the gasket material to be employed rather than overall recommendations. There is also the point to consider that too fine a surface finish can be undesirable since the surfaces may lack grip, especially on a harder gasket material, allowing extrusion to take place. Thus the presence of fine machining marks tangential to the direction of applied fluid pressure can be helpful and some authorities even recommend finishing of the flange surfaces with such 'non-slip' grooves – this is achieved by using a round nosed tool to give grooves of approximately 0.125 mm (0.005 in) depth for gaskets of 0.50 mm (0.020 in) thickness and above or about one half this depth for thinner materials.

Another feature of primary importance is that the flange faces must be parallel and sufficiently rigid to resist distortion on being tightened down and under hydrostatic end loads. Distortion under working loads, often called flange rotation, can appreciably affect

TABLE 2 – MINIMUM GASKET THICKNESS FOR DIFFERENT SURFACE FINISHES

Surface Stress at Working Pressure		Milled, Turned, etc. ∇ (= 160 μ)				Ground, Turned, etc. ∇∇ (= 40 μ)				Ground/Turned ∇∇∇ (= 16 μ)	
kp/cm²	lb/in²	mm		in		mm		in		mm	in
100	1400	5	2.0	1/4	1/8	1.5	1.0	1/16	0.040	0.5	0.020
200	2800	4	2.0	1/4	1/8	1.0	0.75	0.040	1/32	0.5	1/64
500	7000	3	1.5	1/8	1/16	0.75	0.75	1/32	1/32	0.3	1/64
750	10500	–	–	–		0.75		1/32		0.3	1/64
			1.5		1/16		0.5		0.020		
1000	1400					0.5		0.020		0.3	0.008
			1.0		0.040		0.5		0.020		

For various surface stresses (at working pressure) and degree of flange finish. For intermediate surface stress values use the next thicker material.

the working conditions of the gasket. A lot depends on the design of the flange, for bolt spacing should be selected on a basis of providing even distribution of interflange loading, although this may have to compromise with other practical requirements. In many cases, however, this design factor is fixed by using *standard flanges*.

Gasket stress

Basically the pressure rating of a gasket – the value of internal pressure it can withstand without extrusion – is related to the compressive stress the material can withstand, plus its ability to resist creep and cold flow. While it may be possible to combat creep by increasing the hardness of the material, the gasket must still be soft enough to achieve sealing without having to go to unacceptably high bolt and flange loads. A satisfactory gasket material must therefore be capable of retaining a high residual stress along with other desirable properties.

Typical stress relaxation curves, plotted against time, are shown in Figure 2. The upper curve represents a satisfactory material where, after some initial relaxation, the residual stress remains constant. The material designated by the lower curve suffers continuous relaxation and thus a continual loss of pressure rating as a gasket seal.

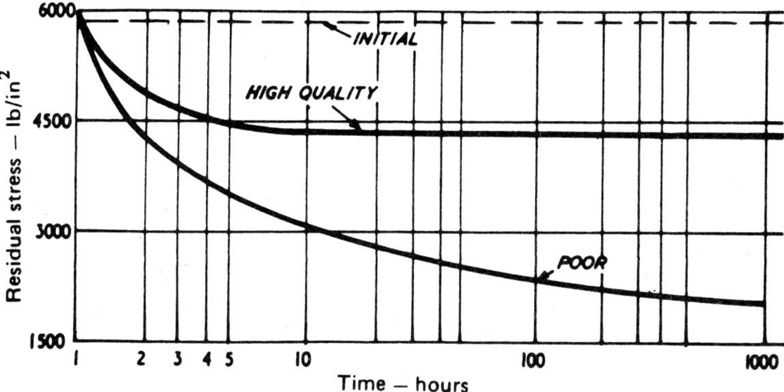

Figure 2

Such curves are essentially specific to a particular material under specific test conditions. Relaxation will also vary with material thickness and temperature. Figure 3 shows a typical plot of residual stress against thickness which emphasizes the fact that for minimum relaxation with any specific material the thinner the gasket the better. Thus for a high pressure high temperature application a gasket thickness of 0.397 mm ($^1/_{64}$ in) would be a logical choice, although in practice it may be necessary to choose a greater thickness to accommodate the flange conditions. Alternatively, for maximum performance from the gasket it may be necessary to improve the flange design and finish to accommodate a minimum thickness gasket or to employ a reinforced gasket material.

Temperature effects

Temperature will have a marked effect on the performance of a gasket since an increase in temperature will both degrade the physical strength of the material and deform it so that the bolt load (and thus the residual stress) is modified. The latter is not accommodated in a residual stress/time curve as in Figure 2, where the test is normally made on a simple

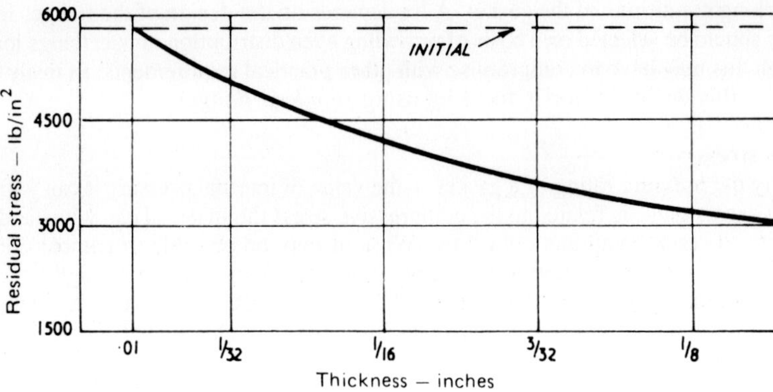

Figure 3

specimen and is usually determined separately by measuring deformation plotted against temperature.

Again such data yield characteristic curves. In the case of a satisfactory gasket material the increase in deformation with increase in temperature is relatively small, indicating that the change in bolt load will also be small. Deformation, it should be noted, represents a decrease in thickness under compressive loading leading to a reduction in bolt load, and consequently loss of residual stress. Thus a poor gasket material which suffers high deformation with increasing temperature will show high relaxation, and in practice will collapse or extrude at higher temperatures under moderate internal pressure.

Deformation plotted against temperature will also show any critical temperature at which the material may become softened or plastic, in which case the curve will have a sharply defined knee. Performance up to the knee may be quite satisfactory, but the knee temperature sets a definite maximum figure for service as a gasket material. In other cases degradation may become progressive rather than sharply defined as the maximum service temperature of the material is approached (Figure 4).

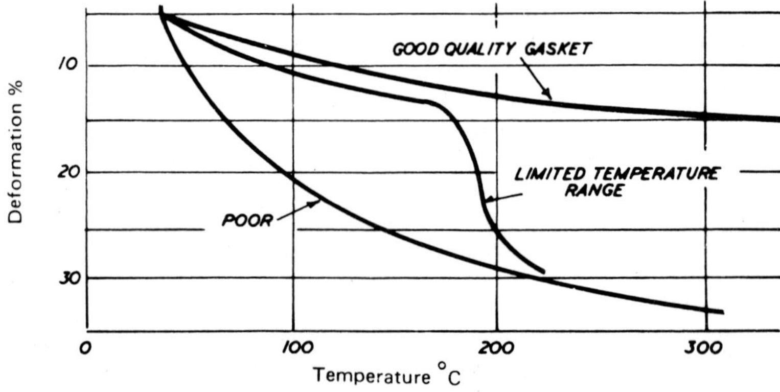

Figure 4

Other properties

The more ordinary physical properties of gasket materials have limited relevance, although they may be quoted as material characteristics. They are mainly useful as a method of production control applied to samples. These include specific gravity or density, tensile strength and compressibility. The loss of tensile strength with increasing temperature may be used as a criterion for establishing a maximum service temperature figure in the absence of knee characteristics in a compression test (Figure 4). Thus in the case of compressed asbestos fibre materials the upper limit of 540°C, usually specified as a maximum for working, corresponds to a loss of tensile strength of about 40%. The material would actually be usable at an even higher temperature – at 650°C the loss of tensile strength would be about 70% – and it is not until about 760°C that this material suffers a 100% loss of tensile strength. Normally, however, maximum recommended operating temperatures rather than strength/temperature characteristics are quoted for specific materials. It can then be assumed that a material can be operated up to this temperature without significant loss of properties (also assuming that relaxation and deformation characteristics are satisfactory).

Compression test figures are useful since these indicate directly the initial residual stress produced by compression and, if fully completed, also indicate recovery characteristics (Figure 5). Normally such data are measured on a cold specimen, and for different thicknesses. Such data may also be presented in terms of a general compressibility figure, indicating the percentage compression desirable or recommended to produce an optimum residual stress for sealing, with which may be applied a corresponding compression pressure. Strictly speaking residual stress will go on increasing

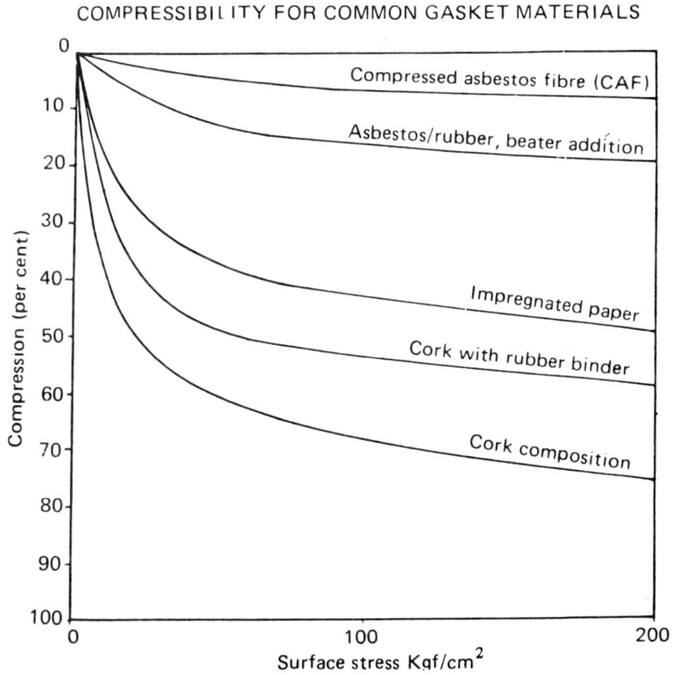

Figure 5

with increasing compression and a compressibility figure does not necessarily indicate the maximum amount of compression a material will tolerate. Usually, it will either indicate actual compression achieved at a given compressive load, in which case the actual closure pressure may be higher or lower in practice, or a maximum recommended compressibility representing a limit to the closure load. Thus the latter is typical for elastomeric materials in sheet form. *Elastomeric ring* seals have actual compression controlled or limited by the matching groove dimensions.

Simple gaskets

Many gasket users produce their own requirements from sheeting stock but increasing use is being made of pre-cut gaskets manufactured by specialist cutters who have the expertise and equipment for cutting even the most complicated shapes within extremely close dimensional tolerances.

Although shapes and dimensions vary enormously there are certain standard shapes common to most industries. Chief among these are flange gaskets produced to British or International standard configurations. Flanges in common use are shown in Figure 6.

The two most common types of gasket are:

(i) *Inside bolt circle* or *ring* type where the periphery of the gasket is generally located by the bolts.

(ii) *Full faced* type where the outside diameter is similar to that of the flange and has a series of holes corresponding to the number and diameter of the bolts.

These are illustrated in Figure 7.

Materials

The following identifies the more general gasket materials in use (see also Tables 5, 6 and 7 for more specific data).

Paper

Papers, which include Kraft and Manilla, are low cost materials, also available stabilized with wax based fillers to reduce absorption of the contained fluids. A further type within this category is cellulose fibre impregnated with a glue/glycerine composition. These materials can be made extremely thin, *ie* from 0.1 mm (0.004 in) upwards. They are used extensively in the automotive industry as gaskets and shims for water, oil and petrol applications. Maximum operating conditions are 120?D2?S0008TgC and 8 bar.

Cork

The qualities of a composition of cork granules bonded with resin will vary according to the size of the granules. Due to the comparatively weak bonding system the material tends to be fragile but this can be improved by the introduction of a fabric interlay. Cork gaskets possess good oil and solvent resistance but can be adversely affected by water.

Their chief application is in situations where mating faces are uneven or where bolting loads are very low. Maximum operating conditions are 50°C and 3.5 bar.

Rubber bonded cork

Rubber bonded cork is a more robust material than resin bonded and is manufactured from high quality fine granule cork bonded with elastomer. The material combines the natural compressibility of cork with the resilience of rubber producing a jointing with high mechanical strength, compressibility and low permanent set. Gaskets from this material provide a leakproof joint under light bolt loading and do not appreciably flow

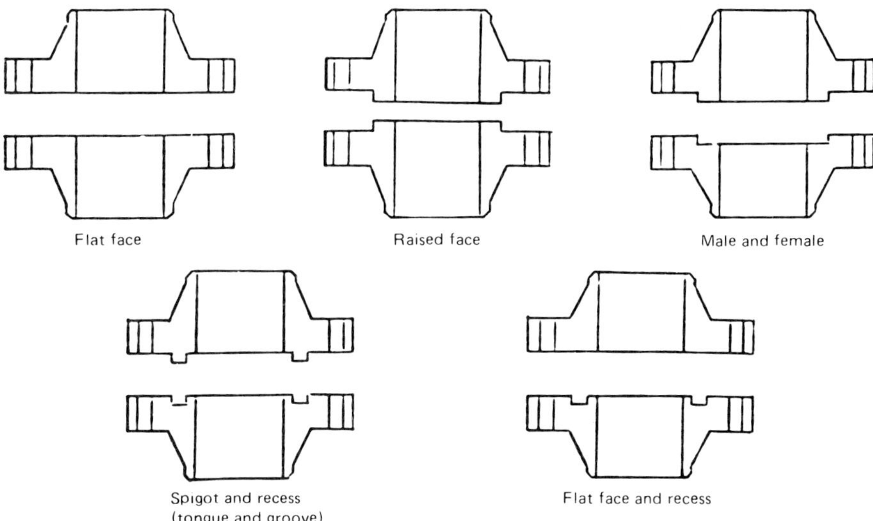

Figure 6
Flanges in common use.

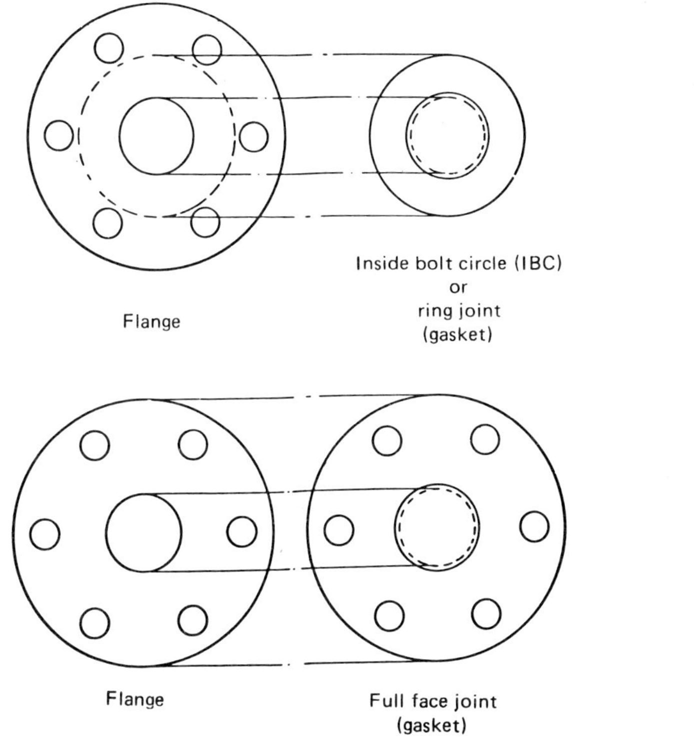

Figure 7
Gaskets in common use.

or extrude. It is a material which is easily cut and due to its strength and flexibility is suitable for gaskets with narrow faces. Chemical resistance is determined largely by the properties of the binder used, normally neoprene or nitrile. Cork products are not recommended for strongly alkaline or acidic conditions but are highly suited for lubricating oils, petrol, air, gas and hot water services, and are widely used as oil seals in electrical transformers. The material generally displays excellent ozone and air ageing characteristics.

Maximum recommended temperature range is from −30°C to +150°C. General pressure range is in the order of 3.5 bar.

Rubber

Among the more traditional materials rubber ranks with the forerunners in versatility and application range.

The property that all elastomers have in common is the ability to deform substantially under the application of stress and then snap back to almost their original shape when the stress is removed.

Most elastomers also possess the following common properties:

(i) They are incompressible. Applied stress changes the shape but the volume remains constant.
(ii) Heat gradually and irreversibly destroys their properties and integrity.
(iii) Cooling causes them to become brittle and rigid. This effect is reversible.
(iv) Prolonged deformation leads to some degree of compression set.

The physical boundaries within which an elastomer can operate effectively are primarily restricted by its chemical make-up but rubber technologists are increasingly able to extend or alter many of those limitations.

An examination of the properties of the basic elastomers clarifies the distinctive features of each type enabling the best selection to be made for a particular application.

Natural rubber (NR)

Natural rubber possesses several excellent mechanical properties. It has good tear and abrasion resistance, high resilience and good low temperature flexibility. Its use in gaskets is limited by low chemical resistance, especially to solvents and oil which greatly reduce its stretch, poor resistance to sunlight and ozone, which cause cracking, and the restricted temperature range.

Alcryn

Halogenated polyolefin, mid-performance, melt processable, rubber which can be processed on both rubber and plastics equipment. Outstanding oil resistance, excellent ozone and weather resistance.

Styrene butadiene (SBR)

A general purpose synthetic rubber which can be regarded as a general replacement for natural rubber except where certain mechanical properties are required.

Polychloroprene (CR)

The best known CR is neoprene. It is superior to NR and SBR in its ozone and weather resistance. It also has a degree of resistance to petrol, mineral oils and other hydrocarbons and can be made flame resistant. It has only moderate low temperature flexibility and tends to suffer from poor compression set.

Nitrile (NBR)
This synthetic rubber is especially resistant to mineral oils, aromatic and aliphatic hydrocarbons and alcohols, depending on the actual nitrile content of the elastomer. High and low temperature performance can be good. Oil resistance can be improved by compounding at the expense of low temperature flexibillity.

Hypalon (CSM)
CSM rubber has excellent resistance to ozone and good resistance to flame, heat, mineral oils, acid and weather, making it ideal for use outdoors or near sparking electrical equipment.

Hytrel (YBPO)
Thermoplastic polyester elastomer with properties similar to those of crosslinked rubbers. Outstanding resistance to solvents, hydrocarbon fluids including petrol and lubricating oils. Excellent abrasion resistance and flex-fatigue performance. Ease of processing by a variety of thermoplastic techniques.

Ethylene propylene (EP)
EP and the diene modified form EPDM has good mechanical properties and is resistant to ageing, weathering, ozone, oxygen, steam and water. It is especially resistant to phosphate ester based hydraulic fluids but is not recommended with petroleum based fluids.

Kalrez (FFKM)
Perfluorinated rubber with excellent all-round chemical resistance. Similar to Teflon PTFE. Best high temperature performance.

Butyl (IR)
Butyl rubber is similar to EP in its ability to cope with phosphate ester based hydraulic fluids. It is highly impermeable to gas and moisture and has good general chemical resistance including mineral acids. Resistance to petroleum based fluids is low.

Viton (FKM)
Fluorinated rubbers cope well with petrol, oils, chlorinated solvents, concentrated alkalis and fuming acids. Ozone and weather resistance are excellent and high temperature performance is good. Fluorinated rubbers have poor low temperature flexibility and should not be used with certain esters or ketones.

Polyurethane (AU/EU)
These polymers provide excellent resistance to oils, solvents, fats, grease, petrol, ozone, sunlight and weather. The mechanical properties are excellent at low temperature but care should be taken at high temperature. Polyurethane rubbers are particularly susceptible to hydrolysis and should not be used with hot water or acids.

Silicone (MQ)
Silicone rubbers are resistant to water, ozone and sunlight and can be made to perform at very low or high temperatures. Silicone is odourless and tasteless and will not support bacterial life. It should not be used with high pressure or steam, nor with oils, petrol or other hydrocarbons.

In general the elastomers are limited to applications of only moderate temperature and pressure and require comparatively low seating stress to make an effective seal. Under conditions of high stress resulting from excessive bolt loading or severely fluctuating operation the materials show a tendency to creep which could be detrimental to sealing efficiency. In certain instances this problem can be overcome by the use of an elastomer with a jute or polyester insertion.

Specific properties of rubbers as a useful guide to material selection are given in Tables 3 and 4.

Compressed asbestos fibre (CAF)

Compressed asbestos fibre is the most universally used gasket material possessing distinctive properties. Compressed asbestos fibre jointing consists essentially of a mass of asbestos fibres bonded together by a relatively small amount of elastomer. It is flexible to handle and although it will adapt itself to surface irregularities in metal faces it will not readily flow or extrude and high loading is required to obtain any appreciable alteration in thickness. When subjected to high temperature it loses much of its flexibility, but nevertheless the material still retains some measure of resilience, and is able to be used in high temperature applications and maintain a pressure-tight seal.

Asbestos itself is a general name. There are four main types, chrysotile, crocidolite, amosite and anthophyllite, which differ in fibre structure and physical properties. Only chrysotile (white asbestos) is used, certainly in the United Kingdom, for the production of compressed asbestos fibre sheet jointing.

The properties of asbestos which make it eminently suitable for jointing materials are:

(i) Ability to withstand high temperature. Up to 550°C without a change in structure, but at this point it begins to lose its water or crystallization and will eventually become quite friable.

TABLE 3 – CHEMICAL RESISTANCE GUIDELINES

		NR	SBR	CR	NBR	CSM	EP	IIR	FPM	AU	Si	FFKM
Acids	Dilute	D	D	D	G	E	G	E	E	P	G	E
	Concentrated	P	P	D	D	G	G	G	G	P	P	E
Alcohols	Ethanol etc	G	G	G	E	G	E	E	G	D	E	E
Alkalis	Dilute	D	G	G	G	G	E	G	G	D	D	E
	Concentrated	D	D	D	D	D	G	D	D	P	P	E
Halogenated Solvents	Trichloroethylene etc	P	P	P	D	P	D	P	G	D	D	G
Aliphatic Hydrocarbons	Petrol, kerosene, etc.	P	P	D	G	D	P	P	E	E	D	E
Aromatic Hydrocarbons	Benzene, toluene etc	P	P	P	D	P	P	P	E	D	P	E
Hydraulic oils	Silicate esters	P	P	G	G	G	D	P	G	G	D	E
	Phosphate esters	P	P	P	P	D	E	E	D	D	D	E
Ketones	Acetone etc.	P	P	P	P	P	G	G	P	P	D	E
Mineral oil		P	P	D	E	G	P	P	E	G	G	E
Synthetic Lubricants (Diesters)		P	P	P	D	P	P	P	G	D	D	E

The information in this chart should only be used as a general guide to the selection of a suitable gasket material

TABLE 4 – RELATIVE PERFORMANCE OF THE COMMON EL

	Elastomer type	General purpose, non oil resistant		General purpose, oil resistant		
		Natural	Styrene Butadiene	Neoprene	Nitrile	
	ASTM designation	NR	SBR	CR	NBR	CSM
	General temp. range °C	–50 to +50	–40 to +70	–40 to +120	–40 to +130	–20 to +150
Resistance to:	Abrasion	Excellent	Excellent	Excellent	Good	Excellent
	Compression set	Good	Good	Good	Good	Fair
	Flame	Poor	Poor	Good	Fair	Good
	Weather	Fair	Fair	Excellent	Fair	Excellent
	Oxidation	Good	Good	Good	Fair	Excellent
	Ozone	Poor	Fair	Excellent	Poor	Excellent
	Radiation	Fair	Good	Fair	Fair	Good
	Water (cold)	Excellent	Excellent	Good	Good	Good
	Steam	Poor	Poor	Fair	Fair	Good
	Gas permeability	Poor	Fair	Fair	Fair	Good
	Electricity	Excellent	Excellent	Fair	Poor	Good

	Elastomer type	Medium temp. non oil resistant		High temp. oil resistant		Oil resistant	High temp. non oil resistant
		Ethylene propylene	Butyl	Viton	Kalrez	Poly-urethane	Silicone
	ASTM designation	EP EPDM	IIR	FPM	FFKM	AU EU	MQ
	General temp. range °C	–40 to +135	–50 to +120	–40 to +250	–45 to +300	–40 to +100	–100 to +250
Resistance to:	Abrasion	Good	Good	Good	Good	Excellent	Poor
	Compression set	Fair	Fair	Good	Good	Good	Excellent
	Flame	Poor	Poor	Excellent	Excellent	Poor	Good
	Weather	Excellent	Excellent	Excellent	Excellent	Excellent	Excellent
	Oxidation	Good	Excellent	Excellent	Excellent	Excellent	Excellent
	Ozone	Excellent	Excellent	Excellent	Excellent	Excellent	Excellent
	Radiation	Poor	Poor	Good	Very good	Good	Excellent
	Water (cold)	Good	Excellent	Good	Good	Good	Good
	Steam	Good	Fair	Good	Excellent	Poor	Good
	Gas permeability	Fair	Good	Good	Good	Good	Good
	Electricity	Excellent	Excellent	Good	Good	Excellent	Excellent

(ii) Virtually chemically inert and is attacked only by strong mineral acids.

(iii) The use of the longer fibre qualities will result in a material with higher tensile strength, better compressibility and recovery characteristics and greater resistance to flow under conditions of heat and stress.

Asbestos fibres are not in themselves self bonding and so are bonded with elastomer to produce the sheet material. Various elastomers are used according to the eventual gasket application. The selection is to a certain extent overlapping and tends to be based on the resistance to the effects of liquid hydrocarbons and mineral acids.

Manufacture

The asbestos is initially subjected to a process for separating and opening up the fibres to ensure good distribution throughout the subsequent mix. Meanwhile, the polymer, in combination with an appropriate solvent, is reduced to the consistency of a smooth solution. The two are then brought together in a mixing vessel where each asbestos fibre becomes coated with a thin layer of polymer. The resulting mash is then fed between the precision rollers of a calender where, under high pressure and temperature, a homogeneous sheet of uniform thickness is produced.

From the foregoing it will be seen that compressed asbestos fibre sheeting is produced in single sheets according to the dimensions of the calender bowl. An alternative manufacturing process, but using the same basic materials, *ie* asbestos fibre and polymer, allows for the production of sheeting in continuous rolls.

This method, known as *beater addition,* is similar to that of the paper making industry where the product, in the form of water based slurry, is transported along a moving wire gauze conveyor during which time the bulk of the water drains away. The solidifying mass continues *via* a series of rollers through drying ovens and on to the finishing and reeling plant. The length of sheeting on the finished roll is governed by the weight which can be conveniently handled during subsequent transport. Beater addition materials can be produced in various thicknesses and within very close dimensional tolerances. They are generally softer and of lower specific gravity than compressed asbestos fibre and require lower bolting loads to make an effective seal. They are generally suited for applications within the automotive and domestic equipment industries.

Woven asbestos

A further material within the asbestos group is proofed woven cloth. These cloths are proofed by impressing into their surface a layer of natural or synthetic rubber. The materials are highly compressible and conformable, being particularly suitable for sealing between uneven faces. Sheeting is produced in various thicknesses by plying up the basic cloth and gaskets can be cut from these sheets using conventional tooling.

An alternative method for producing thick gaskets uses strip material and special folding techniques to build up the gasket to the required shape and thickness.An example of this method is found in manhole door joints for boilers.

The temperature range of the material varies according to the application. The general range is up to 180°C but this can be increased to about 500°C in certain circumstances.

For additional mechanical strength in situations of thermal cycling or vibration, gasket materials of this type may be reinforced with metallic (*eg* brass) wire.

Non-asbestos millboard

This is a relatively new product developed as an asbestos-free material but capable of withstanding similar temperatures (*eg* up to 1000°C. It is intended as a direct alternative to asbestos millboard, with good strength and thermal properties. It is a material which can be rendered flexible by moistening on both sides, hardening to a rigid form when dry. It is basically produced as lagging material, but also has limited application for gaskets.

Plastics

Although the development of plastics has introduced a wide range of materials onto the market, very few have found ready application as *static flat* seals. This is due to the limited temperature range over which they can be reliably used, together with comparatively poor resilience and, in some cases, a tendency to flow under load.

A major exception to this is PTFE (polytetrafluoroethylene) which has such a remarkable combination of properties that it takes a special place in the sealing of aggressive and toxic fluids. PTFE is virtually chemically inert, being attacked only by molten alkali metals and certain fluorine compounds at elevated temperature and pressure.

The working temperature range is very wide for a thermoplastic material being from −190°C to +250°C. It does, however, suffer from a degree of cold flow or creep under load which limits its use as gaskets when cut from base sheet. This problem is readily overcome by the use of a very thin sheet or envelope fitted to the bore of a gasket cut from one of the more conventional materials, *ie* compressed asbestos fibre or rubber (Figures 8a, 8b, 8c, 8d).

Other filler materials used in envelope gaskets include felted or compressed asbestos sheets, cork composition sheets, beater-addition sheets, corrugated metal inserts and sandwich constructions.

These sheets protect the gasket from attack by certain aggressive fluids which could otherwise weaken the material to such an extent that its ability to seal would be greatly impaired. Equally, a conventional material, although making a perfect seal could contaminate the contained product, which in some industries (*eg* food and pharmaceutical) could not be tolerated. To meet such situations a *PTFE sheath* gasket serves to protect the product.

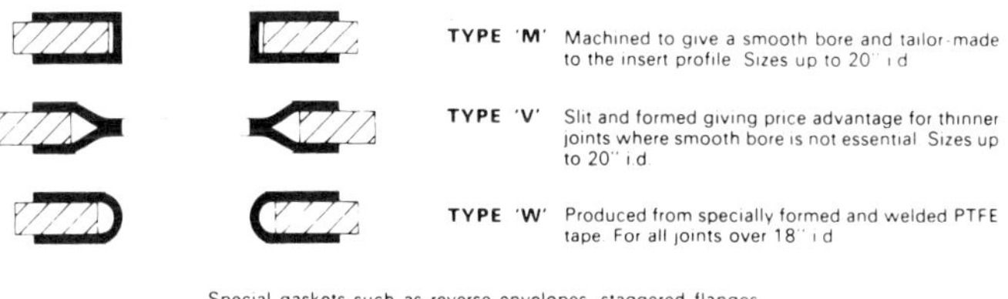

Figure 8a
PTFE envelope gaskets.

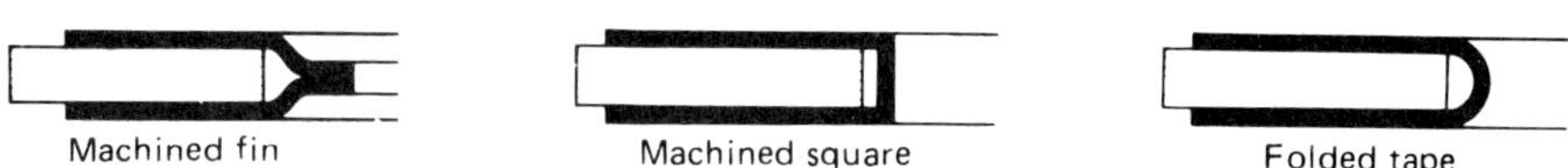

Figure 8b
PTFE envelope gaskets.

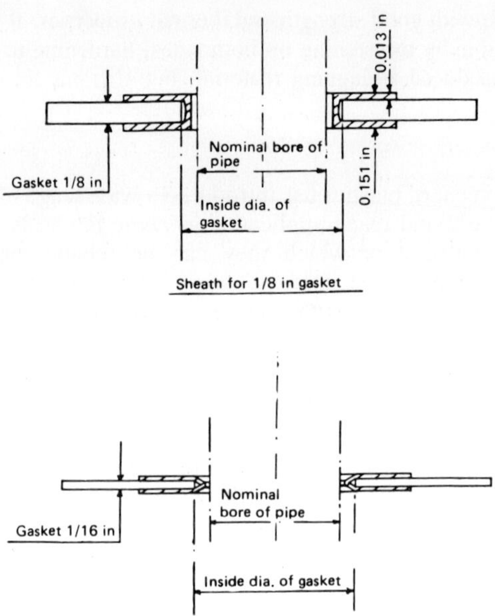

Figure 8c

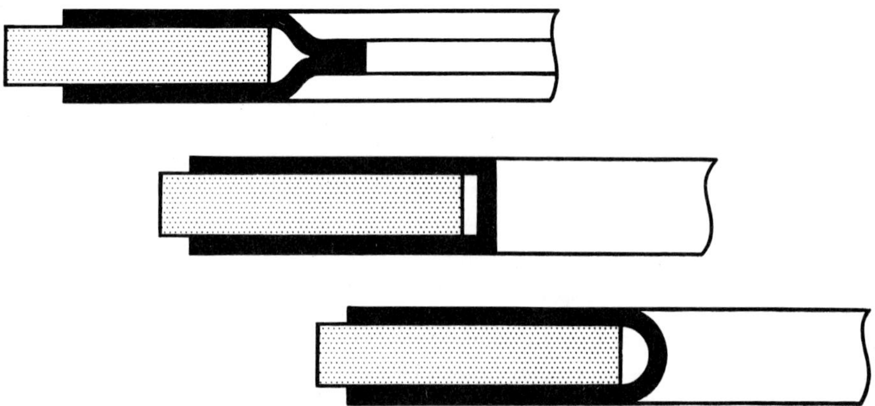

Figure 8d
PTFE gaskets. Machined fin type (top), machined square type (centre), folded tape type (bottom).

Sheaths are produced in the following types:

(i) Split or 'Y' section to fit over a 1.6 mm thick gasket.

(ii) Machined or 'U' section to fit over a 3.2 mm thick gasket.

(iii) *'V' Type* – this is a 'machined' envelope and is probably the most widely used type, available in sizes up to 500 mm (20 in bore).

The method of manufacture is very economic in its use of PTFE, a fact which is reflected in the units' low price. The gasket does however have the disadvantage of creating flow turbulence and sediment trap.

The profile also reduces the effective sealing area.

(iv) *'M' Type* – also a 'machined' envelope gasket available in sizes up to 500 mm (20 in) bore. Due however to the method of manufacture, far more PTFE material is used, which is reflected in the unit price.

The design does, nevertheless, allow for full flange face sealing with smooth bore and uninterrupted product flow.

(v) *'W' Type* – manufactured from special PTFE tape which is formed to the desired shape and welded. Normally available for sizes of not less than 450 mm (18 in) bore.

Types (i) and (ii) are normally produced to BST and ASA dimensions up to 150 mm (6 in) nominal bore, but for larger diameters the sheaths are folded from strip material and welded at the joint. Large diameter welded sheaths are frequently used with *composite* gaskets, *eg* rubber and compressed asbestos fibre for sealing covers of glass lined vessels where faces are not always parallel and a *soft type* gasket becomes necessary.

PTFE in the form of an unsintered cord is used as a ready alternative to *flat* gaskets for certain applications. The cord is simply laid within the bolt circle or laid so as to pass on alternate sides of the bolts. In each case the ends are overlapped and when compressed the material spreads and produces a 'flat' seal, (Figure 9).

Exfoliated graphite (graphite foil)

A comparatively recent addition to the range of static sealing materials is exfoliated graphite. This product differs from ordinary graphite in that its crystal lattice has been expanded by a process which causes the layers of graphite to be held further apart than usual by small electric charges on the surfaces repelling each other like similar poles of magnets.

Graphite foil possesses outstanding physical and chemical properties which render it particularly suitable as a sealing material for the more arduous operating conditions.

In oxidizing atmospheres it can be used in the temperature range of −200°C to +500°C and in reducing or inert atmospheres the range is extended to −200°C to +2500°C.

Figure 9
PTFE/fluorocarbon joint sealant.

As the product has no binder (99.9% graphite) it has excellent chemical resistance, being unaffected by most organic and inorganic chemicals, but exceptions are strongly oxidizing compounds such as concentrated nitric acid. It is also highly resistant to nuclear radiation. At the other extreme it is non-toxic and can be safely used in contact with foodstuffs and potable water. As a gasket material it also possesses excellent stress relaxation characteristics and makes a reliable seal under lower bolt loadings than a number of the more commonly used materials.

Stress relaxation characteristics of graphite foil are shown in Figure 10.

Limited use is also made of foamed materials such as gaskets, *eg* plastic foams and elastomeric sponges for vapour or dust barriers in low pressure applications (see chapter on *Self Adhesive Compression Seals,* Section 7).

Metal and composite materials

The materials considered so far can all be classified as non-metallic and are, to a varying degree, conformable under load and readily compensate for surface irregularities and fluctuations in applied stress. Eventually the point is reached where operating conditions become too severe for these materials, or the assembly stress rises beyond their limits, when it becomes necessary to select a more robust material. It is here that the designer has recourse to a number of semi-metallic and solid metal gaskets.

It should be noted that there is no precise cut-off point where metal automatically supersedes non-metallic, indeed, the decision is sometimes quite arbitrary and it is not uncommon to find metal gaskets employed where a non-metallic material would be more than adequate.

Semi-metallic gaskets are of composite construction in which the distinctive properties of each component are used to advantage.

Corrugated metal gaskets

These are formed in a range of metals including brass, copper, cupro-nickel, steel, monel, aluminium, *etc*, and produced to almost any shape and size required. The thickness of

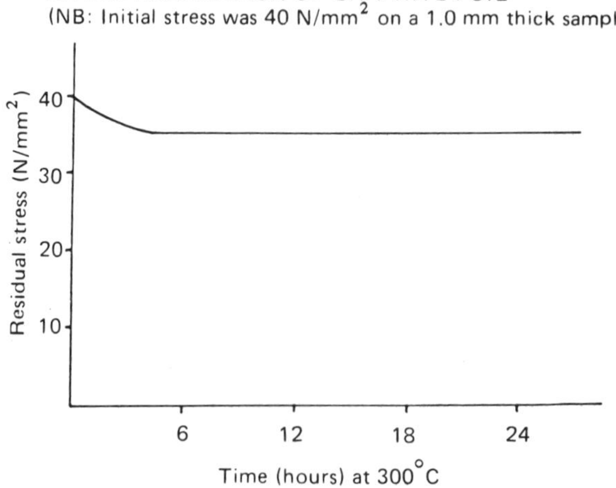

Figure 10
The Flexicarb range of graphite-based products.

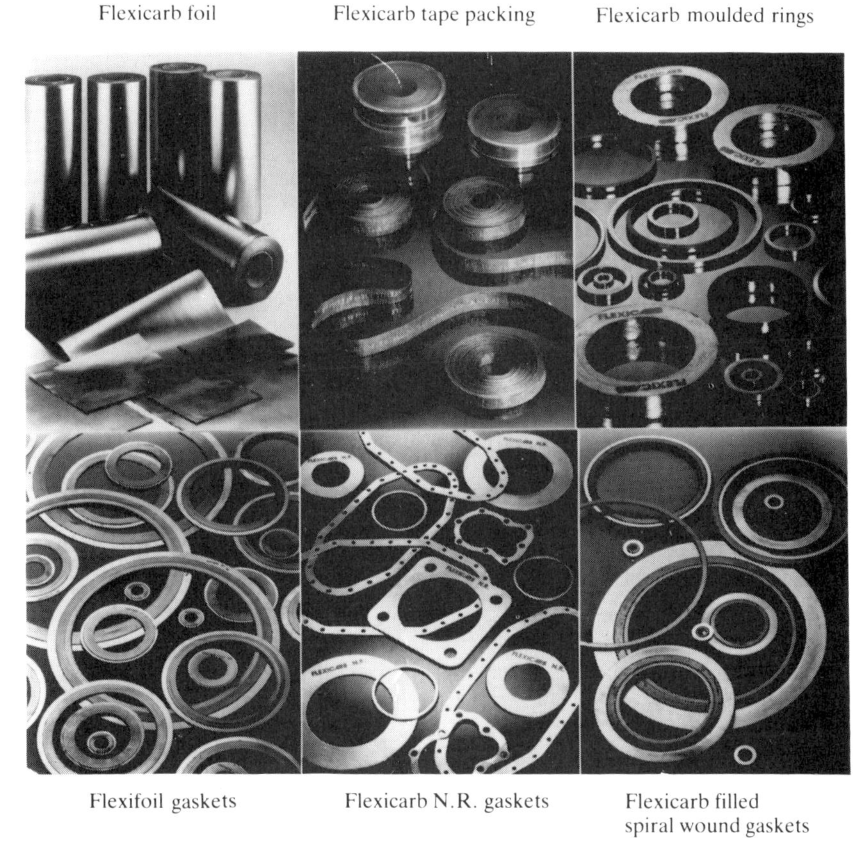

Flexicarb foil | Flexicarb tape packing | Flexicarb moulded rings

Flexifoil gaskets | Flexicarb N.R. gaskets | Flexicarb filled spiral wound gaskets

The Flexicarb range of graphite-based products.

the metal is normally 0.25 mm or 0.30 mm with the corrugations 1.6 mm, 3.2 mm and 6.4 mm pitch.

The sealing mechanism is based on point contact between the peaks of the corrugations and the mating flanges. This produces a relatively high local seating stress which allows some deformation to take place, thus compensating for uneven faces. This type of gasket usually incorporates a filler of jointing compound or asbestos cord located in the well of the corrugations, *eg* Figure 11.

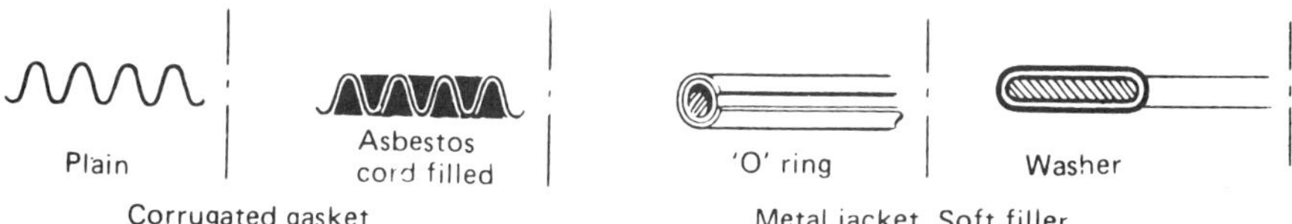

Figure 11
Crush-type gasket featuring triangular corrugations stamped into a metal sheet.

Applications include steam, water, air, chemicals and oils with the metal selection according to the nature of the contained fluid and operating temperature.

Metal clad gaskets
This type of gasket consists of a relatively soft material such as asbestos millboard or compressed asbestos fibre, cut to the required shape and enclosed within a metal covering, *ie* brass, aluminium, soft iron, cupro-nickel, copper, Monel or stainless steel. *Metal clad* gaskets are frequently used where sealing faces are narrow, when the metal cladding provides additional mechanical strength.

They can be produced in complicated shapes, *eg multi-pass heat exchanger* gaskets with the metal casing built up from strip and welded at the joints. It will be seen that with certain types of construction the metal stands proud at the point of overlap. This produces a locally high seating stress which contributes to the effectiveness of the seal in a critical area.

This feature is emphasized in automotive cylinder head gaskets where the bores are metal bound in addition to the metal sheeting covering the softer base material.

Metal clad gaskets are also widely used in the petrochemical industry where high assembly stresses are frequently encountered in pressure vessels.

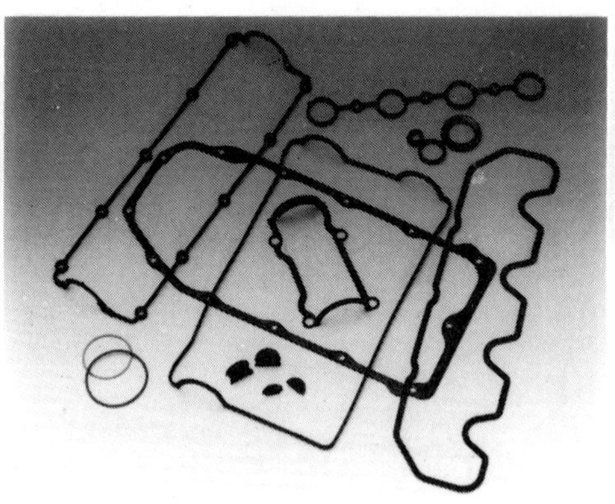

Automotive power train gaskets.

Bonded rubber gaskets
These are of composite construction in which a gasket cut from metal, compressed asbestos fibre or paper is embossed with an elastomer using the screen printing process. The elastomer is deposited in the form of strips extending from one bolt hole to another around the entire flange face. The deposit is frequently thicker mid-way between bolt holes to compensate for the reduction in bolting load which can occur at this point, especially with widely spaced bolts.

These gaskets cover a temperature range from $-60°C$ to $+200°C$ depending on the elastomer. Applications include sealing fuels, lubricants, water and chemicals, especially in conjunction with the smaller type of mass produced components.

Spiral wound gaskets
Spiral wound gaskets are the most versatile of the *semi-metallic* gaskets since they combine considerable mechanical strength with outstanding recovery characteristics.

These gaskets consist of alternate layers of V-section metal strip and a non-metallic filler wound in spiral form into a series of standard shapes (Figure 12). The length of straight sided shapes is limited due to the difficulties in maintaining tension on the material. The filler material does not extend the full length of the metal strip as at the beginning and end of the windings there are several turns of metal alone which are welded together at intervals.

The metal component is normally of stainless steel though other materials such as Monel, nickel, titanium and inconel can be used where operating conditions merit a departure from standard.

Fillers also vary according to application, the dictates of a standards authority or a manufacturer's own specification. Fillers include compressed asbestos fibre, beater addition asbestos, asbestos paper, PTFE lead, ceramic paper and graphite foil.

Style R

Standard construction type. Inner and outer diameters are reinforced with several plies of metal without filler to give greater stability and better compression and sealing characteristics.

Style RIR

Additional inner metal ring acts as a compression stop and fills the annular space between flange bore and inside diameter to prevent accumulation of solids and reduces turbulent flow of fluids.

Style CG

Utilizes an integral external ring which accurately centres gasket on flange face, provides additional radial strength to prevent gasket blow-out and acts as a compression stop.

Style CGI

A Style CG gasket fitted with an internal ring to give an additional compression limiting stop, provide heat and corrosion barrier and prevent flange face erosion.

Figure 12
Examples of spiral wound gaskets.

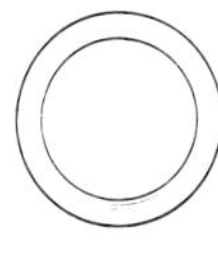

Plain

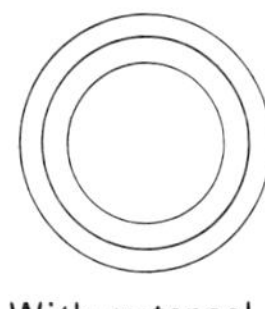

With external centring ring

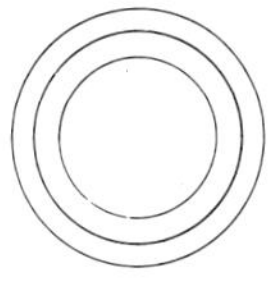

With external centring ring and internal reinforcing ring

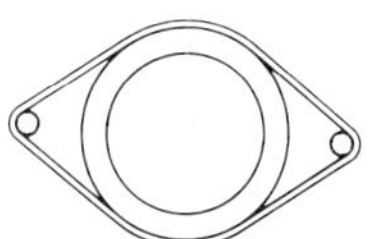

with lugged centre ring.

Types of spiral wound gaskets.

Retaining/supporting rings

Spiral wound gaskets can be used in a variety of flange configurations but to achieve maximum sealing performance they must be compressed by a predetermined amount. In totally enclosed situations, such as spigot and recess, the optimum compression is achieved with metal to metal contact of the flange faces. It is therefore imperative that the spigot and recess be machined within specified dimensions and tolerances.

In unconfined configurations compression is controlled by a metal ring fitted to the outer periphery of the windings, the gasket being compressed almost to this (retaining ring) thickness. In addition to controlling compression these rings serve to locate the gasket centrally within the bolt circle and to support the outer edge of the gasket against radially induced stresses. The rings are normally of carbon steel treated with inhibitor for protection against atmospheric corrosion.

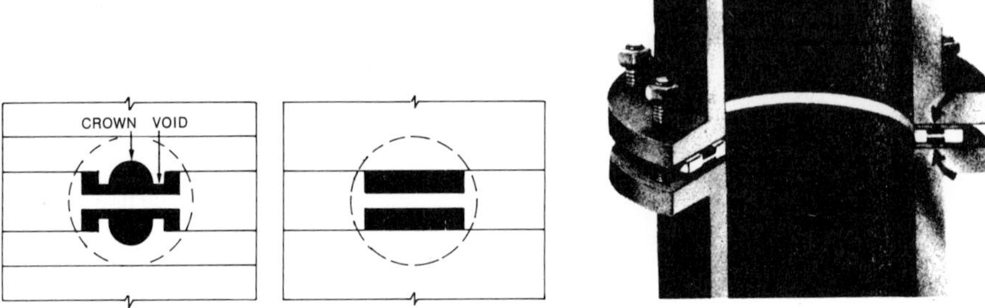

Combined elastomeric O-ring and metal or plastic retainer.

For certain applications inner rings, usually of stainless steel, are also fitted to prevent the build up of solids between the inside diameter of the gasket and the bore of the pipe. They also protect plant from damage in the unlikely event of gasket failure, especially under vacuum conditions when broken components could be drawn into the system.

Table 5 gives details of the recommended compressed thickness for standard spiral wound gaskets.

TABLE 5 – RECOMMENDED COMPRESSED THICKNESS FOR STANDARD SPIRAL WOUND GASKETS

Nominal Thickness		Compound Thickness	
mm	in	mm	in
3.2	0.125	2.4 to 2.6	0.095 to 0.105
4.5	0.175	3.2 to 3.4	0.125 to 0.135
6.4	0.250	4.6 to 4.8	0.180 to 0.190
7.3	0.285	4.7 to 4.9	0.185 to 0.195

Flange surface finish

When a spiral wound gasket is compressed the filler, which stands slightly proud of the metal winding, flows into the slight imperfections in the flange faces while the metal winding provides the mechanical strength and resilience. The more pronounced the asperities in the flange surfaces the greater the load required to effect a reliable seal.

For the majority of service conditions encountered in industry the operating commercial surface texture is acceptable but when extreme operating conditions are involved the need for a superior surface finish becomes essential. Too smooth a surface will reduce the friction between the metal face and the sealing element to such a degree that sealing efficiency could be impaired. A compromise is achieved by a high quality surface texture being machined with a spiral groove of precise configuration. Manufacturers have their own specifications for these finer textures and should be consulted regarding their recommendations.

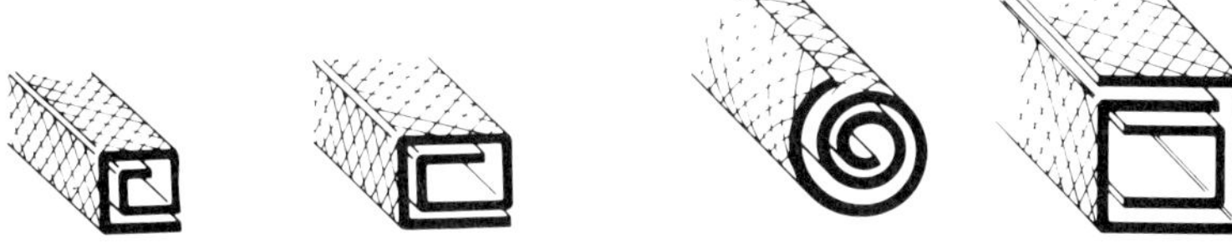

Examples of standard sections produced in solid metallic gaskets.

Applications of spiral wound gaskets

Due to the unique construction of spiral wound gaskets their application range is extremely wide. They can operate at temperatures between −250°C and +1000°C and pressures from vacuum through to 350 bar (5000 lb/in^2). With careful attention to flange design and bolting arrangements, even higher pressures can be accommodated. The selection of filler ensures compatibility with the vast majority of fluids in use over the whole field of industrial activity whether they be liquid or gaseous.

This versatility finds application in service with steam generation and distribution, petrochemicals, nuclear, hydraulics and high pressure gas systems.

Spiral wound gaskets are produced to a number of British and overseas standards and specifications which detail the dimensions and tolerances covering a whole range of operating conditions.

Secure sealing of high performance bolted joints

High performance means not only high pressures and temperatures but the more demanding fluctuations giving rise to thermal shock conditions. Conventional gasketing is unable to cope with such conditions with certainty over a long period, partly because of lack of resilience and partly due to the convention of overloading the joint at cold to compensate for relaxation at temperature. This tends to overstress the bolts which will themselves relax, reducing the closing effort at the very time when it is most needed.

In a high security plant, such as a nuclear pressurised water reactor power station, this is unacceptable.

Expanded graphite, already well accepted in the industry, was the obvious material to provide a solution, but a method of application was needed.

The answer was the LATTY*graf REFLEX concept. This consisted of a tightly wound pure expanded graphite tape held betweeen floating inner and outer stainless steel rings, designed to operate with metal to metal contact (Figure 13).

The system has four main features:-

(a) The low bolt loading removes the disadvantage of highly stressed fasteners (Figure 14).

	Lattygraf Reflex	Spiral wound gaskets Asbestos filler	CAF filler	Graphite filler
Joint closing force required to effect a seal (daN)	19800	40800	30400	22000

Figure 13

Figure 14

(b) The use of expanded graphite as the working seal means that seal relaxation problems are minimised, and that specific pressure on the seal can be calculated and maintained. These are essential features for leak-tight operation. Expanded graphite flows naturally to match with flange deformations or defects.

(c) The precision machined outer stainless steel ring prevents extrusion, and limits compression of the expanded graphite seal. In operation, this ring, in metal to metal contact with the joint faces, protects the seal by absorbing all mechanical stresses from pipework and flange, and also from temperature and pressure changes creating shock conditions. The stability and resilience of the expanded graphite ensure that the seal remains intact.

(d) Exceptional seal resilience is obtained by taking advantage of the elastic recovery of tightly wound expanded graphite working at right angles to the direction of winding. For pre-compressed LATTY*graf E, this recovery is of the order of 10% of the thickness of the limiter.

(e) Production repeatability warrants leaktight operation of each gasket of the same design in the same bolted assembly due to the minimum variation in the closure load requirement. (Not more that + or −5% as against + or −20% or more for a typical spiral wound gasket.

Design elements

The principle of metal to metal contact permits an exact determination of the volume of expanded graphite that will be held between the inner and outer retainer rings. The

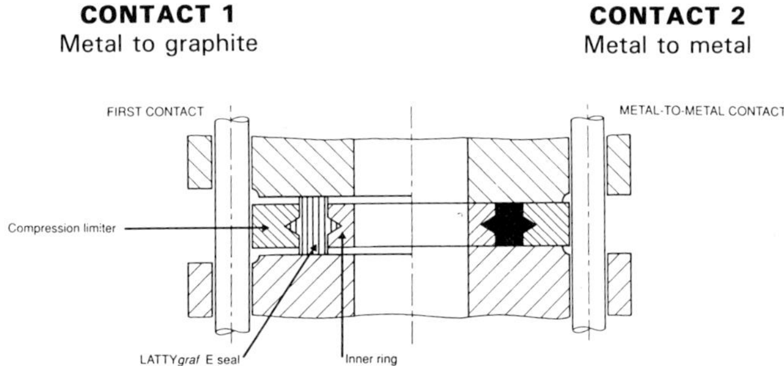

Figure 15
Diagram illustrating the LATTYgraf REFLEX.

load to be applied to the gasket to ensure metal to metal contact under operating conditions is determined by the pressure of the fluid to be sealed, the nature of the fluid, its temperature, the configuration of the gasket housing and any special customer observations.

Figure 16 shows the loading and recovery characteristics ofa Reflex gasket, with recovery a usable 0.4 mm, which is excellent compared to metallic gaskets (0.1/0. mm) or to spiral wound (0.2 mm).

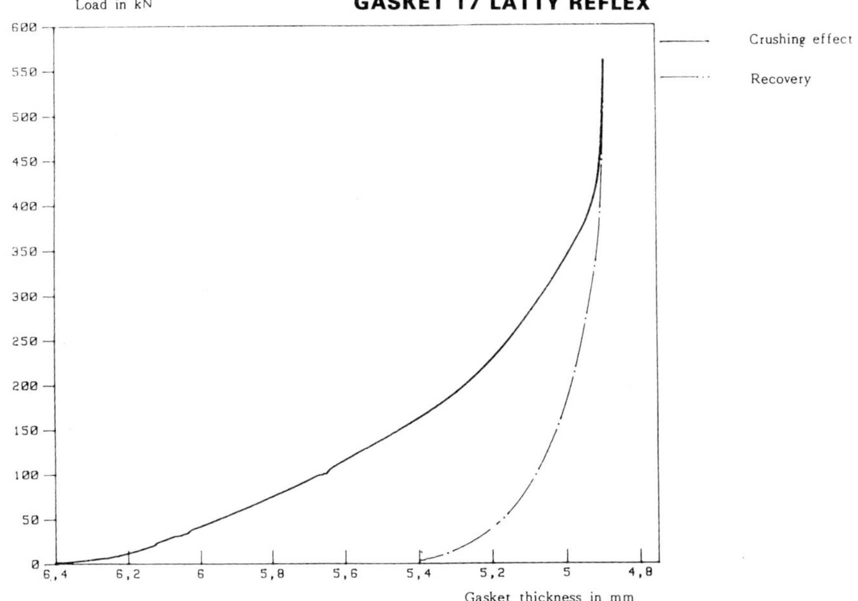

Figure 16
Gasket 17 LATTY REFLEX.

Two gaskets of the same design are compared in Figure 18, illustrating the repeatability of the characteristics.

Application data required for correct design.

Fluid
Nature, temperature, pressure and changes of state.

Working conditions
Particularly the extent and frequency of thermal transients.

Joint
Type, dimensions, material. Position of recesses, shape, sizes and tolerances.

Joint faces
Hardness, flatness, quality of finish.

Closure
Number of bolts, type, thread, material and treatment if any. Method and means of closure. The ideal is to seek the minimum applied load necessary to provide a secure seal.

With the foregoing, a calculation can be made of:-

– the load required for metal to metal contact.
– the load required for guaranteed leaktight operation.
– the characteristics of the LATTY*graf E seal and of the limiter-thickness, surface finish, material (if non-standard).

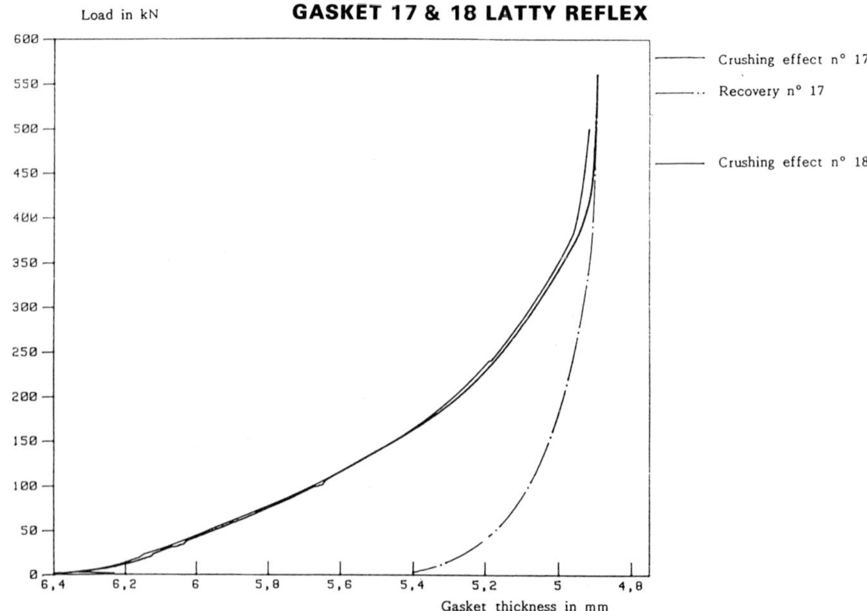

Figure 17
Gasket 17 & 18 LATTY REFLEX.

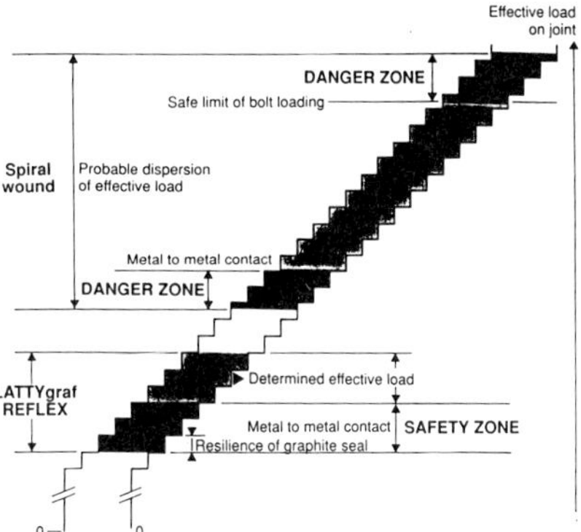

Figure 18
Diagram illustrating the comparison between LATTYgraf REFLEX and a typical spiral wound seal.

The Upload/Download characteristics of the REFLEX gasket is the same at temperature as at cold.

The behaviour of a spiral wound gasket at temperature is quite different from its behaviour at cold. Resilience and recovery are greatly diminished.

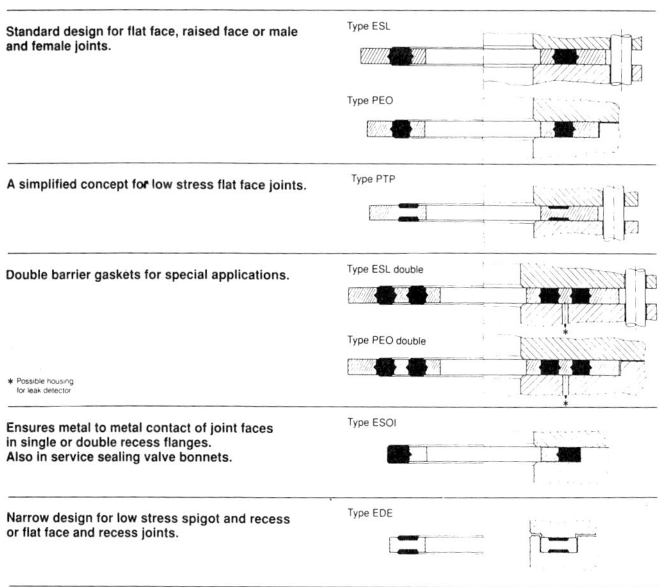

Figure 19
LATTYgraf REFLEX design range.

Solid metal gaskets

The use of *solid metal* gaskets invariably involves higher seating stresses than other gasket types which means that flange faces must be flat and of good finish.

The recovery properties of solid metal are minimal, consequently the design should guarantee that seating stresses will be maintained regardless of fluctuations in operating conditions.

Although flat metal gaskets are frequently used they are normally limited to fairly narrow faces and to help minimize exceptionally high assembly loads. Several types of metal gaskets have a cross section designed to concentrate the load over a small area, thus producing a high seating stress (*eg* Figure 20).

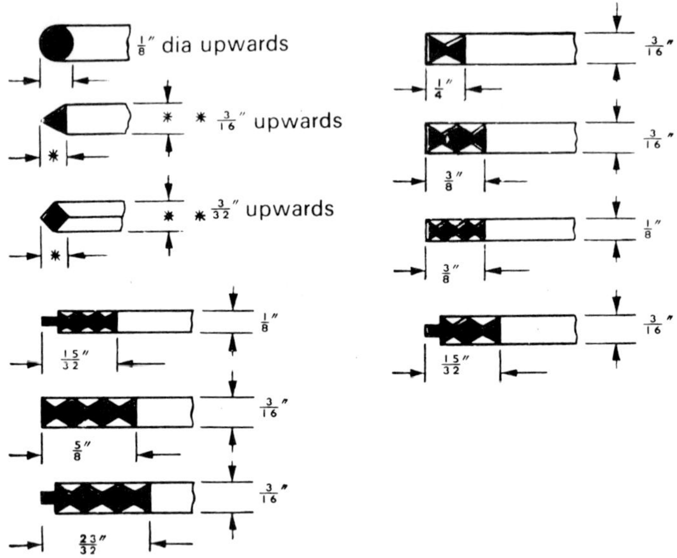

Figure 20
Examples of gaskets fashioned from wire reinforced rubber-asbestos gasket type.

Typical examples of this type are *round, oval, diamond* and *octagonal*, and the materials used include copper, soft iron, lead, aluminium, stainless steel, nickel, monel, inconel and titanium. This type of gasket is often found in the smaller sizes, similar to washers, where sealing faces are narrow and high stress can be induced by the compressive load available.

Larger dimensioned gaskets find application in very high pressure piping systems and pressure vessels. For this purpose it is customary to machine a recess in the flange faces to accommodate the profile of the gasket (Figure 21). A variation on this theme is the *Lens* ring (Figure 22) where the gasket is machined to a slight curvature on the seating surfaces with the mating faces machined at an angle but perfectly flat.

A further type of solid metal gasket has serrated faces with serrations approximately 0.8 mm deep and a pitch of 5.5 per cm ($^1/_{32}$ in deep and 14 per in). The peaks of the serrations are flattened, measuring approximately 0.1 mm and these provide a series of point loads across the sealing faces. During manufacture it is essential that the planes

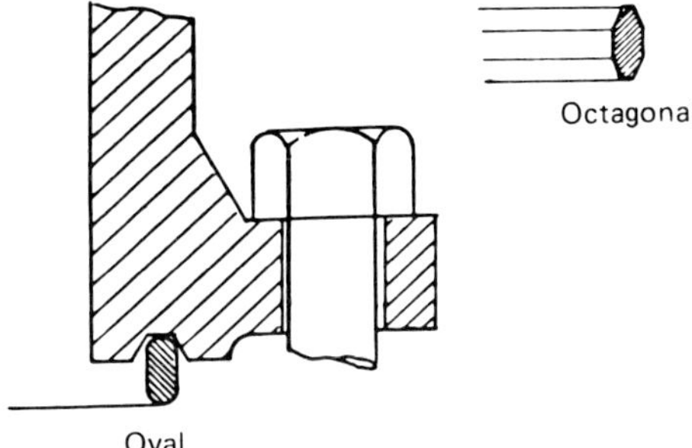

Figure 21
Retained solid metal rings.

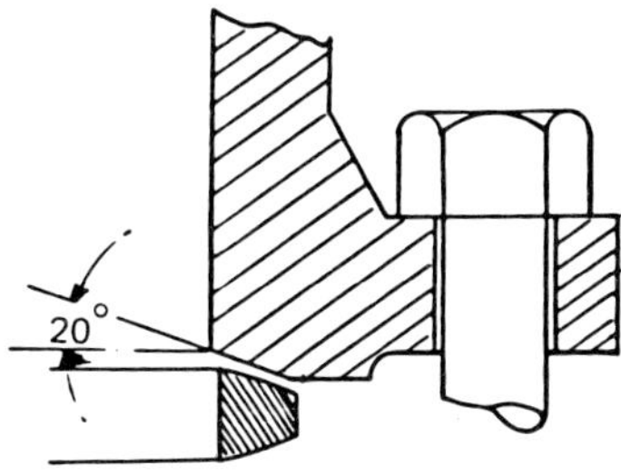

Figure 22
Lens ring.

of the serrations on both faces are parallel and that the contours of the serrations are identical. A variation of this serrated gasket employs a thin compressed asbestos fibre of beater asbestos sheet fitted to each face.

In all these applications with solid metal gaskets the quality of the component faces is critical and every care should be taken to ensure their integrity regarding flatness and smooth texture.

The temperature limitations of most applications are those set by the metal concerned. Pressures up to 1000 bar (approximately 15000 lb/in^2) can be accommodated using these standard types provided attention is given to assembly loads.

Wave ring gaskets

Wave ring gaskets (developed by ICI) are resilient 'wavy rings' which when lightly sprung into a matching groove or cavity are deformed under system pressure acting on the entire internal surface to press the face of the gasket into close contact with the walls of the groove. Metal gaskets of this type (*eg* in stainless steel) can prove particularly effective at high temperatures since their sealing properties are enhanced by expansion of the gasket.

Printed gaskets

The *printed* gasket is a novel approach to this flat form of sealing and is available to a whole spectrum of industry in three different forms:

(i) Metal based high pressure gasket.

(ii) Paper based.

(iii) The pure print soft joint gasket.

Using this technique a gasket may be made with the sealing medium precisely where it is required, incorporating the correct thickness, resilience and environmental resistance to suit the joint under consideration. In addition, because it is possible to produce gaskets unsupported, or deposited on a permanent substrate, it is also possible to extend the role of the gasket to encompass other functions. These are the subject of the *Dowprint* patent.

Metal based or high pressure printed gasket (Figure 23) Group 1

High pressure Dowprints are selective deposits of polyurethane or similar polymers on a metal, phenolic or composite substrate. This type of gasket is used effectively for sealing complex systems of parts or channels normally sealed by O-rings in machined grooves. As the complexity of design or number of parts to be sealed increases, so the cost saving using a high pressure Dowprint also increases. Peak pressures up to 407.5 bar (6000 lbf/in^2) can be sealed effectively, the gasket having resistance to mineral, hydraulic oils and greases within the temperature range of −30°C to +90°C for PVC, −10°C to +120°C for epoxies and −60°C to +200°C for silicone, both standard and conductive.

Paper based Dowprint Group 2

The paper based Dowprint overcomes the handling problems associated with large flexible gaskets. Performance advantages are obtained by selectively printing compressible sealants in problem areas, thus locally increasing gasket thickness. Many applications exist where this type of gasket has replaced the inconvenient and expensive use of 'on site' liquid sealants.

The following significant cost savings can be made when using paper based 'DOWPRINTS':-

(a) Elimination of 'on site' liquid sealants and possible operator error.

(b) Reduced machining on mating components.

(c) Relaxation of bolting patterns and associated hardward bulks. Deflections can be accommodated.

(d) Relaxation in quality of carrier substrate, as substrate may act only as an assembly aid for the flexible seal.

Figure 23
Metal based or high pressure printed gasket as used on a hydraulic circuit for machine tools.

Where micro sealing is required for pneumatic or hydraulic applications, the reverse side of the gasket can be flash coated with about 0.22 mm of the sealant being used for primary sealing operation.

Low pressure pure print gasket (Figure 24) Group 3

The pure print is a selective deposit of cellular or non-cellular elastomeric material, produced without the use of expensive tooling. Fundamentally, material is printed on to a *temporary release carrier* where it is subsequently vulcanized and/or expanded, as required.

The natural domed shape of the *soft joint* gasket obtained provides a seal over a progressively larger area as it is tightened down, giving improved sealing over conventional flat gaskets. The pure print is normally competitive with punched rubber and composite materials.

Normally used for atmospheric sealing the pure print gasket is competitive with punched rubber, composite materials and smaller cork gaskets. Expanded epoxy materials offer higher performance, using the low wastage factors of the process to compensate for higher raw material costs.

Special purpose Dowprint Group 4

Typical of several types developed are these:

(a) There are applications where the assembled thickness of gasket is critical – that is, it must also behave as a 'shim'. To meet these conditions a Group 2 type is used in which the printed sealant consists of intermittent tapered beads between high pressure areas close to the clamping bolts. The correct gap is determined by the thin paper and, by careful adjustment of the thickness and proportions of the bead, distorting components can have a substantially constant sealing pressure all round.

(b) The 'DOWPRINT' process can also be used as a means of improving the performance of existing gaskets of other types.
An example of this benefit is in the application of printed beads on engine cylinder head gaskets to improve sealing at local low pressure areas remote from the bolt positions. The great design freedom of this method enables the designer to achieve an exactly tailored remedy for such gaskets which show a deficiency in service.

Figure 24
Printed gasket plate.

Compression control Dowprint Group 5

Conventional gasket materials only provide a compromise of the desired characteristics of conformability, compressibility and elasticity with good anti-extrusion properties. An old but still common problem is the relaxation effect which occurs when high clamping pressure is accompanied by thermal cycling. To overcome the problem, a range of compression control 'Dowprints' are being developed, and our engineers will be pleased to discuss specific requirements on request.

Epoxy based sealants

These are available in a full hardness range, both in cellular and solid form. Fluid compatibility is very good, being suitable for mineral oils, kerosene, aviation fuels, water, glycols, coal and natural gas. Temperature resistance can be up to 120°C maximum for continuous service.

Epoxy is regarded as the general purpose sealant, and may be used on metal, plastic and paper bases, in addition to the pure print material form discussed earlier.

Urethane based sealants

With higher strength and elongation than the epoxy group, these are normally employed for high pressures. Generally suitable for cycling pressures to 345 bar (5000 lbf/in^2) but static pressures can be considerably in excess of this. Suitable for most mineral and hydraulic oils with operating temperatures up to 100°C maximum. These materials are normally processed on to metal, plastics, or paper substrates. Because of their inherent high abrasion resistance, they can be usefully employed in gaskets where fretting conditions exist due to movements of joint faces.

Silicone sealants

Though of modest strength, silicone can be provided for applications where duties are moderate, except for extremes of working temperatures. This can be from −60°C to +200°C in most hydraulic and mineral oil environments.

Development work is continuing at present aimed at providing an expanded silicone material suitable for manufacturing *pure print* gaskets compatible with the general range of automotive requirements.

PVC based sealant

Available in expanded form and used primarily as a dust and moisture seal where performance requirements are relatively low.

See also chapter on *Liquid Sealants*.

Material selection

Once having established the range of materials available, the selection for a given application depends on certain fundamental criteria:

(i) Nature of fluid.
(ii) Operating temperature range.
(iii) Pressure or vacuum conditions.
(iv) Possibility of cycling conditions, surge pressures, vibration, *etc.*
(v) Shape and dimensions of seating surfaces.
(vi) Texture of seating surfaces.
(vii) Assembly load available.
(viii) Other factors such as radiation, contamination of fluid, *etc.*

TABLE 6 – TYPICAL TEMPERATURE AND PRESSURE LIMITATIONS FOR COMMON GASKET MATERIALS

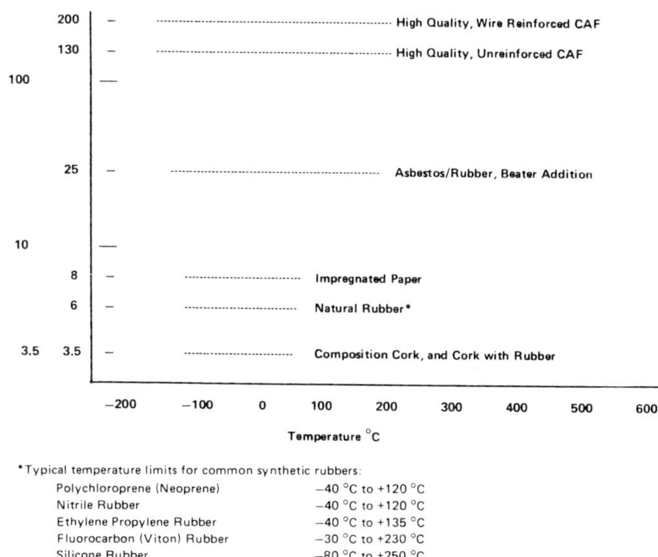

Once these details are ascertained the choice of material could become quite straightforward. Alternatively, it is possible that several materials will emerge, each of equal merit. Selection then becomes centred on wider issues such as:

(a) Current practice in similar, though not identical, applications.

(b) Previous experience where satisfactory performance was achieved.

(c) Standardization and economic considerations in terms of inventory levels.

(d) Safety element where superior performance material is adopted as standard to safeguard against the use of low performance material in a high performance area.

(e) Initial cost.

(f) Availability.

Manufacturers are always willing to provide information and give guidance in respect of particular materials and applications, and certain codes of practice exist which specify the relevant parameters covering materials in general use.

There is also an abundance of information available produced by various standards institutions and industrial technical authorities covering the majority of sealing materials and methods. These provide the designer with ample guidance in respect of standard equipment and services and with due consideration of the factors involved can be interpolated for non-standard applications.

Reliability and failure

The reliability of a seal is as dependent on sound engineering practice in the field as it is on the detailed attention given to the initial design, and any reference to what could be considered elementary engineering procedures is simply to record certain essentials which are occasionally overlooked. The inability to effect or maintain a satisfactory seal can be attributed to a series of factors which can be classified into four distinct groups:

TABLE 7 – GASKET MATERIALS AND CONSTRUCTION

(Based upon the ASME Unfired Pressure Vessels Code, Section VIII, Division 1)

(The details given in this table are suggested only and are not mandatory).

American National Standard Steel Pipe Flanges and Flanged Fittings **ANSI B16.5–1977**

Gasket Group Number	Gasket Material		Gasket Factor m	Minimum Design Seating Stress, γ lb/in^2	Minimum Design Seating Stress, γ MPs	Sketches
1a	Self-energizing types O-rings, metallic, elastomer, other gasket types considered as self-sealing		0	0	0	–
1a	Elastomer without fabric or a high percentage of asbestos fibre: Below 75 Shore Durometer 75 or higher Shore Durometer		0.50 1.00	0 200	0 1.4	
1a	Asbestos with a suitable binder for the operating conditions	3.2 mm (0.12 in) thick 1.6 mm (0.06 in) thick	2.00 2.75	1600 3700	11.0 25.5	
1a	Elastomer with cotton fabric insertion		1.25	400	2.8	
1a	Elastomer with asbestos fabric insertion, with or without wire reinforcement	3-ply	2.25	2200	15.2	
1a		2-ply	2.50	2900	20.0	
1a		1-ply	2.75	3700	25.5	
1a	Vegetable fibre		1.75	1100	7.6	
1b	Spiral-wound metal, with asbestos or other non-metallic filler	carbon steel stainless steel or Monel	3.00	10000	68.9	
1b	Corrugated metal, asbestos inserted or Corrugated metal double jacketed asbestos filled	soft aluminium soft copper or brass iron or soft steel	2.50 2.75 3.00	2900 3700 4500	20.0 25.5 31.0	
1b	Corrugated metal	soft aluminium soft copper or brass	2.75 3.00	3700 4500	25.5 31.0	

	Asbestos with a suitable binder for the operating conditions	0.8 mm (0.03 in) thick	3.50	6500	44.8	
	Corrugated metal, asbestos inserted or Corrugated metal double jacketed asbestos filled	Monel or 4-6% chrome	3.25	5500	37.9	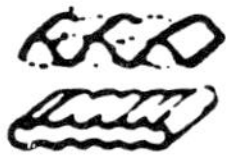
		stainless steels	3.50	6500	44.8	
	Corrugated metal	iron or soft steel	3.25	5500	37.9	
		Monel or 4–6% chrome	3.50	6500	44.8	
11a & 11b		stainless steel	3.75	7600	52.4	
	Flat metal jacketed asbestos filled	soft aluminium	3.25	5500	37.9	
		soft copper or brass	3.50	6500	44.8	
		iron or soft steel	3.75	7600	42.4	
		Monel	3.50	8000	55.2	
		4–6% chrome	3.75	9000	62.0	
		stainless steels	3.75	9000	62.0	
	Grooved metal	soft aluminium	3.25	5500	37.9	
		soft copper or brass	3.50	6500	44.8	
		iron or soft steel	3.75	7600	52.4	
		Monel or 4–6% chrome	3.75	9000	62.0	
		stainless steels	4.25	10100	69.6	
	Solid flat metal	soft aluminium	4.00	8800	60.7	
111a & 111b	Solid flat metal	soft copper or brass	4.75	13000	89.6	
		iron or soft steel	5.50	18000	124.1	
		Monel or 4–6% chrome	6.00	21800	150.3	
		stainless steels	6.50	26000	179.3	
	Ring joint	iron or soft steel	5.50	18000	124.1	
		Monel or 4–6% chrome	6.00	21800	150.3	—
		stainless steels	6.50	26000	179.3	

TABLE 8 – GENERAL GUIDE TO GASKET MATERIALS

Material	Specific Gravity or Density	Tensile Strength lb/in^2	Compressibility %	Minimum Sealing Pressure lb/in^2	Bar	Maximum†† Internal Pressure lb/in^2	Bar	Maximum Service Temp. °F	Maximum Service Temp. °C
Natural cork	9.5–13†	140–170	4–6% @ 100 lb/in^2					250	120
Cork compositions									
Gelatin binder, coarse	20†	150	25–40 @ 100 lb/in^2	Low	Low	–	–	250	120
Gelatin binder, medium	20†	150	25–40 @ 100 lb/in^2	Low	Low	–	–	250	120
Resin binder, fine	23–40†	200–400	10–30* @ 100 lb/in^2	Low	–	–	–	250	120
Impregnated cork	15–20†	–	30–50 @ 100 lb/in^2	100	7	50–100	3.5–7	250	120
			55–70 @ 300 lb/in^2						
	23–30†		20–40 @ 100 lb/in^2	100–200	7–14	50–100	3.5–7	250	
			45–60 @ 300 lb/in^2						
Cork – rubber compositions									
Butyl (typical)	0.7–0.82	200–500	25–40 @ 300 lb/in^2	400	28	50	3.5	250–300	120–150
Sponged butyl	0.4–0.55		25–40 @ 40 lb/in^2	200	14	50	3.5	250–300	120–150
Chloroprene (low density)	0.7–0.8	200–400	25–40 @ 200 lb/in^2	400	28	50	3.5	250–300	120–150
(medium density)	0.9–1.0	200–500	10–20 @ 300 lb/in^2	400	28	50	3.5	250–300	120–150
(high density)	1.2–1.3	200–800	5–15 @ 300 lb/in^2	400	28	50	3.5	250–300	120–150
Sponged chloroprene	0.5–0.6		25–40 @ 50 lb/in^2	200	14	50	3.5	250–300	120–150
Nitrile									
(low density)	0.6–0.75	200–400	30–40 @ 200 lb/in^2	400	28	50	3.5	250–300	120–150
(medium density)	0.95–1.05	200–500	10–20 @ 300 lb/in^2	400	28	50	3.5	250–300	120–150
Sponged nitrile	0.45–0.55		30–40 @ 50 lb/in^2	200	14	50	3.5	250–300	120–150
Sponged styrene	0.4–0.5		10–25 @ 40 lb/in^2	200	14	50	3.5	250–300	120–150
High temperature cork – rubbers (typical)			15–25 @ 400 lb/in^2	400	28	50	3.5	400–450	200–230

Synthetic rubber									
Butyl		1200–1800	20% maximum recommended	500	35	50	3.5	225	107
Chloroprene		1200–1900	20% maximum recommended	500	35	50	3.5	300	150
Nitrile soft		1000–1500	20% maximum recommended	500	35	50	3.5	250	120
Nitrile medium		1200–1500	20% maximum recommended	500	35	50	3.5	250	120
Nitrile hard		1400–1900	20% maximum recommended	500		50	3.5	250	120
Asbestos fibre									
Compressed	1.9	6/12000		1000	70			750–800	400
Wire reinforced	1.95	4/10000		1000–4000	70–250			1000–11000	540–600
Asbestos – rubber									
Styrene (low density)	*	1000	30–40 @ 5000 lb/in^2	High	Low			700–800	370–430
Chloroprene (low density)	*	800	30–40 @ 5000 lb/in^2	High	Low			700–900	370–480
Nitrile (low density)	*	1500–1700	30–40 @ 5000 lb/in^2	High	Low			700–900	370–480
Styrene (precompressed)	*	1700	20–30 @ 5000 lb/in^2	High	Low			800–880	430–470
Chloroprene (precompressed)	*	1200	20–30 @ 5000 lb/in^2	High	Low			800–1000	430–540
Nitrile (precompressed)	*	2000	20–30 @ 5000 lb/in^2	High	Low			800–1000	430–540
Fibre									
Cellulose fibre, cork, styrene	*	700	30–45 @ 1000 lb/in^2	800–2000	59–140	2000	140	300	150
Nitrile	*	800	30–45 @ 1000 lb/in^2	800–2000	55–140	2000	140	300	150
Nitrile (precompressed)	*	1200	12–30 @ 1000 lb/in^2	800–2000	55–140	2000	140	300	150
Cellulose fibre, synthetic Rubber	*	1600–2500	10–30 @ 1000 lb/in^2	2000	140	2000	140	300	150
Cellulose fibre, filler, Synthetic rubber	*	3000	5–15 @ 1000 lb/in^2	2000	140	2000	140	300	150

*Dependent upon composition

† Density – lb/ft^3

†† Higher internal pressures may be sealed with suitable flange design. Figures related to sealing properties of simple gasket only.

(i) Design.
(ii) Assembly.
(iii) Metal faces.
(iv) Gasket material.

Table 9 can be used to identify the fault, its probable cause and the appropriate remedy.

It should be noted that this is by no means exhaustive, as a failure could be traced to something as obviously simple as the omission of a gasket on initial assembly or, at the other extreme, to the effect of some obscure metallurgical process; each, of course, must then be dealt with by whatever straightfoward or complicated remedy best suits the case.

Standards

For production purposes a manufacturer may adhere to a number of standards, *ie* those maintaining the quality of his own individual materials and designs, whilst others are to conform to National standards or specifications laid down by industrial instructions, *etc.* Current National standards are:

(a) Materials

British

BS 1832 : 1972 – Specification for oil resistant compressed asbestos fibre jointing.
BS 2815 : 1973 – Specification for compressed asbestos fibre jointing.
Grade A – For water, inert gases, inert liquids or steam up to 64 bar and 510°C.
Grade B – For water, inert gases, inert liquids or steam up to 16 bar and 230°C.
BSF 125 : 1973 – Specification for compressed asbestos fibre jointing for aerospace purposes.

German

DIN 3754 – Specification for various compressed asbestos fibre grades.

American

ASTM D1170 – Specification for non-metallic gasket materials for general automotive and aeronautical purposes.
ASTM F104 – Classification system for non-metallic gasket materials.

(b) Dimensions

British

BS 3063 : 1965 – Dimensions of gaskets for pipe flanges to BS 10, BS 1770 and BS 2035.
BS 4865 – Dimensions of gaskets for pipe flanges to BS 4505, BS 4622 and BS 4722.
Part 1 : 1972 – Dimensions of non-metallic gaskets for pressures up to 64 bar.
Part 2 : 1973 – Dimensions of metallic spiral wound gaskets for pressures 10 to 250 bar.
BS 3381 : 1973 – Design material and dimensions of metallic spiral wound gaskets for use with flanges to BS 1560.

American

ANSI B16.21 1978 – Dimensions of non-metallic flat gaskets for pipe flanges to ANSI B16.1, ANSI B16.5, ANSI B16.24, API Std 605, MSS SP44 and MSS SP51.
API 601 – Metallic gaskets for piping.

TABLE 9 – TROUBLE-SHOOTING GUIDE

Fault	Cause	Remedy
DESIGN		
Insufficient gasket stress	Insufficient bolt load	Increase number of bolts. Increase diameter of bolts Change to higher tensile material.
	Gasket too thin	Fit thicker gasket.
	Gasket too wide	Reduce area of gasket.
	Wrong gasket type	Fit gasket which requires a lower seating stress
Excessive gasket stress	Excessive bolt load	Reduce number of bolts Change to lower tensile material
	Gasket too thick	Fit thinner gasket
	Gasket too narrow	Increase area of gasket
	Wrong gasket type	Fit gasket which requires a higher seating stress
ASSEMBLY		
Lack of compression	Bolts insufficiently tightened	Apply additional torque
	Incorrect tightening procedure	Bolts should be tightened in sequence, *ie* diametrically opposite and gradually increasing load on each bolt alternately
	Gasket relaxed due to operating temperature	It is recommended that once plant reaches operating temperature all gaskets are 'followed-up' to restore compression
	Bad threads	Ensure nuts are a good running fit over entire length of bolt thread
	Insufficient length of thread	Ensure threads sufficiently long to allow nuts to make contact with metal faces
METAL FACES		
Uneven	Flanges too thin	Flanges should always be sufficiently rigid not to be distorted by the bolt load
	The use of an IBC gasket (ring joint)	With thin flanges and an IBC gasket the bolt load could cause flanges to bow or bend. (Sometimes referred to as 'rotation of flanges'). Flanges should be straightened and full faced gasket used
	Flanges not parallel	Flange faces should always be parallel and bolt load should never be relied on to pull flanges together. *Note:* Spiral wound gaskets can accommodate slight misalignment under certain conditions
Damaged	Mechanical damage while faces exposed	Every attention should be given to ensure faces are clean, flat and free from imperfections too deep for the gasket material to completely fill.
	Qver zealous abraiding during cleaning	

cont...

TABLE 9 – TROUBLE-SHOOTING GUIDE (contd.)

Fault	Cause	Remedy
Dirty or corroded	Previously used jointing compounds frequently harden and form an uneven surface Old gasket not completely removed	Faces should be wire brushed right down to clean metal. Serrations should also be perfectly clean and of sound contour. Spigot and recess faces should be checked for correct fit
Incorrect surface texture		Concentric grooving is ideal for high pressures. Where 'gramophone' or continuous spiral grooving is employed the depth of groove must not be too deep for the gasket to fill
GASKET MATERIAL		
Loss of resilience and interface contact	Re-use of old gasket	The re-use of gaskets is not recommended. The material will have hardened and taken up the contours of flange surface imperfections. It is unlikely that a gasket would be replaced in exactly the same position. Furthermore, downtime costs to replace a faulty gasket are far in excess of initial gasket cost.
	Metal gaskets work hardened	Where possible metal gaskets such as copper should be annealed prior to re-use when they can give further useful service.
Material deteriorates rapidly	Material incompatibility with contained fluid/temperature	Check manufacturer's material recommendations and select a material or gasket type capable of withstanding the conditions
Gasket extrudes from faces	Too high a seating stress	See recommendations under design faults
	Excessive use of jointing compounds	Unless specified by gasket manufacturer the use of compounds and pastes is not recommended. These act as a lubricant which reduces the friction between the gasket and the metal faces thereby reducing the load bearing properties Where a non-stick finish is required it can be applied by manufacture during production
Incorrect dimensions	Design or manufacturing error	Gaskets should always have clean cut edges with the bore slightly larger than that of the vessel or pipe, *ie* the gasket should not protrude into the flow path of the fluid. Protrusion could create turbulence in addition to restricting flow. The gasket could also suffer damage through erosion by the fluid Providing the gasket is compressed over the entire face there is little likelihood of absorption of the fluid. Bolt holes should be sufficiently large to allow a clearance around the bolts

AFNOR (French Standards)

Séries	
A 48-501 à 504	de canalisation en fonte
A 48-802 à 823	
A 84-622, 623	plastiques (Raccords à)
B 35-013	coniques rod (Verrerie de laboratoire)
E 29-531 à 539	(Raccords à)
E 29-591 et 582	(Bouchons à-)
E 29-911 à 956	de tuyauterie industrielle
E 48-041	toriques
E 48-051	pour raccordements hydrauliques
E 20-028	de tampons autoclaves
M 87-615	pour brides
L 17-109	hydrauliques en élastomères
L 46-030 à 033	détanchéité en élastomères

Séries	
L40-114 et L43-603	toriques
M28-245	pour tuyauterie d'air comprim
P 34-403	couvre-joints en zinc
P78-503 et 504	(Fonds del) miroiterie
Sèrie P 85-102à 515	en élastomères et du type plastique
R 93-920	circulairs d'étanchéité
R 99-001	Bague d'étanchéité
R 99-202	(papier pour l)
S 61-825	de lances d'incendie
T 47-301	de canalisations
T54-041	
P 16-343	
T 47-302 et 303	Préformés à base de caoutchouc

Liquid Sealants

LIQUID SEALANTS fall broadly into three categories : general purpose *mastics* or *joint* sealants, *liquid* gaskets, *thread* sealants and *thread-locking* compounds.

Mastics

Mastics are used by the building trade for bedding, sealing frames, joints, *etc.* The majority may be classified as general purpose sealants. Special purpose sealants have been developed for use as caulking compounds for marine use, sealing of cold room doors and other specialized applications.

Joint sealants are best classified by chemical type, (see Table 1). Their use and range of application is well established in non-engineering fields. Depending on their purpose they may be available in tube (paste) or strip form and also in pourable, knife or gun grades. Set properties may range from fully elastic to hard solid.

Liquid gaskets

Liquid gaskets are also sealants but specifically formulated to contain fluids under pressure and provide an alternative to gaskets or static seals. They fall into three general categories: non-hardening, semi-hardening and hardening compounds. Non-hardening compounds are really mastics, with limited adhesive properties. They remain soft or 'wet' in the sense that they contain plasticizers which continually come to the surface. Semi-hardening and hardening compounds are both curing or setting adhesives, the difference being that the final joint is either flexible or rigid, respectively.

Table 2 details typical categories and types of sealing compounds and their application.

Sealing compounds can also be classified under five further headings:

(i) *General jointing* compounds.

(ii) *Specialized gasket* or *static seal replacement* compounds.

(iii) *Retaining* compounds.

(iv) *Hot melt* adhesives.

(v) *Film* adhesives.

General jointing compounds

For many years general jointing compounds were mainly based on shellac and linseed oil. These have now been largely replaced by polyester urethane products. These are

TABLE 1 – TYPES OF JOINT SEALANT

Chemical Type	Base	Pretreatment	Remarks
Oleo-Resins	Drying and non-drying oils, resins and fillers.	Require primers to give adhesion to porous surfaces	Best suited to lap joints, but can be tamped into butt joints. Available in strip form, or for knife or gun application
Rubber-Bitumen	Natural or synthetic rubber compounded with bitumen or pitch.	Movement accommodation of the order of 10%	Cold poured (set by solvent action) or hot poured.
Butyl	Butyl rubbers or degraded butyl rubbers, used alone or in combination with solvents, oils, extenders, etc.	Can be skin-forming, non-skinning, non-drying or curing (curing types cure by solvent evaporation).	Very wide range of compositions from soft, sticky compounds to hard rubber strips.
Polysulphide	Single or two-part mixtures based on polysulphide polymers. Two-part mixtures cure by chemical reaction fast.	Primers required on most porous and friable surfaces.	Wide range of compositions from soft elastic materials to hard rubbers. Available in a wide range of colours. Usual base of caulking compounds.
Silicone	Single-part chemically curing silicone elastomer.	Require primers on most porous and friable surfaces.	Remain fully elastic. Available in a range of colours
Polyurethane	Two-part self-curing mixtures based on polyurethane polymers.	Primers required on most surfaces for adequate adhesion.	Remain fully elastic.
Acrylic	Mixtures of acrylic polymers.		Plastic in nature.

TABLE 2 – SEALANT TYPES, APPLICATIONS AND RELATIVE COSTS

	Material	Application(s)	Relative Cost
NON-HARDENING	Bituminous	Caulking, general purpose sealant	1 – 4
	Butyl	Metal-to-metal, metal-to-glass, caulking	5 – 10
	Acrylic	Caulking metalwork, fibre joints, gaskets	8 – 10
	Polybutane	General purpose construction caulking and sealers	3 – 7
HARDENING FLEXIBLE	Neoprene (solvent-release)	General purpose sealant; also for dissimilar metals	8 – 12
	Butyl (solvent-release)	Sealant for metal, glass, masonry, etc	5 – 9
	Acrylic	General purpose construction sealant, caulking and glazing	8 – 10
	Silicone (one part)	General purpose construction sealant, caulking and glazing	20 – 40
	Silicone (two part)	Sealant for heat shields, etc; porting and encapsulation	40 – 100
	Polysulphide (one part)	General purpose sealant, caulking, glazing	17 – 25
	Polysulphide (two part)	Caulking, porting and encapsulation	12 – 15
	Polysulphine/Bitumen	Flame and fuel resistant caulking medium, sealant for expansion joints	4 – 6
	Polyurethane (one part)	General purpose sealant, caulking, glazing	12 – 15
	Polyurethane (two part)	Caulking, porting and encapsulation	12 – 15
	Epoxy (two-part modified)	Caulking, porting and encapsulation	7 – 15
	Fluorinated hydrocarbon	Oil and heat resistant sealant	80 –120
HARDENING RIGID	Oleo-Resin	Caulking, general purpose sealant	1 – 6
	Epoxy	Pipe sealing, porting, encapsulation of electrical compounds	7 – 14
	Polyester	Gaskets, pipe thread sealant, porting, moulding, encapsulation	6 – 12

TABLE 3 – OTHER SEALANT TYPES – GASKETS AND PIPE THREADS

Material	Application	Viscosity
Cyanoacrylates	Plastics, rubber, metals, ceramic	75 to 125 mPas
Strength – up to 36 MN/m^2 (350 bar) Cure rate – seconds at room temperature Chemical resistance – good Unsuitable for PTFE, polyethylene, polypropylene, silicone		
Material	**Application**	**Temperature**
Single part Thixotropic Solvent free with PTFE	Pipe threads Locknuts Retaining bushes, bearings, liners, studs, splines	–55 °C to +150 °C –67 °F to +300 °F
Adjustment time – up to 30 minutes from application. Unsuitable for use in contact with pure oxygen or high pressure steam		

used, for example, on face-to-face joints on motor pump units, on threaded plugs or threaded connections in hydraulics, as well as on joint faces on hydraulic motors and gearboxes and, in conjunction with gaskets, on geared motors. Much original development work on these was carried out to combat the problems posed by high additive diester-based fuels and lubricants, operating at high temperatures on small land widths. The polyester urethane is produced under carefully controlled conditions, so that it has the correct rheological properties to enable excess compound to flow to the edges of a joint during fastener loading, leaving no more than the minimum amount necessary to fill up the surface imperfections and local gaps. The high tack of the compound prevents this residual minimum from being squeezed out.

Plastic flow occurs during assembly and the absence of subsequent creep phenomenon ensures the retention of fastener loading in service. Therefore, as the compound is applied and air drying takes place, tackiness and adhesion to metal are sufficiently developed to ensure retention of the minimum amount of compound necessary when the joint is placed under load. The joint faces may be left as long as desired before assembly and, although the sealant need only be applied to one face in the vast majority of cases, in some critical areas the coating of both faces may be desirable in order to ensure thorough wetting of the surfaces and to avoid pockets of air at the metal/compound interfaces. Air trapped between the two coated surfaces is forced out by the flow of excess compound during the tightening operation, the rate of flow being greatest midway between the approaching surfaces. The product has a grease-like consistency which gives good distribution.

Polyester urethanes possess good adhesion to metal, yet their tack is of the short type which assists greatly in application as the threads break easily at low elongation, instead of drawing out in long unwieldy thread as in the case of adhesives. They are also mechanically weak in tension. A joint can be easily parted as rupture of the film occurs because the adhesion of the compound to metal is greater than its tensile strength. Its removal is easily carried out by means of a proprietary solvent. The viscosity of polyester urethanes enables them to be applied in any place, including vertically or on the underside of components.

Gasket/O-ring replacement materials

Products for specialized gasket/O-ring replacing compounds are mainly confined to silicone materials and, with the development of one part silicones, the application of these products has become much simpler. The power industry in particular is leading research into low cost gasket technology and *form-in-place* gaskets and O-rings are now rapidly being evaluated. This is because there are many applications on power equipment for *silicone* sealants, such as on hyudraulic motors, pneumatic equipment and compressors, in fact any assembly where the leakage or release of the fuel, the lubricant, the cooling medium or the air would affect the efficiency of the power unit. The method of application can be varied according to production requirements, ranging from low cost automatic dispensing machines to application by hand from cartridges or tubes for low volume or short run production. The *form-in-place* gasket or O-ring is made by applying a thin bead of silicone sealant to one of the two metal surfaces to be mated; when the parts are fastened together, the sealant squeezes out to coat both surfaces. Within fifteen minutes, the thin layer undergoes a preliminary cure and subsequently forms a light and durable seal.

An important attribute is that, in many instances, machining operations can be eliminated and pressure ridges and close metal-to-metal tolerances become unnecessary. When the two parts are fastened together before curing, the resultant joint is essentially metal-to-metal. This eliminates the problem of a bolted joint that loosens because of the gradual relaxation of the conventional gasket. Material usage is cut because the form-in-place gasket system maximizes sealing while minimizing gasket material requirements.

Hot and cold cycling tests involving water and oil have little effect on the sealant's good swell characteristics, low compression set and resistance. Silicones do not rely on compression for sealing, thus the stress factors encountered when using conventional gaskets do not occur and the possibility of scrap is less likely. The product flows into the spaces left at initial mating and, because of its adhesion to both joint surfaces, it 'unitizes' the joint. Thus, true torque is achieved without any reduction and, because no further movement of the compound takes place, no joint relaxation occurs. Additionally its flexibility enables any vibration or impact on the joint to be readily absorbed.

Because silicones are resistant to most industrial fluids, except petrol, and do not flow under their own weight, the joint should be made at 180° rather than sliding surfaces together. It can be easily broken by tapping the joint sideways or rotating it. One part silicones are known as *RTV* compounds, the RTV denoting room temperature vulcanizing.

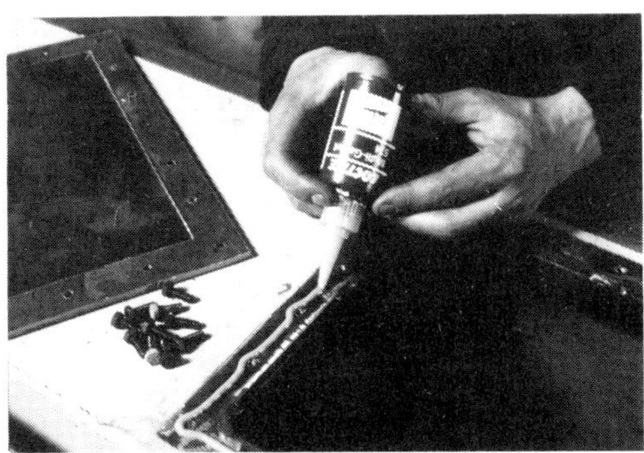

Application of liquid gasket material.

The base compound consists of a chain of alternating silicon and oxygen atoms, the free valencies on the silicon atoms being occupied by methyl groups. Humidity in the atmosphere activates a catalyst and the chains cross link to form a rubberlike material. The catalyst in one instance exhibits the properties of a vapour phase corrosion inhibitor; the other method is to release acetic acid gas. Where small land widths, complex joint designs and poor fastening capacity are encountered on power units, form-in-place silicone compounds can be very effective.

Retaining compounds

Modern retaining compounds are normally *anaerobic* adhesives of *cyanoacrylate* type. They are used on threaded connections and also to retain bearings, liners, bushes, keys and splines. They can also replace welding, brazing, soldering, riveting and bolting. They are available in a variety of viscosities, will cope with extreme conditions and are resistant to solvents, mild acids and alkalis.

Retaining compounds cure at room temperature and all are then inert. They are easily applied and, when deprived of air between the two surfaces, convert from a liquid form to a tough structural solid. They are used mainly to retain liners, bushes and threaded components in the sealing sense, particularly on hydraulic and pneumatic assemblies. In the assembly of metal parts, there are always microscopic irregularities and voids between mating surfaces. The actual contact between faces is restricted to high spots or peaks, only visible when the machine's surface is magnified and even the finest of interference fits is estimated to touch over only 20 to 25% of the surface area. During use, micro-movements take place, wearing off the peaks; consequently, parts loosen and cause fretting, leakage and corrosion. Anaerobic sealants fill the voids, then chemically harden in the absence of air into a tough insoluble plastic to provide total surface contact. The assembly becomes proof against shock and vibration, as well as being leak-proof and protecting the mating surfaces against corrosion.

Thread-sealing adhesives

The chief property required of a *thread-sealing* adhesive is a gap-filling ability to provide a gas- or liquid-tight seal. At the same time it will usually provide a varying degree of thread-locking, depending on the type of adhesive used. It is more important, however, that the properties of the adhesive be tailored so that it will not extrude and block the pipe when assembled, or extrude or disintegrate under pressure. It must be fully compatible with the type of fluid carried by the pipe.

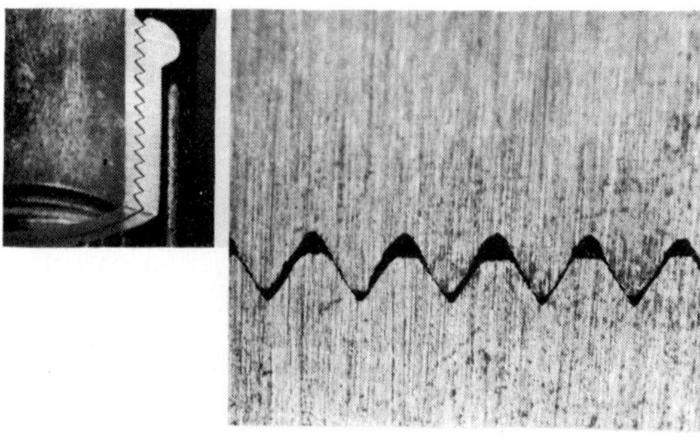

Fully effective thread sealant.

Thread sealants for general pipe sealing applications are usually medium strength, medium viscosity adhesives. Low strength, high viscosity adhesives may be used for high temperature services (*eg* sealing piping on internal combustion engines). The type of adhesive now most widely favoured is a thixotropic sealing paste based on PTFE, which can be effective for sealing pressures up to 700 bar (10 000 lb/in^2) under favourable conditions, with a working temperature range of $-240°C$ to $+260°C$.

Specific recommendations for thread-sealed pipe joints are given in ISO/R7 and ISO/R65.

Thread-locking adhesives

Thread-locking adhesives are formulated primarily with the necessary strength to provide the required break-out torque on a screwed assembly. Their gap-filling properties, and thus sealing properties, are largely governed by the viscosity of the adhesives used.

Pipe sealant being applied to the threads of a steel pipe.

Low strength adhesives are normally used for locking screws and nuts where the components must be removable. Gap-filling requirements are met by using an adhesive with a low viscosity for small screws and nuts, *eg* 6 mm (¼ in) or less, and a medium viscosity for large sizes. Alternatively, since a slightly greater degree of gap-filling is required for locking female (nut) threads, than male (screw) threads of the same size, individual manufacturers may specify different viscosity adhesives for screws and nuts.

Higher strengtrh adhesives are required for locking studs, and here again a medium viscosity is usually used for optimum gap-filling properties. High strength adhesives are used for permanent locking of threads (*eg* nuts and studs).

Adhesive types used may be:

(i) One-part, self-curing containing solvent.

(ii) One-part, anaerobic.

(iii) Two-part, curing by the admixture of a catalyst or accelerator.

(iv) 'Passivated', pre-applied adhesive.

Curing or setting time depends on the size and nature of the joint, and the temperature. Setting time in the case of self-setting and anaerobic adhesives may be accelerated by

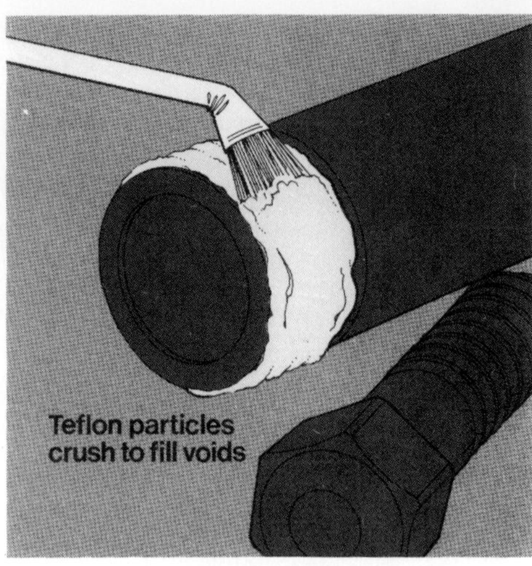

Non/hardening mouldable
Teflon thread sealant and lubricant.*

* Dupont Registered Trademark

the application of catalyst, either as an additive, or directly to the surface to be coated prior to the application of an anaerobic adhesive.

Most thread-locking adhesives are fully resistant to vibration, oils, solvents and heat. They will, however, have a maximum service temperature above which degradation of the bond will occur. Most types have a limited shelf life and should be stored in a cool place, when a shelf life in excess of one year can be expected.

'Passivated' pre-applied adhesives are the latest development in thread-locking adhesives. These are micro-encapsulated products which can be applied to screw threads, the adhesive then remaining dormant until the twisting action of applying the fastener causes the adhesive to cure. The adhesive begins to develop strength immediately at room temperature and achieves almost 90% of its ultimate strength in 24 hours. Complete curing is usually possible in fifteen minutes at a temperature of the order of 70°C (160°F), but would be dependent on the actual heating rate of the assembly and the particular adhesive formulation.

A specific advantage of this type of thread-locking adhesive is that pre-coated fasteners and threaded parts can be packed and shipped in normal fashion, and require no special storage. Shelf life of such adhesive is stated to be at least 4 years.

Self-Sealing Fasteners

FASTENERS SUCH as bolts and screws can be required to provide sealing as well as fastening in specific applications. The obvious method is the use of *sealing washers* under the bolt head or nut, in a suitable resilient material such as soft aluminium, lead, or synthetic rubber. The washer then acts as a simple gasket-type seal when the fastener is tightened.

There are many variations on this theme. For high pressure or high temperature working, or where metal-to-metal sealing is called for, all-metal *sealing rings* may be employed. Instead of using plain sections these may have face points or lipped sections to reduce the closure load necessary to complete the seal as well as rendering the surface finish of the mating surfaces less critical. For even better conformability, a metal ring may be combined with an elastomeric section. Closure then compresses the elastomeric section until the seal is finally tightened down onto the metal ring. Alternatively, the sealing ring may be basically a solid elastomeric section with an internal metal reinforcement spring. This has the effect of providing a self-locking action when the seal is compressed, making it particularly suitable for use as a sealing washer in a bolted or threaded assembly.

Self-sealing fasteners incorporate their own sealing element. This may be a pre-assembled sealing washer in rubber, plastic or composite material, under the head of a self-sealing bolt or screw, or a *moulded-in* seal (either in rubber or nylon). There are numerous proprietary designs, including those with special head designs to accommodate a particular type of sealing washer or sealing ring which may or may not be pre-assembled.

Similar principles can be applied to nuts. Thus *sealing nuts* include a variety of proprietary designs incorporating a separate element which is compressed to produce

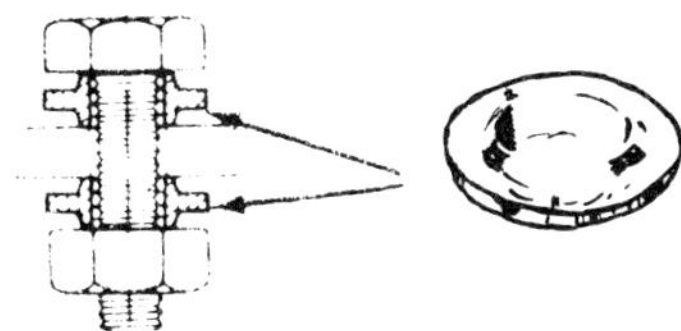

Plastic washers for sealing bolted assemblies.

Micro-encapsulated thread sealing adhesive can be bonded to any make threaded fastener.

Omnicote micro-encapsulated thead-locking adhesive for precoating threads, actuated when threaded parts are screwed manually.

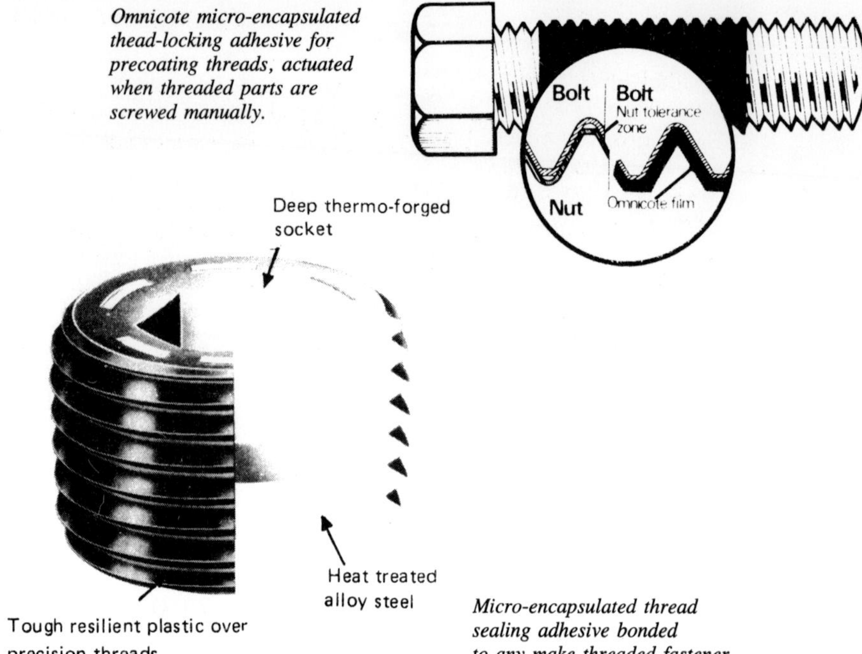

Micro-encapsulated thread sealing adhesive bonded to any make threaded fastener.

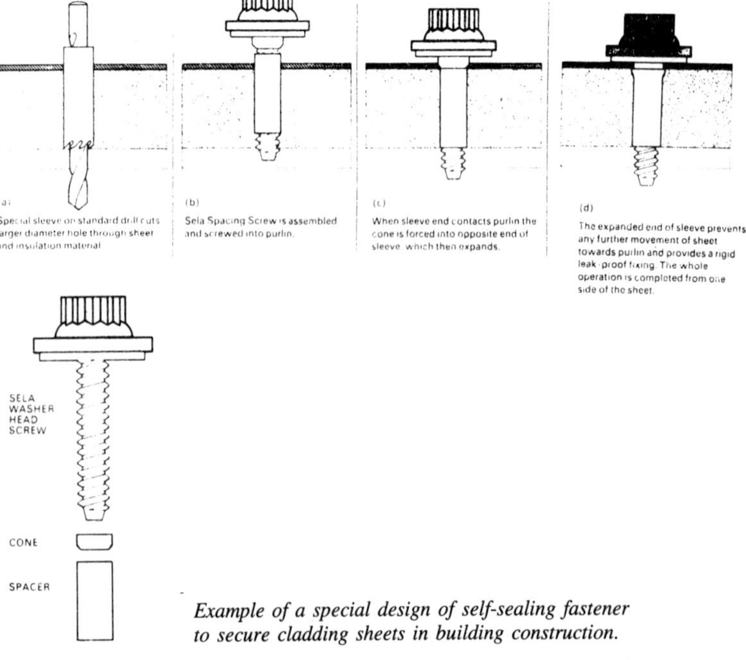

Example of a special design of self-sealing fastener to secure cladding sheets in building construction.

a seal when the nut is tightened. The sealing element may be of rubber, plastic or soft metal.

Alternative sealing treatments for fasteners include:

(i) *Flared-in* sealants, which are liquid or semi-liquid compounds for use with washers where a conventional type of sealing washer may not be suitable or is not easily fitted.

(ii) *Interference fit*, the principle of sealing provided by *Dryseal threads* for pipe couplings and *sealing rivets* which expand to an interference fit when set. Interference fits may also be provided by rubber or nylon sleeves, or special designs of fasteners which incorporate sleeve sections which are deformed elastically on tightening.

(iii) *Liquid* sealants applied to the threads prior to assembly which set to a semi-hard or hard state sealing the clearance space between mating threads.

Miscellaneous Static Seals

SOLID WIRE ring seals may be used for static sealing in applications involving extremes of temperature, pressure, or radioactive or extremely corrosive services and where high flange loadings can be employed. These rings may be circular in section, oval or polygonal (usually octagonal). They can prove effective seals over a pressure range from high vacuum to 1750 bar (25000 lb/in^2).

Circular wire rings can be accommodated in a counterbore or rectangular groove section formed by a retainer plate. Tongue and groove assembly may also be used with softer metal rings. These alternative installations are shown in Figure 1. Typical groove depths and flange loadings required are given in Table 1.

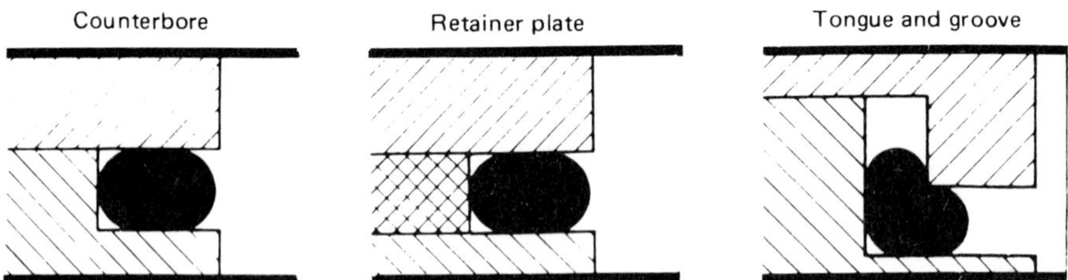

Figure 1

Actual flange load required for effective sealing does, however, vary with the ring material. Wire rings may also be installed in V shaped grooves, when higher flange loading is required. Surface finish requirements are not particularly severe. A finish of the order of 64 μin Ra is usually adequate; or up to 124 μin Ra for V grooves. The main requirement is that the sealing surfaces should be true and parallel.

Other shapes of solid metal rings are shown in Figure 2. These are employed with matching grooves. They may also be designed as pressure-energized rings, *eg* using an octagonal or deepened octagonal section with a single pressure passage hole drilled vertically through the section (Figure 3). Besides increasing the pressure rating, it is claimed that such rings are more resistant to vibration and shock loads.

Solid metal rings can be made from drawn wire in any metal that is weldable; or directly machined. Metals commonly used include aluminium, copper, stainless steel and monel.

TABLE I – TYPICAL WIRE RING SEAL DATA

FOR WIRE/RING SEALS INSTALLED IN COUNTERBORE				
A Wire Diameter (inches)	B Groove Depth (inches)	C Ring Dia. Range* (inches)	Minimum Flange Load (lbs/inch of circ)	Pressure Range lb/in^2
0.035 Al.	0.023/0.027	0.750 – 1.500	1000	Vac. – 4000
0.062 Al.	0.042/0.047	1.000 – 2.000	1400	Vac. – 5000
0.094 Al.	0.074/0.079	1.500 – 4.000	1400	Vac. – 5000
0.125 Al.	0.105/0.110	3.000 – 10.000	1400	Vac. – 5000
0.035 Cu.	0.025/0.029	0.750 – 1.5000	2300	Vac. – 10000
0.062 Cu.	0.046/0.051	1.000 – 2.000	3200	Vac. – 12000
0.094 Cu.	0.078/0.083	1.500 – 4.000	3200	Vac. – 12000
0.125 Cu.	0.109/0.115	3.000 – 10.000	3200	Vac. – 12000
0.035 S.S.	0.028/0.032	0.750 – 1.500	4200	0 – 20000
0.062 S.S.	0.050/0.055	1.000 – 2.000	6000	0 – 25000
0.094 S.S.	0.082/0.087	1.500 – 4.000	6000	0 – 25000
0.125 S.S.	0.113/0.118	3.000 – 10.000	6000	0 – 25000

FOR GOLD WIRE/RING SEALS ONLY – INSTALLED IN TONGUE AND GROOVE						
A Wire Dia. (inches)	D Tongue Dia. (inches) Ring o.d. Minus	E Tongue Depth (inches)	F Groove Depth (inches)	C Ring Dia. Range* (inches)	Minimum Flange Load (lbs/in of circ.)	Pressure Range (mm Hg)
0.020 Au.	0.013/0.017	0.041/0.043	0.047/0.049	0.500 – 2.000	1200	10^{-6} – 10^{-10}
0.030 Au.	0.020/0.025	0.061/0.063	0.071/0.073	1.000 – 3.000	1800	10^{-6} – 10^{-10}
0.040 Au.	0.027/0.032	0.082/0.084	0.095/0.097	2.000 – 6.000	2400	10^{-6} – 10^{-10}
0.062 Au	0.044/0.049	0.127/0.129	0.148/0.150	3.000 – 20.000	3600	10^{-6} – 10^{-10}

* Tolerance on outside diameter of ring is +0.005/–0.000.

GROOVE DIAMETER (Ring o.d. (C) + 0.005in/0.010 in)

GROOVE DIAMETER (Ring o.d. (C) +0.005 in/0.010 in)

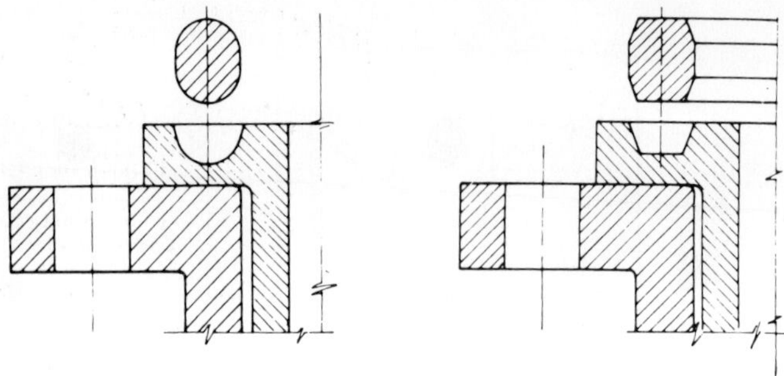

Figure 2
Solid metal ring seals, oval (left) and octagonal (right).

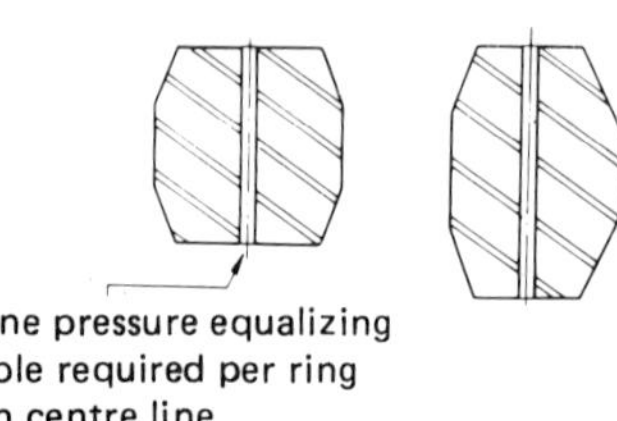

Figure 3
Pressure-energized rings.

Soft coatings may be used on hard metal rings to improve conformability. Such coatings include silver, gold, metal, copper and PTFE.

Chamfer rings

Increasing operational requirements for high temperature static seals employed in military and civil aircraft systems have strained the performance of elastomeric seals to their ultimate limits. The metal wedge seal is designed to meet these requirements and has been developed as a replacement for sealing AGS 3018 or American MS 33649 trapped O-ring housings which are currently used for pipe to component connections AGS 3017 or MS 33648 (*boss* seals).

The metal wedge seal consists of a simple metallic chamfer ring made from either zinc plated S96 steel, or silver plated S80 stainless steel. It is suitable for a wide range of high temperature fluids and resists electrolytic action.

On tightening the union, the wedge-shaped metal ring is compressed and as a result the internal diameter tends to reduce within the elastic limits of the material, forming an effective seal. If thread stretch occurs the internal diameter tends to increase but maintains a high degree of pressure between the angled faces providing positive sealing under all conditions. There must be no damage to the mating faces on tightening the union and it must be possible to re-use the seals after strip-down.

All-metal seals

The all-metal seal as shown in Figure 4 has been developed for face sealing of oils and gases at working temperature outside the effective sealing range of elastomeric materials. Seals of this type have been tested and proved effective under suitable assembly conditions,

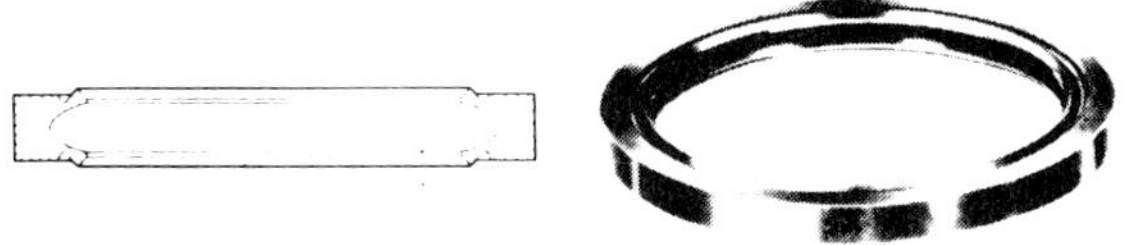

Figure 4
All metal seal.

at static pressures of 690 bar (10000 lbf/in^2) and pulsating pressures of 345 bar (5000 lbf/in^2) over the working temperature range of $-65 \cdot C$ *to* $+300°C$.

The metal lips are mechanically energized by deflection during assembly and in addition are energized by the fluid pressure. The outer rim limits the 'squeeze' on the lips, preventing over-stressing of the seal. It should be noted that this seal is not recommended for use with hardened faces.

The seals should be accurately located by a spigot on the underside of the screwed plug or coupling. This allows re-positioning without imparing the sealing efficiency.

Contacting metal surfaces should be of a good machined finish but need not be ground or lapped. They may be steel or light alloy, and are only lightly marked by the lips of the seal. Care must be taken to ensure that the lips are not damaged prior to assembly.

Bonded seal

As shown in Figure 5 the sealing element of the *bonded seal pressure* gasket is bonded to the i.d. of the metal outer ring which is chemically treated with a bonding agent. The bonding operation is carried out by applying heat and pressure to the rubber during the production cycle.

Today most elastomeric materials can be bonded to metal but after long term immersion in some fluids at high temperature the bond becomes very weak or fails. As an alternative to the bonded item, which is preferred, it is possible for a rectangular section of rubber to be machined and freely inserted into the metal outer ring. Static composite seals can thus be manufactured in a large number of elastomeric materials which can operate over

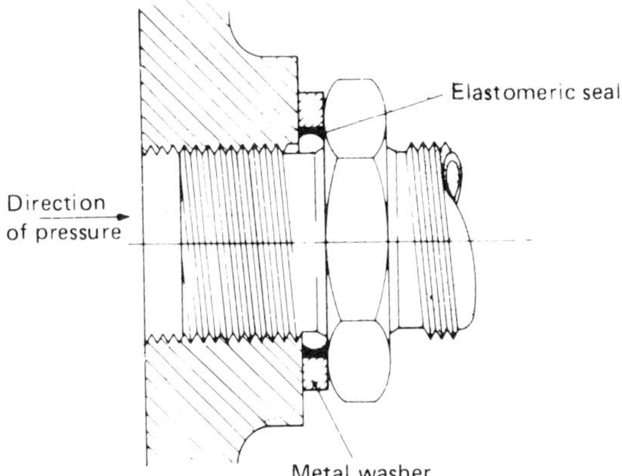

Figure 5
Bonded seal.

a wide temperature range and in a number of exotic fluids for long periods of time. However, as the rubber is not bonded to the metal and is comparatively weak, because of the small section, extra care must be taken on assembly.

Special static gland seals (Figure 6)

In applications where side loads applied to the powered set leg of mining machinery are severe, and local distortion of the collar around the top of the cylinder has occurred, it is possible for leakage to occur from an O-ring. To ensure that under the severest mechanical distortion positive sealing is maintained, a solid lip seal has been introduced which stops the leakage. The shape of the seal gives a higher cross sectional 'squeeze' than that obtained with an O-ring, without causing assembly problems which could damage the seal. The solid lip seals are normally manufactured in a high nitrile elastomeric material of 90 IRHD.

Bonded pressure joint rings (Figure 7)

For sealing pipe unions, bolt heads and flanges at pressures greater than 140 bar (2000 lbf/in^2) use is made of a simple rubber to metal bonded sealing washer. The bonded washer consists of a metal outer ring with a trapezoidal shaped rubber sealing ring bonded to the inside diameter. The bonded washer is assembled under the head of a bolt or union, the final tightening load being taken by the metal outer ring. Thus on tightening, as the rubber height is greater than that of the metal thickness, it is compressed to form a seal. The flexing action of the lips under pressure provides a positive and leakproof seal against water, gases, oils and a wide range of other fluids.

The standard form of *bonded* seal, with a steel outer ring, gives a minimum burst pressure ranging from 670 bar to 2430 bar depending on size. The metal outer rings

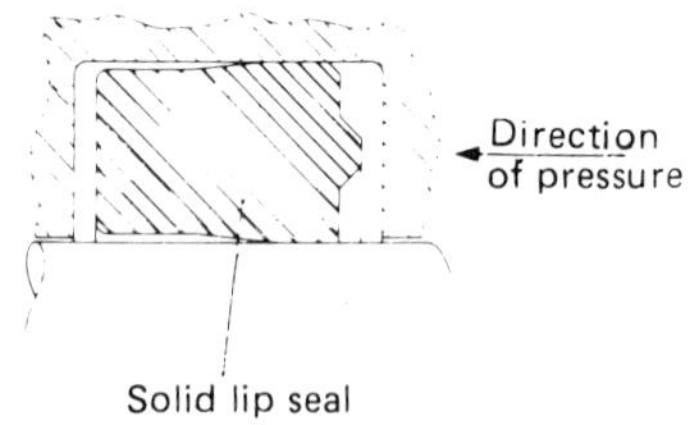

Figure 6
Special static gland seal.

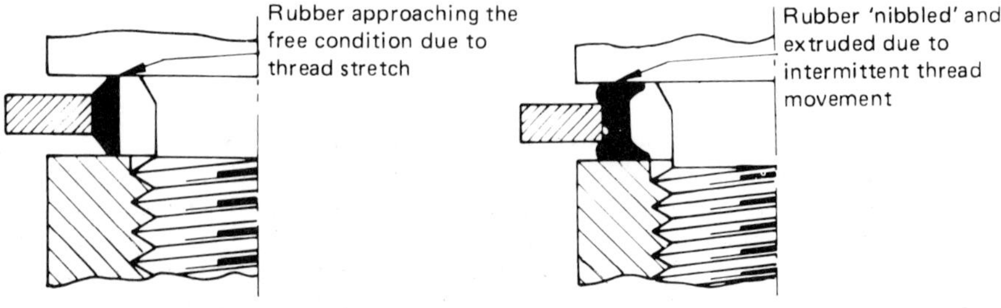

Figure 7
Joint ring.

can be manufactured in mild steel, stainless steel, light alloy or brass. They are suitably corrosion protected, but because of environmental problems the cadmium plated type are being replaced by steels which are zinc plated and then colour passivated.

Pipeline metallic ring joints

Metallic ring joint gaskets are in widespread use with API wellhead equipment, line pipe and refinery flanged fittings. Both the Type R oval design and the later Type R octagonal design rely on bolt generated compressive loads to provide the necessary sealing force and to coin the material against the mating seal groove flange. The resulting flange face stand-off is typically of the order of 5 mm ($^{3}/_{16}$ in) *eg* Figure 8.

The *Type RX* joint fits the same seal groove as *Type R,* but it is designed to react to line pressure as well as to initial bolt compression. This demands greater flange face stand-off, which may exceed 12 mm ($^{15}/_{32}$ in) *eg* Figure 9.

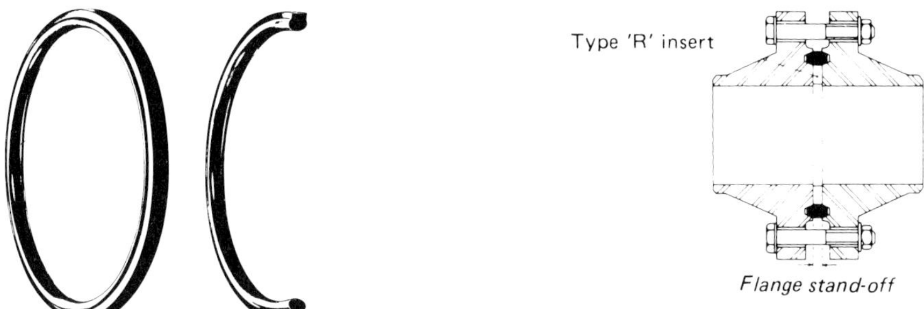

Flange seal for pressures up to 500 bar (7500 lb/in^2).

Figure 8

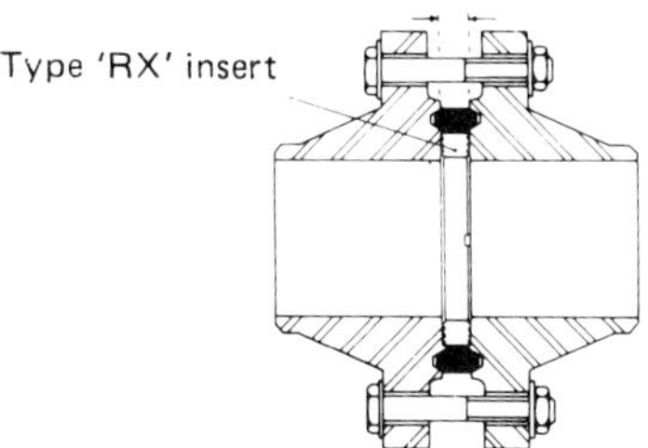

Figure 9

With either type of joint the necessary stand-off can result in a substantial discontinuity in the bore, affecting flow and possibly promoting erosion and/or accumulation of sediment. A solution now commonly adopted is to fit PTFE inserts, suitably proportional to fill the cavity between flange bore and ring joint (Figure 10). Type R inserts have a coarse machined finish. Type RX inserts have a deeper, grooved profile to control spread under compression within the clearance allowance from the flange bore. Top and bottom faces also incorporate ports to provide pressure access to the joint ring, and also to assist venting in the event of sudden system compression.

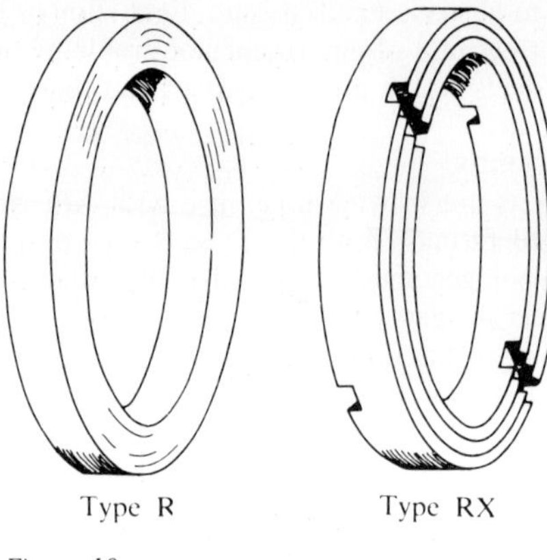

Figure 10
PTFE inserts.

FE tape

PTFE tape is now widely used as a thread sealant on *pipe* joints, *etc.* It is simple and quick to apply, ready for immediate use without any setting time and extrudes into the thread contours, filling voids through which the medium being sealed might otherwise escape. PTFE tape is suitable for all types of pipes and is non-galling and corrosion-resistant (*ie* connections can readily be broken if necessary, even after long periods of service).

Fibrillated PTFE

Fibrillated PTFE in tape form (Lattyflon 3215) is an even more versatile sealing material, suitable for both dynamic and static sealing applications. Apart from the ease with which it can be used to make gasket-type seals and its versatility in this respect, it is suitable for use with all media including aggressive chemicals.

Another (patented) form of PTFE (Lattyflon 3206) which can find particular application for static seals in chemical services, *etc,* is pure PTFE fibre pre-treated under pressure with PTFE then plaited and reimpregnated to a core of the same material. Basically this is a gland packing material for rotary seals in square or rectangular sections, or available in die-pressed rings. It is available with or without lubricant. Rectangular sections without oil are particularly suitable for static seals such as cover and body flanges on equipment handling oxygen or strong oxidizing agents and other chemicals.

Tank lid seals

Tanks used for the bulk transport of spirits and volatile liquids, and particularly inflammable, toxic or corrosive liquids can present severe sealing problems which are often not fully appreciated at the original design stage. These can be aggravated by lack of uniformity and poor surface finish on the sealing faces, as manufactured.

Materials commonly used for *tank lid* seals for hazardous substances include compressed asbestos, PTFE/asbestos packing, pure PTFE yarn packing, machined PTFE and various elastomers. All can show limitations, notably in the rupture of fibres under shear loading with repeated opening and closure of the lid and creep under deformation stress.

A recent development which appears to overcome these limitations is the Lattyflon tank lid packings which are based on a core of elastomer or fibrillated PTFE with an outer cover of PTFE plaited yarn. The former construction relies on the integrity of the cover for chemical resistance and is intended for applications where the lids are in good condition, with even tightening. The packing with a fibrillated PTFE core, however, continues to provide a safe seal whatever the state of the lid or coaming, even if the outer cover becomes worn and badly damaged. It is, in fact, a plastic packing which moulds without creep or risk of shear when it comes into contact with the coaming, regardless of the condition of the latter.

Lattyflon tank lid packings are produced (mostly) in rectangular sections to fit to the tank lid either in a machined groove of U section, or into a welded-on U channel. If necessary the packing can be secured in its seating with stainless steel clips fitted over the channel.

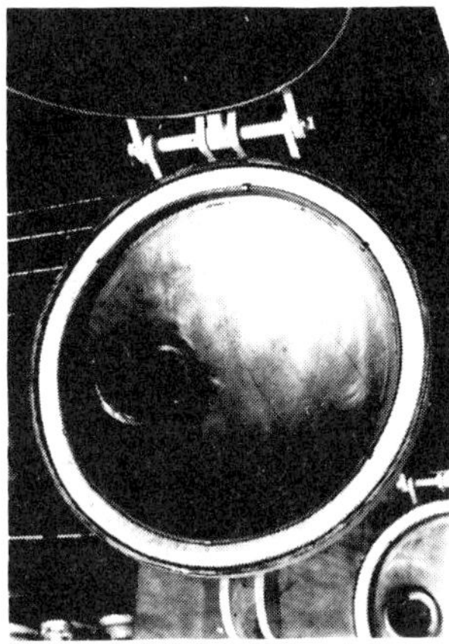

Tank lid packing with fibrillated PTFE core and PTFE plaited cover.

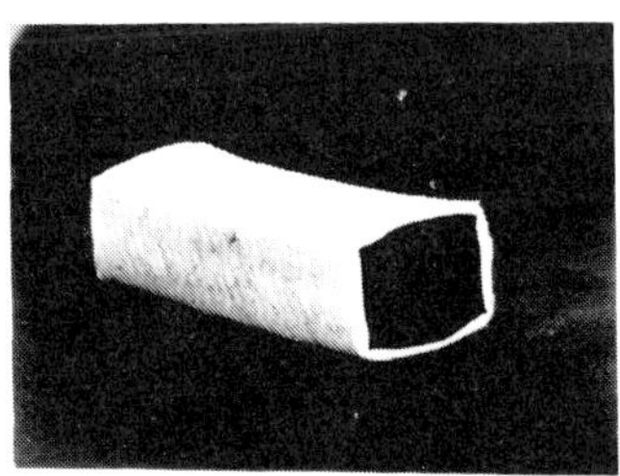

Tank lid packing with rubber core and PTFE plaited cover.

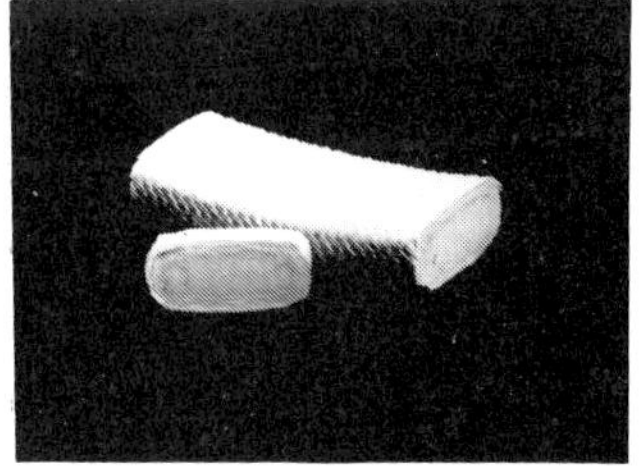

Typical installation with stainless steel retaining clips.

SECTION 4

Dynamic Seals

O-RINGS
HOLLOW METAL RING SEALS
OTHER SOLID ELASTOMERIC RINGS
SPLIT RING SEALS
FLEXIBLE LIP SEALS
COMPOSITE SEALS
FELT SEALS
WIPERS AND SCRAPERS
COMPRESSION PACKINGS

O-Rings

THE O-RING is one of the simplest and most versatile types of seal with a wide range of applications for both static and dynamic sealing. Satisfactory performance, however, depends on the material characteristics, suitable matching groove geometry with close tolerances and smooth surface finish on mating surfaces.

Basically, an O-ring seal comprises a moulded elastomeric circular-section ring nipped in a cavity in which the ring is located. Such seals can be made completely leakproof for static seals up to extremely high pressures; and satisfactorily leakproof for dynamic applications and fluid pressures up to 350 bar (5000 lb/in^2), with particular operating limits. Provided such seals are properly designed and applied they are generally extremely dependable because of their basic simplicity and ruggedness. They are generally worth considering as a simple low-cost possibility for most seal applications except high pressure dynamic seals, rotary seals with a rubbing speed of more than 7.5 m/s (1500 ft/min) and heavier duty dynamic seals. It does not necessarily follow that because an O-ring is suitable for a particular application its performance will be superior to that of an alternative type of seal. Where its sealing performance is satisfactory, however, it will usually have lower friction for dynamic applications than most alternative lip type seals.

The principle of the O-ring seal is shown in Figure 1. On assembly in its groove the ring is subject to an initial squeeze or nip controlled by the relative sizes of the ring, groove depth and diametral clearance. This provides the initial sealing pressure. Since the clearance for the O-ring is now less than its free outer diameter, the O-ring cross-section is squeezed diametrically out-of-round, even before pressure is applied. Thus contact with the inner and outer walls of the passage under static conditions is assured. When internal pressure is applied through the clearance gap to one side of the O-ring, the O-ring further deforms in contact with the opposite side of the groove. This pressure is transmitted to the surfaces to be sealed. Consequently, the sealing pressure is actually higher than the applied fluid pressure, by an amount equal to the initial interference

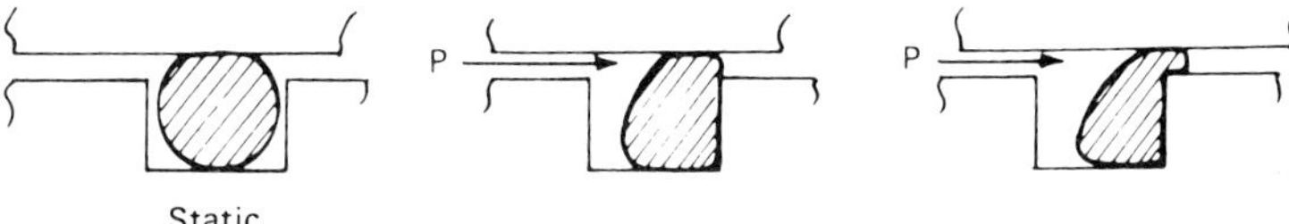

Figure 1

pressure. As the volume of an O-ring cannot change, the applied diametrical squeeze will cause an increase in length of the rubber across the groove. This increase will be even greater as a result of expansion or swell of the rubber, due to heat from the fluid being sealed.

The effectiveness of such a seal depends very much on the initial geometry. With increasing pressure, deformation of the ring will be exaggerated, ultimately tending to extrude a section of the ring into the clearance gap. If the clearance gap is large enough the seal could fail completely by extrusion under static pressure of a moderate order; or, alternatively, if the static pressure is high enough the seal would ultimately fail by extrusion through any clearance gap. In the case of a dynamic application any tendency towards extrusion will obviously localize wear on the O-ring, leading to 'nibbling' and premature failure of the ring (Figure 2).

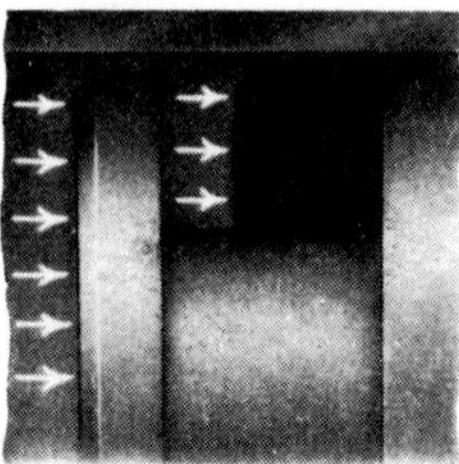

Figure 2

It should be noted that the possibility of seal extrusion is not limited to dynamic applications. In static applications stretching of assembly bolts under high pressure can open up an extrusion gap.

Limits for internal pressure are governed basically by the clearance gap and O-ring hardness – some quantitative data is given in Figure 3. In practice, the gap is normally specified for a given ring size and application, establishing a pressure limit for a particular (or typical) ring hardness. This can be presented only as a design figure, however. Under actual working conditions the usual clearance gap involved may be modified by expansion or contraction of the elements involved under change of temperature or of pressure. Overlooking this factor can lead to extrusion troubles. Thus for operating at low temperature it may be necessary to reduce the groove depth to allow for contraction of the ring and provide the required squeeze of the *contracted* size. Similarly, for high temperature working it may be advisable to increase the groove depth slightly to avoid excessive squeeze on the *expanded* ring at the working temperature. This effect can be significant at extremes of temperature because the coefficient of thermal expansion of elastomers is appreciably higher than that of metals.

A *pressure* effect, which may be overlooked, is the possibility of the external member, such as a cylinder, being expanded slightly under pressure or high pressure surges and thus increasing the clearance gap to provide extrusion of the ring. In extreme cases it may be necessary to stiffen the member to obviate this.

Back-up washers *(anti-extrusion rings)*
The possibility of O-ring extrusion can be eliminated by the use of back-up washers. At the same time this will increase the pressure rating of the seal. A back-up washer

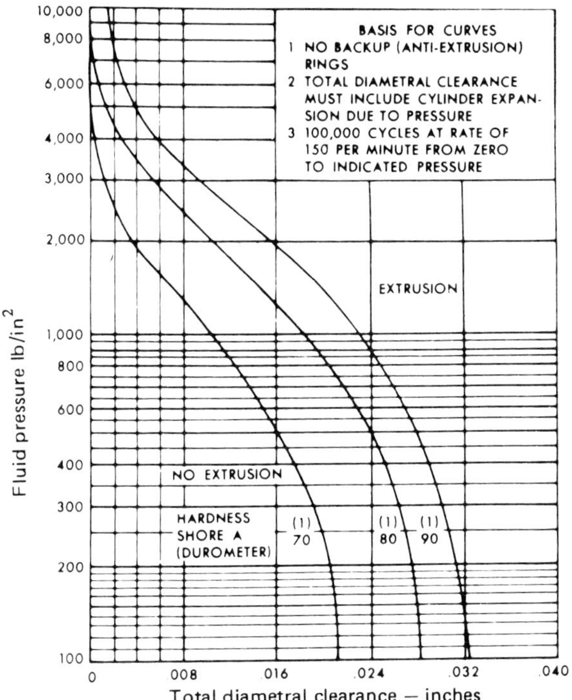

Figure 3

is a ring of harder material than the O-ring itself, but sufficiently resilient to deform under pressure to close the clearance gap. It is fitted in the groove on the opposite side of the ring to the applied pressure; or on both sides if the ring has to seal against pressure from both directions (Figure 4). The groove width dimension is correspondingly increased to accommodate the back-up ring or rings.

Various materials may be used for back-up rings, such as leather, hard rubber and PTFE, the latter being particularly favoured for small sizes and having the advantage of offering very low friction for dynamic applications.

The main requirement of a back-up ring is that it shall not collapse or cold flow, this in turn depending to a large extent on a suitable size of ring being used with proper groove dimensions.

Back-up washers may be single turn scarf cut acetal on PTFE rings, or spiral PTFE, leather or fabric-reinforced rubber rings, *eg* Figure 5. Diametrically 'elastic' rings are necessary to make it possible to stretch or compress the material to fit into the groove. At the same time it is desirable that a back-up washer should have no bias cut or other forms of fabrication which could cause local weakness leading to collapse under pressure, or allow the O-ring to extrude into the surface. Contoured back-up rings may also be used with a concave side facing the O-ring but do not appear to have any specific advantage over smooth flat rings.

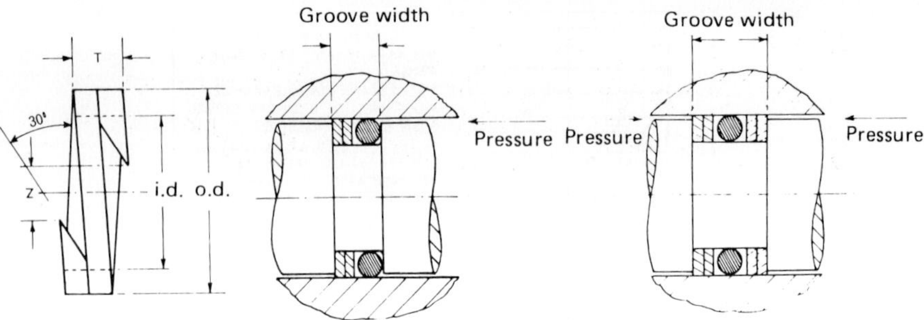

Figure 4
Typical groove widths for O-rings used with spiral back-up rings.

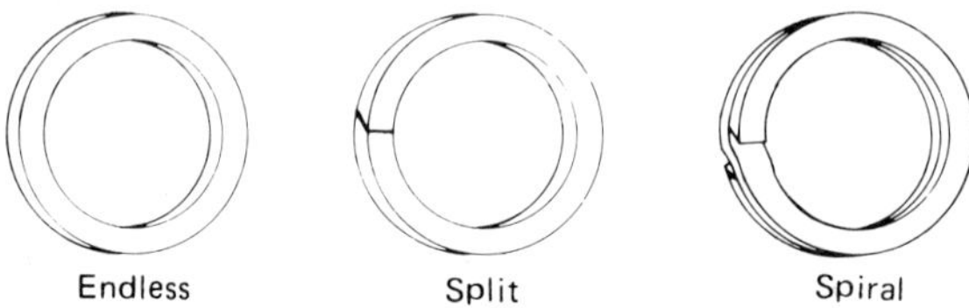

Figure 5
Back-up ring configurations.

Groove geometry

The ideal groove form is purely rectangular although a taper of not more than 5° is permissible for ease of machining. A radius of 0.50 to 0.75 mm (0.020 to 0.030 in) is normally specified at the internal recesses and a top edge radius of 0.12 to 0.25 mm (0.005 to 0.010 in) to eliminate a cutting edge (Figure 6a). This radius must be kept small as otherwise it will assist extrusion.

Specific groove dimensions for standard sizes of O-rings are specified in BS 4518 : 1974 for metric size O-rings; and BS 1806 : 1962 for inch size O-rings. (See Tables 1, 2 and 3; also Figure 6b).

Ultimately the groove depth is the important factor because this controls the compression or nip essential for satisfactory sealing and the clearance gap which affects the pressure rating of the O-ring.

Standard O-rings are designed to fit into grooves of nominal depth, which can set problems in deciding suitable machining tolerances on groove diameter while retaining adequate nip. Specific recommendations are given in Table 4.

Surface finish

Groove finish is important as it affects the wear and life of the seal, although the requirements are not too onerous. A finish of better than 0.75 μm (37 μin) is recommended. An ultra-fine finish is not necessary, but obviously burrs, scores or definite machining marks should be removed.

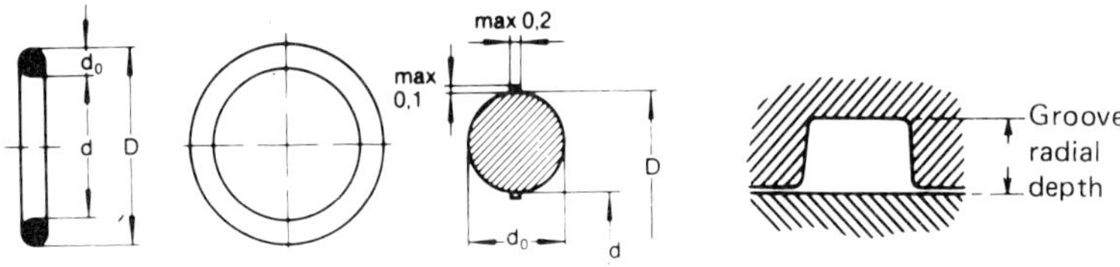

Figure 6a
Basic dimensions.

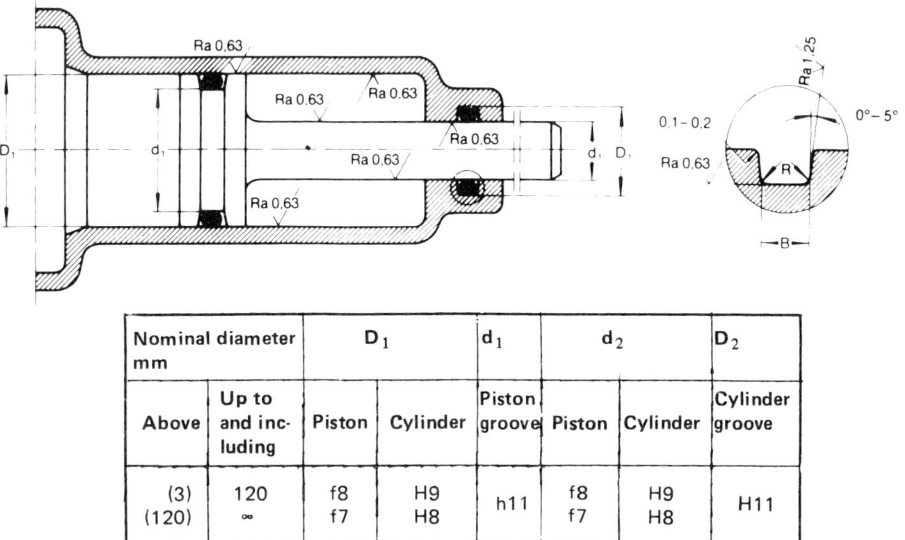

Nominal diameter mm		D1		d1	d2		D2
Above	Up to and including	Piston	Cylinder	Piston groove	Piston	Cylinder	Cylinder groove
(3) (120)	120 ∞	f8 f7	H9 H8	h11	f8 f7	H9 H8	H11

Figure 6b
Groove dimensioning
and tolerance standards.

The finish of the surface over which the O-ring rubs in dynamic applications (*eg* a *rod* or *cylinder bore*) must be of a high order, preferably better than 0.4 μm (16 μin) Ra. The surface must also be free from longitudinal scratch or machining marks and chatter marks. Circumferential marks may be acceptable, but polishing or hard chrome plating are recommended finishes for maximum O-ring life in dynamic applications.

In the case of static applications a very much rougher surface finish, *eg* 0.75 to 1.25 μm (30 to 50 μin) Ra is acceptable, although the lower figure is desirable for high pressure working. One of the advantages of the O-ring used as a static seal is that it is effective with a much rougher, and therefore cheaper, surface finish than is required for gaskets.

The materials in contact with the O-ring also have a bearing on O-ring life. In general the softer metals, such as aluminium, brass and bronze should be avoided for dynamic applications, although they may be suitable for low pressure assemblies. Stainless steel and monel are two further materials which are generally best avoided for dynamic application of O-ring seals, although again there are exceptions.

TABLE IA – STANDARD INCH SIZES OF O-RINGS (BS1806:1962)

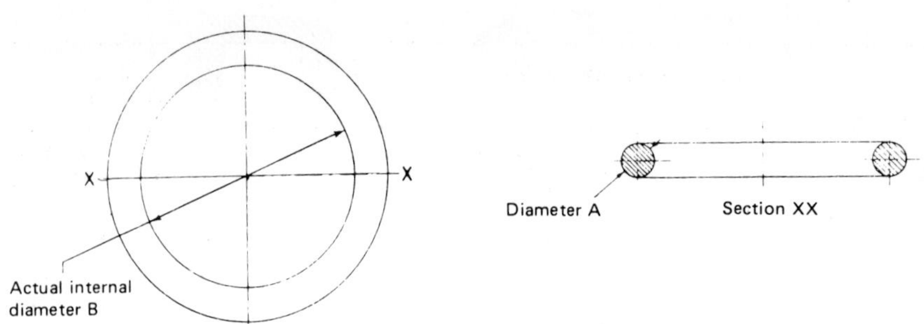

NOMINAL CROSS SECTION 1/16 in

O-Ring Ref No	Actual Diameters				Nominal Housing Dimensions	
	Cross-section A in		Internal B in		Shaft diameter in	Cylinder diameter in
004	0.070	±0.003	0.070	±0.005	5/64	13/64
005	0.070		).101		7/64	15/64
006	0.070		'.114		1/8	1/4
007	0.070		.145		5/32	9/32
008	0.070		176		3/16	5/16
009	0.070		0.208		7/32	11/32
010	0.070		0.239		1/4	3/8
011	0.070		0.301		5/16	7/16
012	0.070		0.364		3/8	1/2
013	0.070		0.426		7/16	9/16
014	0.070		0.489		1/2	5/8
015	0.070		0.551		9/16	11/16
016	0.070		0.614		5/8	3/4
017	0.070		0.676		11/16	13/16
018	0.070		0.739		3/4	7/8
019	0.070		0.801	±0.006	13/16	15/16
020	0.070		0.864		7/8	1
021	0.070		0.926		15/16	1.1/16
022	0.070		0.989		1	1.1/8
023	0.070		1.051		1.1/16	1.3/16
024	0.070		1.114		1.1/8	1.1/4
025	0.070		1.176		1.3/16	1.5/16
026	0.070		1.239		1.1/4	1.3/8
027	0.070		1.301		1.5/16	1.7/16
028	0.070		1.364		1.3/8	1.1/2

cont...

NOMINAL CROSS SECTION 3/32 in

O-Seal size No.	Actual Diameters				Nominal Housing Dimensions	
	Cross-section A in		Internal B in		Shaft diameter in	Cylinder diameter in
110	0.103	±0.003	0.362	±0.005	3/8	9/16
111	0.103		0.424		7/16	5/8
112	0.103		0.487		1/2	11/16
113	0.103		0.549		9/16	3/4
114	0.103		0.612		5/8	13/16
115	0.103		0.674		11/16	7/8
116	0.103		0.737		3/4	15/16
117	0.103		0.799	±0.006	13/16	1
118	0.103		0.862		7/8	1.1/16
119	0.103		0.924		15/16	1.1/8
120	0.103		0.987		1	1.3/16
121	0.103		1.049		1.1/16	1.1/4
122	0.103		1.112		1.1/8	1.5/16
123	0.103		1.174		1.3/16	1.3/8
124	0.103		1.237		1.1/4	1.7/16
125	0.103		1.299		1.5/16	1.1/2
126	0.103		1.362		1.3/8	1.9/16
127	0.103		1.424		1.7/16	1.5/8
128	0.103		1.487		1.1/2	1.11/16
129	0.103		1.549	±0.010	1.9/16	1.3/4
130	0.103		1.612		1.5/8	1.13/16
131	0.103		1.674		1.11/16	1.7/8
132	0.103		1.737		1.3/4	1.15/16
133	0.103		1.799		1.13/16	2
134	0.103		1.862		1.7/8	2.1/16
135	0.103		1.925		1.15/16	2.1/8
136	0.103		1.987		2	2.3/16
137	0.103		2.050		2.1/16	2.1/4
138	0.103		2.112		2.1/8	2.5/16
139	0.103		2.175		2.3/16	2.3/8
140	0.103		2.237		2.1/4	2.7/16
141	0.103		2.300		2.5/16	2.1/2
142	0.103		2.362		2.3/8	2.9/16
143	0.103		2.425		2.7/16	2.5/8
144	0.103		2.487		2.1/2	2.11/16
145	0.103		2.550		2.9/16	2.3/4
146	0.103		2.612		2.5/8	2.13/16
147	0.103		2.675	±0.015	2.11/16	2.7/8
148	0.103		2.737		2.3/4	2.15/16
149	0.103		2.800		2.13/16	3

cont...

NOMINAL CROSS SECTION 1/8 in

O-Seal size No. No	Actual Diameters				Nominal Housing Dimensions	
	Cross-section A in		Internal B in		Shaft diameter in	Cylinder diameter in
210	0.139	±0.004	0.734	±0.006	3/4	1
211	0.139		0.796		13/16	1.1/16
212	0.139		0.859		7/8	1.1/8
213	0.139		0.921		15/16	1.3/16
214	0.139		0.984		1	1.1/4
215	0.139		1.046		1.1/16	1.5/16
216	0.139		1.109		1.1/8	1.3/8
217	0.139		1.171		1.3/16	1.7/16
218	0.139		1.234		1.1/4	1.1/2
219	0.139		1.296		1.5/16	1.9/16
220	0.139		1.359		1.3/8	1.5/8
221	0.139		1.421		1.7/16	1.11/16
222	0.139		1.484		1.1/2	1.3/4
223	0.139		1.609	±0.010	1.5/8	1.7/8
224	0.139		1.734		1.3/4	2
225	0.139		1.859		1.7/8	2.1/8
226	0.139		1.984		2	2.1/4
227	0.139		2.109		2.1/8	2.3/8
228	0.139		2.234		2.1/4	2.1/2
229	0.139		2.359		2.3/8	2.5/8
230	0.139		2.484		2.1/2	2.3/4
231	0.139		2.609		2.5/8	2.7/8
232	0.139		2.734	±0.015	2.3/4	3
233	0.139		2.859		2.7/8	3.1/8
234	0.139		2.984		3	3.1/4
235	0.139		3.109		3.1/8	3.3/8
236	0.139		3.234		3.1/4	3.1/2
237	0.139		3.359		3.3/8	3.5/8
238	0.139		3.484		3.1/2	3.3/4
239	0.139		3.609		3.5/8	3.7/8
240	0.139		3.734		3.3/4	4
241	0.139		3.859		3.7/8	4.1/8
242	0.139		3.984		4	4.1/4
243	0.139		4.109		4.1/8	4.3/8

cont...

NOMINAL CROSS SECTION 1/8 in

O-Ring Ref No	Actual Diameters				Nominal Housing Dimensions	
	Cross-section A in		Internal B in		Shaft diameter in	Cylinder diameter in
244	0.139	±0.004	4.234	±0.015	4.1/4	4.1/2
245	0.139		4.359		4.3/8	4.5/8
246	0.139		4.484		4.1/2	4.3/4
247	0.139		4.609		4.5/8	4.7/8
248	0.139		4.734		4.3/4	5
249	0.139		4.859		4.7/8	5.1/8
250	0.139		4.984		5	5.1/4
251	0.139		5.109	±0.023	5.1/8	5.3/8
252	0.139		5.234		5.1/4	5.1/2
253	0.139		5.359		5.3/8	5.5/8
254	0.139		5.484		5.1/2	5.3/4
255	0.139		5.609		5.5/8	5.7/8
256	0.139		5.734		5.3/4	6
257	0.139		5.859		5.7/8	6.1/8
258	0.139		5.984		6	6.1/4
259	0. 139		6.234		6.1/4	6.1/2
260	0.139		6.484		6.1/2	6.3/4
261	0.139		6.734		6.3/4	7
262	0.139		6.984		7	7.1/4
263	0.139		7.234	±0.030	7.1/4	7.1/2
264	0.139		7.484		7.1/2	7.3/4
265	0.139		7.734		7.3/4	8
266	0.139		7.984		8	8.1/4
267	0.139		8.234		8.1/4	8.1/2
268	0.139		8.484		8.1/2	8.3/4
269	0.139		8.734		8.3/4	9
270	0.139		8.984		9	9.1/4
271	0.139		9.234		9.1/4	9.1/2
272	0.139		9.484		9.1/2	9.3/4
273	0.139		9.734		9.3/4	10
274	0.139		9.984		10	10.1/4

NOMINAL CROSS SECTION 3/16 in

O-Ring Ref No	Actual Diameters				Nominal Housing Dimensions	
	Cross-section A in		Internal B in		Shaft diameter in	Cylinder diameter in
325	0.210	±0.005	1.475	±0.010	1.1/2	1.7/8
326	0.210		1.600		1.5/8	2
327	0.210		1.725		1.3/4	2.1/8
328	0.210		1.850		1.7/8	2.1/4
329	0.210		1.975		2	2.3/8
330	0.210		2.100		2.1/8	2.1/2
331	0.210		2.225		2.1/4	2.5/8
332	0.210		2.350		2.3/8	2.3/4
333	0.210		2.475		2.1/2	2.7/8
334	0.210		2.600		2.5/8	3
335	0.210		2.725	±0.015	2.3/4	3.1/8
336	0.210		2.850		2.7/8	3.1/4
337	0.210		2.975		3	3.3/8
338	0.210		3.100		3.1/8	3.1/2
339	0.210		3.225		3.1/4	3.5/8
340	0.210		3.350		3.3/8	3.3/4
341	0.210		3.475		3.1/2	3.7/8
342	0.210		3.600		3.5/8	4
343	0.210		3.725		3.3/4	4.1/8
344	0.210		3.850		3.7/8	4.1/4
345	0.210		3.975		4	4.3/8
346	0.210		4.100		4.1/8	4.1/2
347	0.210		4.225		4.1/4	4.5/8
348	0.210		4.350		4.3/8	4.3/4
349	0.210		4.475		4.1/2	4.7/8

NOMINAL CROSS SECTION 1/4 in

O-Seal size No	Actual Diameters				Nominal Housing Dimensions	
	Cross-section A in		Internal B in		Shaft diameter in	Cylinder diameter in
425	0.275	±0.006	4.475	±0.015	4.1/2	5
426	0.275		4.600		4.5/8	5.1/8
427	0.275		4.725		4.3/4	5.1/4
428	0.275		4.850		4.7/8	5.3/8
429	0.275		4.975		5	5.1/2
430	0.275		5.100	±0.023	5.1/8	5.5/8
431	0.275		5.225		5.1/4	5.3/4
432	0.275		5.350		5.3/8	5.7/8
433	0.275		5.475		5.1/2	6
434	0.275		5.600		5.5/8	6.1/8
435	0.275		5.725		5.3/4	6.1/4
436	0.275		5.850		5.7/8	6.3/8
437	0.275		5.975		6	6.1/2
438	0.275		6.225		6.1/4	6.3/4
439	0.275		6.475		6.1/2	7
440	0.275		6.725		6.3/4	7.1/4
441	0.275		6.975		7	7.1/2
442	0.275		7.225	±0.030	7.1/4	7.3/4
443	0.275		7.475		7.1/2	8
444	0.275		7.725		7.3/4	8.1/4
445	0.275		7.975		8	8.1/2
445A	0.275		8.225		8.1/4	8.3/4
446	0.275		8.475		8.1/2	9
446A	0.275	±0.006	8.725		8.3/4	9.1/4
447	0.275		8.975		9	9.1/2
447A	0.275		9.225		9.1/4	9.3/4
448	0.275		9.475		9.1/2	10
448A	0.275		9.725		9.3/4	10.1/4
449	0.275		9.975		10	10.1/2
449A	0.275		10.225	±0.030	10.1/4	10.3/4
450	0.275		10.475		10.1/2	11
450A	0.275		10.725		10.3/4	11.1/4
451	0.275		10.975		11	11.1/2
451A	0.275		11.225		11.1/4	11.3/4
452	0.275		11.475		11.1/2	12
452A	0.275		11.725		11.3/4	12.1/4
453	0.275		11.975		12	12.1/2
454	0.275		12.475		12.1/2	13
455	0.275		12.975		13	13.1/2
456	0.275		13.475		13.1/2	14
457	0.275		13.975		14	14.1/2
458	0.275		14.475		14.1/2	15
459	0.275		14.975		15	15.1/2
460	0.275		15.475		15.1/2	16

TABLE 2 – CONDENSED LISTING : STANDARD INCH SIZE O-RING SIZES

British Standard (BS 1806 : 1962) Sizes in bold, others to American Standard A5568

SECTION i.d.	1/16	3/32	1/8	3/16
1/32	001*			
3/64	002**			
1/16	003	102		
5/64	**004**			
3/32	**005**	103		
1/8	**006**	104		
5/32	**007**	105		
3/16	**008**	106	201	
7/32	**009**	107		
1/4	**010**	108	202	
5/16	**011**	109	203	
3/8	**012**	**110**	204	
7/16	**013**	**111**	205	309
1/2	**014**	**112**	206	310
9/16	**015**	**113**	207	311
5/8	**016**	**114**	208	312
11/16	**017**	**115**	209	313
3/4	**018**	**116**	**210**	314
13/16	**019**	**117**	**211**	315
7/8	**020**	**118**	**212**	
15/16	**021**	**119**	**213**	317
1	**022**	**120**	**214**	318
1 1/16	**023**	**121**	**215**	319
1 1/8	**024**	**122**	**216**	320
1 3/16	**025**	**123**	**217**	321
1 1/4	**026**	**124**	**218**	322
1 5/16	**027**	**125**	**219**	323
1 3/8	**028**	**126**	**220**	324
1 7/16		**127**	**221**	
1 1/2	029	**128**	**222**	**325**
1 9/16		**129**		
1 5/8	030	**130**	**223**	**326**
1 11/16		**131**		
1 3/4	031	**132**	**224**	**327**
1 13/16		**133**		
1 7/8	032	**134**	**225**	**328**
1 15/16		**135**		
2	033	**136**	**226**	**329**
2 1/16		**137**		
2 1/8	034	**138**	**227**	**330**

SECTION i.d.	1/16	3/32	1/8	3/16	1/4
2 3/16		**139**			
2 1/4	035	**140**	**228**	**331**	
2 5/16		**141**			
2 3/8	036	**142**	**229**	**332**	
2 7/16		**143**			
2 1/2	037	**144**	**230**	**333**	
2 9/16		**145**			
2 3/8	038	**146**	**231**	**334**	
2 11/16		**147**			
2 1/4	039	**148**	**232**	**335**	
2 13/16		**149**			
2 7/8	040	150	**233**	**336**	
3	041	151	**234**	**337**	
3 1/8			**235**	**338**	
3 1/4	042	152	**236**	**339**	
3 1/8			**237**	**340**	
3 1/2	043	153	**238**	**341**	
3 5/8			**239**	**342**	
3 3/4	044	154	**240**	**343**	
3 7/8			**241**	**344**	
4	045	155	**242**	**345**	
4 1/8			**243**	**346**	
4 1/4	046	156	**244**	**347**	
4 3/8			**245**	**348**	
4 1/2	047	157	**246**	**349**	**425**
4 5/8			**247**	350	**426**
4 3/4	048	158	**248**	351	**427**
4 7/8			**249**	352	**428**
5	049	159	**250**	353	**429**
5 1/8			**251**	354	**430**
5 1/4	050	160	**252**	355	**431**
5 3/8			**253**	356	**432**
5 1/2		161	**254**	357	**433**
5 5/8			**255**	358	**434**
5 3/4		162	**256**	359	**435**
5 7/8			**257**	360	**436**
6		163	**258**	361	**437**
6 1/4		164	**259**	362	**438**
6 1/2		165	**260**	363	**439**
6 3/4		166	**261**	364	**440**

SECTION i.d.	3/32	1/8	3/16	1/4
7	167	**262**	365	**441**
7 1/4	168	**263**	366	**442**
7 1/2	169	**264**	367	**443**
7 3/4	170	**265**	368	**444**
8	171	**266**	369	**445**
8 1/4	172	**267**	370	**445A**
8 1/2	173	**268**	371	**446**
8 3/4	174	**269**	372	**446A**
9	175	**270**	373	**447**
9 1/4	176	**271**	374	**447A**
9 1/2	177	**272**	375	**448**
9 1/4	178	**273**	376	**448A**
10		**274**	377	**449**
10 1/4				**449A**
10 1/2		275	378	**450**
10 1/4				**450A**
11		276	379	**451**
11 1/4				**451A**
11 1/2		277	380	**452**
11 3/4				**452A**
12		278	381	**453**
12 1/2				**454**
13		279	382	**455**
13 1/2				**456**
14		280	383	**457**
14 1/2				**458**
15		281	384	**459**
15 1/2				**460**
16		282	385	461
16 1/2				462
17		283	386	463
17 1/2				464
18		284	387	465
18 1/2				466
19			388	467
19 1/2				468
20			389	469
21			390	470
22			391	471
23			392	472
24			393	473
25			394	474
26			395	475

*1/32 section **3/64 section

TABLE 3 – STANDARD METRIC SIZES OF O-RINGS (BS 4518 : 1974) (contd)......

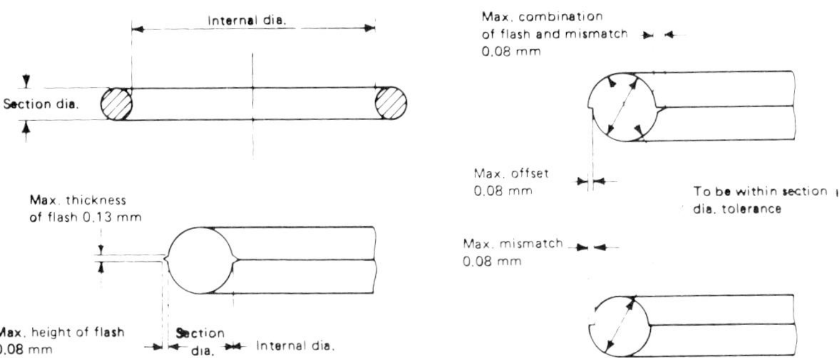

all dimensions in millimetres

O-Ring Ref No.*	O-Ring Dimensions				Nominal Housing Dimensions	
	Internal Diameter	Int Dia Tolerance	Section Diameter	Sect Dia Tolerance	Shaft diameter	Cylinder diameter
0031-16	3.1	±0.15	1.6	±0.08	3.5	6
0041-16	4.1		1.6		4.5	7
0051-16	5.1		1.6		5.5	8
0061-16	6.1		1.6		6.5	9
0071-16	7.1		1.6		7.5	10
0081-16	8.1		1.6		8.5	11
0091-16	9.1		1.6		9.5	12
0101-16	10.1	±0.2	1.6	±0.08	10.5	13
0111-16	11.1		1.6		11.5	14
0121-16	12.1		1.6		12.5	15
0131-16	13.1		1.6		13.5	16
0141-16	14.1		1.6		14.5	17
0151-16	15.1		1.6		15.5	18
0161-16	16.1		1.6		16.5	19
0171-16	17.1		.1.6		17.5	20
0181-16	18.1	±0.25	1.6	±0.08	18.5	21
0191-16	19.1		1.6		19.5	22
0221-16	22.1		1.6		22.5	25
0251-16	25.1		1.6		25.5	28
0271-16	27.1		1.6		27.5	30
0291-16	29.1		1.6		29.5	32
0321-16	32.1	±0.3	1.6	±0.08	32.5	35
0351-16	35.1		1.6		35.5	38
0371-16	37.1		1.6		37.5	40

*NOTE: Reference number consists of ring internal diameter followed by ring section diameter with decimal points omitted.

cont...

TABLE 3 – STANDARD OF METRIC SIZES OF O-RINGS (BS 4518:1974) (contd)......

All dimensions in mm

O-Ring Ref No	O-Ring Dimensions				Nominal Housing Dimensions	
	Internal Diameter	Int Dia Tolerance	Section Diameter	Sect Dia Tolerance	Shaft diameter	Cylinder diameter
0036-24*	3.6	±0.15	2.4	±0.08	4	8
0046-24*	4.6		2.4		5	9
0056-24*	5.6		2.4		6	10
0066-24*	6.6		2.4		7	11
0076-24*	7.6		2.4		8	12
0086-24*	8.6		2.4		9	13
0096-24*	9.6		2.4		10	14
0106-24*	10.6	±0.2	2.4	±0.08	11	15
0116-24*	11.6		2.4		12	16
0126-24*	12.6		2.4		13	17
0136-24*	13.6		2.4		14	18
0146-24*	14.6		2.4		15	19
0156-24*	15.6		2.4		16	20
0166-24*	16.6		2.4		17	21
0176-24*	17.6		2.4		18	22
0186-24	18.6	±0.25	2.4	±0.08	19	23
0196-24	19.6		2.4		20	24
0216-24	21.6		2.4		22	26
0246-24	24.6		2.4		25	29
0276-24	27.6		2.4		28	32
0296-24	29.6		2.4		30	34
0316-24	31.6	±0.3	2.4	±0.08	32	36
0346-24	34.6		2.4		35	39
0376-24	37.6		2.4		38	42
0396-24	39.6		2.4		40	44
0416-24	41.6		2.4		42	46
0446-24	44.6		2.4		45	49
0476-24	47.6		2.4		48	52
0496-24	49.6		2.4		50	54
0516-24	51.6	±0.4	2.4	±0.08	52	56
0546-24	54.6		2.4		55	59
0576-24	57.6		2.4		58	62
0596-24	59.6		2.4		60	64
0616-24	61.6		2.4		62	66
0646-24	64.6		2.4		65	69
0676-24	67.6		2.4		68	72
0696-24	69.6		2.4		70	74

*Only O-rings marked with an asterisk are recommended for dynamic sealing.

cont...

TABLE 3 – STANDARD OF METRIC SIZES OF O-RINGS (BS 4518;1974) (contd)......

All dimensions in mm

O-Ring Ref No.	O-Ring Dimensions				Nominal Housing Dimensions	
	Internal Diameter	Int Dia Tolerance	Section Diameter	Sect Dia Tolerance	Shaft diameter	Cylinder diameter
0195-30*	19.5	±0.25	3.0	±0.1	20	25
0215-30*	21.5		3.0		22	27
0225-30*	22.5		3.0		23	28
0245-30*	24.5		3.0		25	30
0255-30*	25.5		3.0		26	31
0265-30*	26.5		3.0		27	32
0275-30*	27.5		3.0		28	33
0295-30*	29.5		3.0		30	35
0315-30*	31.5	±0.3	3.0	±0.1	32	37
0325-30*	32.5		3.0		33	38
0345-30*	34.5		3.0		35	40
0355-30*	35.5		3.0		36	41
0365-30*	36.5		3.0		37	42
0375-30*	37.5		3.0		38	43
0395-30*	39.5		3.0		40	45
0415-30*	41.5		3.0		42	47
0425-30*	42.5		3.0		43	48
0445-30*	44.5		3.0		45	50
0495-30	49.5		3.0		50	55
0545-30	54.5	±0.4	3.0	±0.1	55	60
0595-30	59.5		3.0		60	65
0645-30	64.5		3.0		65	70
0695-30	69.5		3.0		70	75
0745-30	74.5		3.0		75	80
0795-30	79.5		3.0		80	85

*Only O-rings marked with an asterisk are recommended for dynamic sealing

cont....

TABLE 3 – STANDARD OF METRIC SIZES OF O-RINGS (BS 4518:1974) (contd)......

All dimensions in mm

O-Ring Ref No	O-Ring Dimensions				Nominal Housing Dimensions	
	Internal Diameter	Int Dia Tolerance	Section Diameter	Sect Dia Tolerance	Shaft diameter	Cylinder diameter
0845-30	84.5	±0.5	3.0	±0.1	85	90
0895-30	89.5		3.0		90	95
0945-30	94.5		3.0		95	100
0995-30	99.5		3.0		100	105
1045-30	104.5		3.0		105	110
1095-30	109.5		3.0		110	115
1145-30	114.5		3.0		115	120
1195-30	119.5		3.0		120	125
1245-30	124.5	±0.6	3.0	±0.1	125	130
1295-30	129.5		3.0		130	135
1345-30	134.5		3.0		135	140
1395-30	139.5		3.0		140	145
1445-30	144.5		3.0		145	150
1495-30	149.5		3.0		150	155
1545-30	154.5		3.0		155	160
1595-30	159.5		3.0		160	165
1645-30	164.5		3.0		165	170
1695-30	169.5		3.0		170	175
1745-30	174.5		3.0		175	180
1795-30	179.5		3.0		180	185
1845-30	184.5	±0.8	3.0	±0.1	185	190
1895-30	189.5		3.0		190	195
1945-30	194.5		3.0		195	200
1995-30	199.5		3.0		200	205
2095-30	209.5		3.0		210	215
2195-30	219.5		3.0		220	225
2295-30	229.5		3.0		230	235
2395-30	239.5		3.0		240	245
2495-30	249.5		3.0		250	255

*Only O-rings marked with an asterisk are recommended for dynamic sealing.

cont...

TABLE 3 – STANDARD OF METRIC SIZES OF O-RINGS (BS 4518:1974) (contd)......

All dimensions in mm

O-Ring Ref No	O-Ring Dimensions				Nominal Housing Dimensions	
	Internal Diameter	Int Dia Tolerance	Section Diameter	Section Dia Tolerance	Shaft diameter	Cylinder diameter
0294-41*	29.4	±0.3	4.1	±0.12	30	37
0314-41*	31.4		4.1		32	39
0324-41*	32.4		4.1		33	40
0344-41*	34.4		4.1		35	42
0354-41*	35.4		4.1		36	43
0364-41*	36.4		4.1		37	44
0374-41*	37.4		4.1		38	45
0394-41*	39.4		4.1		40	47
0414-41*	41.4		4.1		42	49
0424-41*	42.4		4.1		43	50
0444-41*	44.4		4.1		45	52
0454-41*	45.4		4.1		46	53
0494-41*	49.4		4.1		50	57
0524-41*	52.4	±0.4	4.1	±0.12	53	60
0544-41*	54.4		4.1		55	62
0554-41*	55.4		4.1		56	63
0594-41*	59.4		4.1		60	67
0624-41*	62.4		4.1		63	70
0644-41*	64.4		4.1		65	72
0694-41*	69.4		4.1		70	77
0724-41*	72.4		4.1		73	80
0744-41*	74.4		4.1		75	82
0794-41*	79.4		4.1		80	87
0844-41*	84.4	±0.5	4.1	±0.12	85	92
0894-41*	89.4		4.1		90	97
0924-41*	92.4		4.1		93	100
0944-41*	94.4		4.1		95	102
0994-41*	99.4		4.1		100	107
1044-41	104.4		4.1		105	112
1094-41	109.4		4.1		110	117
1144-41	114.4		4.1		115	122
1174-41	117.4		4.1		118	125
1194-41	119.4		4.1		120	127

*Only O-rings marked with an asterisk are recommended for dynamic sealing.

cont...

TABLE 3 – STANDARD OF METRIC SIZES OF O-RINGS (BS 4518:1974) (contd)......

All dimensions in mm

O-Ring Ref No	O-Ring Dimensions				Nominal Housing Dimensions	
	Internal Diameter	Int Dia Tolerance	Section Diameter	Sect Dia Tolerance	Shaft diameter	Cylinder diameter
1244–41	124.4	±0.6	4.1	±0.12	125	132
1294-41	129.4		4.1		130	137
1344-41	134.4		4.1		135	142
1394-41	139.4		4.1		140	147
1444-41	144.4		4.1		145	152
1494-41	149.4		4.1		150	157
1524-41	152.4		4.1		153	160
1544-41	154.4		4.1		155	162
1594-41	159.4		4.1		160	167
1644-41	164.4		4.1		165	172
1694-41	169.4		4.1		170	177
1744-41	174.4		4.1		175	182
1794-41	179.4		4.1		180	187
1844-41	184.4	±0.8	4.1	±0.12	185	192
1894-41	189.4		4.1		190	197
1924-41	192.4		4.1		193	200
1944-41	194.4		4.1		195	202
1994-41	199.4		4.1		200	207
2094-41	209.4		4.1		210	217
2124-41	212.4		4.1		213	220
2194-41	219.4		4.1		220	227
2294-41	229.4		4.1		230	237
2394-41	239.4		4.1		240	247
2424-41	242.4		4.1		243	250
2494-41	249.4		4.1		250	257
2594-41	259.4	±1.0	4.1	±0.12	260	267
2694-41	269.4		4.1		270	277
2724-41	272.4		4.1		273	280
2794-41	279.4		4.1		280	287
2894-41	289.4		4.1		290	297
2994-41	299.4		4.1		300	307

*Only O-rings marked with an asterisk are recommended for dynamic sealing.

cont...

TABLE 3 – STANDARD OF METRIC SIZES OF O-RINGS (BS 4518:1974) (contd)......

All dimensions in mm

O-Ring Ref No	O-Ring Dimensions				Nominal Housing Dimensions	
	Internal Diameter	Int Dia Tolerance	Section Diameter	Sect Dia Tolerance	Shaft diameter	Cylinder diameter
0443-57*	44.3	±0.3	5.7	±0.12	45	55
0453-57*	45.3		5.7		46	56
0493-57*	49.3		5.7		50	60
0523-57*	52.3	±0.4	5.7	±0.12	53	63
0543-57*	54.3		5.7		55	65
0553-57*	55.3		5.7		56	66
0593-57*	59.3		5.7		60	70
0623-57*	62.3		5.7		63	73
0643-57*	64.3		5.7		65	75
0693-57*	69.3		5.7		70	80
0743-57*	74.3		5.7		75	85
0793-57*	79.3		5.7		80	90
0843-57*	84.3	±0.5	5.7	±0.12	85	95
0893-57*	89.3		5.7		90	100
0943-57*	94.3		5.7		95	105
0993-57*	99.3		5.7		100	110
1043-57*	104.3		5.7		105	115
1093-57*	109.3		5.7		110	120
1143-57*	114.3		5.7		115	125
1193-57*	119.3		5.7		120	130
1243-57*	124.3	±0.6	5.7	±0.12	125	135
1293-57*	129.3		5.7		130	140
1343-57*	134.3		5.7		135	145
1393-57*	139.3		5.7		140	150
1443-57*	144.3		5.7		145	155
1493-57	149.3		5.7		150	160
1543-57	154.3		5.7		155	165
1593-57	159.3		5.7		160	170
1643-57	164.3		5.7		165	175
1693-57	169.3		5.7		170	180
1743-57	174.3		5.7		175	185
1793-57	179.3		5.7		180	190

*Only O-rings marked with an asterisk are recommended for dynamic sealing.

cont...

TABLE 3 – STANDARD OF METRIC SIZES OF O-RINGS (BS 4518:1974) (contd)......

All dimensions in mm

O-Ring Ref No	O-Ring Dimensions				Nominal Housing Dimensions	
	Internal Diameter	Int Dia Tolerance	Section Diameter	Sect Dia Tolerance	Shaft diameter	Cylinder diameter
1843-57	184.3	± 0.8	5.7	± 0.12	185	195
1893-57	189.3		5.7		190	200
1943-57	194.3		5.7		195	205
1993-57	199.3		5.7		200	210
2093-57	209.3		5.7		210	220
2193-57	219.3		5.7		220	230
2293-57	229.3		5.7		230	240
2393-57	239.3		5.7		240	250
2493-57	249.3		5.7		250	260
2593-57	259.3	± 1.0	5.7	± 0.12	260	270
2693-57	269.3		5.7		270	280
2793-57	279.3		5.7		280	290
2893-57	289.3		5.7		290	300
2993-57	299.3		5.7		300	310
3193-57	319.3	± 1.5	5.7	± 0.12	320	330
3393-57	339.3		5.7		340	350
3593-57	359.3		5.7		360	370
3793-57	379.3		5.7		380	390
3993-57	399.3		5.7		400	410
4193-57	419.3	± 2.0	5.7	± 0.12	420	430
4393-57	439.3		5.7		440	450
4593-57	459.3		5.7		460	470
4793-57	479.3		5.7		480	490
4993-57	499.3		5.7		500	510

*Only O-rings marked with an asterisk are recommended for dynamic sealing.

cont...

TABLE 3 – STANDARD OF METRIC SIZES OF O-RINGS (BS 4518:1974) (contd)......

All dimensions in mm

Ref No	O-Ring Dimensions				Nominal Housing Dimensions	
	Internal Diameter	Int Dia Tolerance	Section Diameter	Sect Dia Tolerance	Shaft diameter	Cylinder diameter
1441-84*	144.1	±0.6	8.4	± 0.15	145	160
1491-84*	149.1		8.4		150	165
1541-84*	154.1		8.4		155	170
1591-84*	159.1		8.4		160	175
1641-84*	164.1		8.4		165	180
1691-84*	169.1		8.4		170	185
1741-84*	174.1		8.4		175	190
1791-84*	179.1		8.4		180	195
1841-84*	184.1	±0.8	8.4	±0.15	185	200
1891-84*	189.1		8.4		190	205
1941-84*	194.1		8.4		195	210
1991-84*	199.1		8.4		200	215
2041-84*	204.1		8.4		205	220
2091-84*	209.1		8.4		210	225
2191-84*	219.1		8.4		220	235
2291-84*	229.1		8.4		230	245
2341-84*	234.1		8.4		235	250
2391-84*	239.1		8.4		240	255
2491-84*	249.1		8.4		250	265

*Only O-rings marked with an asterisk are recommended for dynamic sealing.

TABLE 4 – O-RING GROOVE DIMENSIONS FOR DIAMETRAL SEALING

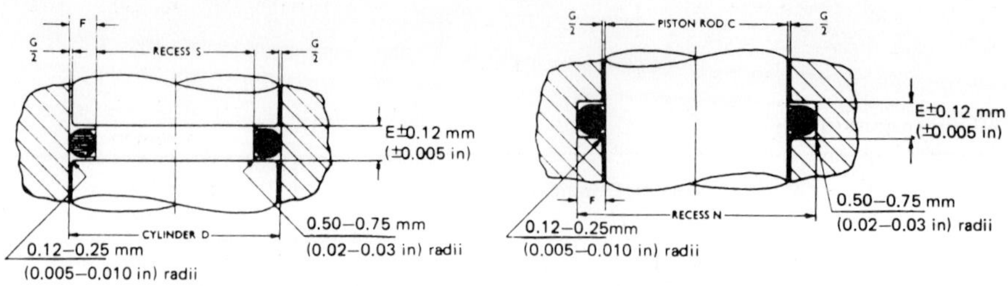

GROOVE IN PISTON GROOVE IN ROD GLAND

Suitable limits must be applied to C or D and mating groove diameter to provide a diametral clearance not in excess of dimension G specified by O-ring standards.

O-Ring section diameter in	Radial Dimension		Minimum cross-sectional squeeze of ring in
	Maximum in	Minimum in	
0.070	0.062	0.060	0.005
0.103	0.094	0.091	0.006
0.139	0.125	0.122	0.010
0.210	0.188	0.184	0.017
0.275	0.250	0.245	0.019

O-ring materials

The majority of O-rings are produced in *Nitrile* rubbers of various grades and hardnesses (*eg* from 40 to 90 degrees). A typical general purpose material is a low/medium nitrile rubber with a hardness of *circa* 75 degrees. Nitrile rubbers have good resistance to heat up to 100 to 120°C, and excellent resistance to mineral oils and greases and a wide range of fluids including water and ethylene glycol. Special grades of nitrile rubbers have also been developed for particular applications.

Various other materials may be used for applications where nitrile rubbers show limitations in performance, or lack of compatibility in contact with specific fluids – see Tables 5,6,7,8 and 9. These also include solid PTFE rings for extreme chemical resistance.

A point to bear in mind is that where O-rings in non-standard materials are produced in the same moulds as *nitrile* rubber rings, some variations in size tolerances can occur because of differences in mould shrinkage. However, such variation outside standard tolerance limits is not normally significant when the rings are fitted in standard grooves because all O-rings are deformed slightly on fitting and thus the functioning of the ring is not likely to be affected.

The exception is the solid PTFE ring. While this can be fitted in standard grooves, it is more effective located in a rounded groove (Figure 7).

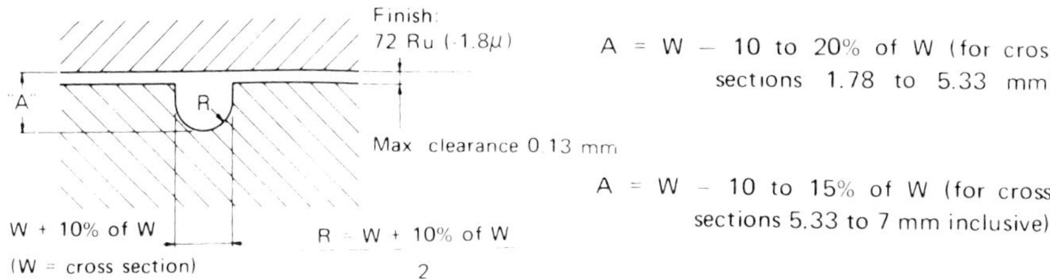

Figure 7
Groove dimensions for PTFE O-rings.

O-ring swell

Compatibility between the O-ring material and the fluid with which it is in contact is a major factor affecting selection, but a point of some significance is that even when the O-ring material is accepted as compatible it may still be subject to a volume change when immersed in the fluid. This can affect the effective squeeze.

Figure 8 plots the linear change in cross section related to volume change both for a free ring and one constrained in a typical assembly. Knowing the volume change (possibly from quoted compatibility test figures for the material) one can determine the change in cross section (diameter). This will always be smaller than the volume change and for low values of volume changes accepted for normal compatibility can often be ignored entirely.

Lubrication

Lubrication is a further factor of some significance. In many dynamic applications the O-ring is lubricated by the fluid with which it is in contact and may also be lubricated with the same fluid to assist initial assembly. In other cases where the fluid concerned lacks lubricity, definite lubrication of the O-ring is necessary as wear on a dry seal will be high. The same applies to initial assembly. All metal parts which will contact the seal, and the O-ring itself, should be lubricated before assembly. This applies both to static and dynamic seals.

COMPARATIVE CHARACTERISTICS OF O-RINGS AND OTHER SEAL TYPES

Type of seal	O-Ring	Rectangular ring	Slipper	U-Ring	Combination seals	
θ	Zero	Zero	Zero	60° to 70° (typical)	20° to 30° (typical)	10° to 15° (typical)
Compression force	High	High	High	Low	Medium to high	Medium to high
Low pressure sealing ability	High	High	Medium	Low	High	High
Width of sealing line	Medium	Wide	Very wide	Varies	Very thin	Very thin

TABLE 5 – O-RING MATERIALS TO SWEDISH STANDARD SMS 1587

Property	Requirement	Test	Test Specimen	Remarks
Hardness	70 ± 3° IRH	Microtest accord. to SIS 16 22 01	O-ring	During testing the ball shall be pressed in the direction of the cross-sectional radius
Tensile strength	14 N/mm², min (140 kp/cm²), min	SIS 16 22 02	2 mm thick sheet	
Elongation	300% min			
Compression set at 373 K (100 °C) cross-section 1.6 cross-section 2.4; 3 cross-section 5.7; 8.4	 30% max 25% max 20% max	SIS 16 22 04	O-ring	The dimensions of the spacers shall be chosen so that the test specimens are compressed 22–28%
Increase in hardness after ageing in air for 72 h at 373 K (100 °C) 72 h at 398 K (125 °C)	 8° IRH, max 10° IRH, max	SIS 16 22 05	O-ring	
Decrease in tensile strength after ageing in air for 72 h at 373 K (100 °C) 72 h at 398 K (125 °C)	 15% max 20% max	SIS 16 22 05	2 mm thick sheet	
Decrease in elongation after ageing in air for 72 h at 373 K (100 °C) 72 h at 398 K (125 °C)	 25% max 35% max			
Change in volume for 71 h in test liquid B at 296 K (23 °C)	+ 35% 0%	SIS 16 22 08	O-ring min. weight 0.3 g	
Change in volume for 71 h in test liquid C at 296 K (23 °C)	+ 60% 0%	SIS 16 22 08	O-ring min. weight 0.3 g	
Change in volume for 71 h of oil No.1 at 373 K (100 °C)	+ 10% – 6%	SIS 16 22 08	O-ring min. weight 0.3 g	Oil No.1 : Aniline point 396.9 ± 1 K (123.9 °C)
Change in volume for 71 h in oil No.3 at 373 K (100 °C)	+ 15% 0%	SIS 16 22 08	O-ring min. weight 0.3 g	Oil No.3 : Aniline point 342.5 ± 1 K (69.5 °C)
Change in hardness for 71 h in oil No.1 at 373 K (100 °C)	+ 10° IRH – 5° IRH	SIS 16 22 08	O-ring	
Change in hardness for 71 h in oil No.3 at 373 K (100 °C)	+ 5° IRH –10° IRH			
Effect on metal (corrosion) (Steel SIS 13 11-00)	None	SIS 16 22 13	O-ring	
Temperature setraction TR 10 TR 30	 243 K (–30 °C) max 248 K (–25 °C) max	SIS 16 22 21	2 mm thick sheet	

TABLE 6 – COMPARISON OF GENERAL PROPERTIES OF O-RING ELASTOMERS

Elastomer	Mechanical properties		Chemical properties* resistance to				Application temp for ageing in air	
							°C	°C
	Wear resistance	Tensile strength MPa	Weathering and ozone	Hot water, steam, weak acids & alkalis	Strong acids and oxidizing acids	Petrol, oils, etc	1 week	1 year
Butyl	F to G	10 to 18	G	E	G	P	160	80
Chloroprene	G	10 to 25	G	E	F to G	F to G	150	70
Ethylene-propylene	G	7 to 18	E	E	G	P	160	90
Fluoroelastomer	E	15 to 20	E	E	G	E	300	210
Natural rubber	G	15 to 30	F	E	F	P	100	50
Nitrile	G	10 to 25	F to G	G	F to G	G	150	70
Silicone	P	4 to 10	E	F	P	F to G	300	200
Urethane	E	30 to 50	E	P	P	G		
Perfluorelastomer	G	13 to 20	E	E	E	E	310	290

*See also Table 8

Key:
P = Poor
F = Fair
G = Good
E = Excellent

TABLE 7 – SIMPLIFIED GUIDE TO O-RING MATERIALS

Material	Working temperature range °C	Remarks
Nitrile rubber	–40 to +100	Suitable for the majority of applications.
Acrylate ester	–10 to +150	Better heat resistance than nitrile, otherwise similar.
Butyl rubber	–35 to +100	Resistant to synthetic fluids.
Polyurethane	–50 to +90	High resistance to abrasive wear.
Ethylene-propylene	–65 to +150	Resistant to synthetic fluids (not resistant to mineral oils).
Fluorocarbon	–40 to +230	Excellent resistance to heat and most fluids.
Silicone	–40 to +230	High temperature applications (static seals only)
Chloroprene	–40 to +100	Good resistance to low temperatures, ozone and acids.
Perfluorelastomer	–55 to +310	Excellent chemical resistance and very high temperature service.

TABLE 8 — CONDENSED MATERIAL SELECTION GUIDE FOR O-RINGS

	Natural rubber	Buna-S	Buna-N bitrile	Ethylene propylene	Butyl	Neoprene	Urethane	Thiokil	Silicone	Fluoro-silicone	Fluoroelastomer (Viton*)	Poly-acrylate	Perfluoro-elastomer
ASTM code	NR	SBR	NBR	EP	IIR	CR	AU EU		Si	FSi	FKM	ACM	FFKM
Hardness (degrees)	30 to 90	40 to 90	40 to 95	40 to 80	20 to 80	20to90	50 to 90	50 to 80	10 to 90	40-80	55 to 95	50-90	65 to 95
Tensile strength (lb/in^2)	3000	3000	2500	2500	2000	3000	5000	1000	1400	900	2500	2000	2000
Tearing strength	E	VG	G	G	VG	VG	E	R	N	N	G	G	G
Wear resistance	E	E	G	G	G	G	VG	R	N	N	G	VG	G
Fireproof hyd fluids	N	N	N	VG	VG	R	N	R	G	VG	RN	N	E
Lubricating oils	N	N	G	N	N	G	VG	E	G	E	E	VG	E
Fuel oils	N	N	G	N	N	R	VG	E	R	E	E	VG	E
Hydraulic oils	N	N	E	N	N	G	VG	E	G	E	E	U	E
Vegetable oils	N	R	E	R	E	G	VG	R	G	E	E	U	E
Animal fats	N	R	E	R	E	G	VG	R	G	E	E	U	E
Petrol (normal)	N	N	VG	N	N	R	VG	E	R	E	E	VG	E
Petrol (high octane)	N	N	G	N	N	N	VG	E	N	E	E	G	E
Kerosene	N	N	VG	N	N	R	VG	E	R	E	E	VG	E
Aromatic hydrocarbons	N	N	R	N	N	N	G	E	N	E	E	R	E
Aliphatic hydrocarbons	N	N	VG	N	N	G	VG	E	R	E	E	VG	E
Water (under 80 °C)	G	VG	VG	E	E	R	R-N	R	E	E	E	N	E
Water (above 80 °C)	N	R	R	E	G	N	N	N	E	VG	G	N	E
Alcohols	G	G	VG	E	E	E	R	R	E	E	E	N	E
Ketones	N	N	N	G	VG	N	N	VG	G	N	N	N	E
Conc. acids	N	N	N	R	G	N	N	N	R	G	VG	N	E
Diluted acids	R	R	R	G	E	G	N	R	G	VG	E	N	E
Alkalis	G	G	R	VG	E	G	N	N	G	R	G	N	E
Chlorinated solvents	N	N	R	N	N	N	R	VG	N	E	E	R	VG
Ozone and sunlight	R	R	R	E	E	VG	VG	VG	E	E	E	VG	E
Max temp at continuous service	70 °C	105 °C	105 °C	135 °C	120 °C	105 °C	70 °C	65 °C	230 °C	205 °C	250 °C	175 °C	300 °C
Electrical properties	E	VG	N	E	E	VG	G	R	E	E	E	N	G
Compression set	VG	VG	VG	G	G	VG	N	N	E	VG	G	G	G
Flame resistant	No	No	No	No	No	Yes	No	No	No	No	Yes	No	Yes

Key:
E = Excellent VG = Very good G = Good R = Reasonable N = Not recommended U = Unsuitable *Du Pont's Registered Trademark

TABLE 9 – STANDARDS FOR O-RING RUBBERS

Specification	Scope
MIL/P/5516	Nitrile rubbers
ASTM D 200) SAE J 200)	Line call-out system covering all elastomers, hardnesses, etc.
BS746	Rubber parts for gas meter unions and adaptors
BS1010	O-rings for draw-off taps
BS2494 Part 2	Rubber joint rings
BS2751	General-purpose nitrile grades
BS2752	General purpose chloroprene grades
BS3222	Low compression set nitrile rubbers
BS3515	General-purpose SBR grades
BS3526 Part 2	O-rings for wet cylinder liners
DTD 560	Rubber grades for use in aviation fuels
DTD 818	Silicone rubber grades
DTD 5509	Nitrile grades for oils, greases and fuels
DTD 5514A	Chloroprene grades for aircraft
DTD 5531	General-purpose silicone grades
DTD 5543A	Fluorocarbon rubber grades
DTD 5582	Silicone rubber grades (oil-resistant)
DTD 5583	Fluorosilicone rubber grades
DTD 5594	Nitrile rubber grades for mineral oils
DTD 5595	Nitrile rubber grades for synthetic oils and fuels
DTD 5596	General purpose ethylene-propylene grades
DTD 5597	Fluid-resistant ethylene-propylene grades
DTD 5603	Fluorocarbon O-rings
DTD 5605	Silicone O-rings
DTD 5606	Nitrile O-rings
DTD 5607	Nitrile O-rings – mineral oil-resistant
DTD 5608	Ethylene-propylene O-rings
DTD 5612	Viton-fluoroelastomer

For general applications, lubrication by smearing the O-ring and groove with high melting point grease is usually adequate. For specific applications it may be necessary to select a lubricant compatible with the system and service involved. Typical recommendations in this respect are summarized in Table 10.

Static O-ring seals

The O-ring is the most commonly used seal for static applications because of its ready availability in a wide range of sizes and materials, low cost, ease of assembly and excellent sealing characteristics.

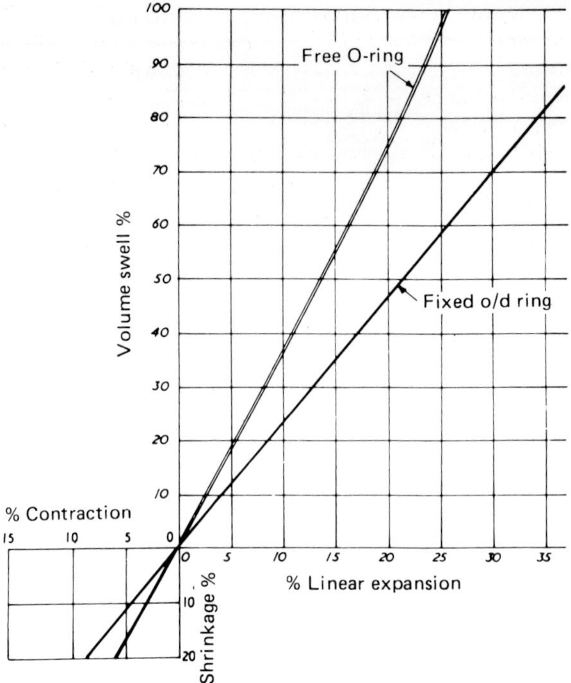

Figure 8

TABLE 10 – LUBRICANTS FOR O-RINGS

O-Ring material	Service	Lubricant
Nitrile rubber	Fuel, oil or hydraulic systems pneumatics, general high pressure, high speed pneumatics.	System fluid Barium grease
	Vacuum systems.	Silicone oil Castor oil or Cellulose ester
Silicone	Vacuum systems.	Cellulose ester
Fluoroelastomer	High temperature hydraulics.	System fluid
	High temperature pneumatics.	Silicone grease
Perfluoroelastomer	High temperature pneumatics and vacuum	Silicone oil/grease

In general, O-ring static seals can be used in standard groove dimensions shown in Table 11 for sealing pressures up to 100 bar (1500 lb/in^2), giving an effective compression or squeeze of 15 to 30%, depending on the O-ring section diameter. If vibration or pressure fluctuations are present, however, either the clearance gap should be reduced, or back-up rings included in the assembly. For sealing pressures greater than 100 bar (1500 lb/in^2) more squeeze, or back-up rings, are essential.

Theoretically, at least, the pressure rating of an O-ring static seal will go on increasing as the clearance gap is closed. Provided the mating surfaces are reasonably smooth and can be closed down to nominal metal-to-metal contact extremely high pressures can be sealed in this manner. Where tight closure is not possible, or there is a possibility of the surfaces breathing, back-up rings may be resorted to for high pressure applications.

For sealing air and gases the permeability of the elastomer employed should be investigated. This is particularly important in the case of static O-ring seals on vacuum systems where a special groove design may also be required to accommodate rubber shrinkage and the reluctance of rubbers to seal at very low pressures. It is generally necessary to aim for the O-ring section to fill the groove section so that no seal movement can take place, which could cause an undesirable pumping loss of seating action.

TABLE 11 – GROOVE DATA FOR STATIC O-RING SEALS

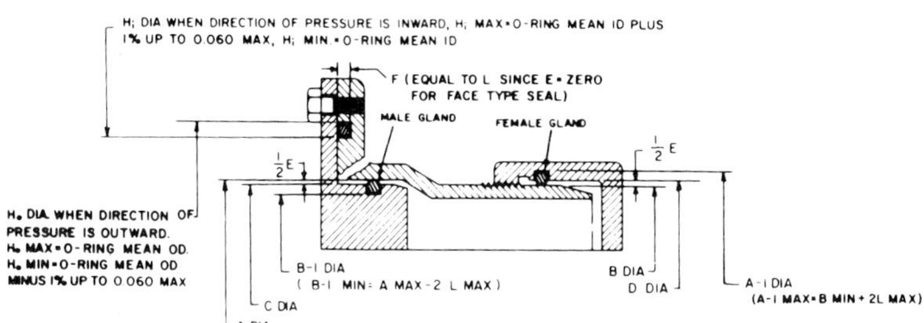

W Cross Section		L	Squeeze		E#	G Groove Width			R	
Nominal	Actual	Gland Depth	Actual	%	Diametrical Clearance	No Back-Up Rings	One Back-Up Ring	Two Back-Up Rings	Groove Radius	Eccentricity Max. T.I.R*
1/16	0.070 ±.003	0.050 0.052	0.015 to 0.023	22 to 32	0.002 to 0.005	0.093 to 0.098	0.138 to 0.143	0.205 to 0.210	0.005 to 0.015	0.002
3/32	0.103 ±.003	0.081 to 0.083	0.017 to 0.025	17 to 24	0.002 to 0.005	0.140 to 0.145	0.171 to 0.176	0.238 to 0.243	0.005 to 0.015	0.002
1/8	0.139 ±.004	0.111 to 0.113	0.022 to 0.032	16 to 23	0.003 to 0.006	0.187 to 0.192	0.208 to 0.213	0.275 to 0.280	0.010 to 0.025	0.003
3/16	0.210 ±.005	0.170 to 0.173	0.032 to 0.045	15 to 21	0.003 to 0.006	0.281 to 0.286	0.311 to 0.316	0.410 to 0.415	0.020 to 0.035	0.004
1/4	0.275 ±.006	0.226 to 0.229	0.040 to 0.055	15 to 20	0.004 to 0.007	0.375 to 0.380	0.408 to 0.413	0.538 to 0.543	0.020 to 0.035	0.005

#This dimension can be reduced by 50% when using silicone rings.

*Total Indicator Reading between groove and adjacent bearing surface.

All dimensions in inches

A point to bear in mind with O-rings (and other solid section rubber seals) used as static seals is that elastomers are incompressible, *ie* they deform elastically but maintain a constant volume. Thus if the groove volume is *less* than the ring volume, no amount of clamping pressure will completely close the joint, only extrude the difference in rubber volume between the joint faces.

Provided the compression of an O-ring static seal (or most other solid rubber sections) does not exceed 20%, the ring should be re-usable if the joint is broken and then remade. Higher clamping pressures – and thus higher compression of the ring – may be used in static seals (*eg* in *crush* seals). In this case, the rings are not re-usable.

Conventional gasket type seals for flanges, *etc,* present no particular problems. The groove can be cut in either component and the necessary squeeze applied by the tightening bolts, making sure that squeeze is applied uniformly around the circumference of the ring. End caps and unions may be sealed in a similar way, or the ring mounted in a groove in the female or male part so that it is squeezed radially when the two parts are assembled. (Figure 9).

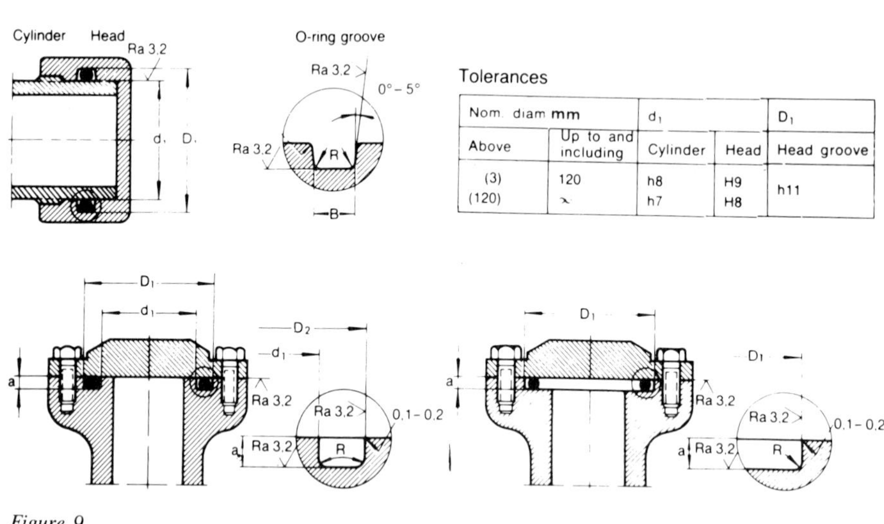

Tolerances

Nom. diam mm		d_1		D_1
Above	Up to and including	Cylinder	Head	Head groove
(3)	120	h8	H9	h11
(120)	∞	h7	H8	

Figure 9
O-ring static seals 1
dimensional standards and tolerances.

Basically the groove overall diameter should be made equal to the ring overall diameter if the pressure is applied from the inside as this eliminates ring movement when pressurized and prevents abrasive wear or even pumping action developing. It is also good design, where practical, to use the system pressure to close the clearance gap.

Where required for static seals, basic non-standard O-rings can be made up from extruded cord material, generally available in the same cross section diameters as standard metric and inch size O-rings.

Certain types of static seals, such as *valve seat* seals, require a modified form of groove in order to eliminate the possibility of blow out. This can occur at relatively low pressures where the actual pressure force under certain conditions, or certain positions of the assembly, is directed radially outwards to lift the ring out of a conventional groove (Figure 10). Alternatively the cause may be gas which has permeated the O-ring causing it to

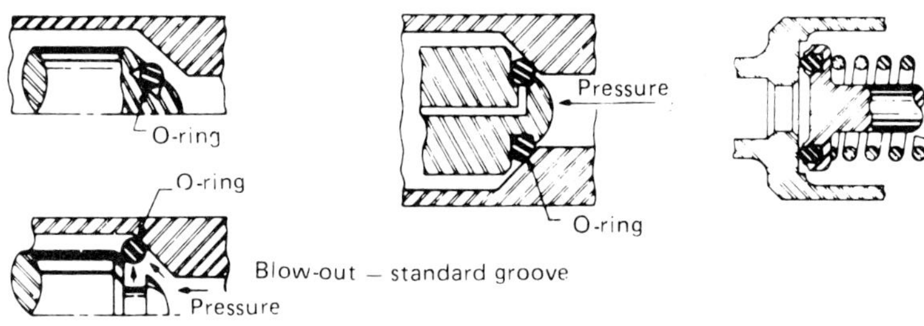

Figure 10

swell or balloon momentarily on release of pressure, and thus pop out of the groove. The simplest solution in such cases is usually to employ a form of groove which mechanically restrains the O-ring against outward radial movement. Alternatively it may be possible to vent the groove so that a negative pressure differential is always applied to the i.d. of the O-ring under conditions which would otherwise promote blow out.

A trapezoidal groove may also be used to accommodate an O-ring with axial compression. In this case recommended groove dimensions are as in Table 12.

TABLE 12 – TRAPEZOIDAL GROOVE DIMENSIONAL DATA

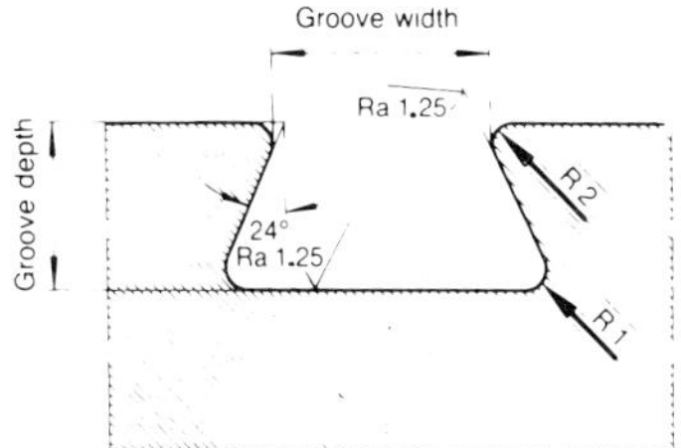

Recessing dimensions

Ring Diameter (mm)	Groove Depth 0 −0.05	R1	R2	Groove Width 0 −0.05
3.53	2.80	0.80	0.25	3.05
5.00	4.15	0.80	0.25	4.10
5.34	4.40	0.80	0.25	4.35
5.70	4.80	0.80	0.40	4.75
6.99	5.95	0.15	0.40	5.65
8.00	6.85	0.15	0.50	6.50
8.40	7.25	0.15	0.50	6.80
9.00	7.80	0.15	0.50	7.25
10.00	8.70	0.15	0.50	7.95

Crush seals

A further possibility with static seals is to dispense with the groove, locate the O-ring against a flange or shoulder and close down on a chamfered edge on the mating member (Figure 11). This is known normally as a crush seal or chamfer seal.

The taper end of the mate fitting should be slightly less (about 10%) than the i.d. of the O-ring selected, when a 45° taper will normally give an adequate gland volume. If preferred, the gland volume required can be calculated, estimating this as 0.9 times the O-ring volume. Once used as a crush seal the O-ring is permanently deformed.

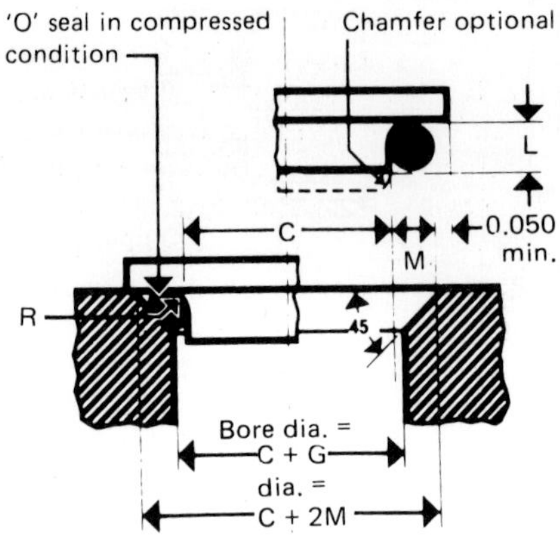

Figure 11

Seat seals

A further type of static application for O-rings is the *seat* seal, where the O-ring is used as a sealing washer on the face of one part which closes against another. Sealing is produced by the closing pressure distorting the O-ring and thus producing a seal against the surface. The degree of distortion is governed by the hardness of the ring material and the closure pressure.

The same form of O-ring mounting may be used to absoprb impact rather than act as a seal or combine both the duties of a seat seal with shock-absorption. Mounting is a simple, straightforward matter, but the design of groove or retainer must be such that the O-ring is trapped and cannot fall out, while still allowing enough of the section to protrude to provide the necessary distortion and face contact for sealing or shock absorption.

Radiator seals

O-rings have replaced the traditional type of washers for sealing taps, mixing valves, swing check valves, radiator valves, *etc*. Special rubber grades are, however, desirable to minimize swell and hardness change over a period of use. Swelling produced by hot water is a process that continues over a very long period, unlike the swelling produced by hydraulic oils which reaches a maximum in about 70 to 100 hours. The preferred rubber for hot water and steam is EPDM, offering good ageing resistance and low swell.

Dynamic O-ring seals

Considerable use is made of O-rings as seals for reciprocating assemblies, *eg* hydraulic cylinder seals and pneumatic spool valve seals. In hydraulic applications, however, the employment of O-rings as primary dynamic seals is normally limited to short strokes and moderate pressures of the order of 100 bar (1500 lb/in^2). In fact O-ring reciprocating seals are most suitable for small diameter, short stroke applications and moderate pressures, although this does not preclude their use for other systems. A rather more significant factor is that O-rings are not particularly suitable for very slow speed reciprocating seals, primarily because the friction under such conditions tends to be high, resulting in early seal failure. Satisfactory performance in any reciprocating application also depends on employing the seal within its rating or capabilities, and with a suitable installation.

General recommendations for groove diameters for O-ring reciprocating seals and clearances are summarized in Table 13, which shows that the squeeze is appreciably less than for static seals. Basically, squeeze is required only to take up differential expansion or contraction and provide sealing at very low pressures or low temperatures, although this in turn will depend on the type of elastomer involved. Too much squeeze will serve only to produce excessive friction and wear, and may also promote spiral failure.

Spiral failure is caused by conditions which allow some parts of the ring to slide and others to roll, simultaneously. The result is an excessive amount of twisting applied to the ring, ultimately leading to its breaking, the break appearing as a tight spiral cut around the ring when removed, although the ring may continue to seal satisfactorily up to the point of breaking.

Properly installed, and operating under suitable conditions, an O-ring should not normally tend to roll or twist under reciprocating motions because its contact area with the groove is greater than the rubbing contact area on the sliding surface and the torsional stiffness of the ring section itself inherently resists twisting. The disposition of frictional forces also tends to hold the ring static in its groove (as static friction is greater than sliding friction and the finish of the groove surface is usually rougher than that of the sliding surface).

TABLE 13 – GROOVE DIMENSIONAL DATA FOR RECIPROCATING O-RING SEALS

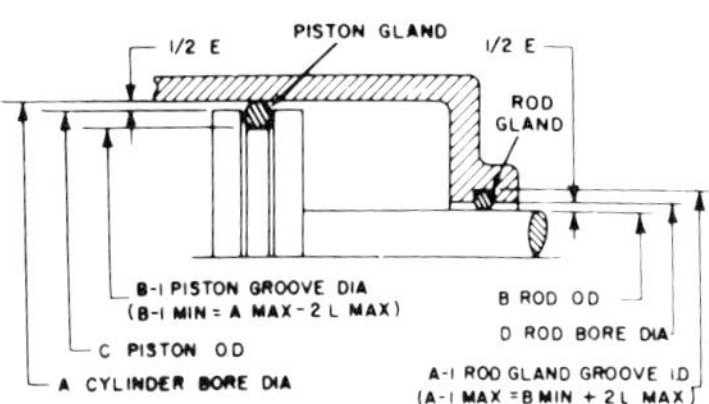

W Cross section		L	Squeeze		E†	G Groove width			R	
Nom.	Actual	Gland depth	Actual	%	Dia-metri-cal clear-ance	No back-up rings	1 back-up rings	2 back-up rings	Groove radius	Eccen-tricity max TIR*
1/16	0.070 ±.003	0.055 to 0.057	0.010 to 0.018	15 to 25	0.002 to 0.005	0.093 to 0.098	0.138 to 0.143	0.205 to 0.210	0.005 to 0.015	0.002
3/32	0.103 ±.003	0.088 to 0.090	0.010 to 0.018	10 to 17	0.002 to 0.005	0.140 to 0.145	0.171 to 0.176	0.238 to 0.243	0.005 to 0.015	0.002
1/8	0.139 ±.004	0.121 to 0.123	0.012 to 0.022	9 to 16	0.003 to 0.006	0.187 to 0.192	0.208 to 0.213	0.275 to 0.280	0.010 to 0.025	0.003
3/16	0.210 ±.005	0.185 to 0.188	0.017 to 0.030	8 to 14	0.003 to 0.006	0.281 to 0.286	0.311 to 0.316	0.410 to 0.415	0.020 to 0.035	0.004
1/4	0.275 ±.006	0.237 to 0.240	0.029 to 0.044	11 to 16	0.004 to 0.007	0.375 to 0.380	0.408 to 0.413	0.538 to 0.543	0.020 to 0.035	0.005

†It is desirable to hold this to a minimum consistent with maintaining positive clearance over full working temperature range.

All dimensions in in

*Total Indicator Reading between groove and adjacent bearing surface

Note: for 1500 lb/in^2 maximum without back-up rings

This system can be upset, and spiral failure brought on, however, by conditions such as excessive sliding friction due to too much squeeze or lack of lubrication; an unsatisfactory groove shape allowing the ring freedom to roll or reducing contact area; high temperatures increasing squeeze and friction; a poor or damaged rubbing surface finish promoting uneven friction and wear; eccentricity; the presence of side or bending loads; and so on. When spiral failure does occur the cause is more likely to be a combination of factors rather than a single factor, but the basic cause is twisting of the ring.

In particular applications, working pressures up to 350 bar (5000 lb/in^2) may be accommodated with back-up rings, with particular attention to material choice and tolerances.

As far as possible the seal should be employed so that the direction of pressure and rubbing friction are in opposition. If the two act in the same direction the tendency for the O-ring to extrude will be much more marked and could reduce the effective pressure rating quite substantially. If the groove cannot be located to reverse the direction of friction, then a back-up ring might be employed to advantage at pressures of 70 to 100 bar (1000 to 1500 lb/in^2).Back-up rings can also assist in maintaining good lubrication of the O-ring.

Shock pressures can be particularly damaging as the momentary pressure may be many times greater than the normal maximum system pressure, with distinct possibilities of damage to the seal. Again this can be guarded against by fitting a back-up ring or rings; although mechanical locks or a relief valve in the fluid system may be alternative solutions.

It is often not appreciated that a low pressure can sometimes be as damaging as excessive pressure, particularly if accompanied by a slow leakage. This can give the equivalent of very slow motion and high friction, as far as the O-ring is concerned, which can lead to twisting and to spiral failure. In general, a high pressure of above 35 bar (500 lb/in^2) tends to eliminate any twisting of the O-ring.

Short stroke applications are generally to be preferred for O-ring seals because these will minimize side and bending loads and eccentricity of the moving system, any of which can result in high unit loads. Alternatively, design may be tackled on the basis of employing a *floating gland* or a definite metallic bearing surface.

A typical floating gland is shown in Figure 12 applied to both the piston and rod. This allows a degree of flexibility or float to counteract eccentricity or misalignment with a comparatively simple and inexpensive solution. Metal-to-metal contact is avoided and high bearing loads are eliminated.

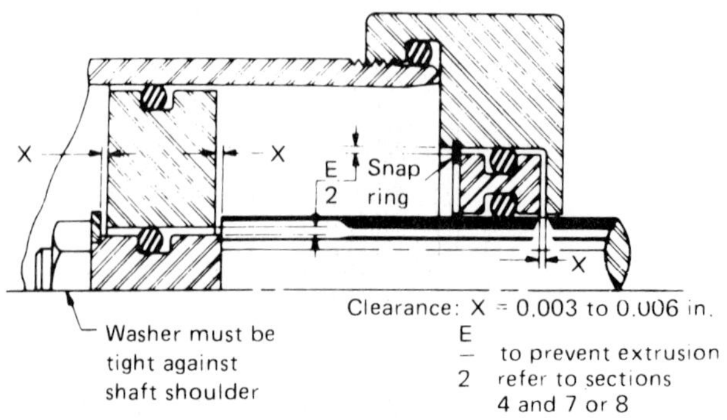

Figure 12

Figure 13 shows an alternative solution where both the piston and rod gland are provided with whitemetal bearings and the O-rings are retained for sealing. This can provide a particularly suitable arrangement for heavy duty applications or high pressure work (in which case the O-rings would be used with back-up rings).

Low temperature working presents a particular problem for reciprocating O-ring seals because the contraction of the ring at working temperature may be considerable and to close the clearance to compensate could cause metal-to-metal rubbing contact.

Choosing an O-ring material with less contraction or increasing the squeeze are the simplest solutions to eliminate leakage at very low temperatures, although spring loading the ring (Figure 14) may be a better solution.

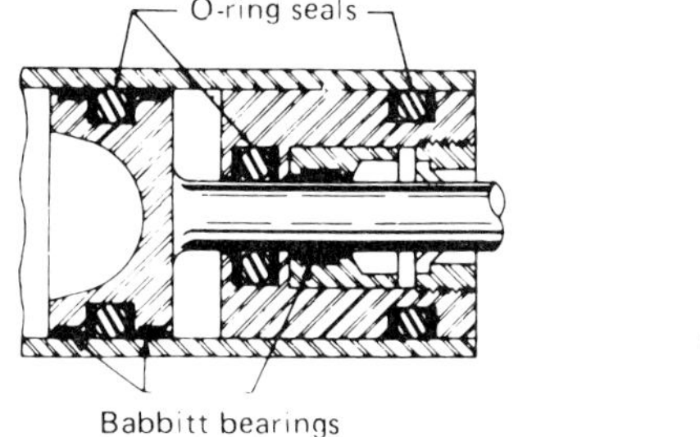

Figure 13

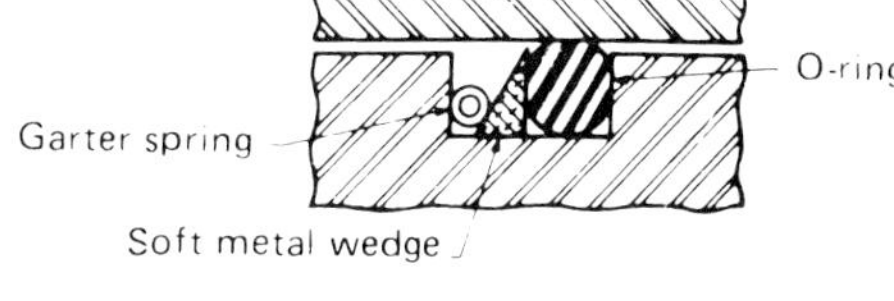

Figure 14

Other common faults with reciprocating O-ring seals, and their cures, are summarized in Table 14. It should be remembered, however, that a small amount of leakage with a reciprocating seal is generally desirable when sealing a fluid since this indicates that the seal is not running (with consequent high friction and wear).

Oscillating seals

An *oscillating* seal is distinguished from a *rotary* seal in that the moving elements describe an arc in one direction of rotation and then the opposite direction, rather than a continuous rotary motion. This oscillation may or may not be accompanied by a sliding or longitudinal motion, but this can generally be ignored as far as seal design requirements are concerned. Here it should be noted that any axial motion, for example, that is caused by a spiral thread, does not change the classification of the seal, if the motion is only of a minor nature.

Where the surface speeds involved are under 1 m/s (200 ft/min), groove requirements can be taken as the same as for reciprocating O-ring seals. For higher speeds (for which O-rings are not generally suitable) it may be necessary to regard the O-ring as a rotary seal. The same considerations apply when the oscillation is continuous motion, although usually the requirements are far less stringent than those of true rotary seals.

By contrast, a seal combining extensive oscillating and reciprocating motion, would be classified as an oscillating-reciprocating seal (Figure 15c).

Slip-ring seals (slipper seals)

A *slipper* seal comprises a standard O-ring associated with an anti-friction ring located on the rubbing side. The ring is normally solid PTFE and can be of various shapes,

TABLE 14 – O-RING SEAL FAULTS AND CAUSES

Fault	Possible cause and/or remedy
Small leakage	Cut or damage during installation. Insufficient squeeze – increase squeeze. Wrong groove dimensions – check for excessive groove width. Poor surface finish on rubbing surface. Side loading – check for eccentricity or side loading.
Large leakage	O-ring failure in service. Badly scored rubbing surface. Deterioration of O-ring due to fluid attack. Excessive shrinkage of O-ring. Excessive differential expansion.
Leakage at low temperatures	Wrong O-ring compound used. Insufficient squeeze – increase squeeze to compensate for thermal contraction.
Excessive friction	Excessive squeeze – check groove proportions. Excessive swell – check compatibility. Metal-to-metal contact – check for out of alignment or excessive differential expansion. Extrusion – O-ring may need back-up ring(s).
Early failure of ring	Poor groove design – check particularly that ring is not unduly stretched. Damage during assembly. Excessive stretch during assembly. Excessive squeeze – check clearance. Wrong O-ring size – use larger cross-section.

eg Figure 15. Matching groove depth is increased slightly over standard size, both the O-ring and the slipper ring being retained by the groove. This means that the slipper ring has to be stretched into position, either with a special tool or by heating to cause it to expand, returning to its normal size on cooling.

Slip-ring seals are suitable for hydraulic applications up to 350 bar (5000 lb/in^2) as a more compact alternative to back-up rings. They also have particular application to pneumatic systems for rod and cylinder seals because of their low friction, *eg* Figure 16.

O-ring rotary seals

O-rings have only a limited application as rotary (shaft) seals, although used within their limitations and with particular attention to clearance, side-thrust and end play they can give satisfactory performance up to surface speeds of the order of 8 m/sec (1500 ft/min).

Specifically, the O-ring needs to be located in a groove in the mating member to the rotating shaft. If the shaft is grooved the O-ring will tend to expand as the shaft rotates and develop high rubbing wear by slipping in the groove. The same applies to other rotary applications. The O-ring groove should be located in the stationary member (Figure 17).

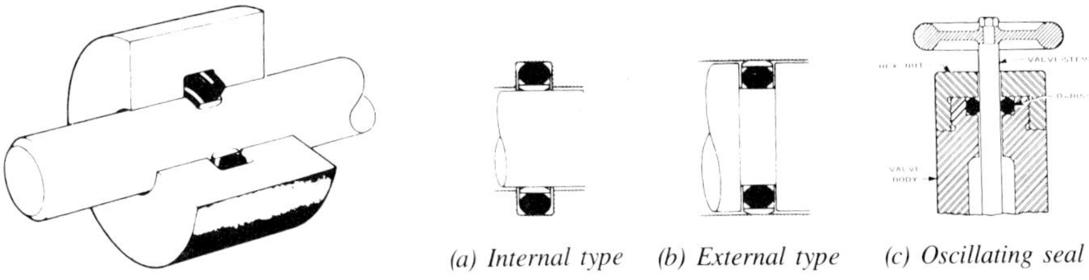

(a) Internal type (b) External type (c) Oscillating seal

Figure 15
Slipper O-ring.

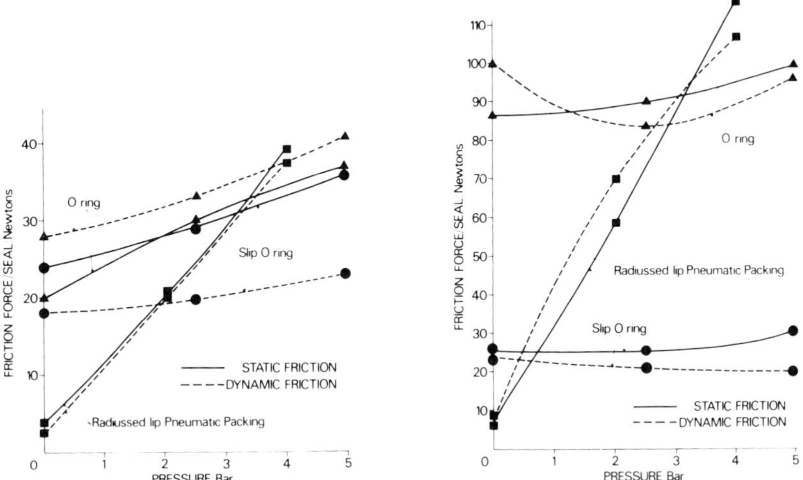

Figure 16
Comparative friction values of differing static and pneumatic lubricated (left) and dry (right).

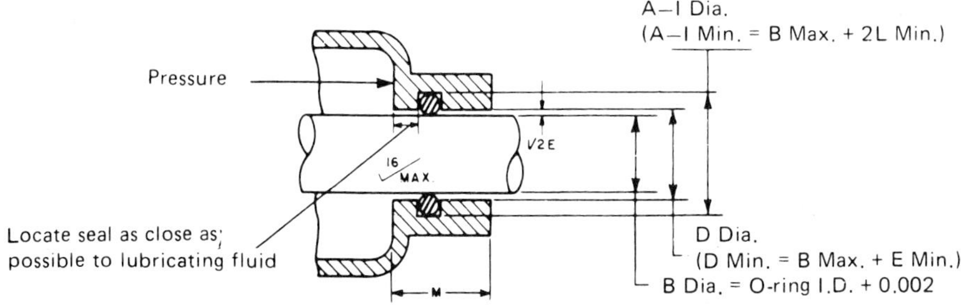

Figure 17

It is then important that this groove or method of retaining secures the O-ring with no tensional stress as, if heated in a stressed condition (by friction or similar method) the ring will tend to contract. This could cause it to seize on the shaft, although this is not always applicable at surface speeds below 1 m/s (200 ft/min).

To reduce seal friction, and minimize wear, the O-ring should be fitted only as a seal and bearing loads should be carried by separate bearing surfaces. This will enable the O-ring to operate with the necessary limited squeeze for satisfactory performance. On the other hand this places a premium on bearing and shaft tolerances to eliminate eccentricity. O-ring rotary shaft seals can prove troublesome or impossible to maintain when lack of concentricity is greater than about 0.127 mm (0.005 in).

Recommended fitting data are summarized in Table 15. Groove size is designed to prevent rotation of the seal, and as a general guide the most satisfactory performance is usually given by the smallest O-ring section (diameter) available for the diameter of seal required – although this is not usually critical for surface speeds of less than 1 m/s (200 ft/min). For surface speeds of 1 to 2 m/s (200 to 400 ft/min) a maximum O-ring section (diameter) of 3.55 mm (0.139 in) is recommended; decreasing to 2.65 mm (0.103 in) for higher speeds or even less, 1.78 mm (0.070 in) for speeds above 3 m/s (600 ft/min).

TABLE 15 – O-RING GROUP DESIGN DATA FOR ROTARY APPLICATIONS
(See also Figure 17)

W Cross-Section			L	G	E†		M	R
Nominal	Actual	Maximum Rubbing Speed ft/min	Gland Depth	Groove Width	Diametrical Clearance	Eccentricity Maximum TIR*	Bearing Length Minimum	Groove Radius
1/16	0.070 ± 0.003	1500	0.065 to 0.067	0.075 to 0.079	0.012 to 0.016	0.002	0.700	0.005 to 0.015
3/32	0.103 ±0.003	600	0.097 to 0.099	0.108 to 0.112	0.012 to 0.016	0.002	1.030	0.005 to 0.015
1/8	0.139 ±0.004	400	0.133 to 0.135	0.144 to 0.148	0.016 to 0.020	0.003	1.390	0.010 to 0.025

† For maximum working pressure of 500–750 lb/in^2. For higher pressures clearances would have to be reduced.
* Total Indicator Reading between the groove o.d., shaft and adjacent bearing surface.

The satisfactory performance of the O-ring as a rotary seal will depend almost entirely on its working temperature. Regardless of material choice, it is unlikely to be suitable for ambient temperatures below about −30°C (−20°F) or above 95°C (200°F), especially for higher surface speeds. The fact that in addition the O-ring may receive heating from adjacent bearings as well as frictional heating must also be considered, as this may call for separation of the bearing from the O-ring as much as possible or, if this is impractical, for the use of a bearing long enough to promote adequate heat dissipation.

In some cases it may be desirable, or necessary, to spring load a rotary seal in order to obtain satisfactory sealing. Two possible methods are shown in Figure 18 using, respectively, a soft rubber back-up ring and a soft O-ring. Either method is preferable

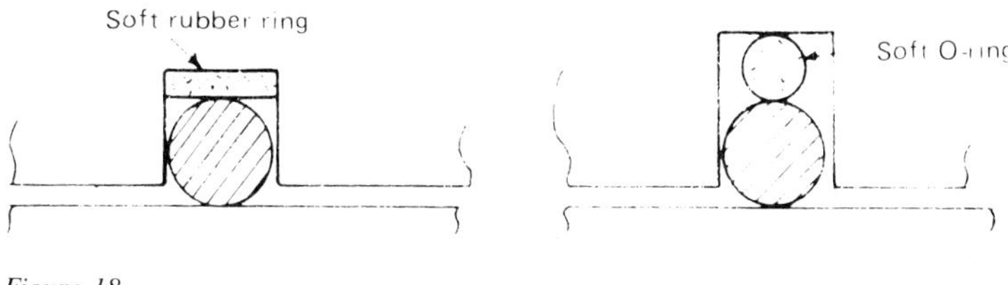

Figure 18

to using a mechanical type spring. Some success has also been achieved using an inclined groove to locate the O-ring, where the preload pressure provides side-thrust.

Encapsulated O-rings

Covering a rubber O-ring with PTFE can provide two advantages. The ring is now given the chemical resistance of PTFE, but retains the flexibility and resilience of the original rubber ring. A high rubber to PTFE ratio (*ie* a thin PTFE covering film) increases the contact area in both static and dynamic operating conditions, reducing the possibility of compression set failure. Used as a dynamic seal, the PTFE surface also materially reduces friction, and particularly breakout friction.

The PTFE covering may be applied in several ways, *eg* as a bonded on split sleeve, overlapping sleeve, or as a flowed on coating. In larger ring sizes (*eg* more than 400 mm diameter), the covering may take the form of endless welded PTFE tubing (see also Figure 19). A further, distinctive type of PTFE enveloped O-ring is shown in Figure 20.

Physical properties of O-ring compounds

Elastomeric compounds now available are vastly better than natural rubber or Neoprene, and as a result, almost all liquids or gases can be handled.

There are, however, still some fluids that are chemically destructive to elastomers. Others have adverse effects on physical properties, so that an acceptable seal life cannot be guaranteed.

Hardness

This is measured using a standard instrument called a SHORE A DUROMETER.

Hardness

This is measured using a standard instrument called a Shore A Durometer.

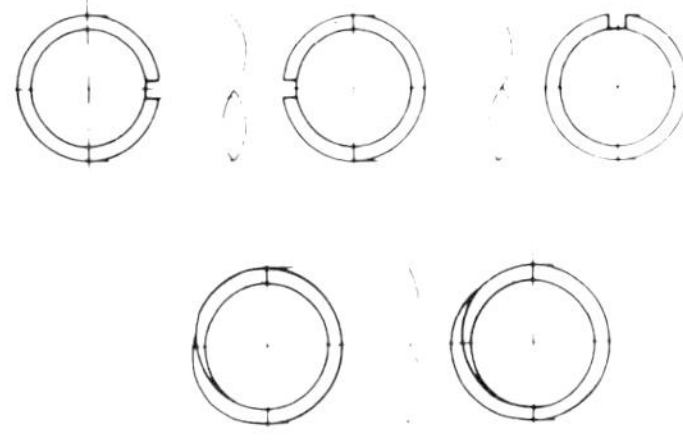

Figure 19
Cross sections of typical PTFE enveloped O-rings.

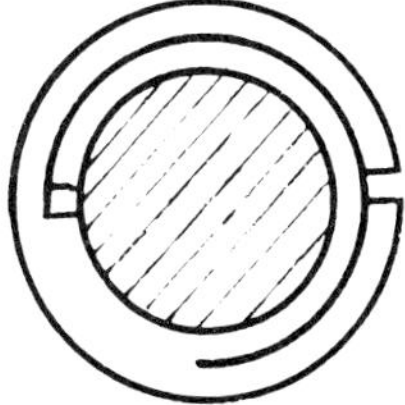

Figure 20
M1 type O-ring with PTFE envelope.

Durometer

The instrument measures indentation hardness, or resistance of the elastomer to the penetrating force of a spring-loaded pin. Durometer hardness is given in increments of 10 points, with a tolerance of 5 points.

Figure 21 shows the force necessary to compress 3.53 mm (00.139 in) cross-section rings per 25 mm (1 in) of circumference.

Figure 22 shows how the hardness of various polymer compounds decreases on initial heating.

Continuous service at a given temperature will increase the hardness.

In some cases, back-up rings will be needed.

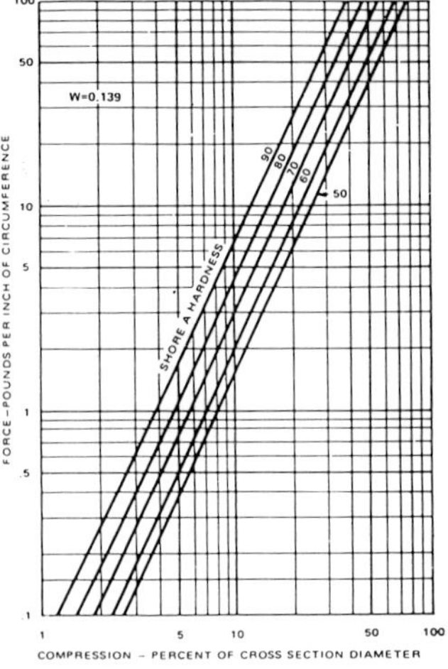

Figure 21
O-ring compression force

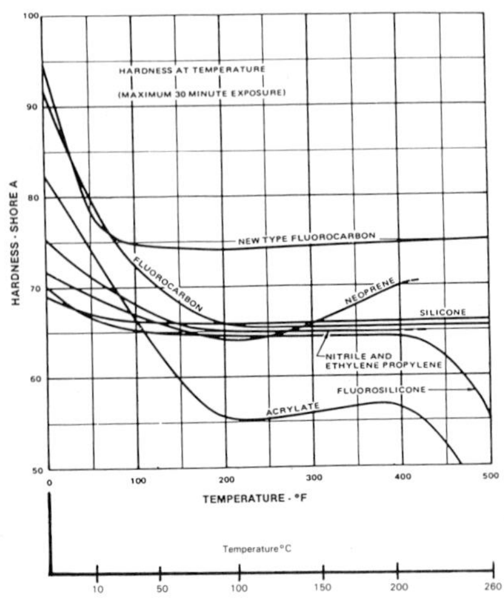

Figure 22
Hardness of elastomers

Hollow Metal Ring Seals

HOLLOW METAL O-rings are made from thin-walled seamless tubing lengths bent to circular form and the two ends butt welded. They are particularly effective as static seals at temperatures higher than those which can be accommodated by non-metallic seals or gaskets and are produced in three different forms:

(i) *Plain* rings which are suitable for groove fitting (fully confined) or flange fitting (semi-confined).

(ii) *Self-energizing* (or *system energized*) rings incorporating holes or slots so that the internal pressure is the same as the system pressure sealed by the ring.

(iii) *Pressure-filled* rings which are plain rings filled with an inert gas (usually nitrogen) at pressures of the order of 42 bar (600 lb/in^2).

Sealing action of a hollow metal O-ring is produced by compressing the ring to a predetermined height equivalent to a section compression of 20 to 25%. The resilience of the metal provides some compensation for surface roughness or lack of parallelism of the surfaces, but the actual sealing properties can be enhanced by a softer surface coating applied to the ring. Such coatings may also be used to provide resistance to chemical attack.

Further control of working properties can be provided by using different tubing wall thicknesses. Thus thin walled tubing can be used for rings where low flange loading is required, although at the same time this will limit the pressure rating of the ring to the order of 420 bar (6000 lb/in^2) depending on cross section diameter and actual wall

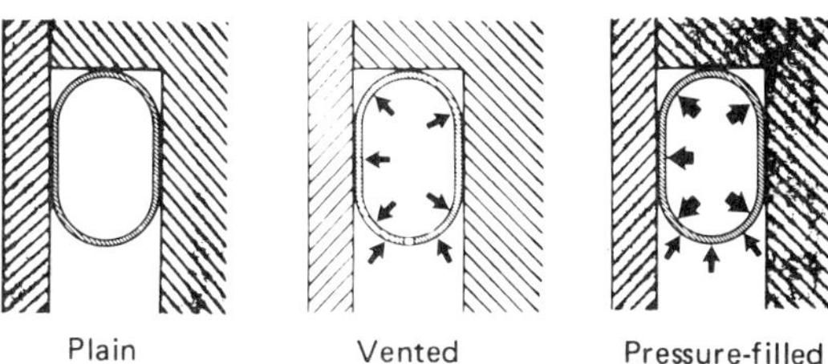

Figure 1
Basic types of hollow metallic O-rings.

thickness. Also some metals (notably aluminium) are not suitable for thin wall tubing rings. Equally, for sealing higher pressure, thick walled tubing can be used for making the rings, at the expense of requiring higher flange loadings. These may be capable of accommodating pressures well above 400 bar (6000 lb/in^2). Normally, however, *vented* rings are preferred to thicker wall rings for sealing pressures above 70 bar (1000 lb/in^2). Some typical production sizes are shown in Table 1.

TABLE 1 – TYPICAL PRODUCTION SIZES OF METALLIC O-RINGS

Ring cross-section		Thick wall section		Thin wall section	
mm	(in)	mm	(in)	mm	(in)
1.59	(1/16)	0.356	(0.014)	0.254	(0.010)
2.38	(3/32)	0.457	(0.018)	0.254	(0.010)
3.18	(1/8)	0.508	(0.020)	0.254	(0.010)
3.97	(5/32)	0.635	(0.025)	0.254	(0.010)
4.76	(3/16)	0.813	(0.032)	0.254	(0.010)
6.35	(1/4)	1.016	(0.040)	0.508	(0.020)
7.94	(5/16)	1.270	(0.050)	–	–
9.53	(3/8)	1.524	(0.060)	0.508	(0.020)

Metals used range from mild steel, copper and aluminium to stainless steels, nickel, monel, inconel and tantalum. Plated coatings commonly used are cadmium, copper, nickel, silver, gold and indium. PTFE coatings may be used for corrosion resistance but in this case the service temperature of the ring is reduced.

Plated or *coated* rings require slightly increased groove dimensions to accommodate the increase in size of the metal ring. Table 2 summarizes maximum service temperature for different ring and plating materials.

Pressure-filled rings are particularly designed for high temperature working, internal pressure increasing with temperature and compensating for loss of ring material strength at such temperatures, as well as increasing the resilience of the seal. Self-energizing (vented) rings are superior to the other two types for high pressure applications.

Uncoated rings are suitable for the majority of liquid sealing applications, but coatings are needed for some liquids and most gas sealing applications. Satisfactory sealing

TABLE 2 – MAXIMUM SERVICE TEMPERATURES FOR METAL O-RINGS

Ring			Plating		
	Maximum temperature			Maximum temperature	
Material	°C	°F	**Coating**	°C	°F
Copper	400	750	Indium	130	270
Mild steel	550	1020	Cadmium	200	390
Cupro-nickel	600	1100	PTFE	300	570
Monel	600	1100	Silver	800	1470
Nickel	700	1290	Gold	850	1560
Stainless steel	800	1470			
Inconel	850	1560			

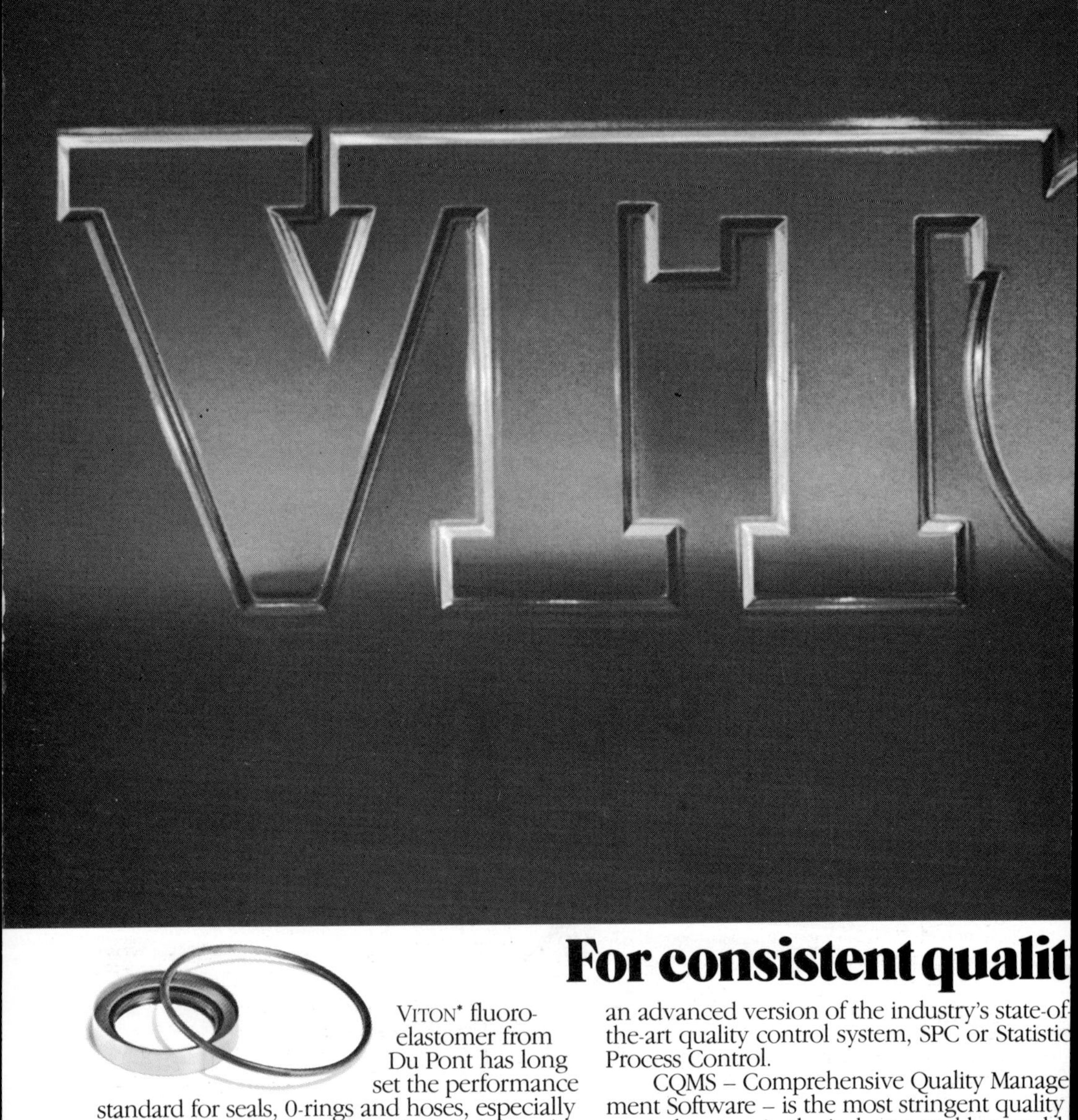

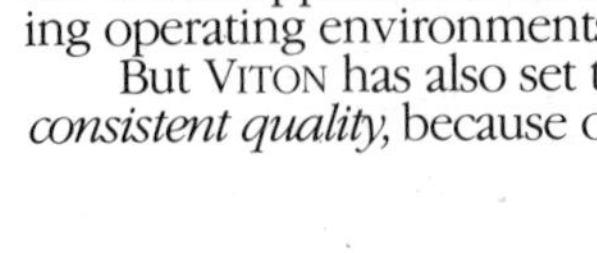

and high performance ... this is the sign.

What this means to you is unsurpassed consistency of performance, part after part, year after year. And that's why VITON remains the natural choice for automotive, aerospace, defence, power generating, chemical and other industries.

If you would like to know more about how products based on VITON high-performance fluoroelastomer can help you achieve (and even exceed) your design goals, call (0442) 61251, or write to Du Pont (U.K.) Ltd., Elastomers and Ethylene Polymers, Maylands Avenue, Hemel Hempstead, Herts HP2 7DP.

*Du Pont's registered trademark.

performance is also related to the surface finish. Typical requirements for surface finish are: 0.8 μm (32 μin) for sealing high viscosity liquids with unplated rings or for sealing medium viscosity liquids with plated rings; 0.8 to 0.4 μm (32 to 16 μin) for sealing low viscosity liquids and gases using plated rings; 0.4 μm (16 μin) or better for sealing hydrogen or helium gas with plated rings.

Hollow metal O-rings can also be produced in a variety of shapes other than purely circular rings, limited only by the corner radii, *eg* see Figure 2. Typically such limits range from about 4 × the tube size for smaller tubes sizes (*eg* 0.035 in) up to 25 × the tube size for tube diameters greater than 0.5 in.

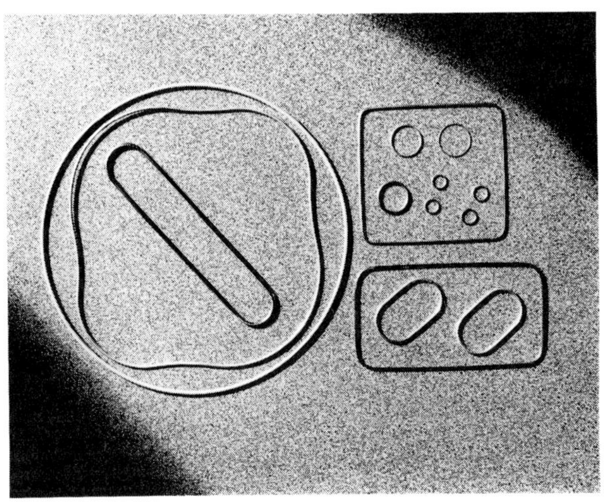

Figure 2
Special hollow O-ring shapes.

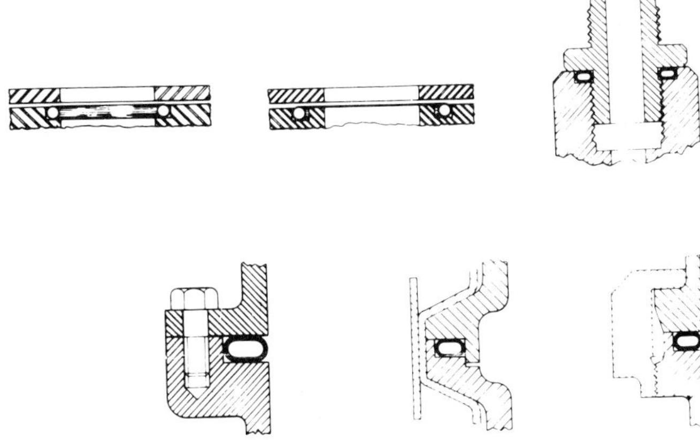

Examples of hollow metallic O-ring installations.

A further feature of this type of configurated seal is that it can be preformed to fit contours which are not in the same plane.

Instead of being installed in rectangular grooves metallic O-rings are often located in opposed triangular grooves. When the joint is closed this provides greater sealing contact area than when a hollow metal O-ring is deformed in a rectangular groove (Figure 3).

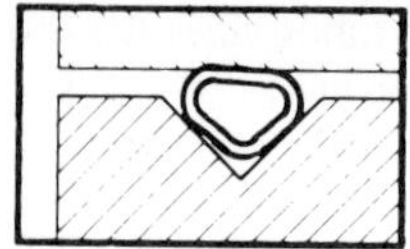

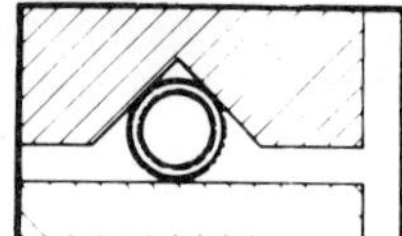

Figure 3

Other metallic rings

Various other sections are employed for *hollow metallic* rings apart from circular (O-rings). They are normally plain rings (*ie* not vented or gas-filled).

Oval metal rings are a simple modification of the hollow metallic O-ring, to fit deeper, narrower grooves. Better sealing in the groove itself is claimed. Diamond rings and double-diamond rings are normally accommodated in deeper grooves than O-rings, the latter providing sealing at six contact points.

Configurated metal rings

Configurated rings are another type of metal gasket with self-energizing properties. They are based on small, resilient cross-sections such as a C, U, V, W, and other variants with the open part of the section facing the system pressure when the joint is closed, *eg* see Figure 4. Seal contact pressure increases with system pressure up to the point where the section yields under hoop stress if unsupported in a groove, or by excessive distortion or extrusion of the unsupported section(s) if housed in a groove.

Configurated metal rings normally require only light clamping forces and can combine the effectiveness of an elastomeric O-ring with the high temperature capabilities of a metal gasket. Materials used are similar to those for hollow metal O-rings, and again coatings are commonly added. Surface finish is not as critical as with hollow metal O-rings, a finish of 0.8 μm (32 μin) being generally suitable for uncoated rings and 0.8 to 2.25 μm (32 to 90 μin) for coated rings.

Figure 4

Metal C-rings (Figure 5)

Metal *C-rings* can be dimensioned to be interchangeable with metallic O-rings of the same free height, provided the groove depth permits a compression of 15 to 20% of the original free height of the C-ring. It is, in effect, a 'fully vented' O-ring, self-energizing under pressure (the open side of the ring always facing the higher pressure side of the system).

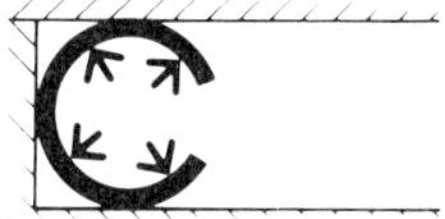

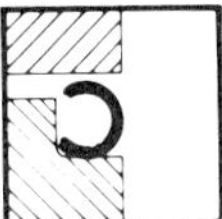

Figure 5

A particular advantage offered by the metallic C-ring is good spring-back, providing a consistently high level of sealing performance under pressure cycling, stretching of the flange bolts, or flange distortion.

C-rings are made in similar materials to metallic O-rings, both in plain metals and with coatings.

Metal V-rings (Figure 6)

A *V-ring* section in metal provides a robust seal capable of withstanding a wide range of temperature and/or pressure extremes. It is particularly suited to applications where flange loadings are relatively low, but for satisfactory sealing this demands a relatively high degree of surface finish on the mating surfaces. The rings themselves are normally precision finished. They are usually available in standard configurations matching recommended groove sizes, but any of the standard configurations can usually be modified in height to fit existing groove depths if necessary.

Metallic U-rings

The distinction between the section geometry which defines a metal ring as a V-ring or a *U-ring* is a fine one. Both types are really similar and have similar properties and applications. In fact, most so-called metal V-rings are more realistically U-ring form in section geometry, characterized by a thickened heel section rather than a plain V-section.

The major application for both metallic V-rings and metallic U-rings is for high temperature or arduous duty sealing which would normally favour the choice of a metallic O-ring, but where it is desirable to use a lower flange load than would be necessary for adequate sealing performance with a metallic O-ring.

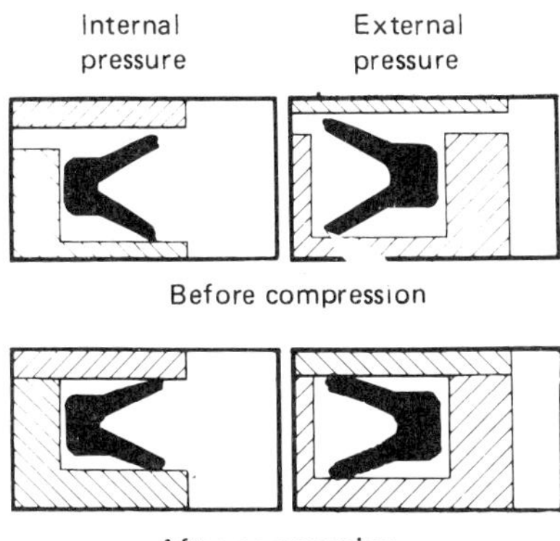

Figure 6

Helicoflex seals

Helicoflex seals are (patented) flexible metal O-rings with an elastic core capable of giving absolute sealing from vacuum pressures of 10^{-10} torr up to 690 bar (10000 lb/in^2) over a temperature range of −250°C to +750°C. Compression and elastic recovery properties of the seal are exceptional, due to the unique construction.

In its basic form, the *Helicoflex* seal comprises one or two metal linings around the toroidal section of a helically wound spring (Figure 7). The sealing principle is based on the plastic deformation of a lining of greater ductility than the materials surrounding it, between a flat surface constituted by the sealing face of the seal and an elastic core of a helical spring with its coils compressed against each other. Permanent contact of the seal against the sealing face is ensured because the compressive stress on the lining is offset by the radial elasticity of the spring. This spring, with its coils compressed against each other, gives total independence to each coil during radial compression of the section, which in turn energizes the coil edgewise.

Helicoflex seals are produced in sizes ranging from 4 mm to 8000 mm nominal diameter and 1.5 mm to 25 mm section diameter. There are two main types: one for assembling in grooves with metal to metal tightening and one for assemblies with plain flanges and used with an external or internal limiter (Figure 8). Various planform shapes can also be produced, each retaining the circular cross section (Figure 9).

Particular applications of *Helicoflex* seals are: in nuclear and conventional power stations, as boiler seals; in petrochemical refineries; and for ultra high vacuum duties.

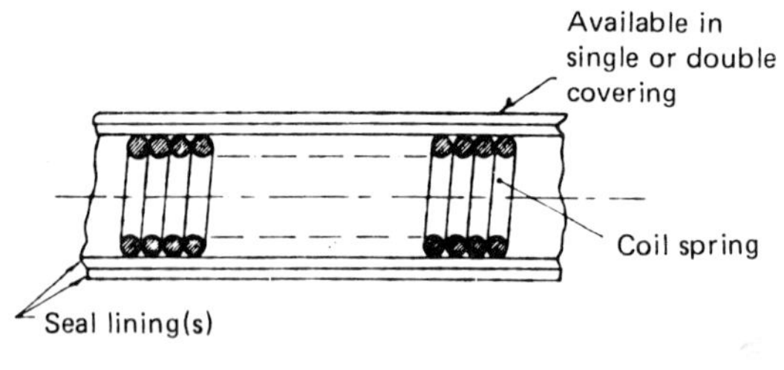

Figure 7

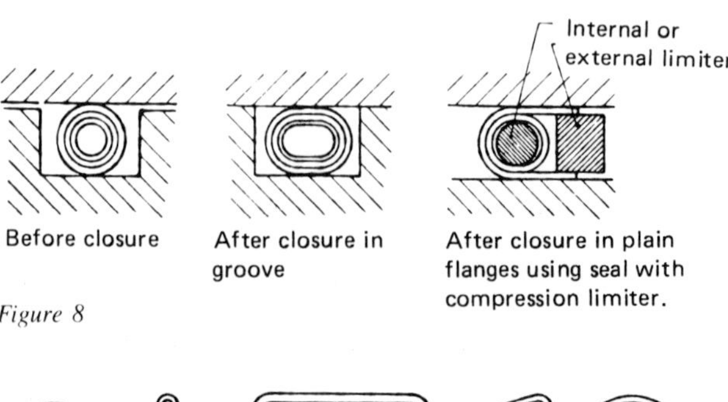

Figure 8

Figure 9

Other Solid Elastomeric Rings

Rectangular rings

THE RECTANGULAR section ring is stable under all pressure conditions, with less mechanical back-lash than an O-ring with pressure cycling (Figure 1). Also it cannot roll or twist like an O-ring, and can resist higher pressures than an O-ring. Actual pressure rating is up to 200 bar (3000 lb/in^2), which may be increased up to 370 bar (5400 lb/in^2) with the addition of back-up rings in certain applications. Back-up rings may, however, be recommended for both static and dynamic applications where the pressure exceeds 100 bar (1500 lb/in^2).

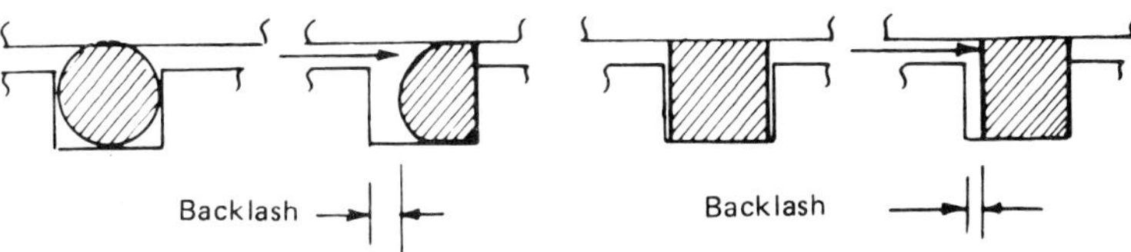

Figure 1
Rectangular ring has less back-lash than O-ring in the same size groove.

Friction is higher than with an O-ring, although this can be reduced if the dynamic sealing edges are chamfered (Figure 2). It is best suited for low speed reciprocating applications. A practical advantage with this form of ring is that it can be cut from sheet or tube stock to adjust the dimensions to any size. Thus, if an O-ring groove has been machined incorrectly a rectangular section ring can be machined to fit the groove.

There is no standard size range for rectangular section rings and the cost compared with O-rings is about 66% for similar sizes. Rectangular rings are normally proportioned to be accommodated in standard O-ring grooves. *Square-section* rings (which are merely a geometric variant) may require slightly different groove proportions for optimum performance.

Rectangular rings can be used as external and internal seals, *eg* for both piston head and rod seals in hydraulic cylinders. The back-up ring configuration for use with a rectangular piston head seal to give an efficient double-acting piston head assembly can vary in a number of ways (Figure 3).

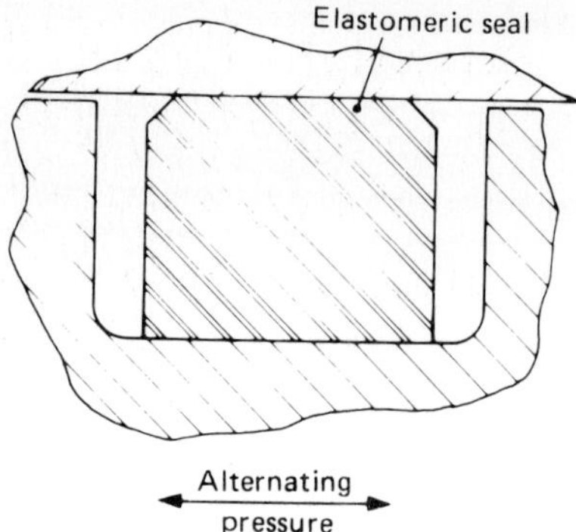

Figure 2
Rectangular ring. Friction can be reduced by chamfering the edges of the rubbing surface, as shown.

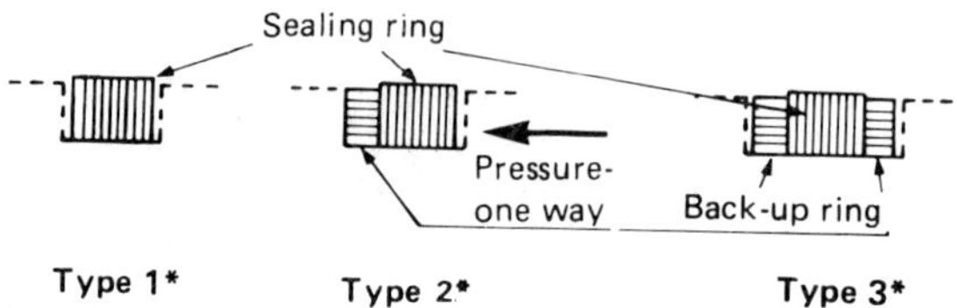

Figure 3
Nominal designations for the purpose of Table 1.

Examples of proprietary standard sizes of rectangular rings with recommended groove proportions and surface finishes are given in Table 1. Table 2 then gives the radial squeeze dimensions with these installations.

Rectangular (and square) *section* rings may also be made in other materials than rubber, *eg* cork or felt, for use as wipers or lubricating rings. The sealing performance with such materials is strictly limited.

Tetraseals

These are lathe-cut, square cross-sectional seals, equivalent to standard AS568A O-ring seals, and are designated by the AS568 dash number. Thus a *Tetraseal-214* is equivalent to an AS568-214 O-ring.

In non-standard form, they may also be used in flange applications and as face seals (Figure 4).

Definition

The majority of compression-type seal applications are static, *ie* the rigid components of the seal gland do not move in relation to each other or to the seal. The only movement of the seal is caused by vibration or fluid pressure within the gland.

TABLE 1 – PRECISION RECTANGULAR SECTION SEALS – GROOVE AND RECESS SIZES

Surface finish numbers represent μ in CLA

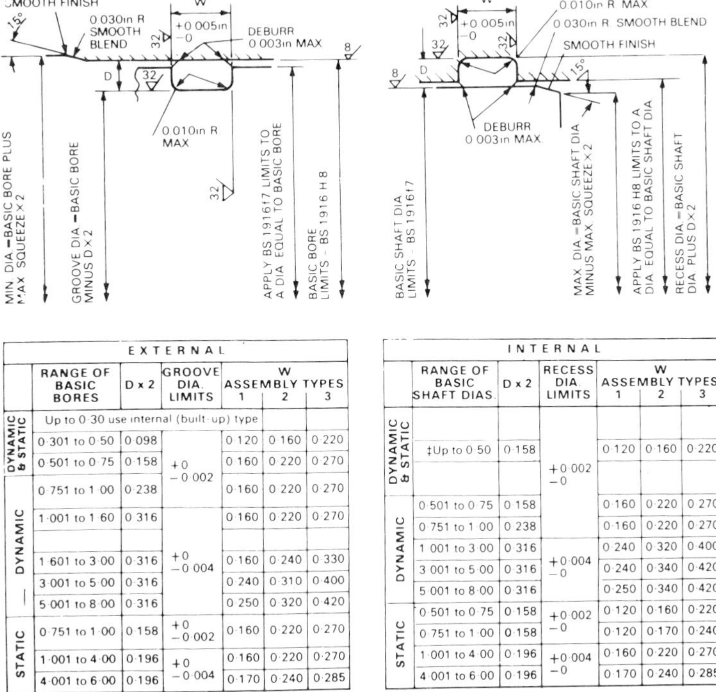

EXTERNAL						
	RANGE OF BASIC BORES	D x 2	GROOVE DIA. LIMITS	W ASSEMBLY TYPES 1	2	3
DYNAMIC & STATIC	Up to 0·30 use internal (built-up) type					
	0·301 to 0·50	0·098	+0 −0·002	0·120	0·160	0·220
	0·501 to 0·75	0·158		0·160	0·220	0·270
DYNAMIC	0·751 to 1·00	0·238		0·160	0·220	0·270
	1·001 to 1·60	0·316	+0 −0·004	0·160	0·220	0·270
	1·601 to 3·00	0·316		0·160	0·240	0·330
	3·001 to 5·00	0·316		0·240	0·310	0·400
	5·001 to 8·00	0·316		0·250	0·320	0·420
STATIC	0·751 to 1·00	0·158	+0 −0·002	0·160	0·220	0·270
	1·001 to 4·00	0·196	+0 −0·004	0·160	0·220	0·270
	4·001 to 6·00	0·196		0·170	0·240	0·285

INTERNAL						
	RANGE OF BASIC SHAFT DIAS.	D x 2	RECESS DIA. LIMITS	W ASSEMBLY TYPES 1	2	3
DYNAMIC & STATIC			+0·002 −0			
	‡Up to 0·50	0·158		0·120	0·160	0·220
DYNAMIC	0·501 to 0·75	0·158		0·160	0·220	0·270
	0·751 to 1·00	0·238		0·160	0·220	0·270
	1·001 to 3·00	0·316	+0·004 −0	0·240	0·320	0·400
	3·001 to 5·00	0·316		0·240	0·340	0·420
	5·001 to 8·00	0·316		0·250	0·340	0·420
STATIC	0·501 to 0·75	0·158	+0·002 −0	0·120	0·160	0·220
	0·751 to 1·00	0·158		0·120	0·170	0·240
	1·001 to 4·00	0·196	+0·004 −0	0·160	0·220	0·270
	4·001 to 6·00	0·196		0·170	0·240	0·285

‡BUILT UP ASSEMBLY REQUIRED

In service, the compressive force exerted by the seal on the mating surfaces ensures a positive seal, even at low pressure. The elastomeric seal material conforms to the mating surfaces, closing off the flow of fluid.

In Figure 4 the Tetraseal is shown installed, compressed to form a seal and under pressure. As system pressure increases, the Tetraseal slides to the low-pressure side of the groove and is supported by the groove wall. The pressure at which the seal moves within the gland is determined by the frictional force of the seal material on the mating surfaces. Only when system pressure exceeds the frictional forces will the seal seat itself on the low-pressure side of the groove.

Under high pressure, the seal acts like a viscous fluid. Following the laws of fluids, the elastomeric material transmits the system pressure to all mating surfaces including the low-pressure side of the groove. In this manner, system pressure ensures consistent sealing.

Face seals

Most Tetraseals are used in static face seal applications, *ie* they are placed in a groove cut into a flange and compressed by a mating flange. Although the seal is circular, it is quite common to fit the seal to non-circular gland applications and thus take advantage

TABLE 2 – RADIAL SQUEEZE ON RECTANGULAR RINGS*

Application	Bore size range	Cross-section (radial) squeeze	
		Min.	Max.
EXTERNAL			
Static and dynamic	0·301 to 0·500	0·0035	0·0080
	0·501 to 0·750	0·0045	0·0090
Dynamic	0·751 to 1·000	0·0050	0·0100
	1·001 to 1·600	0·0085	0·0150
	1·601 to 3·000	0·0100	0·0170
	3·001 to 5·000	0·0130	0·0200
	5·001 to 8·000	0·0160	0·0230
Static	0·751 to 1·000	0·0040	0·0090
	1·001 to 1·600	0·0100	0·0165
	1·601 to 3·000	0·0130	0·0195
	3·001 to 6·000	0·0170	0·0235
INTERNAL			
Static and dynamic	up to 0·500	0·0025	0·0070
Dynamic	0·501 to 0·750	0·0034	0·0077
	0·751 to 1·000	0·0042	0·0086
	1·001 to 3·000	0·0090	0·0145
	3·001 to 5·000	0·0106	0·0163
	5·001 to 8·000	0·0134	0·0191
Static	0·501 to 0·750	0·0034	0·0077
	0·751 to 1·000	0·0052	0·0096
	1·001 to 3·000	0·0088	0·0144
	3·001 to 6·000	0·0134	0·0192

*Conforming to size ranges and dimensions detailed in Table I.

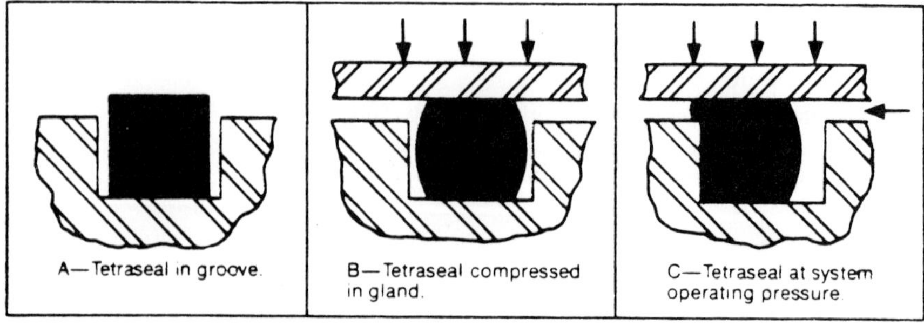

Figure 4

of the seal compound's ability to be deformed within reasonable limits without loss of sealing properties.

To optimize seal effectiveness at all pressures, grooves for static face seals should be designed so that the seal is in contact with the low pressure side of the groove at assembly. Thus, for internal pressure applications, the seal groove is matched to the o.d. of the seal. When pressure is external or from the o.d. side of the seal, the gland should be designed to support the seal on the i.d. (Figure 5).

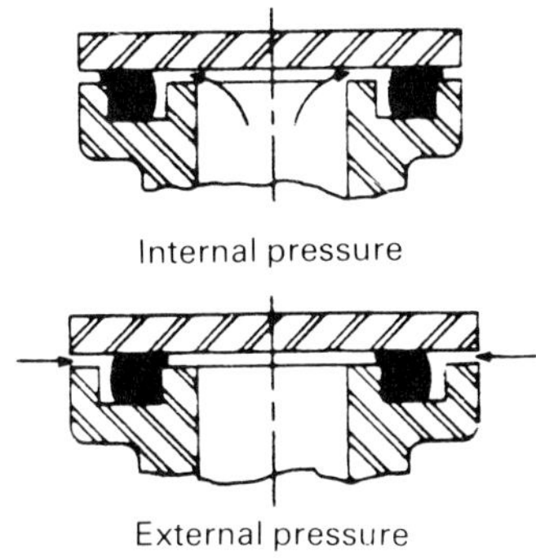

Figure 5

A *flange* seal or *cover* seal on a pressure vessel is typical of internally pressurized seals, as shown in Figure 6.

The general rule for determining the actual groove diameter takes into consideration manufacturing tolerances of the seal amd machining tolerances of the gland itself. Thus, for an internally pressurized seal, the gland o.d. equals the nominal seal o.d. plus a suitable machining tolerance. Conversely, for external pressure sealing, the groove i.d. should equal the nominal seal i.d. minus a machining tolerance.

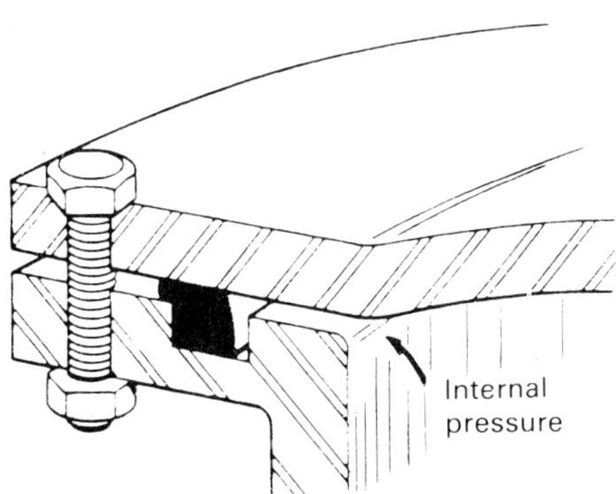

Figure 6
Flange seal.

Trapezoid rings (Figure 7)

Trapezoid-section rings to fit in trapezoid grooves and made in felt are used as lubricating rings for rotary applications, the ring being impregnated or fed with oil. Such rings should not be allowed to dry out since felt is a hygroscopic material which can absorb water and cause corrosion on the shaft on which it bears.

T-section rings (Figure 8)

While still accepted as a type and used in certain applications, the *T-section* ring has limited use. A plain T-section ring is suitable as a light duty rotary seal, but it is more usual to employ a T-ring with back-up rings. The section is then fully supported when deformed under pressure and can accommodate high pressures without extruding, whilst retaining a generous area in contact with the groove for stability. At the same time the rubbing contact area is small so that friction is comparable with that of an O-ring.

Back-up rings used with elastomeric T-rings are best spirally wound to be capable of expanding and contracting with deformation of the T-section under pressure variations.

Delta rings (Figure 9)

The plain *delta-section* ring provides a good contact area in the groove for stability with small rubbing contact area. However, when deformed under pressure, the peak section is relatively thin and prone to extrusion. Thus, although the plain delta-section has low friction its pressure rating is relatively low unless used with a back-up ring (or rings). Although more stable than an O-ring, friction tends to be higher and wear substantially higher.

D-rings (Figure 10a)

The *D-section* ring is basically an O-ring with a square base to provide greater base contact area, thus eliminating twisting and the possibility of spiral failure. Friction is comparable with that of an O-ring, but the section is deformed more at high pressures and requires the use of back-up rings for handling high pressures.

D-rings can make excellent seals for reciprocating motions (*eg* rod seals for both hydraulic and pneumatic applications), or other applications where an O-ring is not satisfactory.

H-rings (Figure 10b)

H-rings have lower friction than O-rings, particularly with low speed reciprocating motions. They are more resistant to spiral failure at moderate pressures, but may require the support of back-up rings at higher pressures to eliminate extrusion. Matching groove sizes for H-rings are normally closer than O-ring fits (or an H-ring to fit an O-ring groove wider than the corresponding O-ring diameter). The ring itself may also be subject to more wear than an O-ring unless the surface finish is better than 10 μin Ra.

Quad rings (Figure 10c)

The *quad* ring is virtually a refinement of the basic H-section ring, with similar comments applying. Quad-section rings have a low back-lash and are particularly effective as an alternative to O-rings for low speed applications. There are further proprietary variants on the H-ring, quad-ring section, such as the *Nu-Lip*, where the section has been 'tailored' to enhance specific performance characteristics.

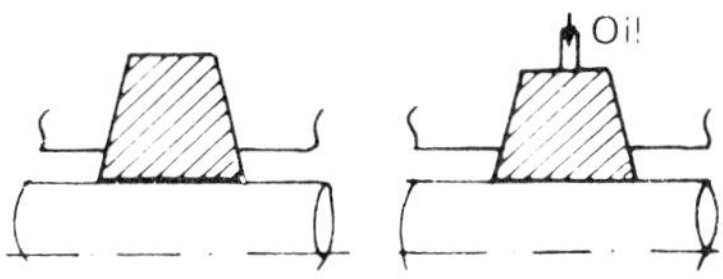

Figure 7
Trapezoid ring.

Figure 8
T-section ring.

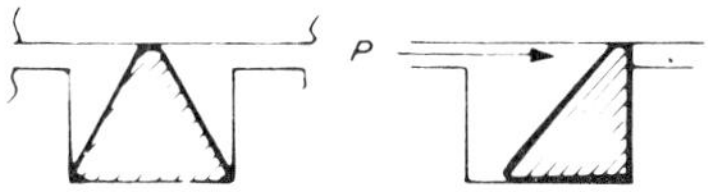

Figure 9
Delta ring.

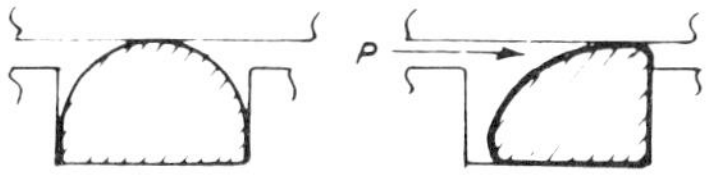

Figure 10a
D-ring.

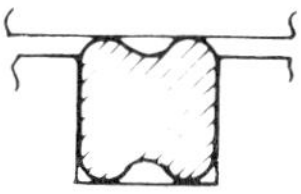

Figure 10b
H-ring.

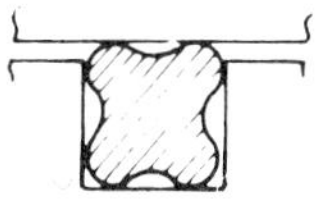

Figure 10c
Quad ring.

Heart rings (Figure 11)

The *heart-section* ring has a very small rubbing contact area and strictly limited sealing ability, although this can be improved by the addition of a back-up ring (or rings). Heart-section rings are only used as a light duty rotary shaft seal with a single back-up ring and possibly a further supporting *wedge-section* ring.

Diamond rings (Figure 12)

The *diamond-section* ring is basically a square or rectangular ring on edge, located in a diamond-shaped groove and has little practical application as an elastomeric ring. It has a low rubbing contact area, but is prone to extrusion under moderate pressure and also subject to spiral failure. The diamond-section is, however, used in *metallic gasket* rings.

Figure 11
Heart seal.

Figure 12
Diamond ring.

Fabric rectangular rings

The Lattytex SG seal (Figure 13) is an interesting variant on the rectangular cross-section seal in that it has a much higher aspect ratio than a conventional rectangular seal and is made in rubber impregnated fabric. The object is to produce a ring seal with controlled expansion and complete resistance to extrusion when fitted in closed grooves. Single seals are suitable for pressures up to 250 bar. For higher pressures two or more rings can be used, fitted at intervals of at least half the seal ring length. Multiple ring sets may also incorporate sintered rings to work as guide rings (and intermediate seals) with the end seals providing the main sealing action.

Lattytex SG seals are normally supplied as cut rings for ease of fitting. These can also be trimmed for smaller diameters. Standard diameter sizes range from 16 mm to 300 mm.

Hollow rings

Hollow rings are sometimes considered as an alternative for O-rings and are usually proportioned to fit standard O-ring grooves. They are basically *open-section* rings with an internal spring, the outer or cover section usually being made of PTFE for low rubbing friction, or fluoroelastomer for maximum chemical resistance. Seals of this type are commonly based on a *V-cup* section with square head, energized by an elastomer (core) or metal spring, *eg* Figures 13 and 14.

†Registered trademark.

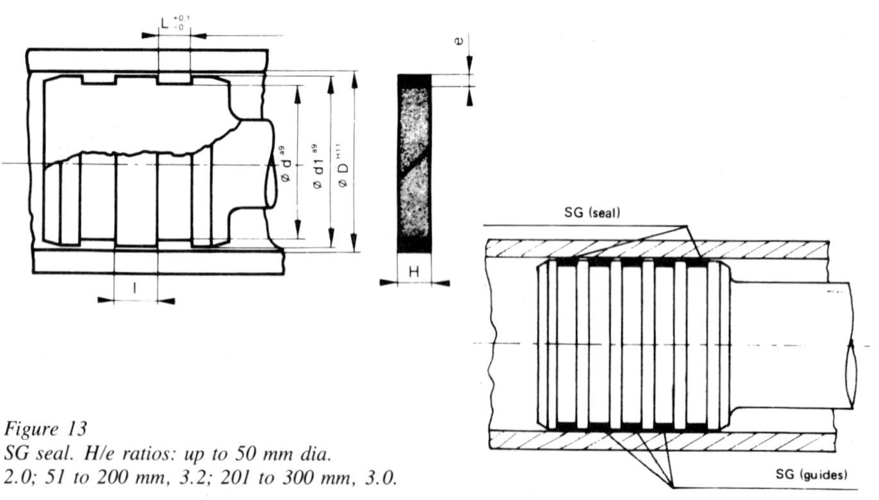

Figure 13
SG seal. H/e ratios: up to 50 mm dia. 2.0; 51 to 200 mm, 3.2; 201 to 300 mm, 3.0.

Figure 14
Tech-Ring spring-energized envelope seal of Teflon†.

Split Ring Seals

SPLIT-RING seals, which are basically a refined piston ring, are suitable for working at much higher temperatures than those possible with elastomeric seals. They are mostly metallic rings (and are also known as metallic ring seals), but increasing interest is shown in split-rings made of PTFE or reinforced PTFE for non-lubricated application and/or where minimal friction is desirable. They can be used both as reciprocating and rotary seals.

It should be noted that the description 'split-ring' may also be used to describe an elastomeric ring seal (*eg* an oil seal or lip seal) which is available in cut or split form to facilitate assembly. Correctly this type is called a split seal, to distinguish from split-rings as a separate category.

Split-rings can be designed to expand or be *outspringing* or to be contracting or *inspringing*. Unlike single lip seals, either can be used as an internal or external seal without inversion of geometry. The three main variants in basic geometry are illustrated in Figures 1a,1b and 1c.

Figure 1a is an outspringing seal working as a piston seal (reciprocating external seal). The seal can be single- or double-acting (as shown). Figure 1b is an inspringing seal as a rod seal (reciprocating internal seal). Figure 1c is an outspringing seal as a rotary shaft seal (rotary external seal).

It is also possible to use an inspringing seal as a rotary shaft seal, assembled in a housing rather than in the shaft (Figure 1d). However, with this configuration centrifugal force will lift the seal at high speeds.

Sealing performance

A ring seals on two faces, the circumferential face pre-loaded by inherent spring pressure (outspringing or inspringing) and one side face bearing against the retaining groove (Figure 2). The latter may be provided by fluid pressure or pressure from a separate spring or similar mechanical means. It should be emphasized that without adequate side pressure, leakage is likely because in its working position there is always a gap between the ring and the bottom of its groove.

Leakage is also inevitable through the gap in the circumference of a single ring. This depends on the size and form of the gap. Gap size needs to be proportioned so that it is the minimum necessary to accommodate expansion when fitted. The gap as manufactured may be positive or negative (*ie* overlap), depending on whether the ring

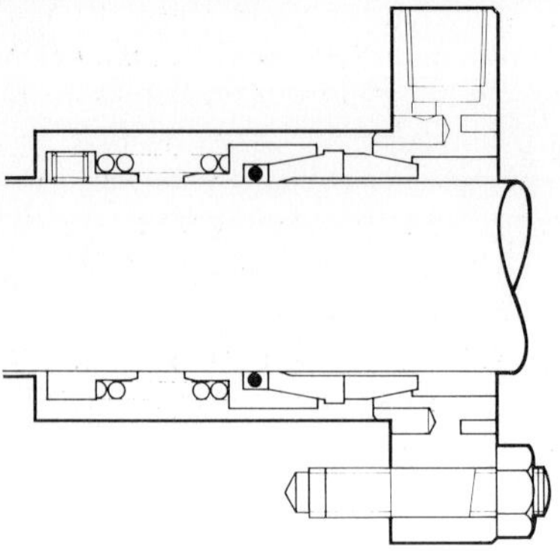

Figure 1a
Split seal assembled sectiona view

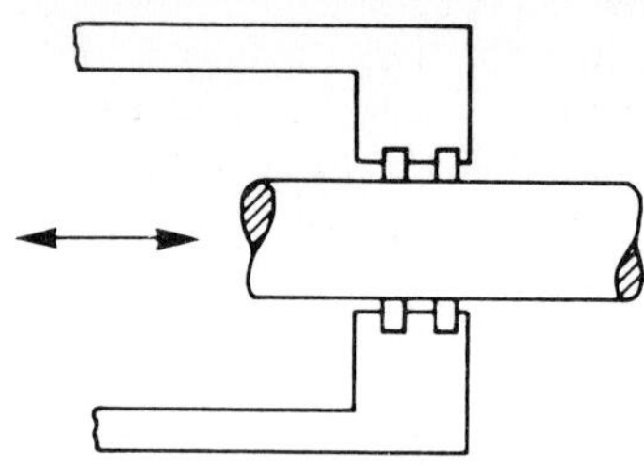

Figure 1b
Inspringing reciprocating seal.

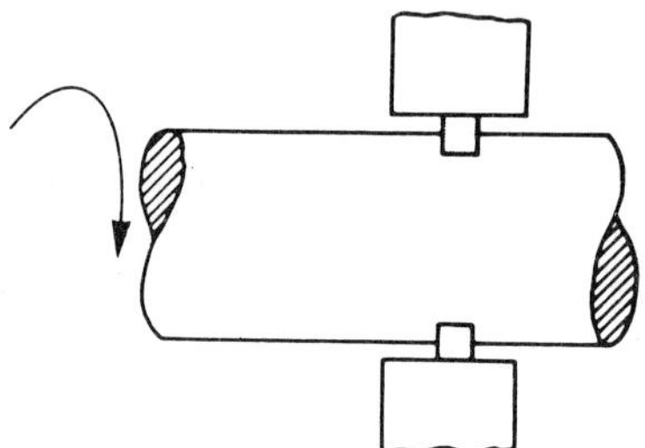

Figure 1c
Outspringing rotary seal.

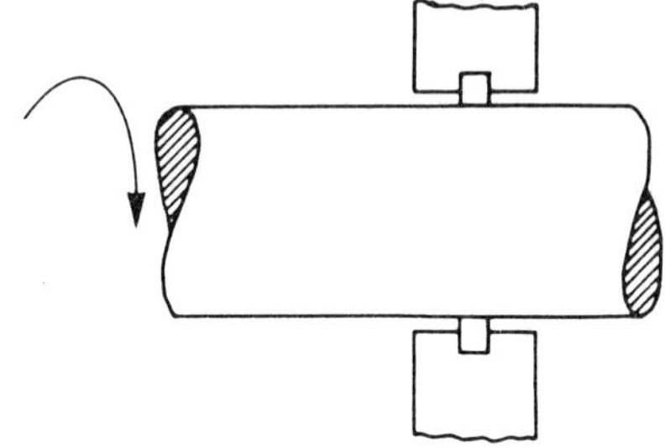

Figure 1d
Inspringing rotary seal.
(Not normally used)

is outspringing or inspringing, respectively. In the case of outspringing seals this gap has to be chosen to balance fitting stress with fitted stress, normally calling for a free gap of the order of three or four times the radial thickness of the ring. In the case of an inspringing seal installed in a housing, the regular gap or overlap is determined by the contracting seal pressure required.

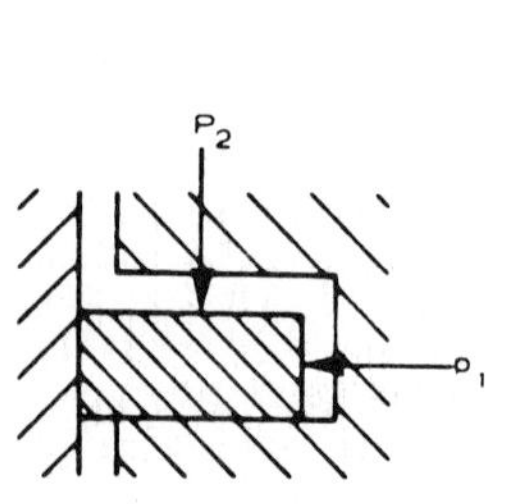

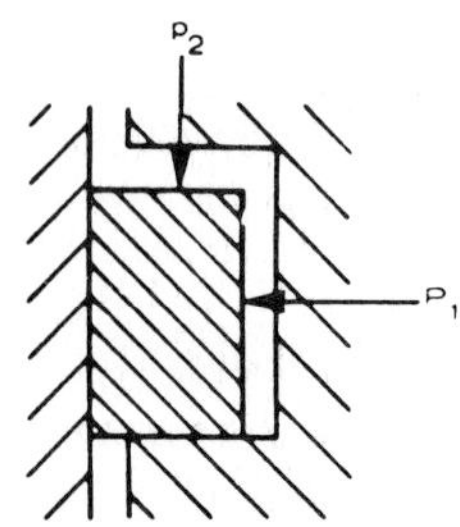

Figure 2

The *closed* gap is a critical dimension because it determines the size of the leakage path. For hydraulic applications it is usually of the order of 1/1000 of the ring diameter (*eg* 0.001 inches per inch diameter), but should not exceed 0.25 mm (0.01 in). Actual leakage is dependent not only on the actual gap but also the viscosity of the fluid being sealed. In the case of an outspringing seal, wear on the ring or mating rubbing surface will tend to increase the gap (and leakage) only slightly. With an inspringing seal wear on the i.d. of the ring will tend to close the gap. If this continues until the gap becomes zero, leakage will then develop across the face of the seal on the i.d.

Outspringing rings seal on the o.d. under the inherent tension infused in the ring material when the diameter is contracted when fitted. Face pressure imparted by such a seal is:

$$P = \frac{E\,G}{7.07D\left(\frac{D}{t} - 1\right)^3}\ \text{lb/in}^2$$

where P = pressure (lb/in^2)
E = modulus of elasticity (lb/in^2)
G = free gap (in)
D = fitted diameter (in)
t = minimum radial thickness (in)

If the tangential load (T) required to closed the seal is known (Figure 3):

$$P = \frac{2T}{Db}\ \text{lb/in}^2$$

where b = width of ring in contact with cylinder (in)

If the diametral load (Q) required to close the seal is known (Figure 4):

$$P = \frac{0.76\,Q}{Db}\ \text{lb/in}^2$$

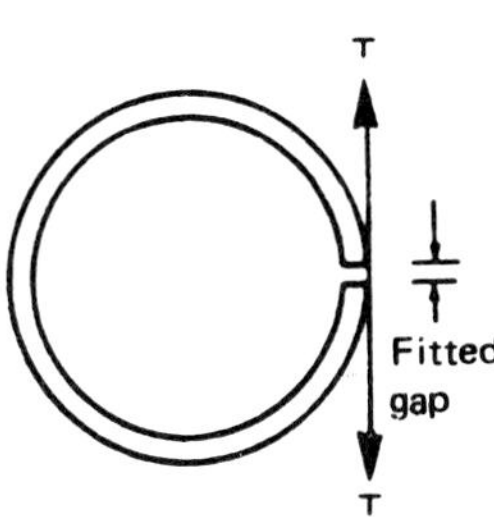

Figure 3

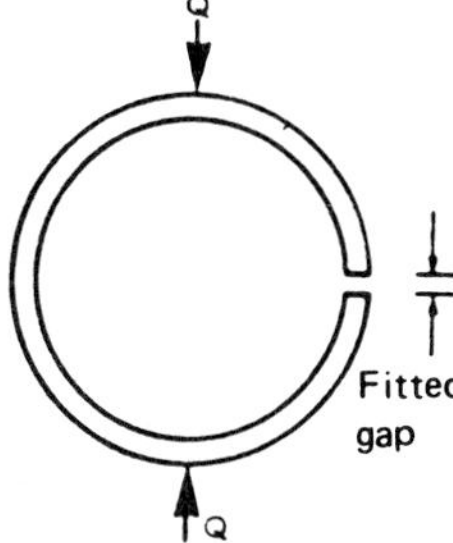

Figure 4

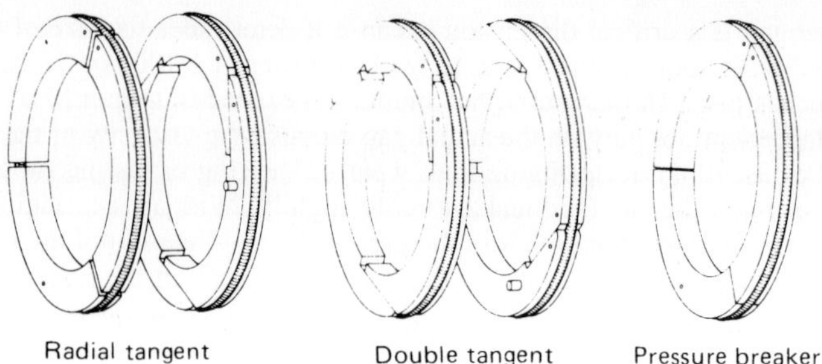

France-type ring seals (packing rings).

Material stress

The material of the ring may be stressed in two ways: once when fitting and then by the stress imparted by the change in diameter when fitted. Values can be determined from these two formulas:

$$\text{Maximum fitting stress} = \frac{4(G - g)t\,E}{3\,\pi\,D^2}\ (\text{lb/in}^2)$$

where g = additional amount by which the free gap has to be extended to fit the ring in place

$$\text{Maximum fitted stress} = \frac{4G\,t\,E}{3\,\pi\,D^2}(\text{lb/in}^2)$$

Ring Section

The radial thickness (t) has a major influence on the face pressure imparted by the ring. With conventional materials the thickness (t) is between D/32 and D/22 to enable the ring to be sprung in place over its own o.d. A thicker ring would require a built-up piston for assembly, but would give improved sealing pressure. Thicknesses less than D/32 become increasingly less effective in sealing. It should be noted that the radial thickness and groove depth also determines the sealing area on the side face of the seal.

The axial width dimension of the ring is significant in determining the working load and rubbing friction of the seal, as well as the conformity of the seal. If necessary, conformability can be improved by the use of backing expanders, *eg* a *coil spring, leaf* spring or *serpent-type* spring. Coil spring expanders are circumferentially limiting, housed in a machined recess behind the seal. Leaf spring expanders are fitted in the bottom of the groove. Serpent-type expanders are fitted in a circumferential housing machined in the back face of the seal.

Ring joints

The simplest forms of ring joints are the straight butt and scarf (Figure 5). In practice both normally have a similar leakage path for the same size of ring. They may be used for piston seals where some leakage is accepted, but are seldom used for rod seals.

A *stepped* joint (Figure 6) is preferable where wear is likely to be high and can greatly reduce wear on cylinder bores. A simple rectangular *lap-step* has about the same leakage rate as a *straight butt* joint, but the *scarf-stepped* joint has a much lower leakage rate.

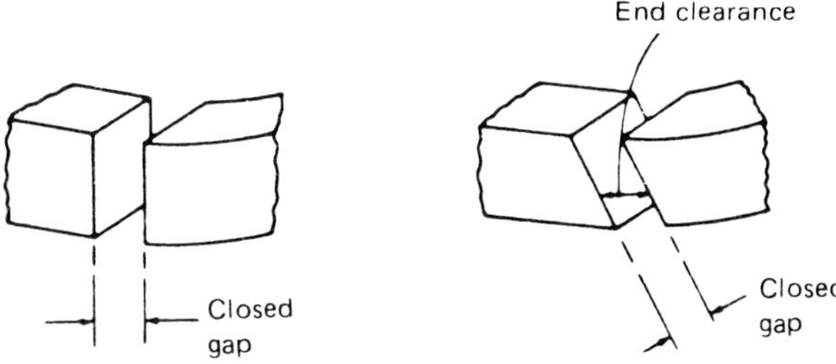

Figure 5

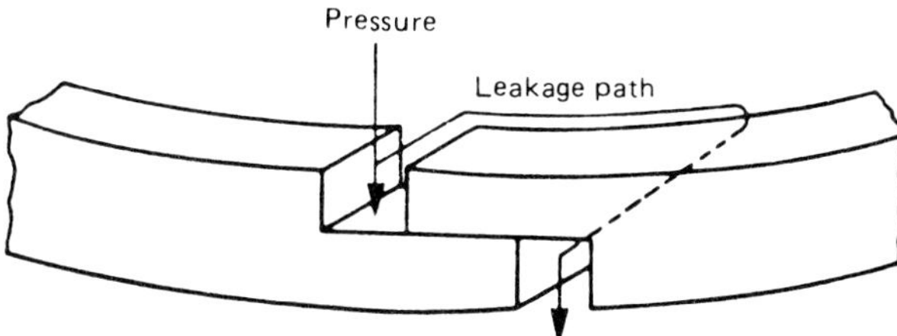

Figure 6

In this case, however, the ring seals effectively in one direction only, so it must be fitted the right way round. It is also more expensive to manufacture. For double-acting sealing duties two scarf-stepped rings must be fitted in separate grooves in back-to-back configuration.

Other joints may be used for specific purposes, *eg overlapping* joints which provide a minimal leakage path but at the same time seal equally effectively in either direction. Again these are a more expensive production, but can also be designed for self-compensation if ring or cylinder bore wear during service.

Interlocking joints (Figure 7a) are used for rings which require a restrained diameter in the free state, *eg* to facilitate blind assembly or limit expansion to facilitate removal. They are best used as *unidirectional* seals. Another method of controlling the free diameter of the seal is to employ an L-shaped cross-section located in a built-up assembly or by a retainer ring (Figure 7b).

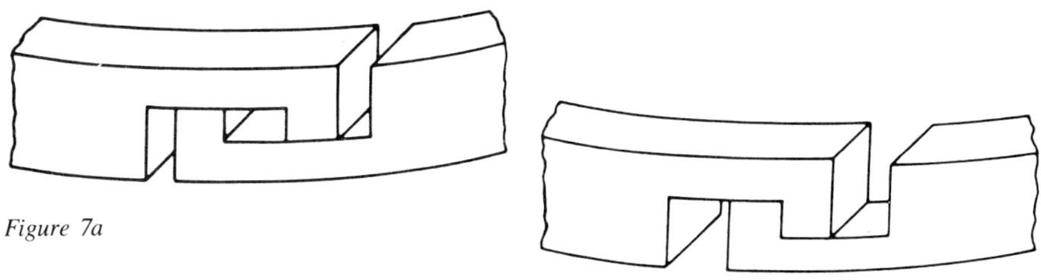

Figure 7a

Figure 7b

Two-piece seals

Two-piece seals are normally used where leakage has to be held to a minimum. A typical example is the use of one ring with a step pad joint and a second concentric ring with a butt joint, the two joints being diametrically opposed. Usual configuration is to employ the *plain butt* joint on the inner ring on an outspring seal (Figure 8a) and on the outer ring (Figure 8b) on an inspringing seal.

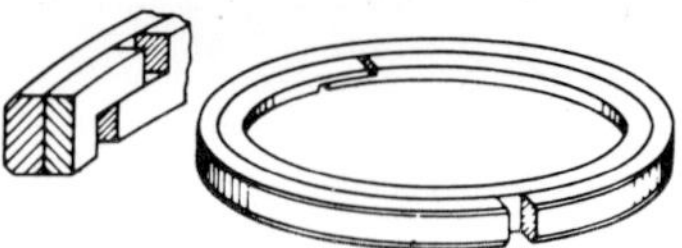

Figure 8a

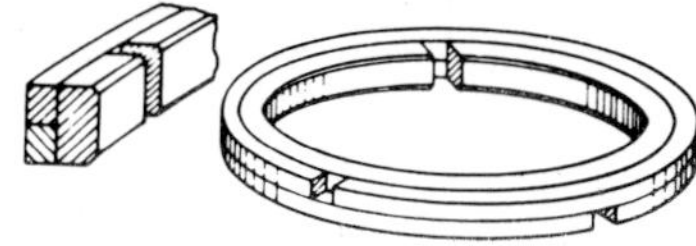

Figure 8b

Seals of this type can seal equally well in either direction. Accuracy of manufacture and fine surface finish is important for proper performance.

The gap leakage path of two plain split-ring seals can be improved by incorporating a stepped joint on the side face of the ring. It is important that the surface of the stepped joint should be accurately mated to ensure efficient sealing. When a seal is likely to be at an angle to the axis of the bore in which it is sealing (due perhaps to misalignment), the combined periphery of the two rings can be spherical, (Figure 9).

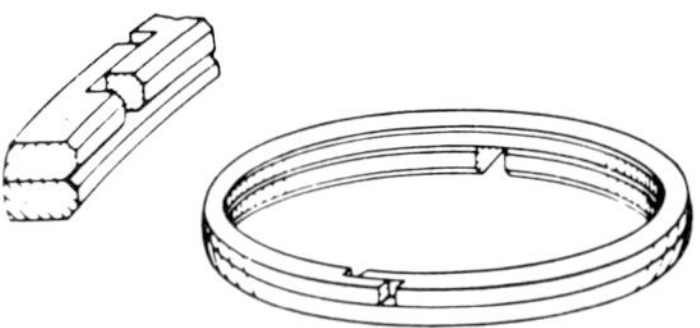

Figure 9

Three-piece seals

Three-piece rings are normally based on an assembly of one full width ring and two half-width rings; Figure 10a shows an outspringing ring that consists of two outer split sealing rings (A) and (B), with a split inner ring (C). The inner member, which is equal in width to the combined widths of the outer two, seals off the radial leakage path through the closed gap on the outer ring farthest from the pressure. With the gaps of the outer rings staggered, there is no axial leakage path, and angled joints prevent them lining up. One disadvantage of these seals is that it is difficult during manufacture to prevent differences in size between the radial dimensions of the outer rings, allowing the full pressure of inner ring (C) to bear on one outer member only, thus leaving a leakage path (D) behind the other. This form of leakage can be minimized by reducing the tolerance of the radial dimension of the outer rings to 0.05 mm (0.02 in).

The three-piece seal, the ideal assembly for small diameter seals, can be made to sizes less than 25 mm (1 in) diameter.

The bearing rings on this type of seal can be segmented, to ease replacement of worn parts. This arrangement is ideal on reciprocating-rod seals of the contracting type.

Alternative three-piece seal

To overcome the disadvantage of the standard three-piece seal, when the variation in the radial thickness can cause a leakage path, a two-piece seal with a side sealing member can be used. It consists of an outer ring (A) with a cut-straight or angled joint, backed by an inner member (B) which provides radial sealing. To block the axial leakage path there is a side sealing member (C) with a radial thickness which is the combined size of the inner and outer rings (Figure 10b).

To effect a complete seal in one direction only it is necessary to lap members A and B together, to ensure that the axial widths are equal.

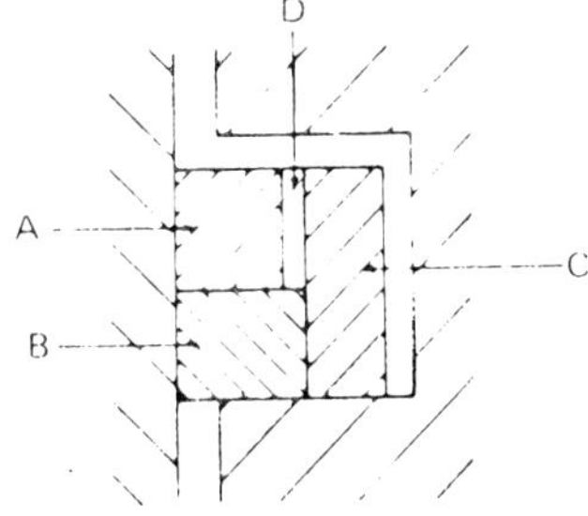

Figure 10a

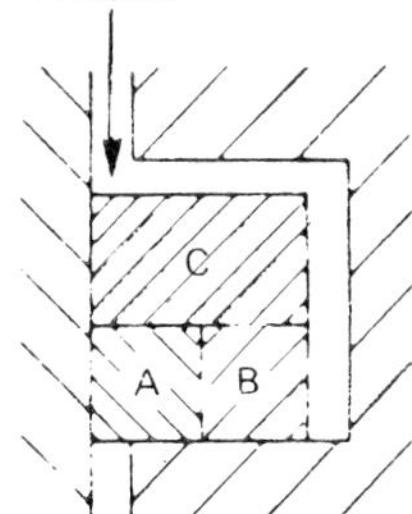

Figure 10b

Materials

Conventional piston ring materials are the most common choice for split-ring seals and are suitable for operation up to about 275°C. For higher surface temperature, or other special requirements such as corrosion resistance, other metals may be used – see Table 1. Various surface coatings may also be used for enhanced performance, *eg plated coatings,* including for corrosion resistance or to improve bedding-in of seals; *cadmium, copper, chromium, silver* and *tin, phosphate coating* for corrosion resistance; *metal spraying* to improve anti-galling characteristics of heat-resistant metals; and *anti-friction coatings* of PTFE or graphite.

TABLE I – METALS FOR SPLIT-RING SEALS

Metal	UTS tons/in^2	UTS kg/mm^2	Modulus of Elasticity x 10^6 lb/in^2	Modulus of Elasticity x 10^5 kg/cm^2	Remarks
Grey iron	16	25	15.5	10.9	Self-lubricating to a degree and wear resistant
Low alloy iron	29	46	20.0	14.0	Wear resistant
Austenitic iron (Ni-resist)	16	25	14.0	10.0	Non-magnetic
High alloy iron	35	55	28.0	20.0	Good heat resistance
Spring steel	30	47	30.0	21.0	High strength spring material
Aluminium bronze	35	55	15.0	10.5	Good bearing properties under marginal lubrication
Nickel alloy (Nimocast 90)	45	70	32.0	22.0	Resistant to heat and corrosion
Stellite	58	90	30.0	21.0	Resistant to heat and corrosion

PTFE split-ring seals

PTFE is a possible alternative to metals, offering extremely low friction (even when rubbing dry), extreme resistance to chemical attack and a service temperature range of −100°C to +250°C. Its main limitation is its relatively low strength and high moduli compared with metals, although these properties can be enhanced by filling with glass, bronze, carbon, *etc*. One particular point to watch in the design of PTFE split-ring seals (particularly in combination with metal rings) is that the coefficient of expansion of PTFE is about ten times that of cast iron or steel.

Limiting PV values for PTFE consistent with 0.125 mm (0.005 in) wear in 1000 hours are:

	lb/in^2 x ft/min	**kg/cm^2 x m/s**
PTFE (plain)	1000 max	0.35
15% glass-filled	3000	1.05
40% glass-filled	8000	2.8
60% bronze-filled	9000	3.15
25% carbon graphite-filled	4500	1.575

(filler percentages by weight)

See also chapter on *Piston Rings*, Section 7.

Flexible Lip Seals

THE U-RING is one of the earliest designs of pressure-energized lip seals, primarily intended for dynamic applications where low friction is required and/or generous tolerances are desirable, *eg* as piston head or gland seals in hydraulic and pneumatic cylinders. Homogeneous rubber *U-rings* are particularly effective for sealing at low to moderate pressures, but produce increasing friction and accelerated wear at higher pressures, as well as being prone to extrude into the clearance space. Performance can be improved by modifying the *U-section* to reduce localized stressing and/or the use of a tougher material (*eg* harder rubbers or rubber-impregnated fabric). Homogeneous

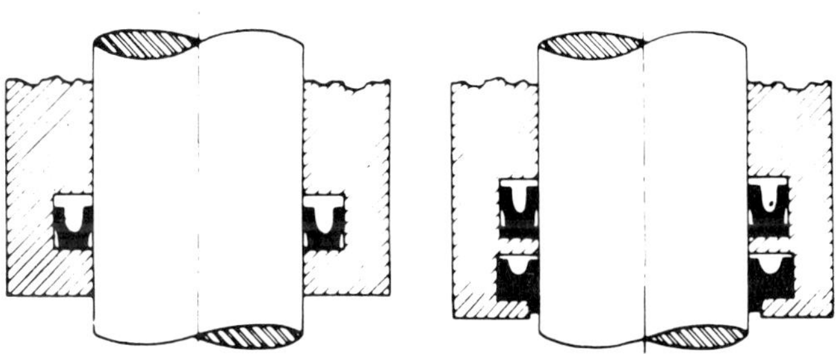

Typical U-ring seals.

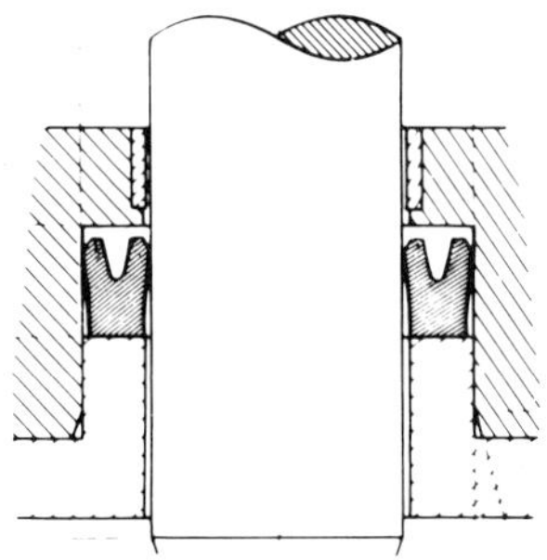

Typical section of polyurethane U-ring.

rubber U-rings designed for pressures in excess of 175 bar (2500 lb/in^2) or where the extrusion gap exceeds 0.65 mm (0.025 in) can incorporate anti-extrusion heels in a semi-flexible or rigid material, or use them in combination with headers and/or back-up rings. They are used only as reciprocating seals.

Basically the U-section chosen has to be a compromise between friction, wear and sealing performance. A thin-walled section will have low friction and favourable sealing characteristics for low pressure working (*eg* pneumatics). Thickening the section will provide better sealing performance at high pressures at the expense of increased friction (Figure 1).

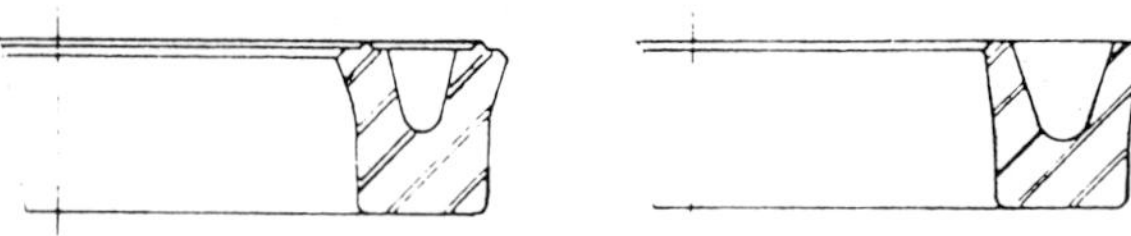

Figure 1
Typical heavy-duty U-ring section for hydraulics (left) and light-duty U-ring section for pneumatics (right).

The heel of the section is a vulnerable point because this is where the section is subject to maximum stress, which makes the material more prone to abrasive wear or cutting. It is also the point at which extrusion can occur. Damage to the heel can also affect the proper functioning of the flexible lip by modifying the stress distribution throughout the section (Figures 2 and 3). Resistance to nibbling and wear can be increased by using a *rounded-heel* section, although some designs specifically incorporate a square base for use in conjunction with a back-up ring. Other basic variations are the use of either chamfered or squared off lips – see also (Figure 4.)

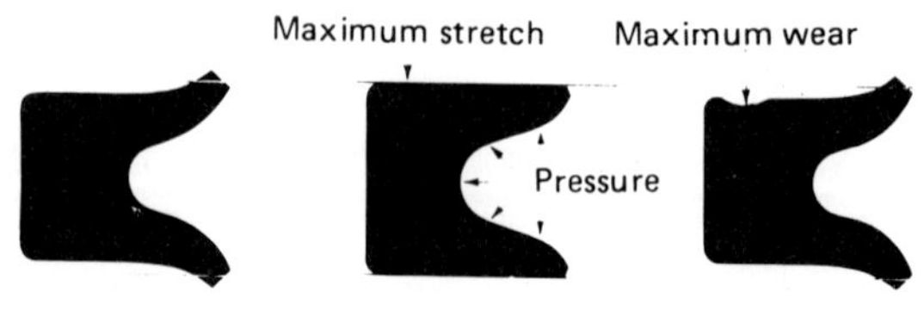

Figure 2

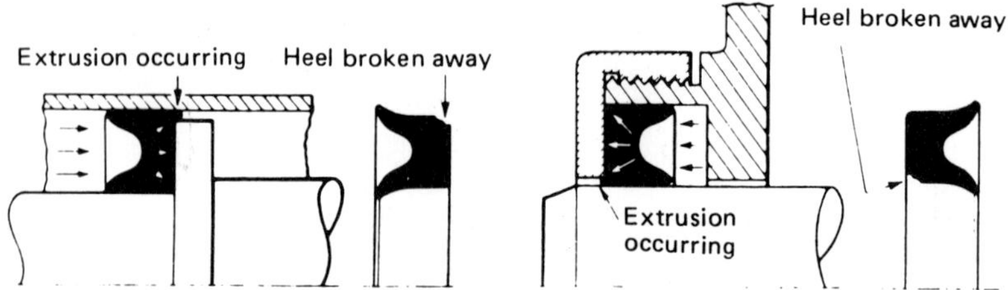

Figure 3
Typical damage to U-rings working at high pressure.

Figure 4
Examples of U-ring sections.

U-rings can be installed in closed grooves, half-open grooves, or open grooves (Figure 5). With a closed or half-open groove it is important that the width of the groove (B) is sufficient to accept the full length of the seal (b) when assembled with an interference fit, otherwise the working section will be distorted. At the same time the groove width must not be excessive, otherwise the seal will tend to float and 'pump' (Figure 6). With a half-open groove it is also necessary to ensure that the shoulder diameter is sufficient to retain the seal in the groove under reciprocating conditions.

With an open groove (or equivalent) mounting it is necessary to incorporate a retainer or lantern ring to hold the seal in position (Figure 7). This retainer must be proportioned so that the base of the seal is not under compression when assembled and the edge in contact with the seal is smooth and rounded. A vent hole may also be necessary in the retainer to ensure through-flow so that both sealing lips are energized with equal pressure. This is not normally necessary with U-rings that incorporate a web.

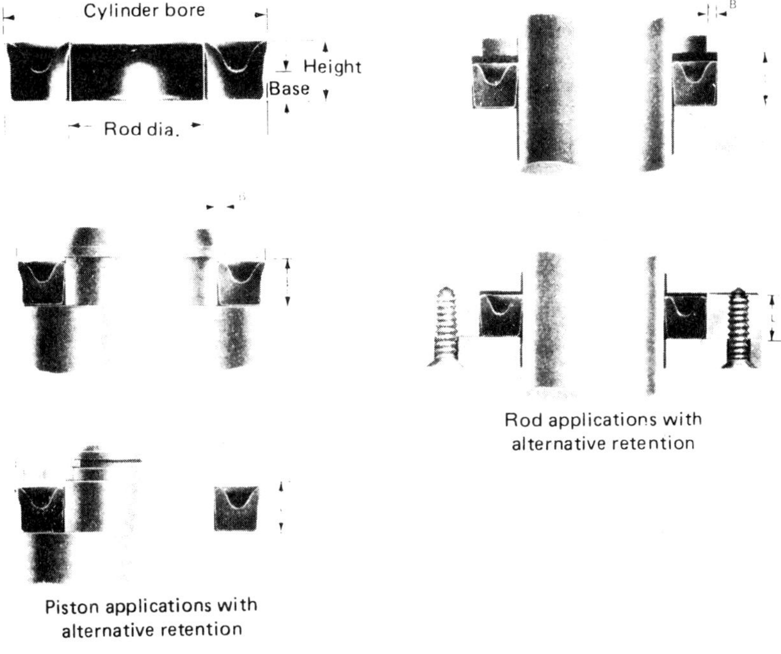

Figure 5

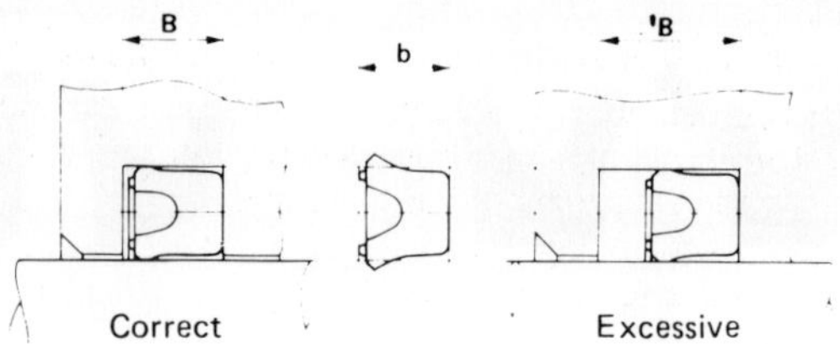

Figure 6
Groove design for U-rings.
Groove length (B) must be sufficient to accept the length of the seal (b) when seal is assembled.

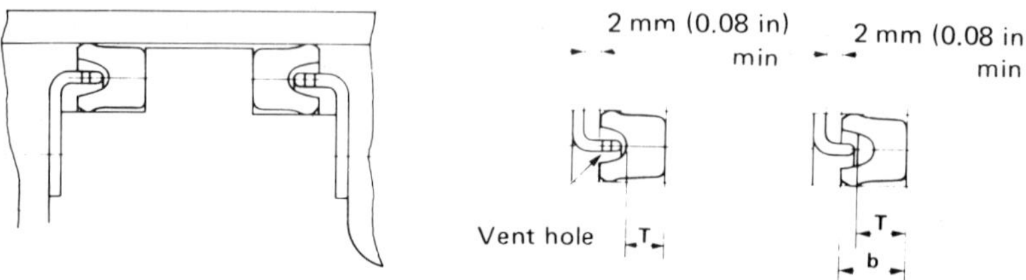

Figure 7
U-rings mounted with retainer rings.
Base of seal (T) must not be under compression when assembled.
Variations on U-ring forms.

U-rings in leather are normally formed with straight side walls and a rounded base and assembled with support rings (back-up ring and *internal support* or *pedestal* ring). In the absence of a pedestal ring, the seal may be filled with hemp, flax or rubber to prevent the side walls collapsing.

The U-ring is a single-acting seal. Two seals can be used back-to-back for double-acting piston applications, where the piston head design is suitable. However, many modern piston heads can be jumped into a one-piece groove and as these can also incorporate bearing rings the use of the U-ring in high pressure double-acting piston head assemblies is decreasing. Similarly, the use of U-rings in sets, with headers, to provide increased pressure rating is now less favoured than single U-rings, with anti-extrusion heels, or other seal forms.

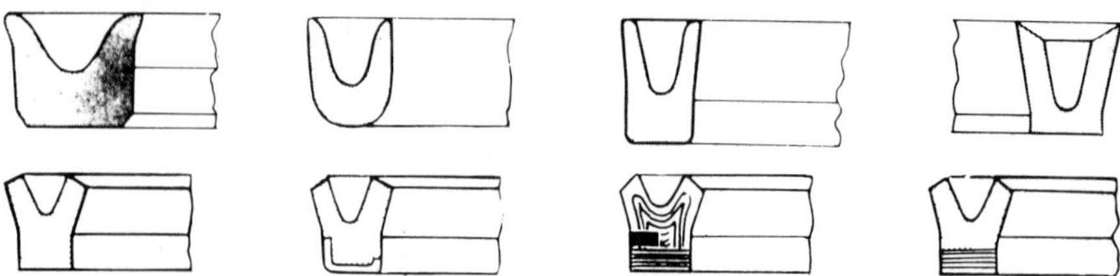

Variations on U-ring forms.

Composite U-rings

The modern composite seal based on a U-ring or similar flexible lip section, together with an integral back-up ring, offers two specific advantages. The back-up ring eliminates the possibility of extrusion of the heel, also enabling the seal to accommodate higher working pressures. At the same time the elastomeric section can be proportioned for progressive pressure energizing, with low initial interference pressure. The effective seal contact area then increases progressively with increasing pressure.

This is shown in Figure 8. Under zero or low pressure only the tips of the seal lips provide sealing contact. Under increasing pressure fluid pressure expands the seal section radially to increase the contact area to maintain sealing. At the same time, the back-up ring may be so shaped that its forward (pressure-side) tip flexes to close any extrusion path. Finally, under high pressure the seal section is fully deformed with the back-up ring completely deflected to fully contact the cylinder wall and presents a substantial base providing resistance to shear and preventing extrusion.

Other types of composite seals may work with little or no initial interference. In this case. sealing contact force and contact area are fully developed by the fluid pressure itself.

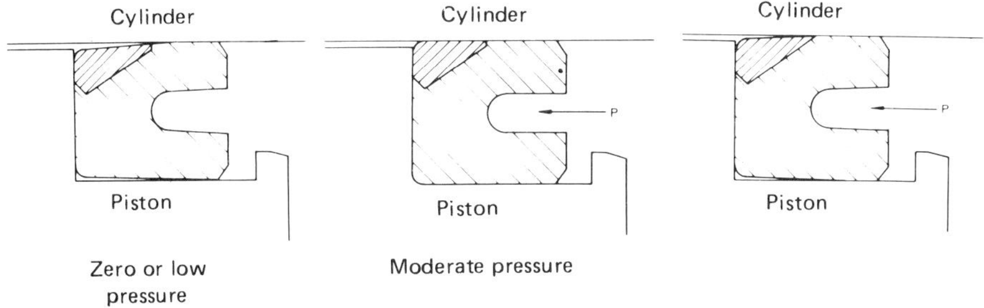

Figure 8
High performance packing rated for 700 bar (10000 lb/in²) maximum pressure.

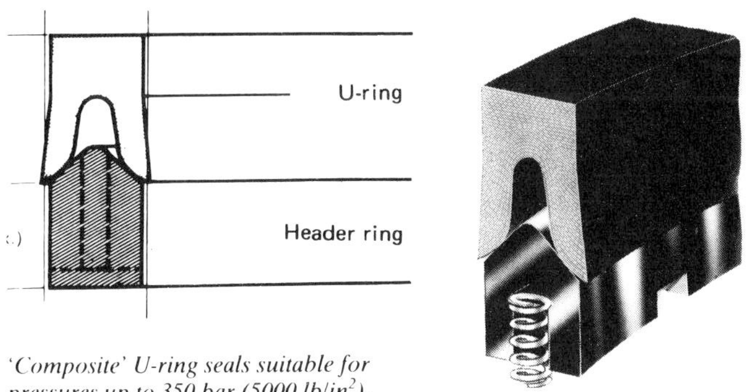

'Composite' U-ring seals suitable for pressures up to 350 bar (5000 lb/in²).

V-rings

V-rings, also known as *chevron* seals, have a more robust section than U-rings and can better withstand both axial and radial loading. Their pressure rating and low pressure sealing capability can both be increased by using two or more rings in a set, together with female and male adapters or headers (*eg* Figures 9a and 9b). This is the normal employment of V-rings (*ie* in sets rather than individual rings). Headers can be solid or vented; and in some cases spring-loaded headers may be used for automatic compensation for wear.

V-rings can be made of leather, homogeneous rubber, or rubber-impregnated fabric. V-ring sections are normally similar, regardless of the material used, although there are some variations in the form of moulded rubber V-rings (*eg* Figure 10). Headers are usually metal, but may be rubber-impregnated fabric, glass-filled nylon, acetal resin, hard rubber, PTFE and impregnated PTFE.

V-ring sets may include rubber rings interposed with fabric rings; the greater elasticity of the rubber ring(s) improving sealing under static conditions or under low pressures. On smaller diameter seals rubber rings may alternate with fabric rings, but with larger diameters the number of rubber rings is usually restricted (*eg* Figure 11).

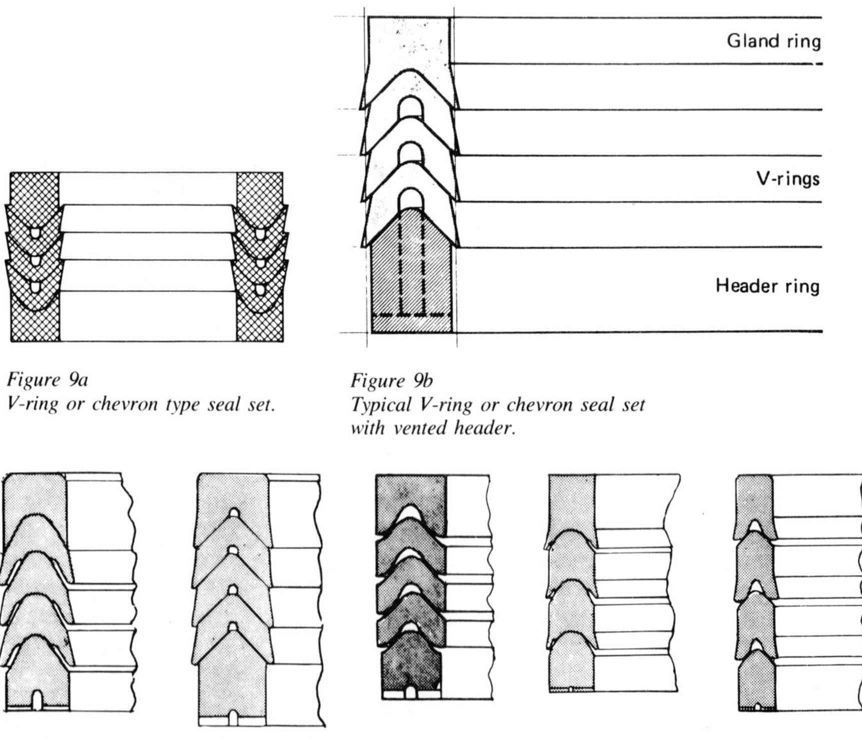

Figure 9a
V-ring or chevron type seal set.

Figure 9b
Typical V-ring or chevron seal set with vented header.

Figure 10
Variations on V-ring profile forms.

Leather- and rubber-impregnated fabric V-rings used in sets may be split to facilitate assembly, in which case joints should be staggered. Moulded V-rings are invariably one-piece (never split rings). Rubber V-rings provide generally tighter sealing, although their maximum pressure rating is somewhat lower than that of leather or fabric V-rings. Rubber and fabric V-rings are sometimes used together in the same set.

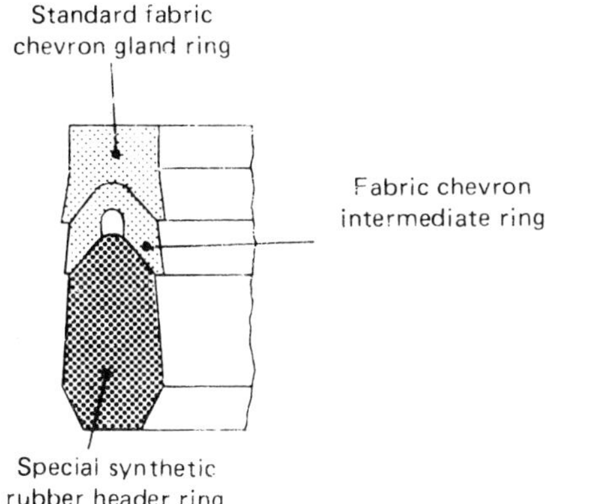

Chevron packings designed as rod seals with optimum film control.

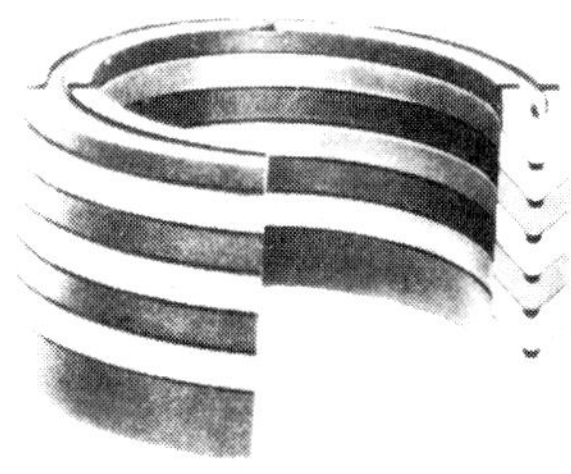

Figure 11
V-ring set alternating laminated fabric and rubber rings, with laminated fabric headers.

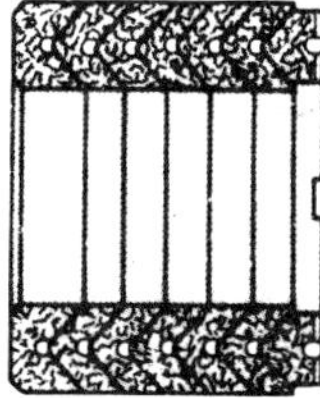

Examples of modern V-ring sets (V-rings and matching headers) in rubber-impregnated fabric. Pressure rating up to 400 bar. Material combinations include cotton/nitrile rubber; asbestos/PTFE; asbestos/Viton; and asbestos/neoprene.

V-ring sets may be classified as *heavy-duty rod* or *cylinder* seals. Initial sealing is provided by pre-load produced by the axial clamping force generated when the gland is tightened to close the seal to the design stack height (including headers). Spacers may be used to adjust the stack height as necessary, *eg* to adjust static sealing, or compensate for seal wear. Under internal pressure the V-rings work as pressure-energized seals.

Depending on the V-ring material and the number of rings in the set, working pressures up to 700 bar (10000 lb/in^2) can be accommodated by rubber-impregnated fabric rings and up to twice this figure with laminated fabric or leather rings. The number of rings required depends both on the operating pressure and the diameter of the seal. Typical recommendations in these respects for sealing hydraulic cylinders are summarized in Tables 1 and 2, with relevant dimensions. It will be noted that with higher pressure ratings and larger seal diameters, the length of the seal is increased, as is seal friction (although the V-ring itself has lower friction than a U-ring). The modern trend is to use more compact seals or seal sets than V-rings, although they still find numerous applications in medium duty as well as heavy duty services. V-ring sets are relatively compact seals in depth, in fact, although seal length can be relatively large with a large number of rings.

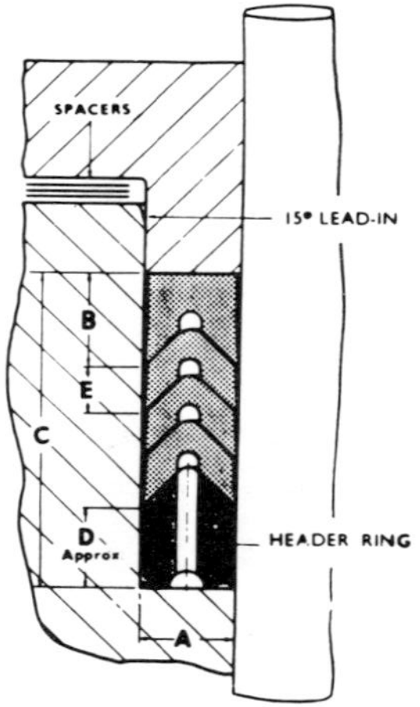

TABLE I – TYPICAL DIMENSIONAL DATA FOR V-RING SETS

Seal Size Cylinder or Rod Diameter		Section A		Depth V-Ring E		Depth Gland Ring B		Header Ring D		Pressures up to 350 bar (3500 lb/in²)			Pressures above 350 bar (3500 lb/in²)		
										Number of V-Rings	Total Length of Seal C		Number of V-Rings	Total Depth of Seal C	
in	mm	in	mm	in	mm	in	mm	in	mm		in	mm		in	mm
up to 1	up to 25	3/16	5	3/32	2.5	3/16	5	1/8	3	3	½	12.5	4	19/32	15
1.1/16 to 1½	27.0–37.6	¼	6	1/8	3	¼	6	¼	6	3	¾	19.0	4	7/8	22
1½–2.5/8	37.6–66.7	5/16	8	5/32	4	5/16	8	9/32	7	4	1	25.4	5	1.5/32	30
2¾–4	69.8–101.6	3/8	10	3/16	5	3/8	10	5/16	8	4	1¼	32.2	5	1.7/16	36.5
4¼–7	107.9–177.8	½	12.5	¼	6	½	12.5	7/16	10	4	1¾	44	5	2	50
7¼–17	184.1–431.8	5/8	16	5/16	8	5/8	16	½	12.5	4	2	50	5	2.5/16	60
17¼–25	438.1–635.0	¾	19	3/8	10	¾	19	5/8	16	5	2.7/8	75	6	3¼	80
25½–37	647.7 939.8	¾	19	3/8	10	¾	19	5/8	16	6	3¼	85	7	3.5/8	90
above 37	above 1000.0	1	25	½	12.9	1	25	¾	20	7	4¼	105	8	5¼	130

TABLE 2 – MAXIMUM PRESSURE RATING OF V-RING SETS (typical values)

Number of V-rings	V-ring Material					
	Rubber		Rubber-Impregnated Fabric		Leather	
	lb/in²	bar	lb/in²	bar	lb/in²	bar
2	500	35	1000	70	–	–
3	1000	70	1500	100	500	35
4	2000	140	3000	210	1500	100
5	3000	210	5000	350	6000	420
6	5000	350	10000	700	20000	1400

C-rings

A C-ring is the general description for a single-sided seal with a finely tapered lip presenting a relatively large contact area on the dynamic side. Unlike a single-sided U-ring section this increases progressively from the tip of the lip to the blunt end of the other (static) leg.

It can be described as a modified U-ring with one arm shortened or cut off. It seals on one surface only, *ie* the lip side, and so can be made as either an external or internal seal. It is similar in mechanical characteristics to the U-ring and like the U-ring can be used singly or in seal sets with headers. It is the basis of a number of proprietary seals aimed at improving the performance of a conventional U-ring seal for specific applications.

Because they are designed specifically to be used in sets, (one or more seals together with top and bottom headers) C-rings are also referred to as gland rings or *automatic gland* packings to emphasize that they are pressure-energized, as opposed to compression packings. Examples are shown in Figures 12 and 13.

Construction is usually in cotton or asbestos cloth moulded and bonded with rubber. A graphited finish is usual to facilitate running in when first fitted. Rings may be integral or cut. In the latter case a minimum of two, or preferably three, rings is recommended with joints staggered. Single (integral) seals are suitable for low speed applications and

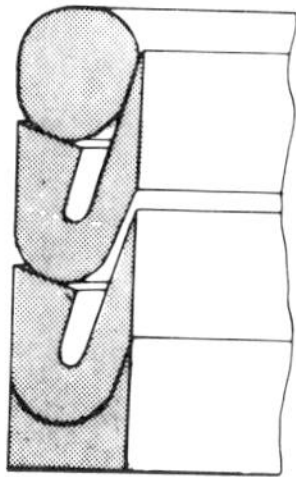

Figure 12
C-rings normally used with headers are particularly suitable for low pressure, long stroke cylinder seals.

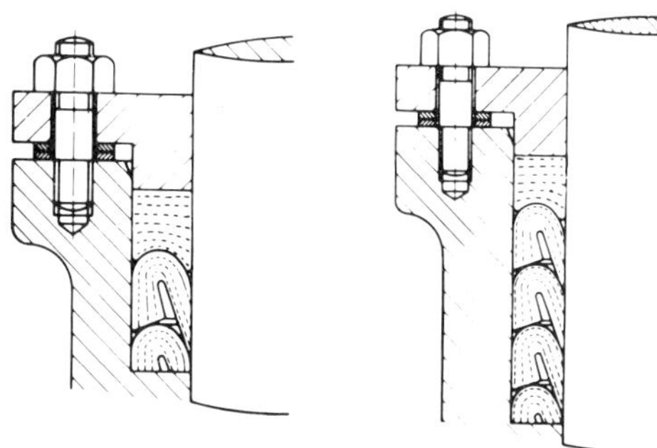

Figure 13
Proprietary C-ring sets and moulded headers.

will give minimum friction. Two or three rings are to be preferred where there is lateral movement or lateral loading.

For light duties up to about 25 bar (350 lb/in^2) headers may be lubricated plaited packings. For other duties the headers can be moulded rings of similar material to the seal rings suitably machined bushes or shaped metal header rings, *eg* Figure 14.

The performance of gland rings of this type is generally excellent, giving maximum sealing efficiency with minimum frictional loss. Satisfactory performance does, however, depend on correct sizing, correct fitting (to avoid any excess of compression pressure) and the use of correct header rings.

C-rings in homogeneous rubber should be suitable for pressures up to 70 bar (1000 lb/in^2); semi-reinforced up to 140 bar (2000 lb/in^2); and in rubber-impregnated fabric up to 350 bar (5000 lb/in^2). C-rings in laminated fabric may be capable of handling pressures above 350 bar (5000 lb/in^2). They are mainly suitable for sealing reciprocating motions but homogeneous rubber C-rings may also be used as light duty rotary shaft seals.

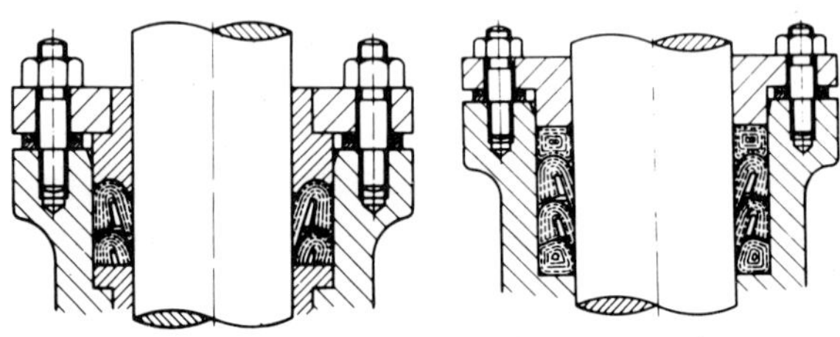

Figure 14
Seal ring sets with machined headers (left) and plaited packing headers (right).

Automatic lip seals

Sections derived generally from the basic U-ring but with a more rigid section and the lip(s) formed by a V-notch are sometimes described as *automatic* seals. This merely implies that they are pressure-energized, but it is a convenient description for sections which are not true U-rings and embraces a considerable variety of proprietary designs. Such seals may be designed to be used singly or in sets. They may also have two sealing lips (like a U-ring) or only one sealing lip (like a C-ring) (Figure 15). In the latter case they may further be categorized as *distributor seals,* (Figure 16).

Most seals of these types are more robust than U-rings and primarily intended for medium and heavy-duty applications, although there is not any clear distinction between light, medium or heavy-duty because the size of the machine is also a factor, as well as the actual loading the seal may be subject to. Thus a more rigid or heavy-duty section may be needed in cases where the seal may be subjected to side loads, for example, bearing the weight of a component on a horizontal machine. As a general rule eccentricity and side loads are to be avoided as far as possible with all flexible lip seals as such factors can interfere with their proper function, leading to leakage and excessive wear.

There can also be notable exceptions to any attempt at precise classification. Thus the distributor type seal may be produced with a relatively thin lip, specifically as a light-action seal for pneumatic services,(*eg* Figure 17).

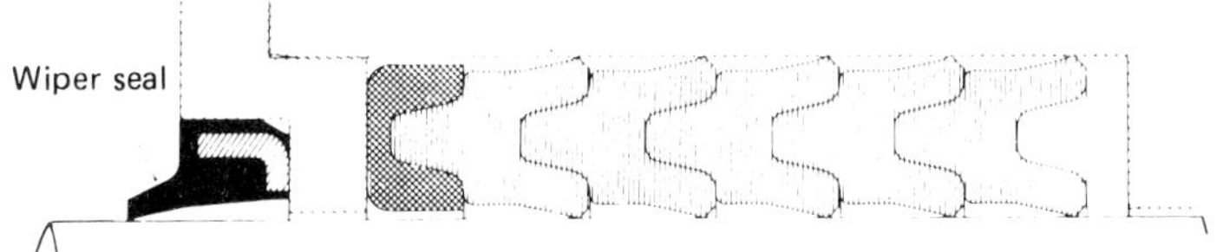

Automatic split-type seal for hydraulic piston seals. Maximum pressure rating 350 bar (5000 lb/in^2).

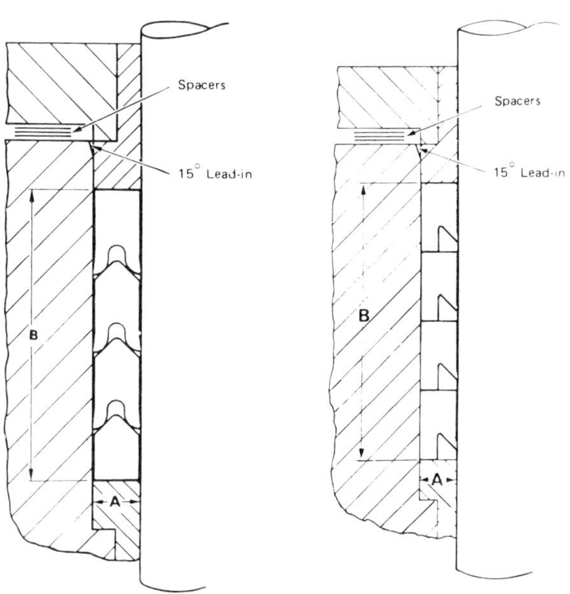

Examples of proprietary ring packings.

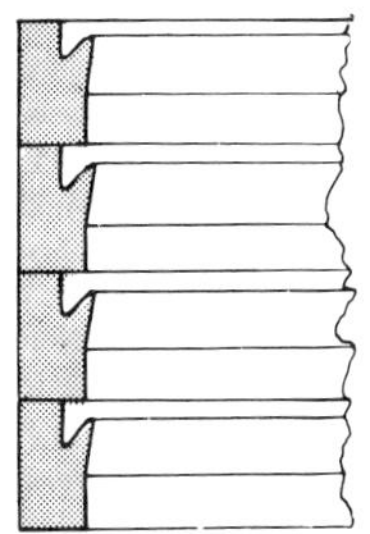

Basic single-lip or automatic packing rings.

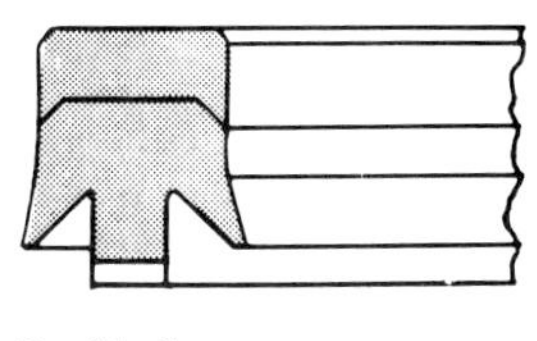

Double-lip automatic packing.

Figure 15

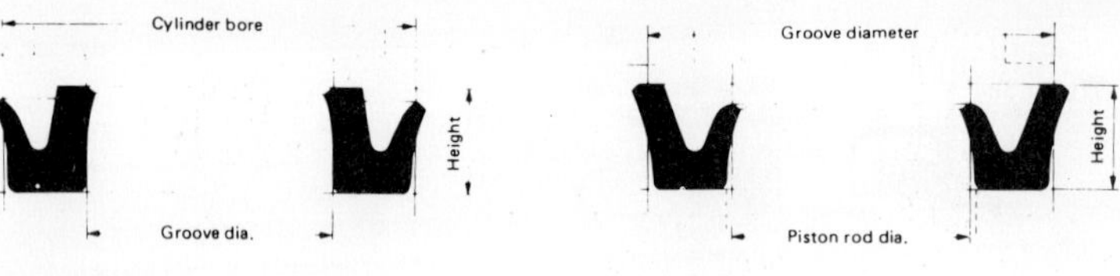

External distributor seal

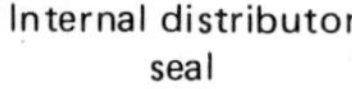
Internal distributor seal

Figure 16

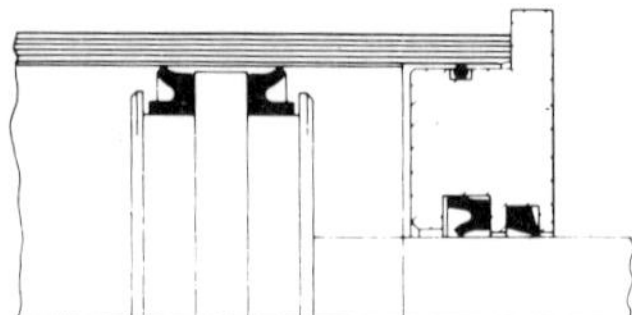

Typical internal and external distributor seals installation illustrating comparative design and machining simplicity.

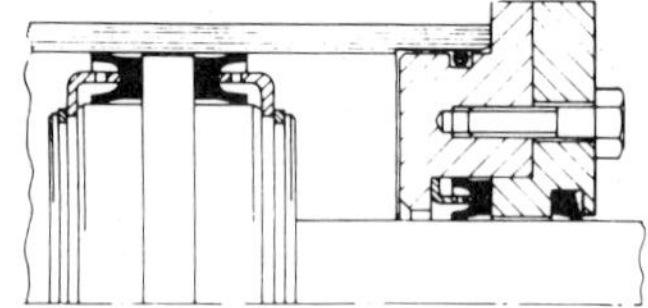

Typical 'U' seal installation illustrating design and machining complexity.

Distributor seals (left) have the advantage over U-rings (right) in that they can provide simpler assembly.

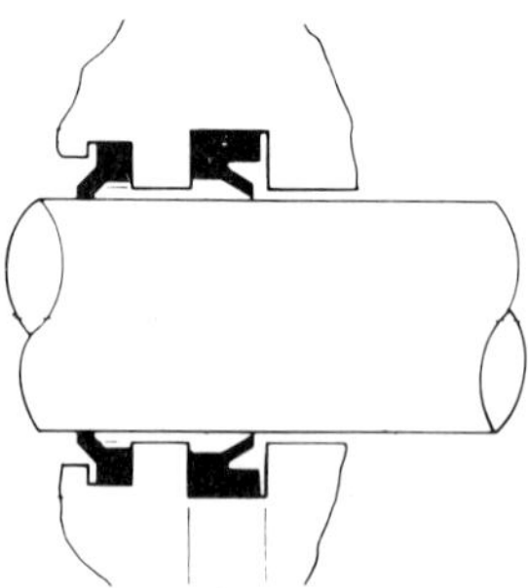

Figure 17
K-seal.

Wedge-Action Seals

The conventional lip seal has a pre-load pressure by virtue or the deflection of the lip(s) when installed. A *squeeze-type* seal also has a pre-load pressure imparted by deformation of the section when installed. A *wedge-action* seal has pre-load further enhanced by spring pressure, (see Figure 18). All three types are then further pressure energized, increasing the face loading when under pressure. The advantage of a wedge-action seal is that the manner of pre-loading prevents the sealing lips from collapsing away from the sealing surface under severe load changes or high cycling rates, as well as providing more positive sealing action (through higher pre-load) at low pressure or even vacuum conditions.

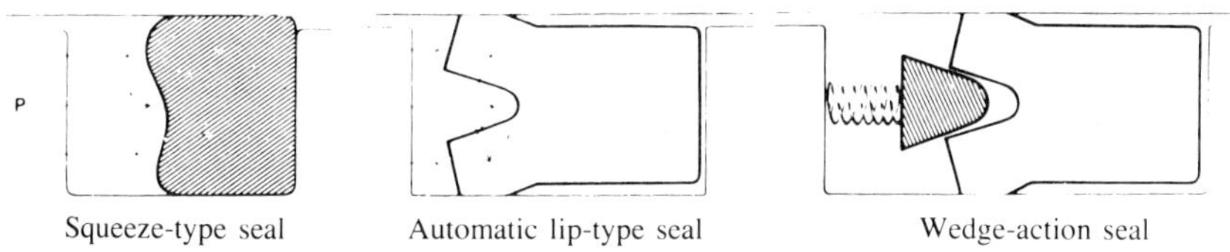

Figure 18

Rubber lip ring seals for pressures up to 160 bar (2400 lb/in^2).

The wedge-action principle has been developed in a number of proprietary seals. A logical combination is a relatively hard or tough rubber lip seal section with the lip opening filled with a soft rubber core. This core then acts as a spring to provide wedge-action. In effect, seals of this type combine the automatic lip sealing performance of a V-ring with the squeeze-sealing force of an O-ring. Both the rigidity of the main section and the form of the core can be tailored to meet specific application requirements as shown in the *Polyseal* designs in Figure 19.

Here the standard seal has a relatively flexible lip with a soft core fully contained within the lips. This provides a low friction seal under lightly lubricated conditions. In the second section the lip shape is modified by bevelling to increase the scraping action of the seal and make it suitable for dry working. The two deeper sections follow the same two movements, but the more rectangular form of the seal provides greater stability for working under heavier loads or more arduous conditions. The fifth section has the core section extended with the base section in a hard, extrusion-resistant material. The softer

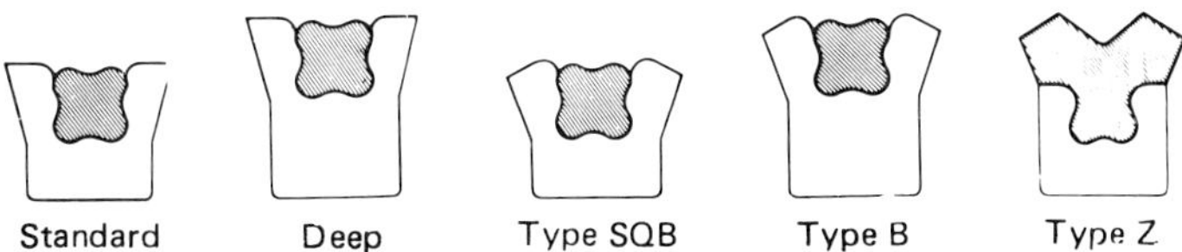

Figure 19
Proprietary wedge-action seals.

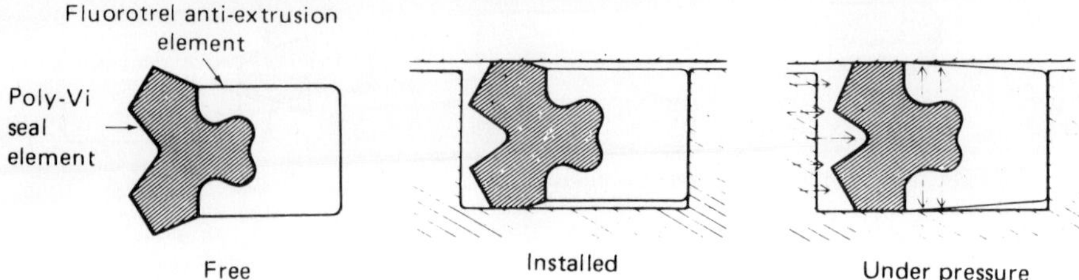

Figure 20
Sealing principle of wedge-action seals.

rubber section then forms the working part of the seal, (Figure 20). This particular type is designed for extreme environmental conditions.

Table 3 details the different materials which may be used in these particular seals.

Single-lip seals
For *single-lip* seals designed specifically as rotary shaft seals – see Section 5 on *Oil Seal Types*. Another class of single-lip seal designed for use on rods is the *wiper* seal – see chapter on *Wipers and Scrapers*.

TABLE 3 – MATERIALS USED IN POLYSEALS

Seal Type	Standard Materials		Alternative Materials	
	Core	Main Section	Core	Main Section
Types: Standard SQB, B and deep seals	nitrile rubber rubber	urethane	urethane ethylene-propylene fluoroelastomer	nitrile PTFE ethylene propylene
Type Z	fluoro-elastomer	polyester elastomer	nitrile neoprene urethane ethylene-propylene	polyester elastomer PTFE polyphnylene-sulphide

Spring-energized rings
The spring-back characteristics which make the metallic C-ring attractive can be enhanced by fitting the ring with an internal coil spring. This makes it more suitable for applications where the finish on the mating surfaces is relatively poor. For lower temperature working, even better conformity can be realized by moulding the outer jacket in PTFE, elastomers, or even better polyamids (*eg* nylon). The use of a non-metallic jacket material in this way also makes the seal suitable for both static and dynamic sealing. With a simple *C-section* for the jacket, however, the jacket itself is relatively weak. It is more usual, therefore, with spring-energized non-metallic seals for a stiffer section resembling a U-cup to be employed (Figure 21).

U-cup rings may be used in the dynamic mode, both in reciprocating and rotary motion applications. Their temperature and pressure ranges are shown in Figure 22. While it

Figure 21
Spring-energized metallic U-ring.

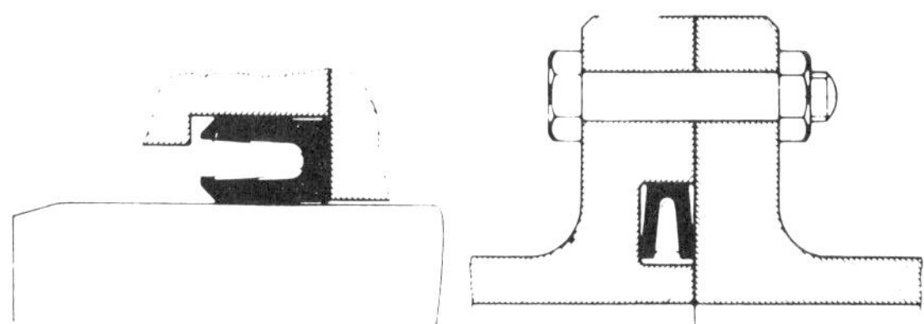

This seal comprises a PTFE or modified PTFE section with an energizing spring. Designed to be interchangeable with O-rings with freedom from O-ring failings.

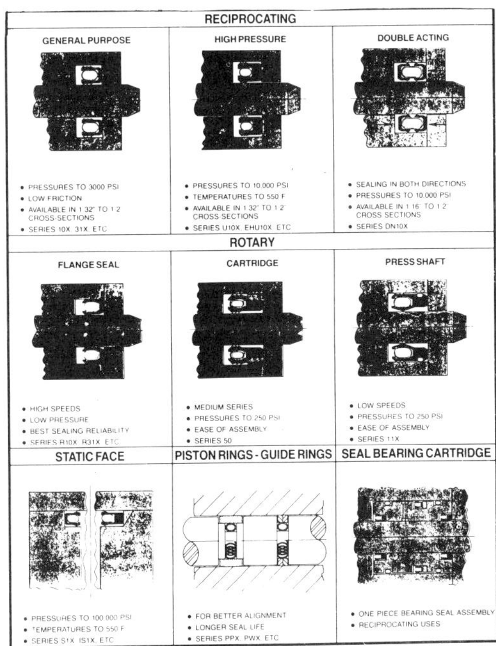

Figure 22
Temperature and pressure ranges for U-cup rings.

is true that the seals will work on relatively poor surfaces, it is also true that improving a surface by grinding and polishing will greatly enhance seal life. This results from the much reduced friction and concomitant build-up.

In many cases, the U-seals must work in a corrosive and/or abrasive atmosphere. Electroless nickel plating has been found to resist these conditions, with the added advantage that it can be applied in deep bores and intricate shapes.

Composite Seals

THE DESCRIPTION composite seal is given to two- or multi-part seals produced as a simple integral unit. They can offer an efficient, compact solution for dynamic sealing applications, with long life. Numerous individual seal designs have been produced on this basis, and complete classification is impossible. A basic classification is by composite geometry, *ie* whether the elements of the seal are co-axial or 'in-line' (axial). The latter provides the greater variety of designs. Here at times it becomes difficult to distinguish between a true composite seal (*ie* integral construction) and a seal *set* comprising (separate) elements, as the name 'composite' may be used to describe both types. A distinction would be to call the latter *compound* seals.

A typical simple co-axial seal is shown in Figure 1. It combines the inherent simplicity of an elastomeric ring seal with a rigid, low friction working face. The basic section is a square or rectangular elastomeric ring in combination with a thinner ring of PTFE to produce a composite seal. The PTFE ring is on the inner diameter in the case of an internal seal and on the outer diameter in the case of an external seal.

The seal is located in a simple groove in a similar manner to an O-ring fitting, but with groove dimensions specific to the size of composite ring. This governs the degree of squeeze and thus the radial pressure on the PTFE ring, and also locates this latter ring axially. Besides providing a bearing surface this PTFE ring also confines the elastomeric section within its groove, eliminating any possibility of extrusion so that no back-up rings are required, even at pressures of the order of 350 bar (5000 lb/in^2).

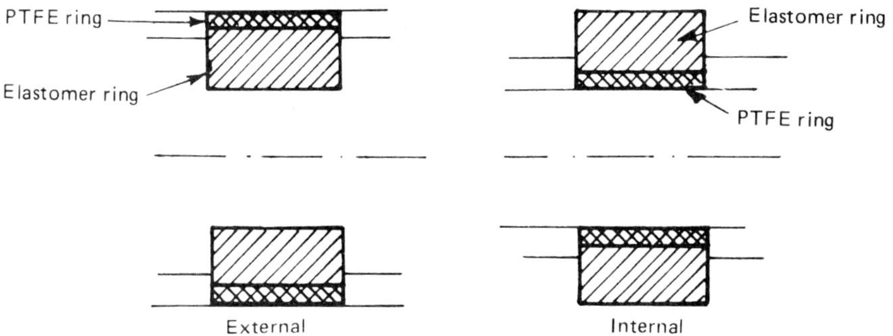

Figure 1

Broadly speaking, there are two main types of *co-axial* seal, namely the *stepped* type, for gland or low pressure applications where no leakage can be tolerated and where extended periods of inactivity are likely, and the *unstepped* type, which is used for piston head applications subject to high alternating pressures, *eg* Figure 2. Some examples of proprietary types are given in Figure 3.

Co-axial seals, because of their small configuration and their use in high pressure applications, have replaced O-ring and rectangular-section seals in servo actuators, aileron jacks and flying controls. The elastomeric part of the seal energizes the PTFE sleeve; thus contact of the PTFE on to the rod or cylinder is maintained under all service conditions. The PTFE element also acts as an anti-extrusion device or back-up ring as well as providing the low friction characteristics. At pressures above 70 bar (1000 lbf/in^2) the use of the co-axial seal proves advantageous if low friction is required. A special low crystalline PTFE can be used to ensure that the ring recovers to its original dimensions after stretching over the piston.

The low friction characteristics of the PTFE contact ring permit a certain relaxation in surface finish requirements on the mating surface; also the static and dynamic frictions of the seal are substantially the same, with marked absence of break-out force. The

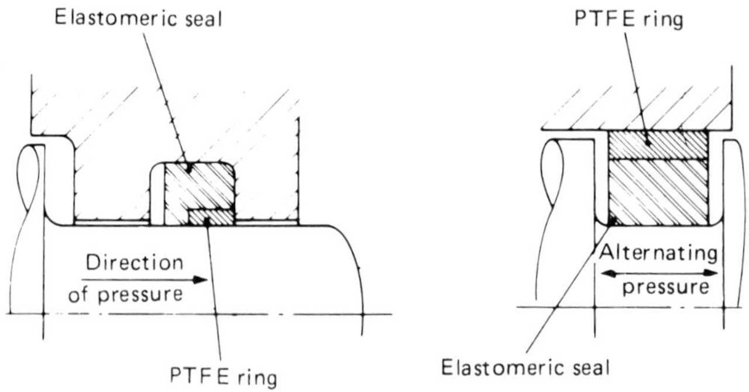

Figure 2
Stepped composite ring (left) and unstepped ring (right).

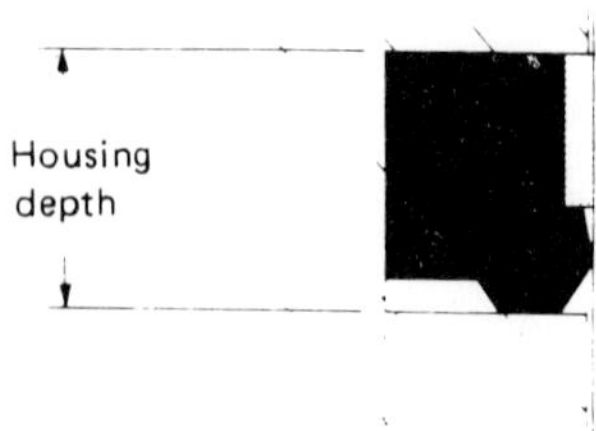

This composite seal comprises prefilled rubber section with PTFE sealing sleeve

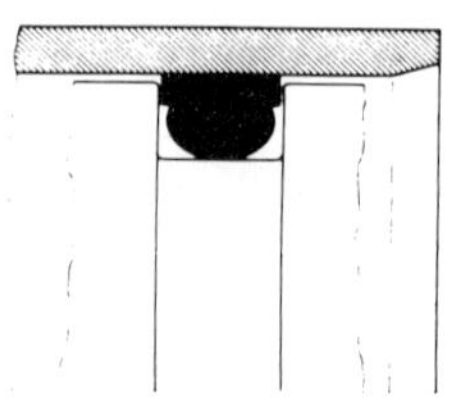

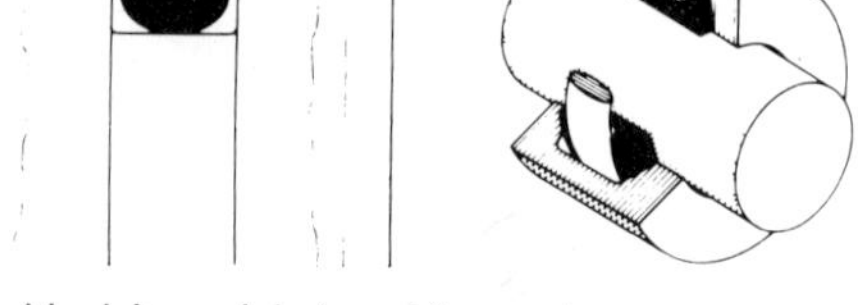

Double-delta seal designed for use in conventional O-ring grooves.

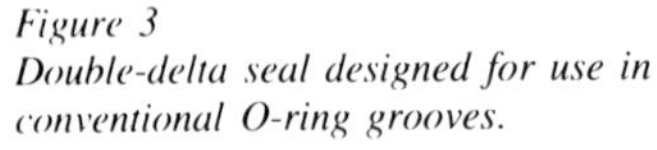

Figure 3
Double-delta seal designed for use in conventional O-ring grooves.

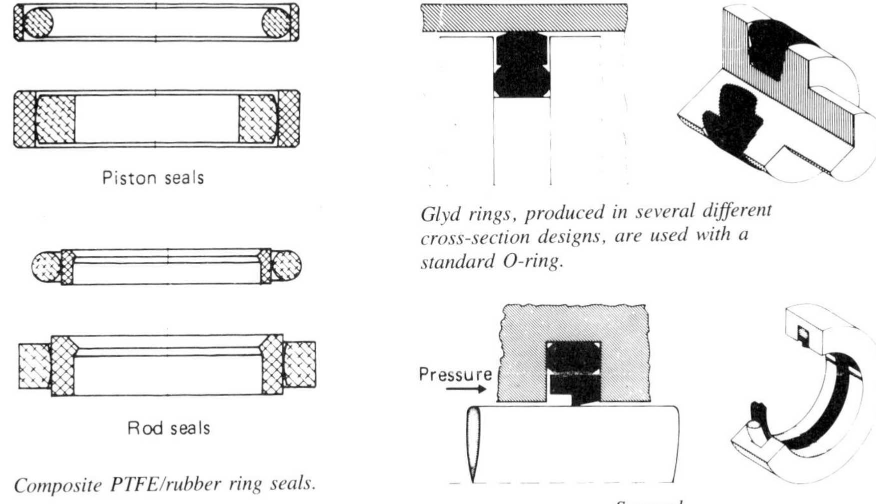

Composite PTFE/rubber ring seals.

Glyd rings, produced in several different cross-section designs, are used with a standard O-ring.

Stepseal.

seal is effective at low as well as at high pressures and the hardness of the elastomeric ring can be selected accordingly. Rubber type is selected on the basis of compatibility with the fluid concerned – hardness and compatibility being the only factors involved affecting seal performance for normal applications. A service temperature range of −60°C to +250°C can be accommodated. Maximum pressure rating is nominally 400 bar (6000 lb/in^2) but may be higher in specific applications with the smaller size of composite seal. The type is particularly suitable for piston and rod seals, installed singly. Usual precautions apply when assembling such a seal, preferably using a fitting tool to avoid excessive stretching of the thin PTFE ring and allowing ample time for this ring to recover and contract after positioning.

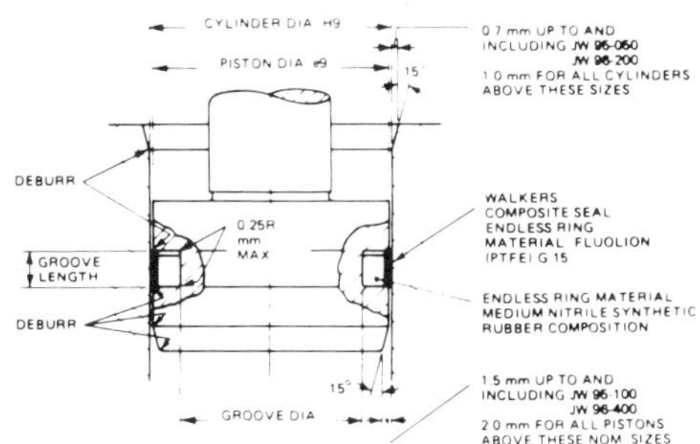

Proprietary composite seal – general detail.

Duo-Cone seals.

Axial-composite (in-line) seals follow the principle of employing an elastomeric sealing element combined with hard anti-extrusion elements. Construction may or may not be integral, and the number of individual elements involved may range from two to seven. For the purpose of description it is only possible to deal with these as individual proprietary types, although many may be similar in geometry, *eg* Figure 4 and Tables 1,2 and 3.

A
A
PISTON DIA
= MIN CYL DIA
− CLEARANCE
B
HOUSING
DIAMETER
HOUSING
DEPTH
SEATING
DIAMETER
NOMINAL SEAL DEPTH

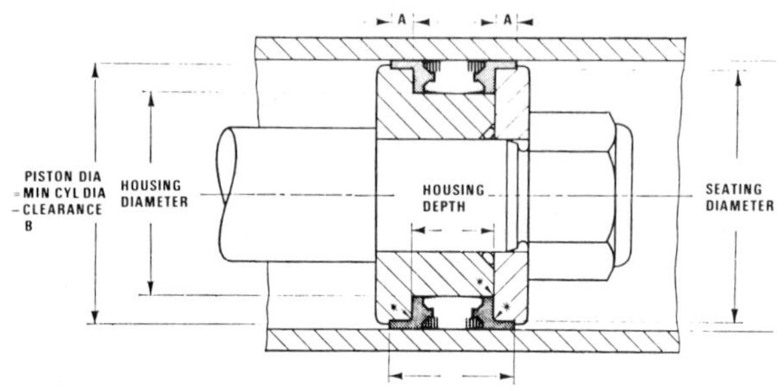

Figure 4
Double-acting composite seal.

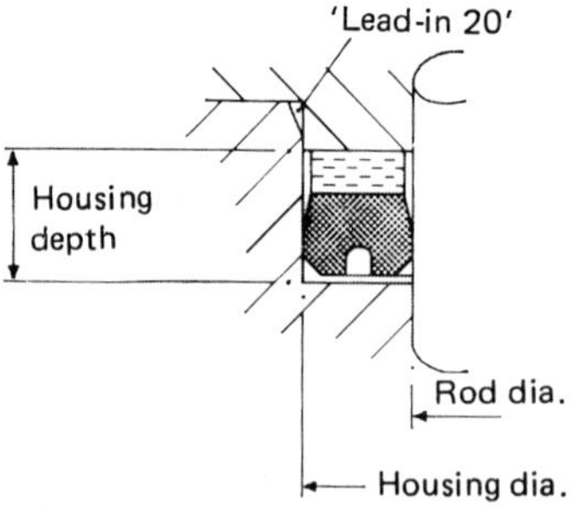

'Lead-in 20'
Housing
depth
Rod dia.
Housing dia.

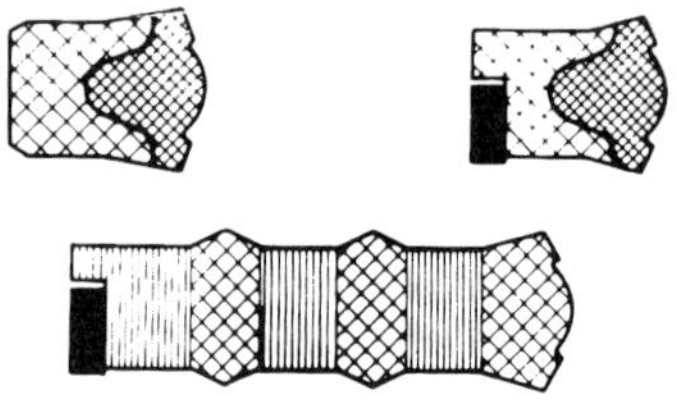

Composite seals.

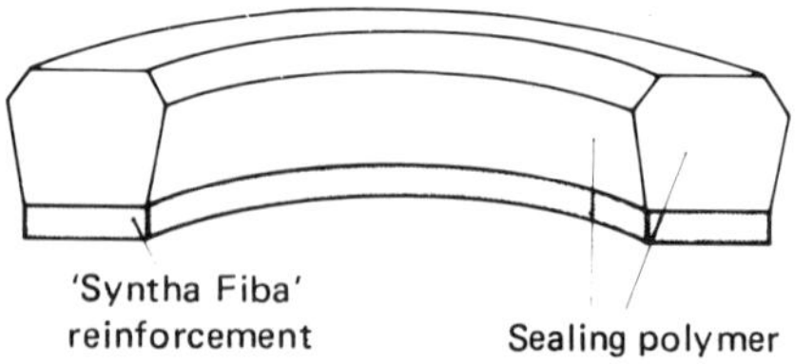

'Syntha Fiba'
reinforcement
Sealing polymer

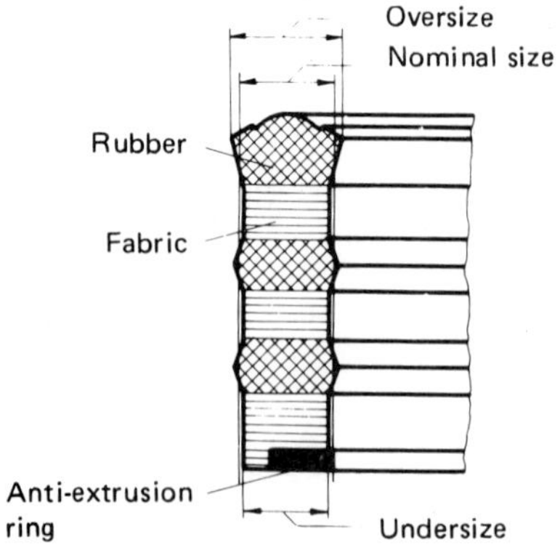

Oversize
Nominal size
Rubber
Fabric
Anti-extrusion
ring
Undersize

Composite seals.

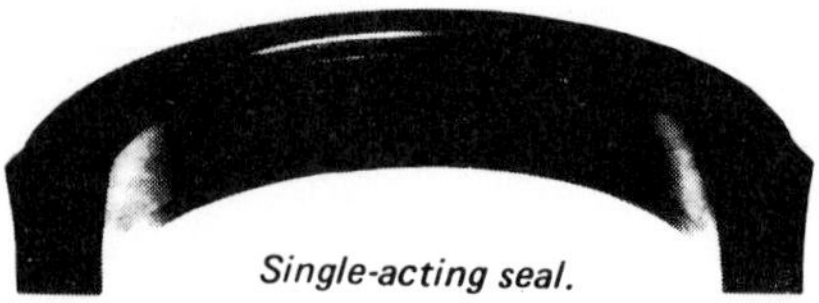

Single-acting seal.

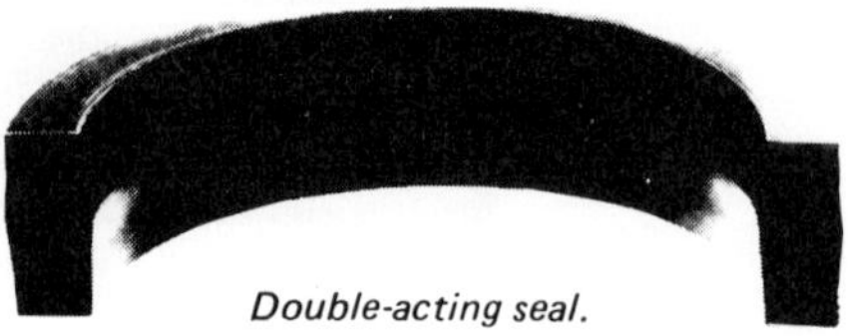

Double-acting seal.

TABLE 1

Type	Geometry	Construction	Application(s)
S11	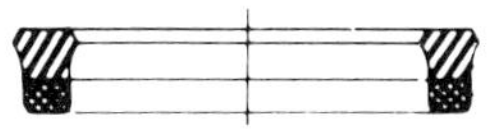	One-piece rubber moulding with fabric reinforcement at one end	Single-acting seal for reciprocating motions and pressure up to 140 bar (2000 lb/in^2)
D11		One-piece rubber moulding with fabric reinforcement at both ends	Double-acting seal for reciprocating motions and pressures up to 140 bar (2000 lb/in^2)
S11EO	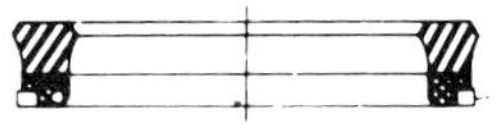	Similar to S11 with addition of nylon or acetal anti-extrusion ring recessed into heel	Single-acting reciprocating seal suitable for higher pressures or working with larger clearance spaces
D11EO		Similar to D11 with addition of nylon or acetal anti-extrusion rings recessed into each heel	Double-acting reciprocating seal suitable for higher pressures or working with larger clearance spaces
S11WO	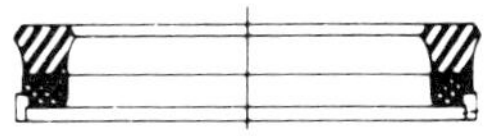	Similar to S11EO but with L-section nylon or acetal combined anti-extrusion and wear ring replacing recessed rings	Single-acting cylinder seals where generous clearances are required
D11WO		Similar to D11EO but with L-section nylon or acetal combined anti-extrusion and wear rings replacing recessed rings	Double-acting cylinder seals where generous clearances are required
S11E1	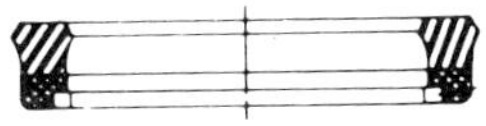	Similar to S11EO but with internal anti-extrusion ring	Similar applications to non-grooved counterparts, but more flexible seal has less friction and can also more readily accommodate any eccentricity between piston and cylinder
Grooved S11		Similar to S11 but with grooved pressure face	
Grooved S11EO		Similar to S11EO but with grooved pressure face	
Grooved S11WO		Similar to S11WO but with grooved pressure face	
Grooved S11E1	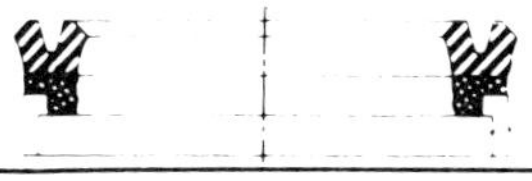	Similar to S11E1 but with grooved pressure face	

TABLE 2

Name	Geometry	Maximum Pressure Rating bar	Maximum Pressure Rating lb/in^2	Service Temp Range °C Continuous	Sizes Metric (mm)	Sizes Imperial (inch)	Remarks/Application
Balmaster Single-acting	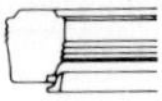	500	7000	−20 to +110	20–140	0.75–6.0	Rod and general gland seal
Balmaster Double-acting	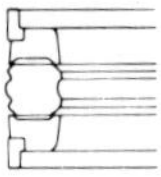	500	7000	−20 to +100	40–200	1.5–8.0	Suitable for one-piece or multi-piece pistons. Also available with split-support and bearing rings.
Selemaster Single-acting		700	10000	−20 to +100	12–500	0.5–20	Heavy duty rod seal.
Selemaster Double-acting	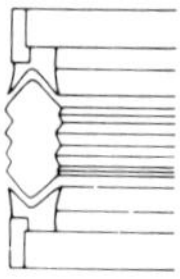	700	10000	−20 to +110	45–250	2–14	Heavy duty piston seal designed to be used in conjunction with single-acting Selemaster.
Balsele Single-acting	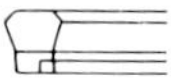	180	2500	−20 to +110	6–800		Compact single-acting gland or piston seal. Pressure rating may be enhanced by fitting anti-extrusion ring.
Balsele Single-acting	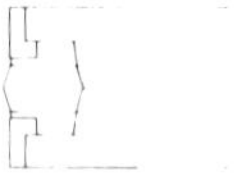	350	5000	−20 to +110	16–380	0.875–15	Hydraulic cylinder seal incorporating built-in bearing rings.
OP70	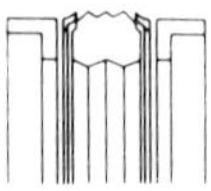	250	3500	−20 to +110	50–250	2–8	Double-acting seal for one-piece pistons; incorporates wear rings.

*Weir Polypac Limited

TABLE 3 – UNIT SEALS (COMPOSITE SEALS)

Description	Geometry	Maximum Pressure Rating bar	Maximum Pressure Rating lb/in^2	Speed m/sec	Speed ft/sec	Sizes Metric mm	Sizes Imperial inch	Applications
Single-acting		350	5000	1	3	16–140	0.5–8.5	For pistons and rods
Single-acting with anti-extrusion ring		700	10000	1	3	16–200	0.312–3.500	For rods
Single-acting with anti-extrusion ring		700	10000	1	3	32–130	1–4	For pistons
Single-acting with wear ring and retaining ring		350	5000	1	3	30–110		For pistons
Double-acting		350	5000	1	3		0.75–3.000	For pistons
Double-acting with anti-extrusion ring		700	10000	1	3		1.250–2.750	For pistons
Double-acting with wear rings		350	5000	1	3	30–90	1.250–7.000	For pistons
Double-acting with wear rings for one-piece pistons		350	5000	1	3	30–200		For pistons
Evco Mercury double-acting with wear rings		700	10000	1	3		2.00–8.000	For pistons
Evco Apollo single-acting		700	10000	1	3	25–160	H	Heavy-duty seals for rods
Evco Apollo double-acting		700	10000	1	3	65–250		Heavy-duty seal for pistons

The *Hallprene Delta* seal (Figure 5) has proved a particularly successful type of composite seal for double-acting hydraulic cylinders. Nominal pressure range is 0 to 210 bar (0 to 3000 lb/in^2) and temperature range -40 to $+100$°C. Surface speeds up to 0.5 m/s (100 ft/min) can be accommodated with recommended bore finish of 0.4 μm (16 μin) Ra or 0.2 μm (8 μin) Ra for speeds above 0.3 m/s (60 ft/min).

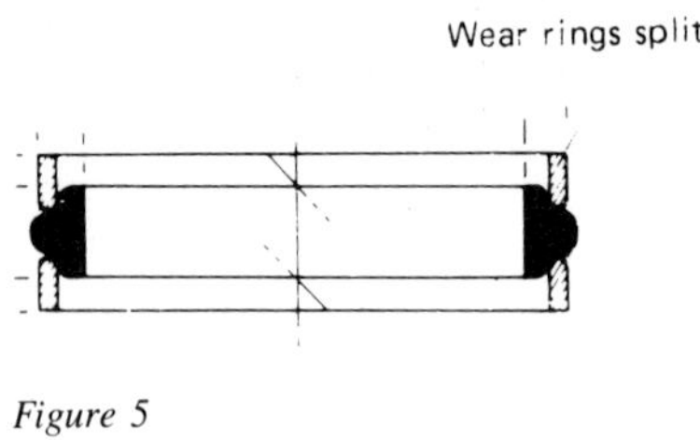

Figure 5
Delta seal.

The Delta seal consists of two acetal-resin bearing rings and a delta-section *Hallprene* synthetic rubber sealing member. Under pressure, the seal exerts a wedging action on the bearing rings, reducing the clearance between the cylinder bore and the bearing rings, thus minimizing the possibility of extrusion of the seal.

Assembly is carried out as follows:

(i) The intermediate seal is stretched over the piston into the central recess.

(ii) The split bearing rings are positioned with the chamfer-side adjacent to the intermediate seal.

Standard materials are compatible with mineral-based hydraulic oil, water-soluble and water glycol fluids.

Special materials (ethylene propylene or fluoroelastomer) are available for synthetic non-flammable fluids (*ie* phosphate ester-based fluid). Fluoroelastomer seals require housing alterations.

The *Hallite Alpha* seal (Figure 6) is a development of the Delta seal, with maximum pressure rating extended to 315 bar (4500 lb/in^2). It has five elements: a rubber seal, two split anti-extrusion rings in acetal resin and two split bearing rings in acetal resin.

The seal proportions are of a robust section but friction characteristics are not excessive. Due to the support of the anti-extrusion rings, the Alpha seal has a great capacity for withstanding transient pressure surges. The anti-extrusion rings are so designed that the

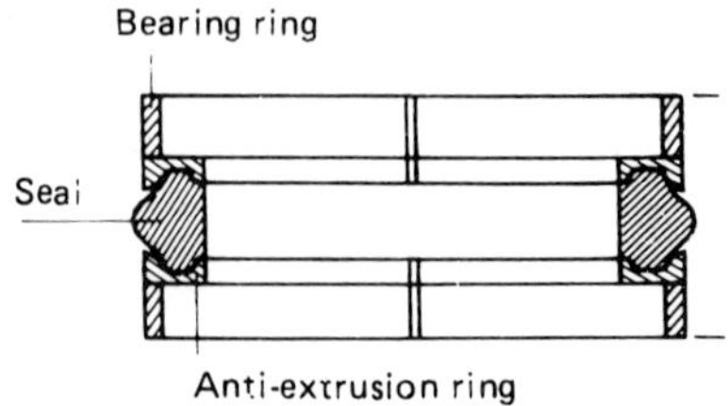

Figure 6
Alpha seal.

seal is protected by the inner and outer lips, which when energized close the extrusion gap. The bearing rings, mounted independently of the seal, are suitable for a wide range of applications. Two lengths of bearing rings are available for most sizes, but if there are heavy side-loads, or there is a possibility of 'dieseling' heavy-duty bearings can be used in place of phenolic fabric material.

The *Crown* seal shown in Figure 7a is another simple double-acting seal, the actual seal element being in polymer elastomer or polyurethane rubber combined with a conventional O-ring. The O-ring acts both to spring-load the seal and as a static seal. At the same time the seal section acts as an anti-extrusion device for the O-ring under pressure energization.

Similar principles have been developed in the seal shown in Figure 7b where the seal is of triangular section elastomer associated with hard triangular back-up rings. These act both as anti-extrusion devices and scraper rings. Due to the hardness of the anti-extrusion elements the Poly Tri-Seal can accommodate larger extrusion gaps than conventional seals, making it particularly suitable for pistons using non-metallic wear rings. A version of this seal is also available with bearing rings included in the assembly. Working characteristics of these seals are given in Figure 8.

The seals listed in Tables 1 and 2 are a specially-designed range of composite seals based on rubber mouldings with fabric reinforcement section(s) and additional anti-extrusion and/or wear rings where appropriate. They are generally suitable for pressures up to 140 bar (2000 lb/in^2) without extrusion devices, and higher pressures when so fitted. The range covers a variety of different sections.

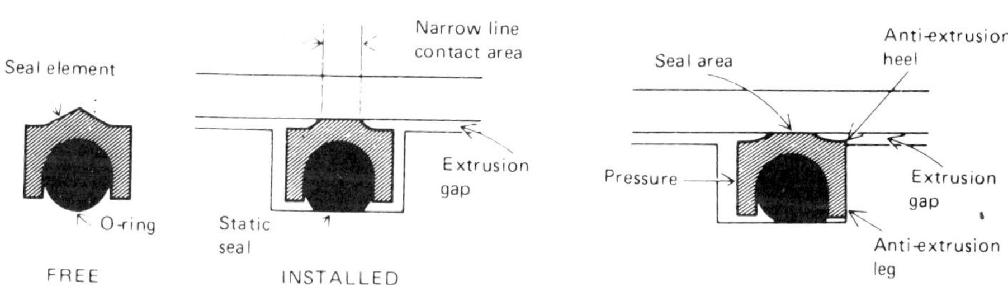

Figure 7a

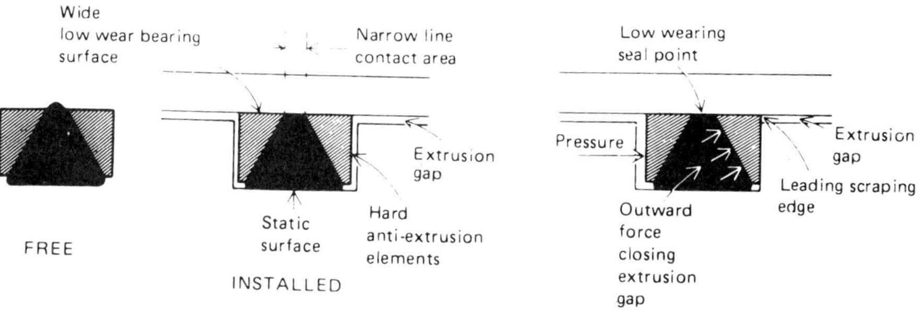

Figure 7b

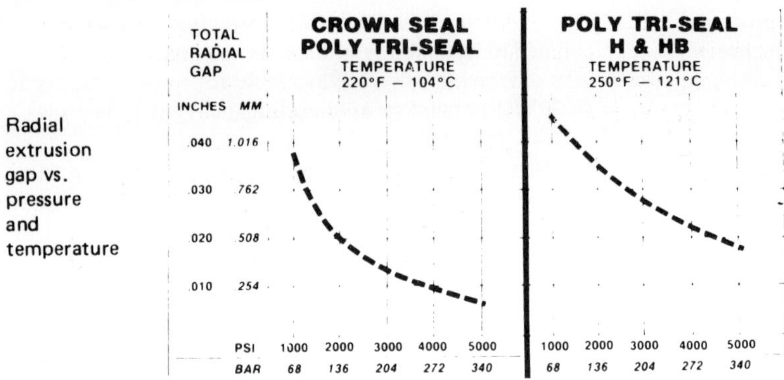

Figure 8

The *Selemaster* shown in Figure 9 is an extra high-duty composite seal specifically intended for very high pressures and arduous working conditions.

The rubber sealing element is designed with a number of sealing lips, offset on the inner and outer faces to form a flexible unit which seals effectively at pressures from zero upwards and which accommodates considerable eccentricity between rod and bush or piston and cylinder.

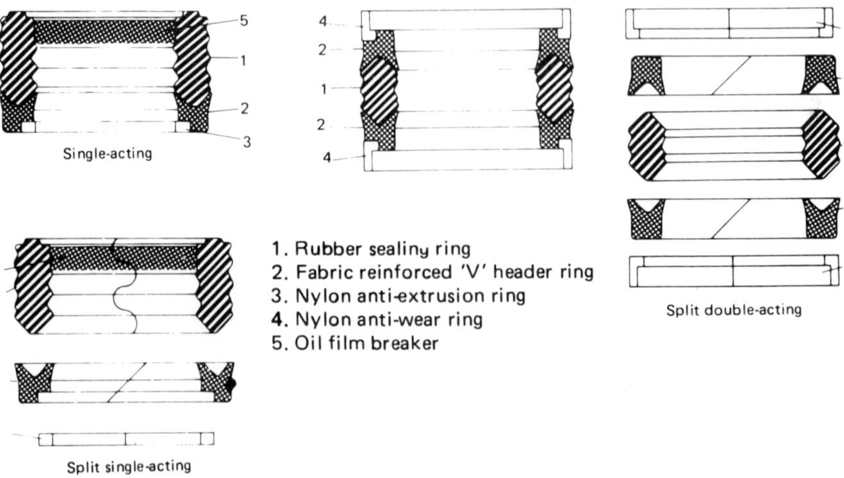

Figure 9
Composite seals

The flexible rubber element is supported by a fabric-reinforced V-shaped header ring at one end of the single-acting seal and at both ends of the double-acting seal. This is reinforced with an acetal resin anti-extrusion ring or wear ring.

A similar range of composite seals are shown in Table 3.

Another common form of composite seal is that based on a fabric U-ring section and an elastomeric section, which can be considered as direct replacements for O-rings and/or U-rings and with superior performance characteristics to either basic types. The principle of combining a U-shaped lip ring in rubber-impregnated fabric with an integral rubber support ring provides good sealing at low pressures (*ie* better than a conventional U-ring), with enhanced pressure energization – and thus better sealing – at higher pressures. The complete seal section is more stable.

Again this leads to a variety of individual designs, *eg* Figure 10, some of which also include anti-extrusion devices. Seals of this type may be true composite (integral) construction, or compound, comprising separate U- and O-ring forms. (See also chapter on *Flexible Lip Seals* for further details of the latter type).

The *Hallprene* range of fluid seals developed from an original combination of a rubber O-ring with either a single rubber-impregnated fabric U-ring (single-acting seal), or two

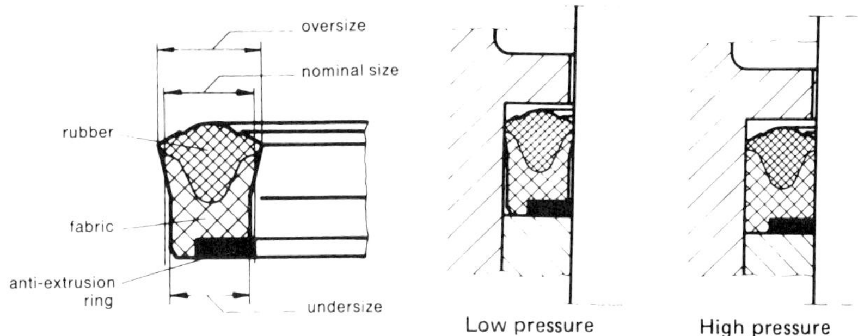

Figure 10
Piston seal.

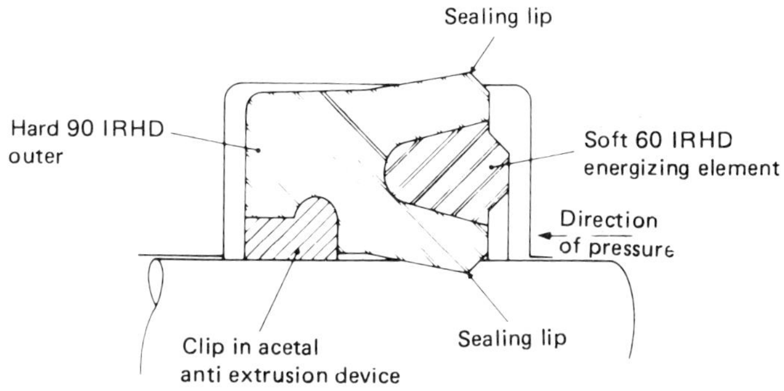

Resilient gland seal.

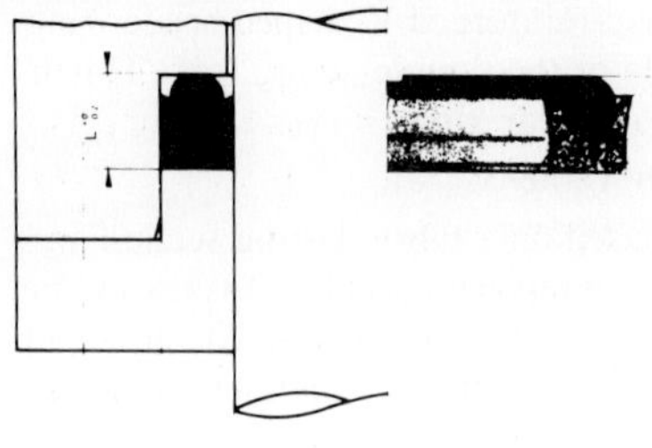

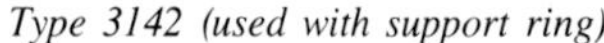

Type 3142 (used with support ring)

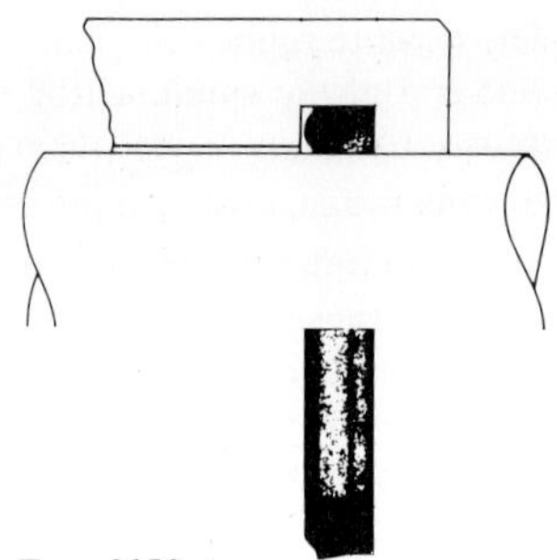

Type 3150

U-rings (double-acting seal). The two configurations arrived at are shown in Figure 11. Further single- and double-acting *Hallprene* seals subsequently appeared with integral bearing rings (Figure 12) to minimize scoring in the event of pick-up between piston bearing and cylinder bore. A further advantage of the bearing rings is that they also act as anti-extrusion devices when high pressures are being sealed.

Hallprene seals can provide efficient sealing up to 700 bar (10000 lb/in^2), but as pressure increases finer surface finishes are called for and reduced extrusion gaps. A surface finish of 20 μin Ra is adequate for pressures up to 70 bar (1000 lb/in^2) but should be of the order of 5 μin Ra for good seal life at higher pressures.

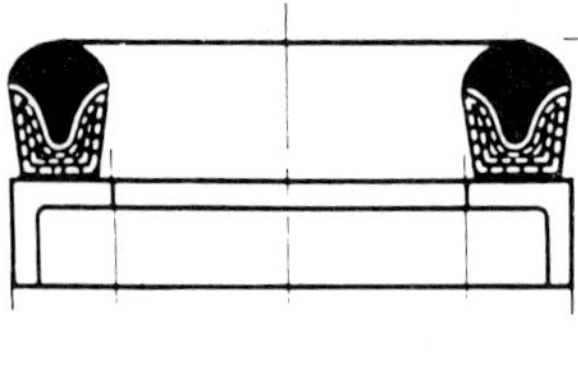

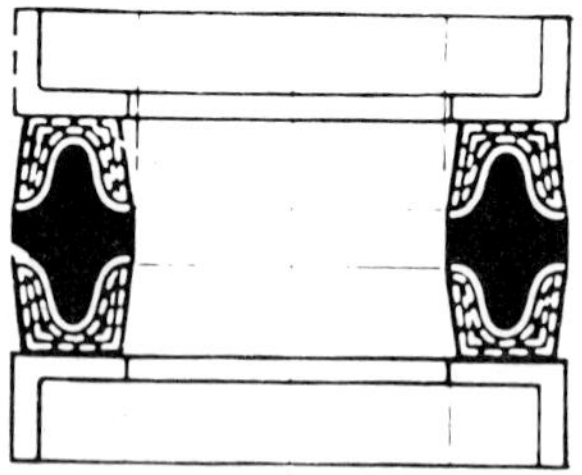

Figure 11
Single-acting and double-acting seals.

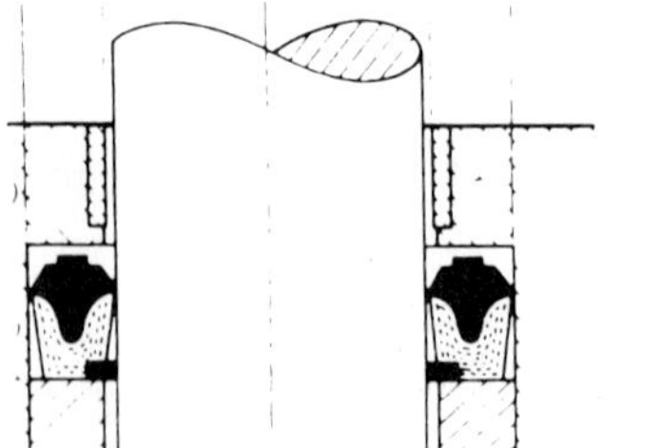

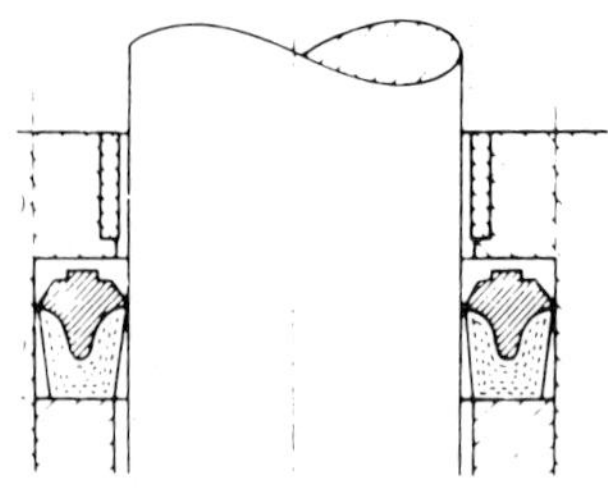

Figure 12
Composite seal with and without anti-extrusion ring.

Servo wedge seals

The *Servo wedge* seal consists of two PTFE members and a metal spring as illustrated in Figure 13. The metal spring radially energizes the wedges against the shaft bore and side walls prior to the application of sufficient system pressure to provide the necessary sealing force on the unsplit wedge.

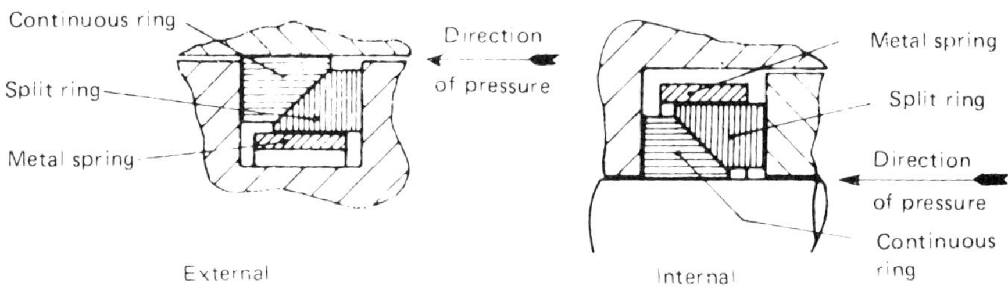

Figure 13
Servo wedge seal.

The *Servo wedge* seal was introduced to replace elastomeric seals operating in temperature environments in excess of 200°C. These seals have been developed for aircraft fuel systems and engine applications over an operating temperature range of −50°C to +230°C at pressures not exceeding 138 bar (2000 lbf/in^2) and give low hysteresis values. The seals are suitable for use as a piston seal (external) and a *shaft* seal (internal) under static or dynamic conditions where pressure is single-acting.

The materials of the seal assembly are such that it can be used with all known fuels, oils and engine lubricants.

Controlled leakage seal

The seal shown in Figure 14 is a composite seal which relies on the natural ability of loaded PTFE to resist axial shear, and assure anti-extrusion qualities at high pressure. Since this material also possesses a very low co-efficient of friction, it is ideal for occasional contact with the shaft or dry-running on start-up.

Essentially the seal consists of a flexible elastomeric element which eliminates leakage round the back of the seal housing, and a loaded PTFE sleeve designed to be a spin fit on the shaft. This sleeve is attached to the elastomeric component by tabs arranged radially to prevent independent motion.

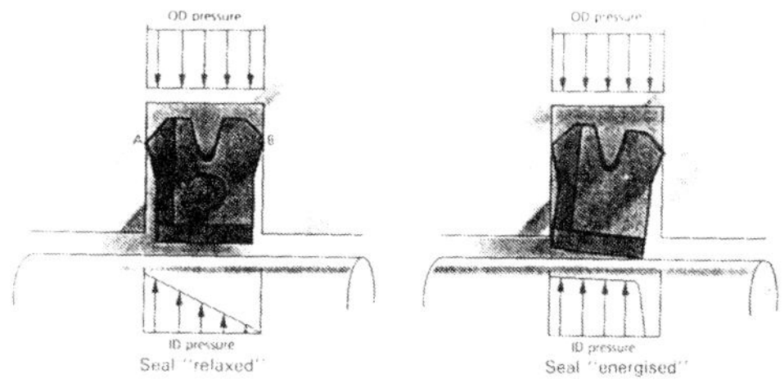

Figure 14
Composite seal showing PTFE resisting axial shear.

The ported pressure face (A) allows system fluid to the outside diameter of the housing, but the unported atmosphere face (B) forms a static seal, preventing the fluid from continuing round the back of the seal to atmosphere.

Because the seal is designed to give a small clearance on the shaft, fluid can pass between the shaft and the PTFE running sleeve.

If the sleeve was rigid, the pressure would fall almost linearly to atmospheric pressure along the sleeve length, and there would be high leakage. Pressure distribution on the inside and outside of the seal produce a moment (M) which causes the seal to tilt, bringing the atmosphere edge of the seal into close proximity to the shaft. This movement stems the leakage.

The undersleeve pressure then "backs up" to produce a distribution approximating to that shown. Most of the pressure drop occurs near the atmosphere edge of the seal.

The hydrodynamic pressure generated in the thin fluid film as a result of shaft rotation prevents loaded contact, and an equilibrium is set up between the closing and the opening forces.

The seal is therefore in virtual pressure balance, thus providing high pressure capability, and running on a hydrodynamic fluid. This gives it its high speed capability.

Since the seal is a clearance device, a certain amount of leakage must be expected, but this often amounts to less than 0.3 L/h (about one drip every three seconds).

A particular compound seal set has been specifically developed to eliminate the damaging effects of severe operating environments created in high frequency, short stroke sealing applications. Under these conditions a standard seal gets no lubrication in the centre and in this area the limiting PV of the dynamic PTFE seal ring is exceeded causing a catastrophic failure of the seal.

The design of this type of seal allows the sealing area to be lubricated by using an interconnecting series of radial, axial, and circumferential –bleed" grooves (see Figure 15) in the PTFE based dynamic sealing element which prevents premature failure. The PTFE ring is statically energized by an elastomeric O-ring. This technique typically extends the functioning service life of the seal set by 300 to 400%.

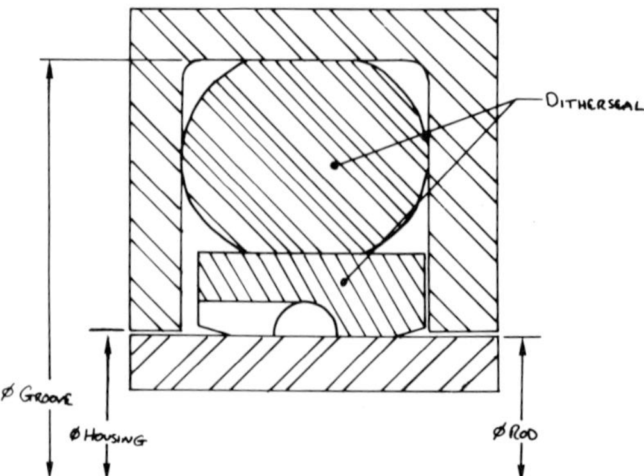

Figure 15
Compound seal with lubrication grooves to prevent damage from operating environment.

The seal shown in Figure 16 combines a separate rubber sealing element which has the cross-section of an O-ring with a harder, low friction, PTFE based sealing element into an overall geometry that optimizes the main advantages of each material and eliminates their main disadvantages.

This seal is designed as a secondary rod seal, used in combination with a uni-directional primary seal, in an unvented, redundant sealing system that requires zero external leakage (refer to Figure 16). Elastomeric contact is required to wipe the rod dry at the low pressures encountered between the seals. In order to minimize the friction, the elastomeric contact of the O-ring is kept to a near line contact with the rod. In the event of a primary seal failure the seal is capable of containing very high system pressures without a significant increase in friction. The seal is also designed with circumferential mini-grooves on the dynamic sealing surface which further aids in reducing friction, eliminating external fluid leakage and maximizing service life.

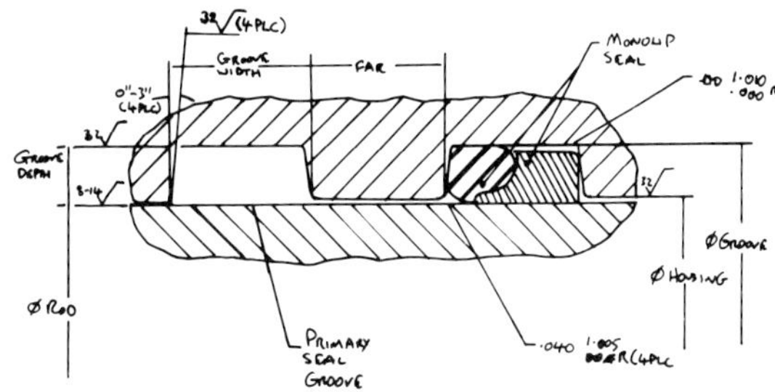

Figure 16
Secondary rod seal developed to reduce friction in an unvented sealing system.

Felt Seals

FELT IS a fabric composed of interlocking fibres of wool used alone or combined with animal, vegetable and synthetic fibres. It is a material which has long been used for sealing duties because of numerous favourable characteristics, notably its oil wicking and oil absorption properties, effectiveness as a fine filter, and resilience. The latter enables a *felt* seal to maintain constant sealing pressure under adverse operating conditions, or wear. With wear, too, the surface of a felt seal remains unchanged, *ie* it does not glaze like many other materials.

Felt seals are used mainly as oil or grease seals on shafts for retaining lubricant, and at the same time preventing dust or dirt from entering the sealed system. They are cut from felt sheet in a variety of ring shapes for installation in grooves or glands, or in casings for cartridge assembly. *Cupped* or *flanged* felt seals are also produced. *Ball-bearing* seals are cut to fit a machined housing.

Felt seal rings are usually pre-saturated with lubricants of slightly greater viscosity than that used in the bearing. Plain felt seals afford positive bearing protection and provide a reservoir for lubricant storage – making it available as needed. If run dry, they tend to protect and polish rather than score a shaft and seldom fail through ageing, embrittlement or disintegration. Under normal bearing temperatures and operating conditions, the felt seal is highly economical and requires replacement only when the machine in which it is used is overhauled. It should be emphasized that these seals are not suitable for use in conjunction with oils of extremely low viscosity, nor to retain pressurized lubricants.

Felt seals properly cut, and properly installed, are effective over a variety of operative conditions and a wide range of speeds. Normal maximum rubbing speed is 10 m/s (2000 ft/min), but can be as high as 20 m/s (4000 ft/min) where shafts are hard and smooth and ample lubricant is present in the seal. Because of their resilience, felt seals provide close contact between the seal and shaft surface without undue pressure. Excessive pressure must be avoided as the structure of felt can be modified or damaged by compression. Equally the structure of felt can be destroyed if stretched excessively; this calls for care in assembly.

Felt seals are normally replaced when a machine is overhauled, regardless of condition. If overhaul periods are frequent, the designer of the seal housing should allow for easy replacement. Where long periods are specified between overhauls, the felt seals may require replacement at more frequent intervals.

Specific points applying to the design and installation of felt seals are:

(i) Avoid excessive stretching when mounting.

(ii) The felt seal must not be fitted too tightly to the shaft and the retainer should not exert an excessive compressive force on the felt.

(iii) In the usual installation, the height of the felt-ring section should be greater than its width. This proportion minimizes seal distortion and permits firm clamping of the felt in its groove.

(iv) When *ring felt* seals are not practical due to design or size, *strip felt* seals can be economical and effective, if properly installed. However, when a strip seal is used the joints should be bevelled at an angle of 30° so that the compression of felt in the carrier groove fills any gap.

Temperature limits for felt seals are normally −50°C to +120°C (−60°F to +250°F), although synthetic fibre felts may have a maximum service temperature of 200°C (+400°F). The co-efficient of friction of felt is low, averaging about 0.22 for dry felt rubbing on steel, falling to 0.15 when the felt is pre-saturated with lubricant. Felt seals are normally saturated with lubricant before assembly and can absorb up to 80% of their volume of oil. The capillary action of the felt also enables the material to retain lubricant well, so felt seals do not dry out during extended idle periods.

As a filter, felt seals can trap and retain particles as small as 0.7 microns. To meet special requirements, felt can be impregnated with one of several materials. Compounds of paraffin, petroleum jelly or colloidal graphite increase the resistance of basic felt to water and mud, improve its resistance to pressurized lubricants, and lower its co-efficient of friction.

Wipers and Scrapers

WIPERS AND scrapers are classfied as exclusion seals (or exclusion devices), used in conjunction with sliding or reciprocating motions. The two descriptions are often used synonymously, but basically a wiping action is gentle, such as that provided by the thin lip of a flexible material. A scraping action is tougher, such as that given by a harder, semi-rigid edge like a flexible metal strip. A further difference is that a wiper will often act also as a seal whereas a true scraper has little or no sealing action as such, although its shape may produce a sealing effect in throwing surface contaminants clear.

Probably the best known application is the *rod wiper* which protects the rod of a pneumatic cylinder from drawing foreign matter into the cylinder on the in-going stroke, the wiper in this case being of ring form encircling the rod. In such cases it is used purely as a protective device and any necessary sealing of the rod would be accomplished by a separate gland or rod seal. In other applications wipers may be employed to traverse straight or curved surfaces, employing straight or profiled wiper blades, respectively, again to remove surface contamination. The same principle of working is involved, although the wiper section employed and the form of construction of the wiper unit may differ appreciably.

Rod cleaning devices, therefore, embrace either *wiper* rings or *scraper* rings, depending on their construction or action and some may combine the two actions of hard wiping and sealing. In addition, they may also be described as wiper seals or scraper seals, although intended as protective rather than sealing devices. The implication is normally that the shape of the wiper (or scraper) is such as to prevent build-up of particles around and under the wiper. Such devices are not intended to perform the dual role of protection and rod sealing. A separate seal is invariably employed for the latter purpose.

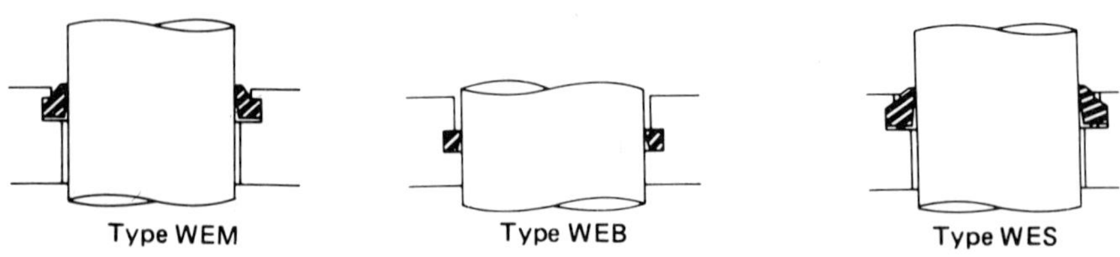

Wiper seals.

Rectangular section wipers

A simple rectangular section of felt, hard rubber or any semi-resilient material mounted with pre-load so that it acts as a simple compression type seal will have a wiping action over any substantially flat surface on which it rubs (Figure 1). Considered in terms of a normal surface motion traversing the face of the wiper backwards and forwards the wiping action will be similar in both directions. The degree of protection provided by the wiper against the passage of surface particles will depend to a large extent on the pre-load pressure and the nature of the wiper face. The latter may absorb and hold particles which tend to roll under its edge (*eg* an *oil-soaked felt* wiper) or deform to the extent that it can ride over such particles completely. An increase in pre-load pressure may improve the protection offered, but only at the expense of increased friction. In neither case is protection satisfactory, and abrasive particles stopped and trapped by the section can often be more damaging than those passed, as far as eventual surface scoring is concerned. Nevertheless, simple felt and other wipers of this type have their uses, although for good wiper action a rather more sophisticated section is to be preferred.

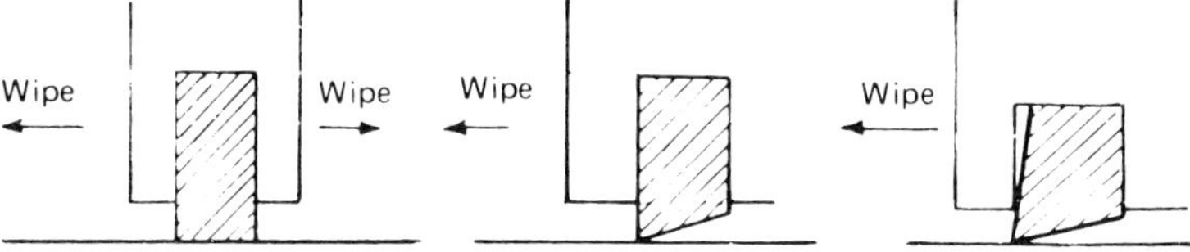

Figure 1

A far more effective wiping action is provided by a *chamfered-section*. This will have the effect of increasing the wiping efficiency in one direction (positive chamfer angle), and decreasing it in the other. Essentially, therefore, a chamfered face makes the wiper unidirectional and its efficiency as a wiper is directly related to the (positive) chamfer angle involved (Figure 2). Tests show that wiping efficiency increases dramatically from a chamfer angle of zero up to about 3°, with a further slight increase in efficiency with increasing chamfer angle up to about 10°. Any further increase in chamfer angle has no further effect, other than that if the angle is considerably increased the bottom section may lack stiffness and deform or collapse under the wiping motion. Thus a chamfer angle of 4° to 10° is entirely adequate for obtaining maximum performance from a simple wiper section.

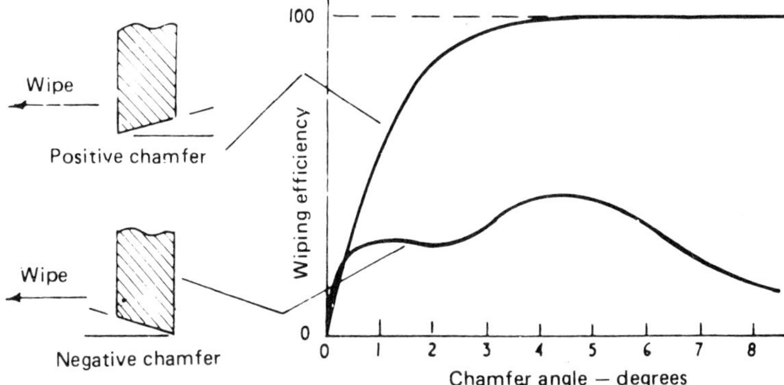

Figure 2

Satisfactory performance from a simple wiper of this type will be dependent on establishing an optimum pre-load for the section and material involved. With plain rectangular-sections an increase in pre-load will normally produce a proportionate increase in wiping efficiency, but this may also be accompanied by high wear. With a chamfered wiper an increase in pre-load above an optimum figure may show a *decrease* in wiping efficiency, as well as increased wear.

Wiper action in all cases is that of scraping over the surface on which the wiper rides, although in fact the two surfaces may remain separated by an oil film of microscopic thickness. The closer the wiper edge approximates to a true knife edge the more efficient the scraping action can be expected to be, although such a thin section may be subject to high wear unless this is countered in the design. At the same time the lip or sealing edge must be sufficiently elastic to avoid breaking the oil film on the rod.

Lip seal wipers

The same is generally true of *notched-* or *specially shaped*-sections (Figure 3). Sections of this type embody the positive chamfer characteristics of the simple chamfered wiper but deliberately aim at a more flexible foot capable of providing high wiping efficiency at moderate pre-loads. This avoids the problem of excessive friction and, more important, excessive wear which can result from high pre-loads. With a reasonably flexible foot the pre-load necessary to keep the wiper in contact with the surface may be only of the order of 0.2 kg per cm (1 to 1.5 lb per in) length of wiper lip. Actual performance as a wiper, however, may be inferior to that of a simple chamfered-section with a substantially higher pre-load, particularly at low speeds, although again much will depend on the wiper material. The high wear rate associated with high pre-loads can rapidly deform and modify an original section so that its efficiency falls rapidly.

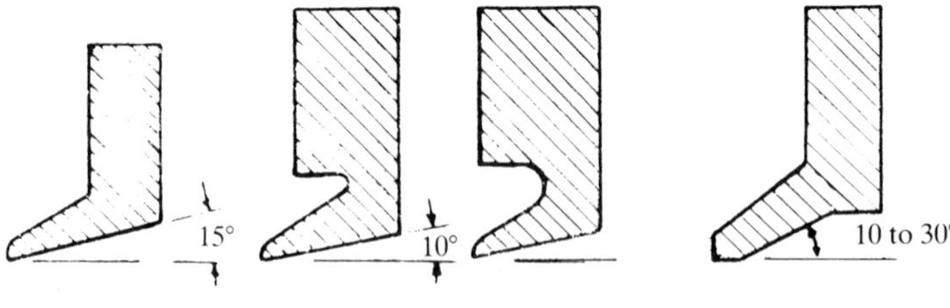

Figure 3

Both from a design and operating point of view a flexible foot section is generally to be preferred as this is effective with minimum pre-load. A reasonably strong leading edge is required, however, both to resist any lifting action due to motion or pre-load, and also to provide uniform wear at the contact face. This may call for reinforcement of the basic moulded section with a metal insert or casing, depending on the size of the wiper and its particular application.

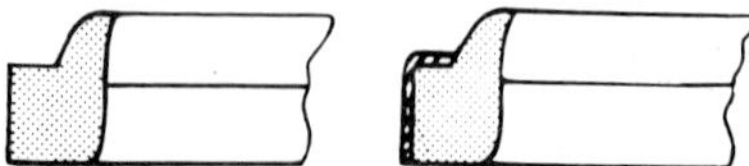

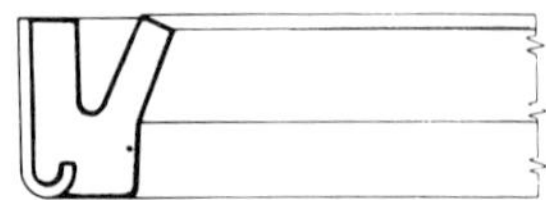

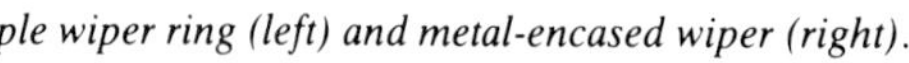

Simple wiper ring (left) and metal-encased wiper (right).

Metal-clad wiper.

Small ring wipers may need little or no stiffening. Large straight wipers may need considerable stiffening, as well as requiring a rigid section for mounting. With all such types of flexible sections the actual shape is a major factor governing performance.

The majority of wiper rings for cylinder rods are simple synthetic mouldings with a semi-flexible foot section and a moderate degree of positive chamfer (Figure 4). These are simply fitted in a matching groove and are essentially unidirectional wipers. Alternatively, the section may be double-sided to provide wiping in both directions if required. Metal-cased rings can be employed for fitting into open-ended housings.

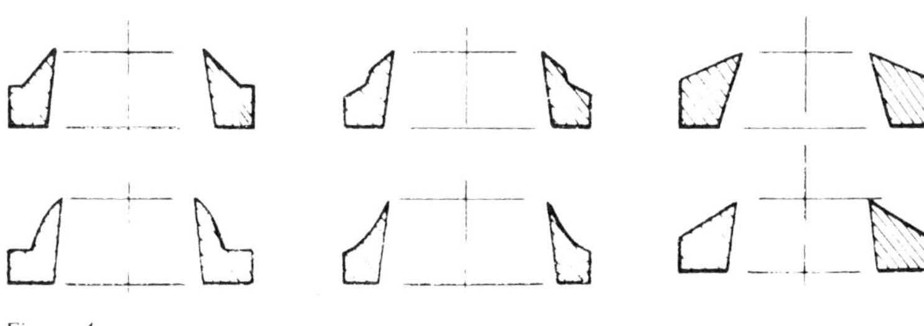

Figure 4

Wipers can be categorized into three types: light-, medium- and heavy-duty (Figure 5).

Light-duty wipers are normally fitted to devices which are operating inside buildings, and thus are only used for preventing damage to seals and bearings from dirt, dust and contaminants found in a normal factory environment.

Medium-duty wipers normally comprise a hard rubber or polyurethane bonded to a metal case. The design incorporates a sharp wiper lip which will remove most contaminants (except ice) and can be a press fit into an open ended recess. This type of wiper is most popular and is highly effective.

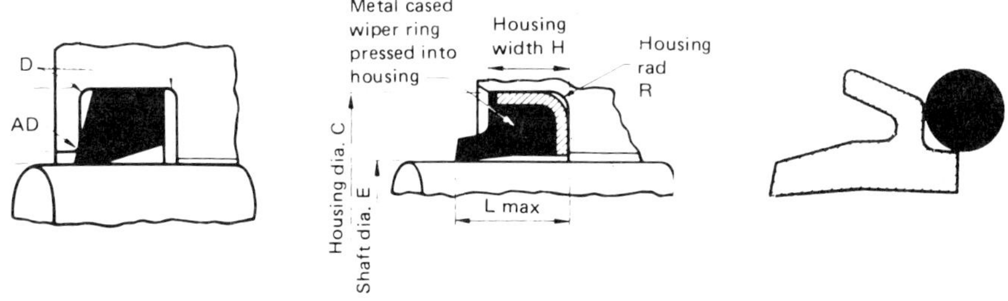

Figure 5
Examples of light-duty (left), medium-duty (centre) and heavy-duty (right) wiper rings.

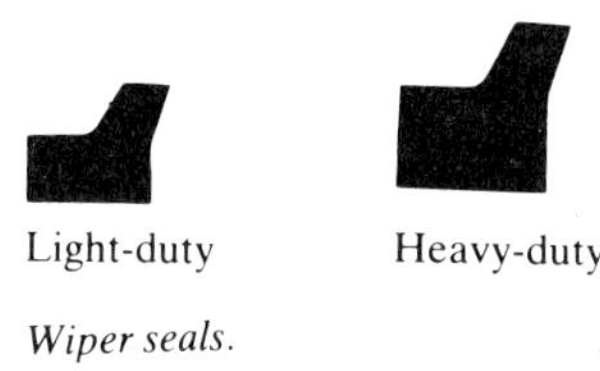

Wiper seals.

The more aggressive contaminants *eg* ice and caked mud, require a *heavy-duty wiper/scraper* such as on earth-moving equipment. One type of heavy-duty wiper scraper incorporates a hard acetal scraping element which is a slight interference fit on the rod backed up by a conventional O-ring to provide a dust and moisture seal around the o.d. and back of the scraper. The o.d. of the scraper has a clearance in its housing groove which allows the assembly to float during working operation. This is to cater for any off-centring or misalignment of the rod. Thus it is necessary to examine the type of wiper seals on the market and select the type that will give improved performance and reliability of the equipment under its normal working environment.

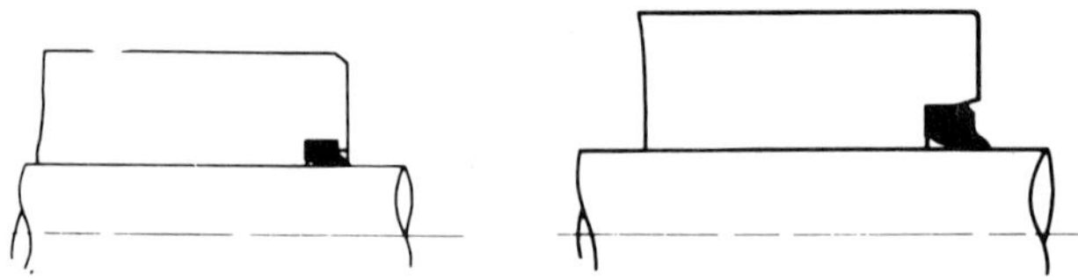

Solid rubber and internally reinforced wiper (scraper) rings.

Scraper rings.

Scraper rings

Most scraper rings take the form of two conical metal rings assembled with a rubber mounting ring, *eg* Figure 6. The conical taper of the rings both effectively scrapes the rod surface and throws any foreign matter clear of both the rod and ring. Positive contact is provided by notches in the rings producing a spring effect as well as allowing for expansion or contraction under temperature differences. The two rings are mounted on the rod with their notches staggered while pre-load is applied by compression of the rubber ring on installation. This rubber ring also provides cushioning and renders the entire assembly self-adjusting for alignment. Typical methods of fitting are shown in Figure 7.

Scraper seals

The description *scraper* seal is given to simple forms of scraper rings the designs of which allow the scraping edge to float effectively so that it remains in contact with the rod regardless of any sideways movement or flexing of the rod. Equally, the lip must

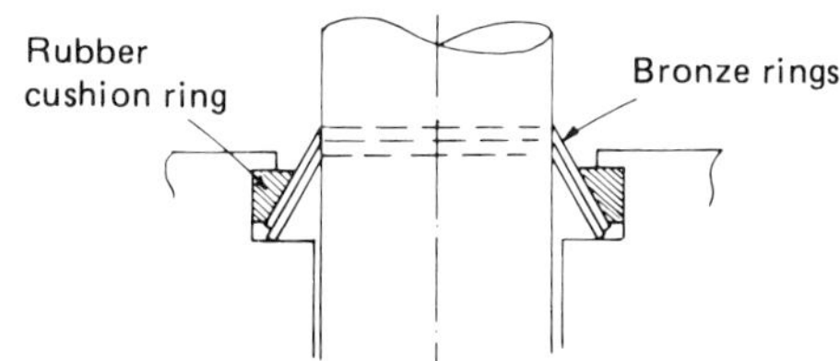

Figure 6

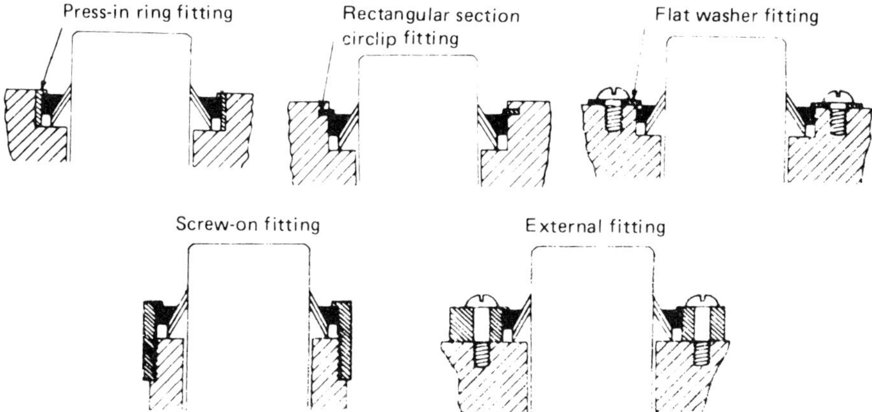

Figure 7

be shaped so that material removed by the leading edge of the ring falls away and cannot become trapped around the ring or fixing point. Conforming with the original definitions, the sealing is of softer material than metal.

A modern design of scraper seal is shown in Figure 8, manufactured in acetal resin and designed to be a snap-on fit in a simple groove. There are numerous other types, often called *scraper-wipers* or wiper/scrapers based on more substantial elastomeric

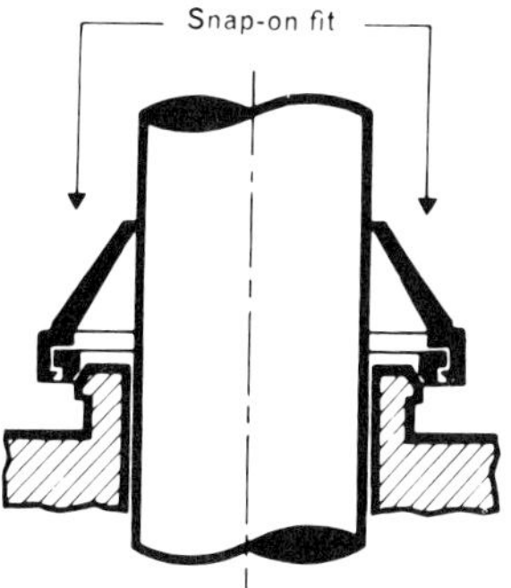

Figure 8
Proprietary scraper.

sections, *eg* Figure 9. These are basically *lip-type* seals resilient enough to maintain positive rod contact and, at the same time, be tough enough to be capable of removing heavier deposits by a scraping action. They are normally designed to be sprung into place in a simple recess machined in the housing.

Materials

Wiper rings are normally moulded from oil-resistant rubber (*eg* nitrile) of around 80 degrees hardness. Polyurethane rubbers of 80 to 90 degrees hardness are more or less standard for heavier duties (and particularly for scraper-wipers). The mechanical properties of polyurethane rubber may also be enhanced by glass-filling for particularly arduous duties.

More limited use is made of felt for wipers, particularly in the presence of more abrasive contaminants. Felt wipers, however, tend to have a high wear rate used under such conditions.

Assembly and pre-load

The necessary pre-load is normally realized simply by choosing a suitable size of wiper, specified for a given rod diameter in the case of rod wipers, and installing in a machined groove or housing of suitable proportions – see Tables 1 and 2. Tolerance on these dimensions is usually rather more generous than in the case of seals, but will normally be specified by the wiper manufacturer, together with any special assembly instructions. For fitting into open ended housings a *metal-caged* ring can be employed which can be simply pressed in place.

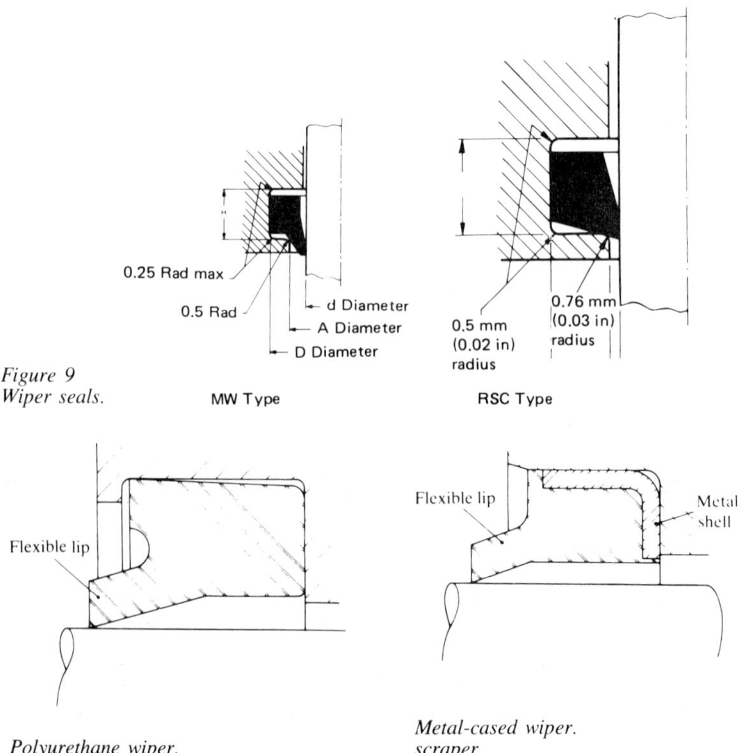

Figure 9
Wiper seals.

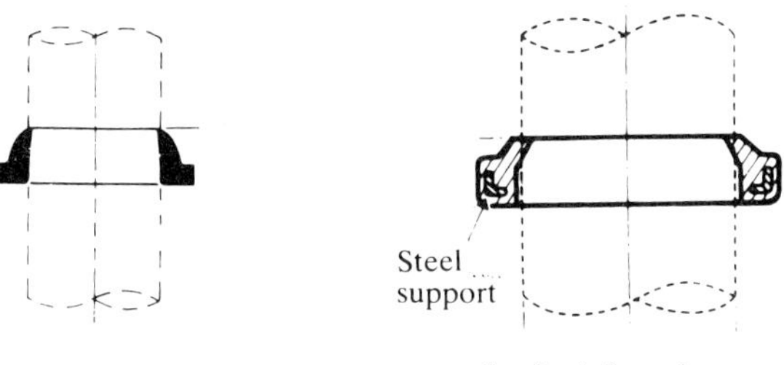

Wiper ring.

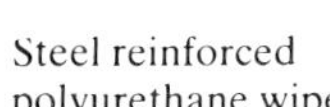

Steel reinforced polyurethane wiper.

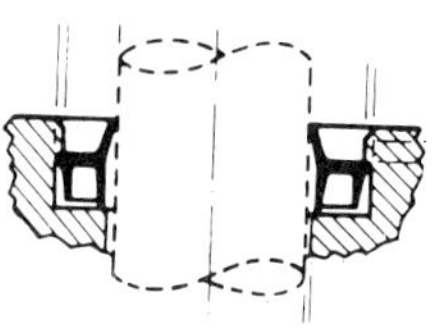

Wiper-scraper.

TABLE I – TYPICAL WIPER AND INSTALLATION GEOMETRY FOR CETOP CYLINDER ROD SIZES 12–200 mm

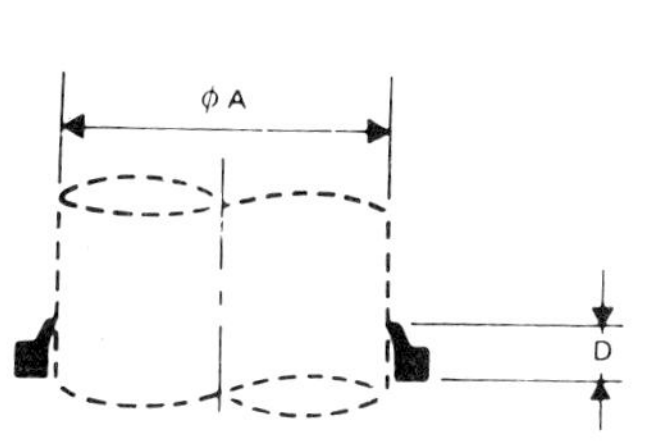

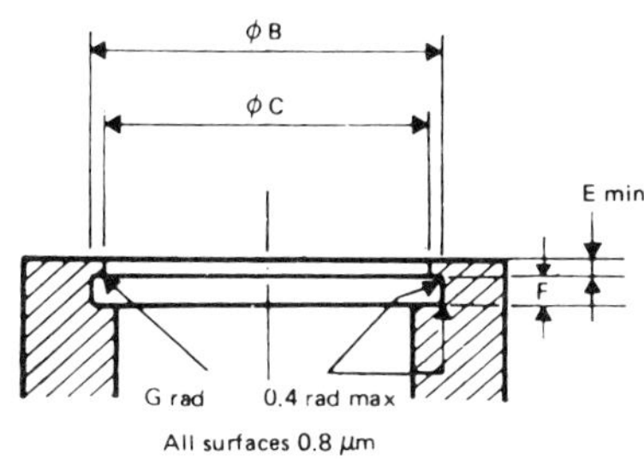

A	B dia max	B dia min	C dia max	C dia min	D	E	F (+0.1 −0.0)	G
12	20.13	20.00	16.11	16.00	6	2	4.0	0.50
14	22.13	22.00	18.11	18.00	6	2	4.0	0.50
15	23.13	23.00	19.13	19.00	6	2	4.0	0.50
16	24.13	24.00	20.13	20.00	6	2	4.0	0.50
18	26.13	26.00	22.13	22.00	6	2	4.0	0.50
20	28.13	28.00	24.13	24.00	6	2	4.0	0.50
20	28.73	28.60	23.13	23.00	7	2	5.3	0.50
22	30.13	30.00	26.13	26.00	6	2	4.0	0.50
22	30.76	30.60	25.13	25.00	7	2	5.3	0.50
25	33.16	33.00	29.13	29.00	6	2	4.0	0.50
25	33.76	33.60	28.13	28.00	7	2	5.3	0.50
28	36.16	36.00	32.16	32.00	6	2	4.0	0.50
28	36.76	36.60	31.16	31.00	7	2	5.3	0.50
30	38.76	38.60	33.16	33.00	7	2	5.3	0.50
30	42.16	42.00	36.16	36.00	9	3	6.0	1.00
32	40.76	40.60	35.16	35.00	7	2	5.3	0.50
32	44.16	44.00	38.16	38.00	9	3	6.0	1.00
35	43.76	43.60	38.16	38.00	7	2	5.3	0.50
35	47.16	47.00	41.16	41.00	9	3	6.0	1.00
36	44.76	44.60	39.16	39.00	7	2	5.3	0.50
36	48.16	48.00	42.16	42.00	9	3	6.0	1.00
40	48.76	48.60	43.16	43.00	7	2	5.3	0.50
40	52.19	52.00	46.16	46.00	9	3	6.0	1.00
42	50.79	50.60	45.16	45.00	7	2	5.3	0.50
42	54.19	54.00	48.16	48.00	9	3	6.0	1.00

All dimensions in mm

cont...

TABLE I – TYPICAL WIPER AND INSTALLATION GEOMETRY FOR CETOP CYLINDER ROD SIZES 12–200 mm (contd) . . .

A	B dia		C dia		D	E	F		G
	max	min	max	min					
45	55.79	55.60	48.16	48.00	7	2	5.3	+0.1 −0.0	0.50
45	57.19	57.00	51.19	51.00	9	3	6.0		1.00
48	60.19	60.00	54.19	54.00	9	3	6.0		1.00
50	60.79	60.60	53.19	53.00	7	2	5.3		0.50
50	62.19	62.00	55.19	55.00	9	3	6.0		1.00
55	65.79	65.60	58.19	58.00	7	2	5.3		0.50
55	67.19	67.00	61.19	61.00	9	3	6.0		1.00
56	66.79	66.60	59.19	59.00	7	2	5.3		0.50
56	68.19	68.00	62.19	62.00	9	3	6.0		1.00
60	70.79	70.60	63.19	63.00	7	2	5.3		0.50
60	72.19	72.00	66.19	66.00	9	3	6.0		1.00
63	73.79	73.60	66.19	66.00	7	2	5.3		0.50
63	75.19	75.00	69.19	69.00	9	3	6.0		1.00
65	75.79	75.60	68.19	68.00	7	2	5.3		0.50
65	77.19	77.00	71.19	71.00	9	3	6.0		1.00
70	80.82	80.60	73.19	73.00	7	2	5.3		0.50
70	82.22	82.00	76.19	76.00	9	3	6.0		1.00
75	87.22	87.00	81.22	81.00	9	3	6.0		1.00
75	87.42	87.20	81.22	81.00	12	5	7.1		1.25
80	92.42	92.20	96.22	86.00	12	5	7.1		1.25
80	96.22	96.00	88.22	88.00	12	4	8.0		1.25
85	97.42	97.20	91.22	91.00	12	5	7.1	+0.1 −0.0	1.25
85	101.22	101.00	93.22	93.00	12	4	8.0		1.25
90	102.42	102.20	96.22	96.00	12	5	7.1		1.25
90	106.22	106.00	98.22	98.00	12	4	8.0		1.25
100	112.42	112.20	106.22	106.00	12	5	7.1		1.25
100	116.22	116.00	108.22	108.00	12	4	8.0		1.25
105	121.25	121.00	113.22	113.00	12	4	8.0		1.25
110	122.45	122.20	116.22	116.00	12	5	7.1		1.25
110	126.25	126.00	118.22	118.00	12	4	8.0		1.25
115	127.45	127.20	121.25	121.00	12	5	7.1		1.25
115	131.25	131.00	123.25	123.00	12	4	8.0		1.25
125	140.25	140.00	132.85	132.60	16	6	10.1		1.25
125	141.25	141.00	133.25	133.00	12	4	8.0		1.25
135	151.25	151.00	143.25	143.00	12	4	8.0		1.25
140	155.25	155.00	147.85	147.60	16	6	10.1		1.25
140	156.25	156.00	148.25	148.00	12	4	8.0		1.25
150	165.25	165.00	157.85	157.60	16	6	10.1		1.25
150	166.25	166.00	158.25	158.00	12	4	8.0		1.25
160	175.25	175.00	167.85	167.60	16	6	10.1		1.25
160	176.25	176.00	168.25	168.00	12	4	8.0		1.25
170	186.29	186.00	178.25	178.00	12	4	8.0		1.25
180	200.29	200.00	190.29	190.00	18	8	10.2		1.50
180	200.29	200.00	190.29	190.00	15	5	10.0		1.50
190	210.29	210.00	200.29	200.00	15	5	10.0		1.50
200	220.29	220.00	210.29	210.00	18	8	10.2		1.50
200	220.29	220.00	210.29	210.00	15	5	10.0		1.50

All dimensions in mm

TABLE 2 – TYPICAL WIPER AND INSTALLATION GEOMETRY FOR HEAVY DUTY WIPER-SCRAPER RINGS

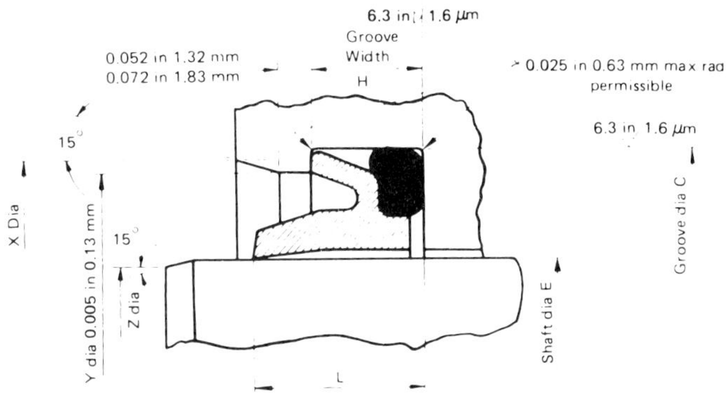

Shaft Diameter E Max mm (in)	Shaft Diameter E Min mm (in)	Groove Diameter C Max mm (in)	Groove Diameter C Min mm (in)	Groove Width L	Groove Width H	Z Dia Max mm (in)	X Dia Min mm (in)	Y Dia mm (in)
12.70 (0.500)	12.57 (0.495)	22.61 (0.890)	22.86 (0.900)	8.63 ± 0.25 mm 0.340 ± 0.010 in	5.08 + 0.12 mm / 0.00 0.200 + 0.005 in / 0.000	11.68 (0.460)	22.35 (0.880)	20.70 (0.815)
15.87 (0.625)	15.47 (0.620)	25.78 (1.015)	26.03 (1.025)			14.86 (0.585)	25.40 (1.000)	23.75 (0.935)
19.05 (0.750)	18.92 (0.745)	28.96 (1.140)	29.21 (1.150)			18.03 (0.710)	28.70 (1.130)	27.05 (1.065)
22.22 (0.875)	22.09 (0.870)	32.13 (1.265)	32.38 (1.275)			21.21 (0.835)	31.75 (1.250)	30.10 (1.185)
25.40 (1.000)	25.15 (0.990)	37.85 (1.490)	38.10 (1.500)	10.92 ± 0.25 mm 0.430 ± 0.010 in	6.48 + 0.12 mm / 0.00 0.255 + 0.005 in / 0.000	24.00 (0.945)	37.59 (1.480)	35.69 (1.405)
28.57 (1.125)	28.32 (1.115)	41.02 (1.615)	41.27 (1.625)			27.18 (1.070)	40.64 (1.600)	38.73 (1.525)
31.75 (1.250)	31.50 (1.240)	44.20 (1.740)	44.45 (1.750)			30.35 (1.195)	43.94 (1.730)	42.04 (1.655)
34.92 (1.375)	34.67 (1.365)	47.37 (1.865)	47.62 (1.875)			33.53 (1.320)	46.99 (1.850)	45.08 (1.775)
38.10 (1.500)	37.85 (1.490)	50.55 (1.990)	50.80 (2.000)			36.70 (1.445)	50.29 (1.980)	48.39 (1.905)
44.45 (1.750)	44.20 (1.740)	56.90 (2.240)	57.15 (2.250)			43.05 (1.695)	56.64 (2.230)	54.74 (2.155)
50.80 (2.000)	50.55 (1.990)	63.25 (2.490)	63.50 (2.500)			49.40 (1.945)	62.99 (2.480)	61.09 (2.405)
57.15 (2.250)	56.90 (2.240)	69.60 (2.740)	69.85 (2.750)			55.75 (2.195)	69.34 (2.730)	67.44 (2.655)
63.50 (2.500)	22.74 (2.470)	82.30 (3.240)	82.55 (3.250)	16.00 ± 0.25 mm 0.630 ± 0.010 in	9.40 + 0.12 mm / 0.00 0.370 + 0.005 in / 0.000	60.96 (2.400)	81.79 (3.220)	78.61 (3.095)
69.80 (2.750)	69.09 (2.720)	88.65 (3.490)	88.90 (3.500)			67.31 (2.650)	88.14 (3.470)	84.96 (3.345)
76.20 (3.000)	75.44 (2.970)	95.00 (3.740)	95.25 (3.750)			73.66 (2.900)	94.49 (3.720)	91.31 (3.595)
82.55 (3.250)	81.79 (3.220)	101.35 (3.990)	101.60 (4.000)			80.01 (3.150)	100.84 (3.970)	97.66 (3.845)
88.90 (3.500)	88.14 (3.470)	107.70 (4.240)	107.95 (4.250)			86.36 (3.400)	107.19 (4.220)	104.01 (4.095)

Compression Packings

Definition of compression packing

A COMPRESSION packing can be described as a deformable material used to prevent or control the passage of a fluid between surfaces that move in relation to each other.

Packings are the traditional form of *gland* seal and can be manufactured from a wide variety of different materials. A packing may be in shredded form (usually mixed with a lubricant) with which the gland or *stuffing box* is packed; in made-up sections (normally rectangular but sometimes circular); wound in position spiral fashion from a suitable cut length; or fabricated sections assembled in the gland as cut rings. In all cases sealing is obtained by tightening the gland, so that the packing is forced against or compressed on to the surface to be sealed.

The operating principle of a packed gland is illustrated in Figure 1. The compressive force generated by tightening of the gland spigot produces a radial pressure providing the sealing effect. The radial pressure distribution varies throughout the packing length in an exponential manner. To keep the packing 'dry' the radial pressure on the inner ring must be at least equal to the internal pressure, implying that the radial pressure of the outer ring will be considerably higher – and excessively high for most purposes (resulting in excessive friction, shaft wear and pneumatic seal failure). Thus in the majority of applications the compressive force is adjusted to give very *slight* leakage past the last ring, *ie* the radial pressure of this ring is very slightly less than the internal pressure. It follows, however, that if the gland is adjusted to *minimum* compression for *no* leakage, there will be some leakage past the majority of the rings.

The question of optimum gland tightening is also complicated by the fact that some packings expand under service conditions, such as may be encountered due to a rise

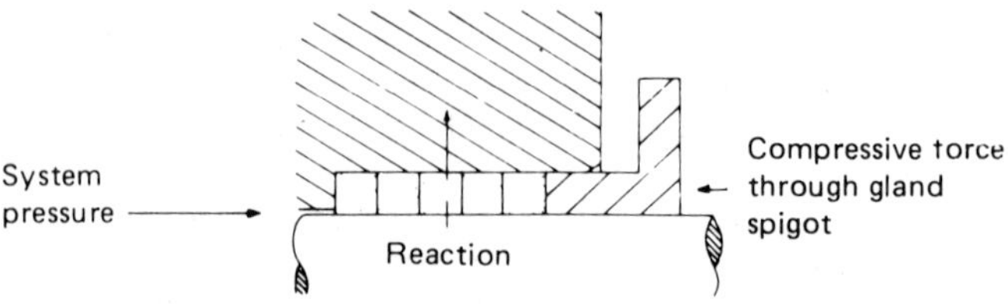

Figure 1

in temperature, and very little initial tightening of the gland may be necessary. In other cases periodic re-tightening of the gland may be necessary to compensate for wear and relaxation of the packing and to maintain a satisfactory seal.

With a typical packing material the ratio of radial pressure generated to applied axial pressure in tightening the gland is of the order of 0.6 to 0.7. Typical radial pressure throughout the gland is then as shown in Figure 2.

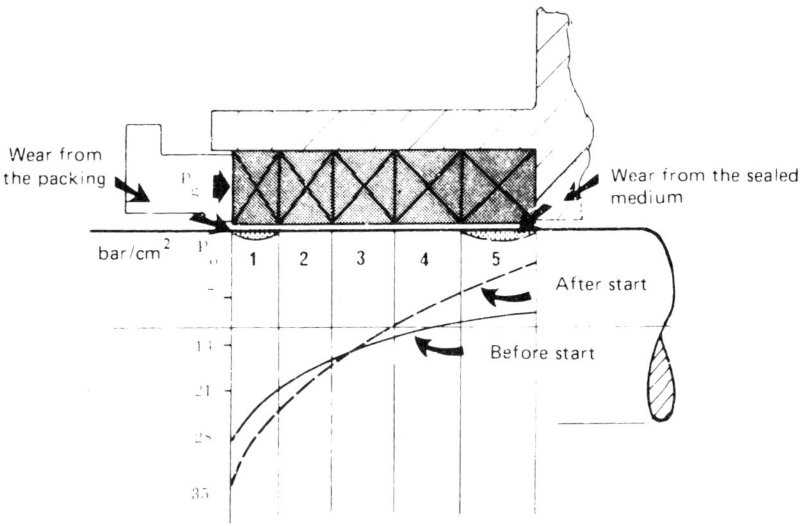

Figure 2

Uses of compression packings

The major uses for compression packings are in the process industries such as petrochemical, paper, steel, *etc* and also in service industries such as marine, water, sewage, food and nuclear. They seal all types of fluid such as steam, water, acids, alkalis, solvents, oils and gases, *etc*. They are also used in rotary, centrifugal and reciprocating shaft pumps, agitators, valves and many other types of mechanical equipment.

Flexible seal sets or single seals may replace packings for many reciprocating duties, particularly in large sizes and heavy duty applications, unless minimum leakage is a primary requirement, when a mechanical shaft seal may be preferred. It is significant, however, that with the widespread use of mechanical seals, the demand for gland packings shows no sign of falling off.

The degree of 'softness' of packings may vary considerably. They usually contain lubricants and excessive compression or over-heating during service can produce loss of lubricant, with subsequent reduction in packing volume and loss of radial pressure leading to leakage.

Where lubrication is a problem, or a degree of gland cooling is required, additional lubricant/coolant can be fed to the centre of the gland (Figure 3). The degree of cooling which can be provided by this is limited and for higher temperature working cooling may have to be applied to the complete gland housing to maintain the operating temperature of the gland within the limits of the packing material. Many of the problems associated with high friction and heating through necessary high compression shrinkage of fibres and/or lack of adequate lubrication have largely been eliminated by the latest packing materials based on PTFE-coated aramid fibres.

Packing sizes

Compression packings are normally of substantially square section (although round braided packings may be used on reciprocating rods and valve stems and loose packings may be used for sealing valves and some pump glands). Thus most packings are made in a standard range of section sizes from about 6 mm (¼ in) square upwards. Section size is largely arbitrary, but as a general rule housing width should be of the order of 25% of the shaft (or rod) diameter for shaft sizes of 12 mm (½ in), reducing to 10% for shaft sizes of the order of 150 mm (6 in).

Again there are no exact rules as to the optimum number of rings, but for average duties a total of four or five square rings is typical – see also Figure 4.

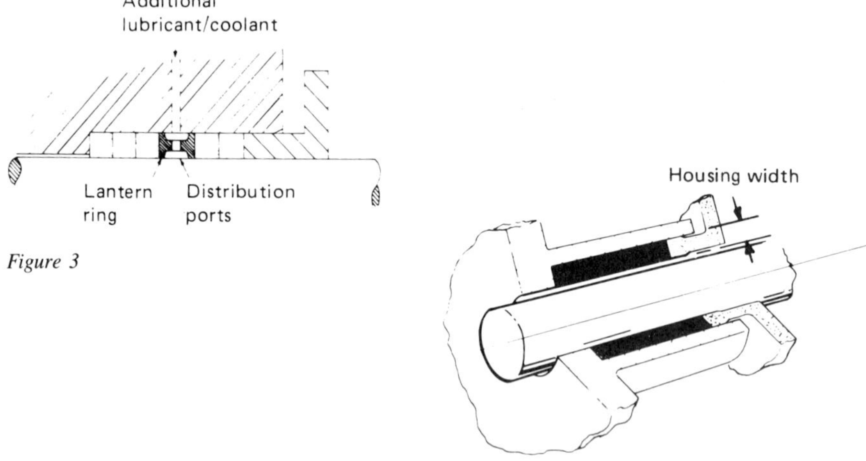

Figure 3

Figure 4

Gland design

Gland design for handling clean, non-abrasive fluids under positive pressure is straightforward (Figure 5a). Detail requirements are the provision of an adequately tapered lead-in at the mouth of the gland to facilitate fitting the packing rings without damage, and a reasonably good surface finish on the gland surface; 2.5 μ mm (64 μ in) Ra is generally considered satisfactory for most applications.

Where the product being sealed contains abrasive, it is highly desirable to keep this out of the packing area. This can be done by injecting a compatible flushing fluid through a lantern ring in the centre of the gland (Figure 5b). It can be noted here that the controlled leakage in this case is that of flushing fluid which, because of the distribution of radial pressure, will also leak back into the product. Where it is not possible to arrange a suitable liquid flush, a grease flush may provide a solution (Figure 5c). The grease in this case must be clean and compatible with the product.

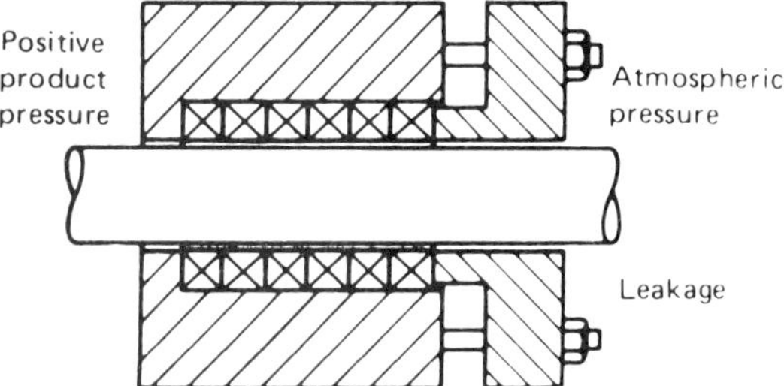

Figure 5a

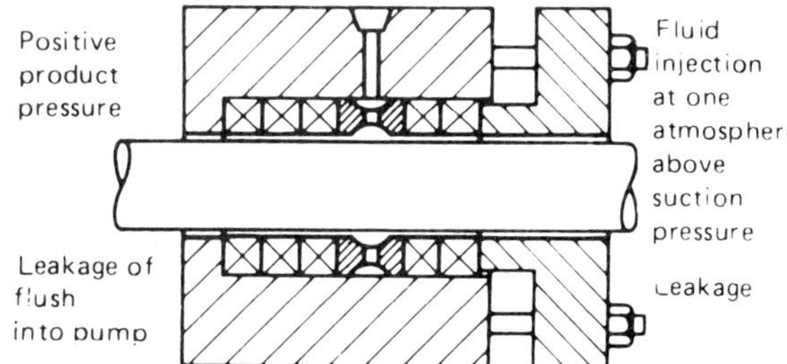

Figure 5b

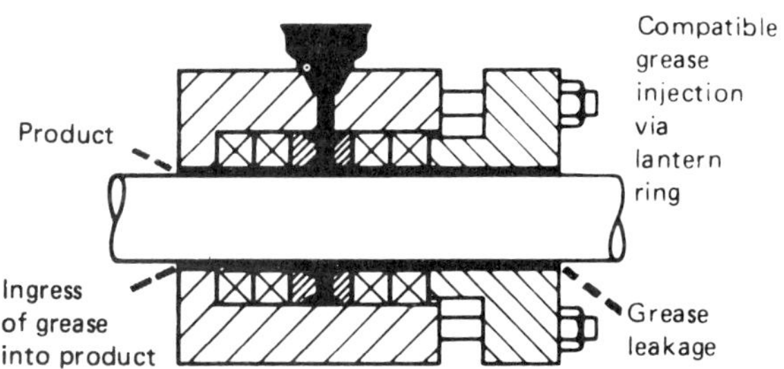

Figure 5c

Two other gland arrangements are shown in Figure 6. In Figure 6a the product being handled is at less than atmospheric pressure, requiring a fluid barrier to prevent air being drawn in through the gland. This barrier is provided by tapping the discharge side of the product and feeding it into the gland *via* a lantern ring. Controlled leakage in this case is that of the product.

In Figure 6b the product being handled is toxic or hazardous, so a flushed gland is again used to provide a primary barrier. This is backed up by a *quench (flushing annulus)* in the *gland follower* and an *auxiliary packed gland* to exclude leakage.

Operating principles

To generate the necessary compressive force it is usual to have a bolted gland follower. Adjustment of the retaining nuts (or studs) enables the axial movement of the follower to be easily controlled.

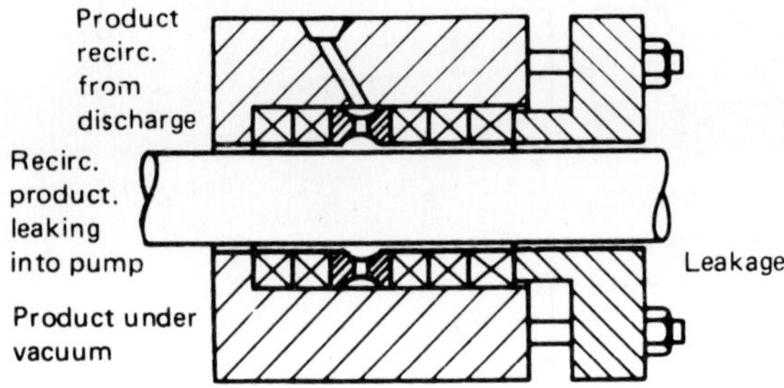

Figure 6a

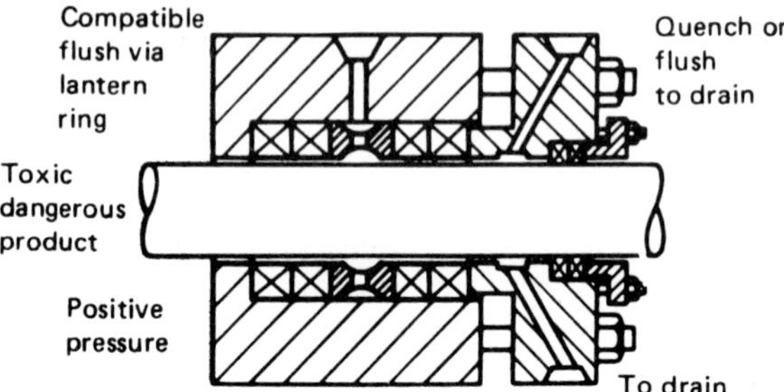

Figure 6b

In hazardous or inaccessible situations spring loading is sometimes used, *eg* Figure 6c shows Belleville washers assembled between the gland nuts and the gland follower. This can prove acceptable where there is slow or infrequent shaft motion, *eg* in valves, but the inherent lack of fine control demanded by some packing types, particularly on high speed shaft applications can cause problems. The range of load capability available from springs is also a limiting factor.

On applications where there is slow or infrequent shaft motion, compression packings are required to seal without leakage. This condition can be achieved by ensuring the compressive force exerted by the gland follower causes the packing radial force to exceed the pressure to be sealed.

For high speed shaft applications fluid leakage is essential for heat dissipation and lubrication purposes and must be controlled. Fluid leakage ensures the frictional heat generated at the packing/shaft interface is dissipated. The sealing force can be adjusted to cater for service wear but care must be taken to avoid over compression and loss of adequate fluid leakage with the consequent friction increase, shaft wear and premature packing failure.

Design guidelines

There seem to be very few international standards defining the design of the installation environment for compression packings. Furthermore, stuffing box depths vary according to specific service conditions such as the pressure of the fluid being sealed, lantern ring inclusion, *etc.* Figure 6d shows general design guidelines.

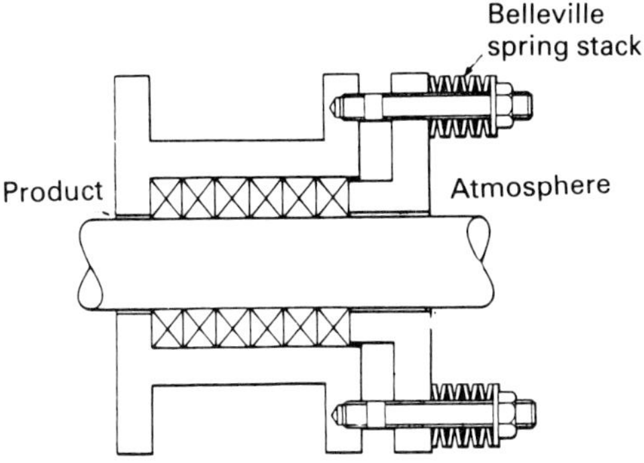

Figure 6c

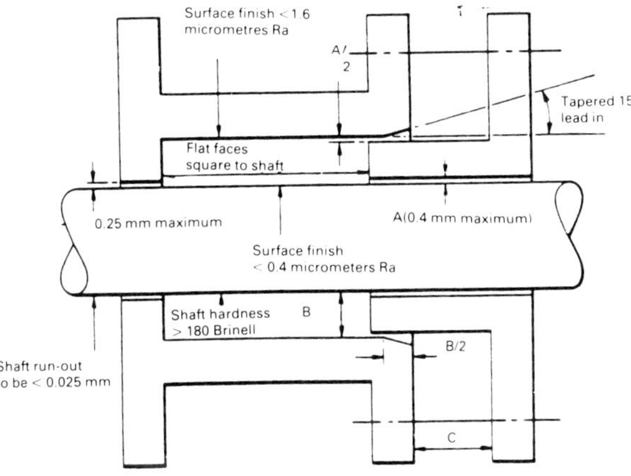

Figure 6d
Surface finish < 1.6 micrometres Ra

Points to note are:

(a) Assuming square section packing is to be used, dimension C should be at least 1.5 times B.

(b) The shaft must be adequately supported. The packings must not be used as a bearing. It must also be free from scores, pits, grooves and ridges.

(c) If the clearances exceed those stated then suitable anti-extrusion rings must be fitted. These can be machined from metal or suitable plastic with appropriate clearances or formed from a suitable packing material.

(d) If metallic packings are to be used the shaft hardness should be at least 500 Brinell. There are a number of techniques and materials available that can be employed to achieve this, *eg* spray deposition of stellite or chrome oxide, *etc.*

(e) The use of a sleeve over the length of packing reduces refurbishment/replacement costs particularly when hardening is necessary.

When a lantern ring is required in the installation its length should be at least 1½ times the packing radial section (dimension B) and the same clearances in the stuffing box bore and over the shaft as the gland follower should be applied in order to minimize the possibility of it contacting, and therefore damaging, the shaft. PTFE is particularly suitable as a lantern ring material because of characteristics such as chemical inertness and relative non-abrasiveness. The cross section of a lantern ring should balance the need for structural strength with minimum restriction of injection fluid flow to the packing and shaft. On installation it should be positioned below the barrier fluid inlet in such a way as to allow for slight compression of the inboard packing rings.

Tables 1 and 2 provide general guidance on:

(a) The cross section of packing to use related to the shaft diameter (all types of equipment).

(b) The number of square section rings required in relation to the pressure of the fluid to be sealed.

With reference to Table 2 it is not possible to be definitive. Decision-making will be influenced by practical experience coupled with the knowledge that a particular packing:

(a) is resistant to deterioration by the fluid being sealed;

(b) is capable of withstanding the fluid temperature; and

(c) can be formed in such a way as to ensure that the fluid pressure alone will not cause loss of packing volume, *eg* through lubricant squeeze out, *etc.*

To exceed the cross section and number of rings recommended is unnecessary and can be detrimental. Too many rings of too large a section means:

(a) Excessive area of packing/shaft contact with excessive heat generation.

(b) Possible loss of packing thermal conductivity efficiency.

(c) As the stuffing box increases so does:

  (i) the difficulty of installing/dismantling and adjusting the packing rings properly; and

  (ii) the chance of shaft malalignment with its inherent adverse effect on packing sealing performance.

TABLE 1

Shaft diameter mm	Packing section mm
Up to 16	3
Above 16 to 25	5
Above 25 to 50	6.5
Above 50 to 90	8
Above 90 to 150	10
Above 150	12.5

TABLE 2

Stuffing box pressure bar	Number of rings
Up to 35	4
Above 35 to 70	6
Above 70 to 140	8
Above 140	10

Furthermore, reference to Figure 6e shows how the fluid pressure is progressively throttled by each successive packing ring. It can be seen that the five rings nearest to the gland follower do the majority of the sealing and most of the wear on the shaft occurs under the outer two to three rings. Unless considered desirable because the fluid being sealed is:

(a) At high temperature.
(b) At high pressure.
(c) Corrosive.
(d) Abrasive.

Four or five rings should be sufficient.

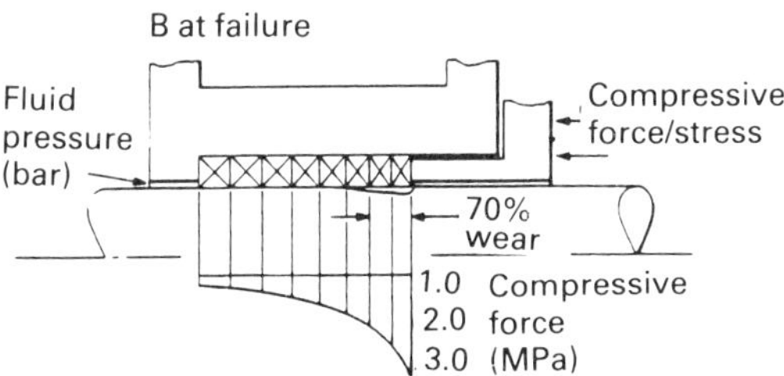

Figure 6e

Traditional materials

The traditional type of packing based on lubricated fibre yarns remain in common and widespread use. The range of materials employed is quite wide (see Table 3, which is only a partial list) and further extended by the introduction of synthetic yarns to upgrade certain performance characteristics, although rayon and nylon have proved to have only limited benefits, (see also Table 4).

Vegetable-based fibre yarns are generally suitable for oil, water and non-corrosive chemical duties where the working temperature does not exceed 90°C and rubbing speeds are moderate (not more than 8 m/s). Cotton and flax are the most widely used fibres, and to a decreasing extent hemp. Ramie, jute and sisal have largely disappeared from the scene.

Asbestos yarns are the traditional choice for higher temperature services (up to 320°C) and higher rubbing speeds (up to 20 m/sec). Asbestos has, of course, been the subject of much concern as a health hazard and the use of blue (crocidolite) has virtually ceased. Blue asbestos, however, has better corrosion resistant properties. There is less realistic objection to white (chrysotile) asbestos, which produces the principal yarns now used for asbestos packings, particularly if the fibres are securely bound by impregnation during the manufacture of packings and will not emit asbestos dust which is the main quoted source of health hazards.

Traditional lubricants

Apart from specialized applications where physically dry packings are required, yarn packings are invariably combined with a lubricant.

TABLE 3 – TRADITIONAL PACKINGS

Material and construction	Maximum pressure lb/in²	Maximum pressure bar	Maximum service temperature °C	Maximum service temperature °F	Main application(s)	Fluid(s)
Asbestos – plaited	1000 to 2000	70 to 140	340 to 540	650 to 1000	Reciprocating and rotary up to 750 ft/min.	General service.
Asbestos – proofed	3000	210	120 to 260	250 to 500	Rotary and-reciprocating.	Oil resistant packings.
Asbestos – resistant binders	2000 to 3000	140 to 210	up to 315	up to 600	Rotary and reciprocating.	Acids, corrosive fluids, *etc.*
Asbestos – braided sleeve			120	250	Rotary	Dilute acids, alkalis, ammonia.
Asbestos – wire reinforced	750 to 2000	50 to 140	260	500	Up to 1250 ft/min.	Steam, hot water, aqueous solutions, oils, *etc.*
Asbestos – plaited – metal braided	1500	105	650	1200	Reciprocating and rotary.	High temperature service.
Asbestos – white metal (inserts)	700	50	260	500	Reciprocating and rotary.	Steam.
Asbestos – wire, plaited	700	50	260	500	Rotary up to 2000 ft/min.	Steam, water, acids, alkalis.
Asbestos cloth – rubber core	1000	70	315	600	Reciprocating.	High pressure steam.
Asbestos – wire – rubber	700 to 1000	50 to 70	260 to 315	500 to 600	Reciprocating	Steam, air, gases, water.
Soft cotton (lubricated)			95	200	1250 ft/min.	High speed centrifugal pumps, suitable with solids in suspension.
Cotton yarn (lubricated)			95	200	Up to 1250 ft/min.	Water, oil, aqueous solutions, ammonia.
Cotton – plaited			65	150	Reciprocating and rotary pumps.	Air, water, refrigerants.
Cotton duck – rubber			140	280	Piston rods, heads of reciprocating pumps.	Hot and cold water, petroleum products.
Cotton duck – metal reinforced	50 (steam) 3000 (water)	3.5 (steam) 210 (water)	150	300	Piston rods, *etc.*	Low pressure steam, water, oils, ammonia.
Canvas covered cotton			120	250	Centrifugal and rotary pumps.	Water, caustic solutions, ammonia.
Cotton cloth – bonded	3000	210	140	280	Reciprocating low-medium speed rotary.	Oils, solvents, hydro-carbons, *etc.*
Flax – plaited (lubricated)			65 to 140	150 to 280	Circulating water pumps.	Fresh and salt water.
Flax – wire reinorced cotton covered	50 (steam) 3000 (water)	3.5 (steam) 210 (water)	140	280	Water feed and circulating pumps.	Fresh and salt water.
Foil – wrapped fibres	500 to 700	35 to 50	65 to 800	150 to 1500	High speed rotary.	Steam, water, hot oils, mild acids and alkalis.
Foil – wrapped, plaited core	500	35	190 to 800	300 to 1500	High speed rotary.	Oil pumps, high temperature steam.
Hemp (lubricated)	500	35	65 to 95	150 to 200	Medium speed rotary pumps.	Water.
Plaited vegetable fibre			65	150	Rotary and reciprocating, medium speeds.	Fresh and salt water.
Plain and plaited leather			37 to 50	100 to 120	Low and medium pressure reciprocating.	Cold water and cold oils oils.
Leather – fibre – metal			37 to 50	100 to 120	Reciprocating (low-medium pressures).	Water, oils, limes, *etc.*

Graphite is the lubricant commonly introduced into packing sections and provides excellent self-lubricating properties for many services where the packing would otherwise have to run dry, or in contact with a fluid which is not a lubricant. Thus graphited lubricant is particularly suitable for steam and water services, and especially salt water. In certain cases, however, the presence of loosened graphite may be objectionable; or in the case of the packing operating against a stainless steel rod, graphite may produce pitting of the steel through electrolytic action. An alternative lubricating impregnant which overcomes these difficulties is mica. These lubricants, together with molybdenum disulphide and PTFE, remain the standard 'dry' lubricants.

Traditional 'wet' lubricants such as tallow, have been superseded by mineral oils, greases, waxes and soaps. Silicone greases are particularly used with asbestos packings for high temperature services, but are now no longer regarded as suitable for use with edible products or drinking water. The lubricant employed for such services is usually an inert mineral oil.

The actual percentage of lubricant employed will vary with the service. Thus a packing intended for high speed movement, and particularly high rotary speeds, would normally be of a softer type to maintain flexibility for long periods and contain a large percentage of lubricant. A packing used only for a static application would not normally contain any lubricant at all. That for a reciprocating motion may contain anti-friction wire reinforcement in preference to lubricant, with perhaps a final dressing of graphite. Others may contain both soft anti-friction wire reinforcement and be impregnated with lubricant. The value of soft anti-friction wires is that they provide continuous lubrication to the shaft and also assist in conducting heat away from the working surface.

With the notable exception of packings constructed from exfoliated and multifilament yarn graphite, those commonly used for high speed shaft applications contain both dry and wet lubricants.

These lubricants perform a number of important functions including:

(a) Provision of initial lubrication and sealing during start up and running in of equipment. The sealed fluid or external injection fluid eventually takes over this function.

(b) Provision of a fluid barrier. By filling voids in braided packings, sealed fluid penetration into the packing is minimized and sealed fluid passage through the packing is prevented.

(c) Provision of inter-fibre lubrication. This minimizes fibre on fibre abrasion/breakage and aids even distribution of the compressive force.

(d) Improvement of thermal conductivity/heat dissipation characteristics of the packing.

Figure 6f shows an installation used when handling fluids at high temperature. Injection of coolant *via* the lantern ring alone may be inadequate. In this case a jacketed stuffing box can be used with coolant passing through it.

If the fluid being handled crystallizes, becomes highly viscous or a solid when cooled, *eg* sugar, bitumen, wax, *etc,* then heaters or a jacketed stuffing box must be used to restore the fluid state before equipment start up. In this case steam may be suitable to pass through the jacket.

N.B. The packing rings should be made from a material which has a high thermal conductivity characteristic.

When the packing installation is required to hold a vacuum, or where a suction lift is involved, the fluid injected into the lantern ring can be a grease. In this case, a screw or spring-loaded Stauffer cap type lubricator is recommended.

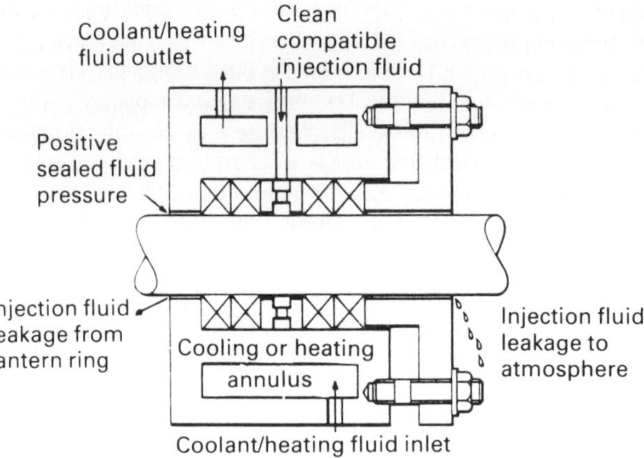

Figure 6f

Plaiting *versus* **Braiding**
Plaited packings are constructed from multiples of yarns interwoven in conventional or modified plaiting forms, with pockets created between each plait to retain lubricant. The lay of the plait may also be matched to specific duties, *eg* following the direction of shaft rotation for rotary seals so that the effect of individual fibres wearing away can be tolerated for some considerable time without detriment to the overall behaviour of the section.

Multi-Lon™ packing.

Long life, non-asbestos gland packing.

Braided packings may be formed in two distinct ways. Continuous braided packings are constructed from individual yarns braided together in the form of a tube, the required section being built up layer upon layer in similar fashion. The alternative form is diagonal braiding (with lattice braiding as a variation). Both produce a more dense form of packing with a higher surface density, but with smaller lubricant retaining pockets and thus more impermeable than plaited fibres (like a plaited packing) without the packing disintegrating. Examples of 'classic' types of construction are given in Figure 7.

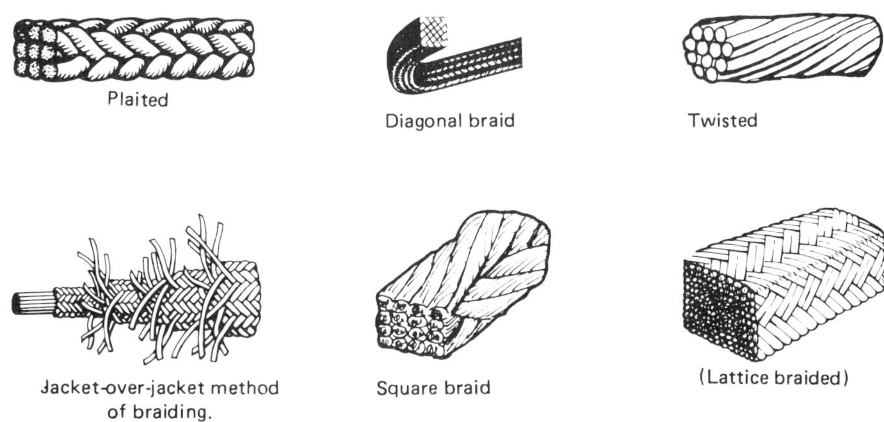

Figure 7
Examples of different types of plaiting and braiding styles.

Braided sections may be laid up square or round. In the latter case they are usually formed to square section after braiding and impregnation with lubricant simply by passing through a roller-die. In practice manufacturers have developed their own particular forms of plaited or braided constructions, *eg cross plait* (Crossley) and *superplaiting* (Latty International), aimed at overcoming the limitations of simple or 'classical' braids. Two examples of well developed sections which provide a robust, homogeneous, non-porous braid with good suppleness are shown in Figure 8.

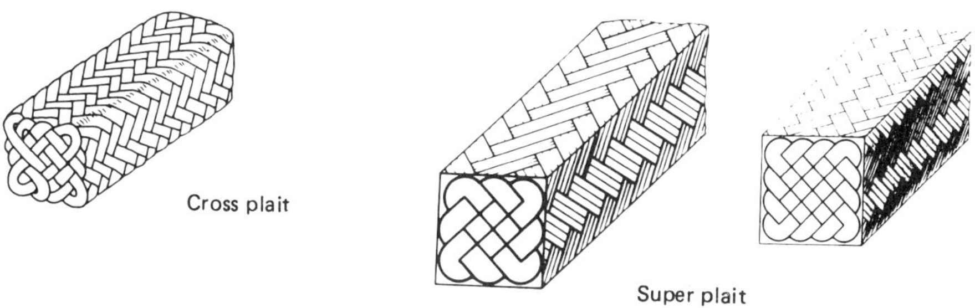

Figure 8

Modern graphited asbestos packings

Recent developments in producing an intimate mix of graphite and asbestos rather than a surface coating have led to the appearance of graphited asbestos packings with superior lubrication for cooler running and lower friction, and with superior high temperature performance. Figure 9 shows a comparison between the performance of an outstanding proprietary packing of this type and a conventional graphited asbestos packing. The test data were obtained with water and the fluid being sealed, at a pressure of 5 bar and a rubbing speed of 10 m/s.

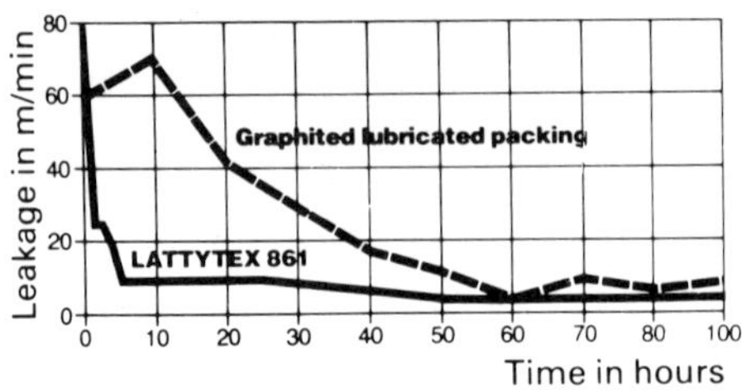

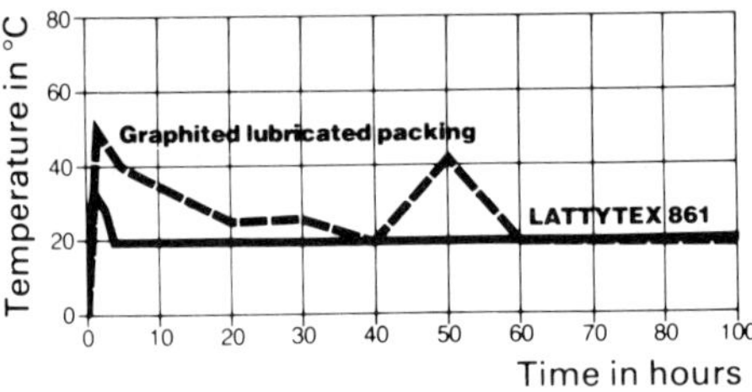

Figure 9

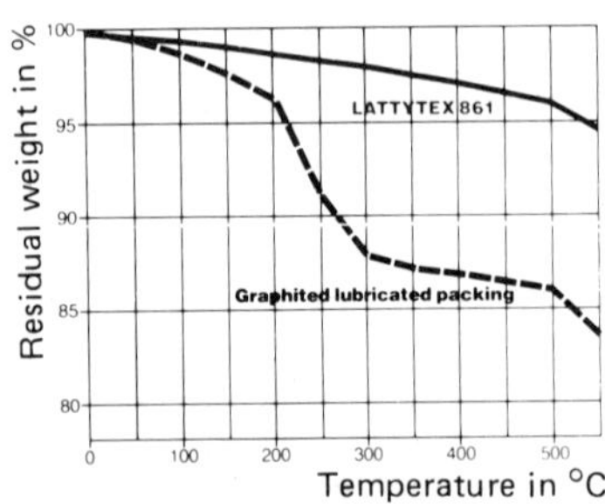

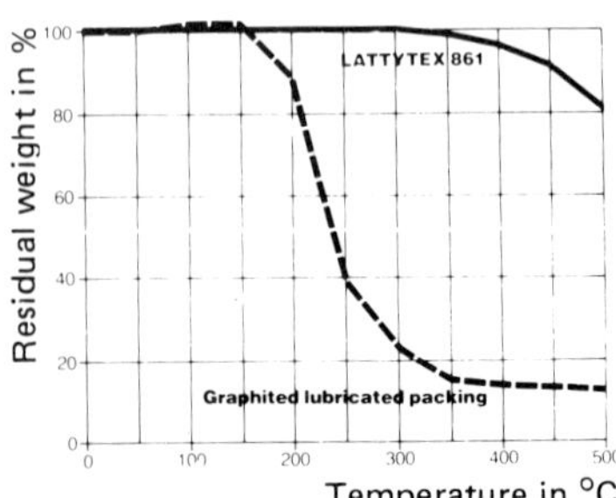

Performance of homogeneous graphited asbestos packing compound with conventional graphited asbestos packing

PTFE packings

The outstanding properties of PTFE as a low friction material with extreme resistance to chemical attack makes it an attractive choice for packings. Its unfavourable characteristics are its low strength, low thermal conductivity and its tendency to shrink with increasing temperature (*ie* it has a negative coefficient of thermal expansion). The latter characteristic, combined with the use of *yarn* packings (usually abestos yarn) as a lubricant, normally limits maximum rubbing speed to the order of 8 to 10 m/s and maximum service temperature to the order of 250 to 290°C (see also Table 3).

Heat transmission can, however, be improved by incorporating graphite, and *PTFE/graphite* packings produced by extrusion are amongst the most attractive and useful of the modern packing types, with superior performance to conventional yarn packings, particularly as regards life and reduction of shaft or rod wear.

Aramid/PTFE packings

This generation of modern packings is based on aramid fibre which in filament form has a tensile strength better than steel with a specific gravity of only 1.44 (substantially lower than cotton (1.50) or asbestos (2.55)). The material is also flame-resistant, self-extinguishing and does not melt, although it chars at a temperature between 425°C and 485°C. Yarns or monofilaments pre-coated with PTFE dispersant, braided and impregnated with PTFE yield exceptionally long life, low friction packings with a similar chemical resistance to PTFE, and similar maximum service temperature. Specific examples of this type of packing are shown in Figure 10 and included in Table 5.

Figure 10

Synthetic fibre/PTFE produced in moulded rings.

Synthetic fibre/PTFE packing for rotary applications.

Material property comparisons

PTFE, PTFE-coated aramid fibre and graphite fibres are all low friction materials with the first two having a coefficient of friction of the order of 0.05 to 0.2. Graphite fibres generally yield a coefficient of friction within the range 0.1 to 1.0. By comparison, the coefficient of friction of lubricated cotton is of the order of 0.4 to 0.6; and that of lubricated asbestos or PTFE/asbestos even more wide ranging than graphite. This is because asbestos is naturally a friction material (*eg* it is used for brake shoes) and so relies heavily on added lubricant to work as a (relatively) low friction packing. Equally, should the lubricant be lost, the asbestos packing will generate high friction and heat.

As regards thermal conductivity, graphite is the best material (25×10^{-3} cal/s/cm/°C) followed by aramid fibre (15×10^{-4} cal/s/cm/°C), and PTFE (6×10^{-4} cal/s/cm/°C). Asbestos at 15×10^{-6} cal/s/cm/°C rates very poorly in this respect.

Both asbestos and aramid fibre have virtually zero coefficient of thermal expansion and so are ideal packing materials in this respect. PTFE decreases in volume with increasing temperature, so packings made from PTFE fibre will tend to shrink onto the shaft. Graphite, on the other hand, expands slightly with increasing temperature.

TABLE 4 – TYPES OF MATERIALS

Type	Construction	Maximum Performance Parameters*			Application	pH*
		Rubbing Speed m/sec	Temp. °C	Pressure bar		
Cotton	Plaited or braided with grease graphite or mica lubricant	7.5	90	–	Seals for water.	6–9
Hemp	Plaited or braided with grease graphite or mica lubricant	–	–	–	Seals for water	5–9
Flax	Plaited or braided with grease graphite or mica lubricant	–	70–120	–	Rotary seals or pumps and slow reciprocating seals; ships' stern glands, etc.	–
Textile	Plaited or braided cotton, nylon, rayon, hemp, etc, with PTFE impregnation	–	–	–	Seals for water services, solvents, weak alkalis, oils, greases, foodstuffs, etc.	–
Asbestos (dry)	Round or square plaited or braided white asbestos	–	500	–	Autoclaves, boiler stop valves, etc.	
Asbestos (lubricated)	Round or square plaited or braided white asbestos with graphite, mica or mineral oil lubricant		–40 – +300		Steam services, evaporators, cannisters, liquors, etc.	5–12
Asbestos (wire reinforced)	Square section plaited or braided with brass or inconel wire reinforcement. Impregnated with graphite or mica lubricant		–50 – +750	250–650	Water, air, steam, solvents, hydrocarbons, weak acids and alkalis	4–11
Asbestos (PTFE)	Plaited or braided asbestos yarn coated with PTFE dispersion	15	280	100	Acids, solvents, oils, hydrocarbons, etc (not caustic solutions)	0–8
Asbestos (Glass fibre)	Plaited or braided asbestos yarn and glass fibre with added lubricant	5	200	250–300	Corrosive duties with equivalent chemical resistance to blue asbestos, except caustic solutions	0–11
Blue asbestos	Braided asbestos fibres with added lubricant	15	280	60	Limited production because of health hazard and not made in many countries	0–12
Asbestos (lead)	Manufactured from asbestos fibres and granulated lead with added lubricant	–	–	–	–	–
Ramie (PTFE)	Ramie yarn impregnated and coated with PTFE. Plaited or braided with added lubricant	15	120	400	Water, mildly aggressive clean fluids and mildly abrasive slurries	4–11
Acrylic	PTFE coated, lubricated and plaited or braided	10	120–200	100	For low temperature, low general service slightly acidic fluids	2–8
PTFE	Yarns or tape	–	200–250		Soft packings, services involving corrosive media	0–14

TABLE 4 – TYPES OF MATERIALS (contd)

Type	Construction	Maximum Performance Parameters* Rubbing Speed m/sec	Temp. °C	Pressure bar	Application	pH*
PTFE and lubricant	PTFE braided and impregnated with graphite and/or molybdenum disulphide lubricant	20	–100 to +250	30+	Water, steam, acids, alkalis, solvents, oils, greases, hydrocarbons, etc	0–14
	Braided pure PTFE fibres treated with PTFE dispersion	10	–200 to +300	100	Sealing all media	0–14
PTFE (graphite)	Extruded fibrillated PTFE and graphite with mineral oil lubricant	10	–100 to +250	100	Water, acids, alkalis, solvents, oils, degreasing fluids, oxidizing agents, foodstuffs, etc.	0–14
Aramid – PTFE	Braided PTFE coated Kevlar yarn impregnated with lubricant and PTFE	10–20	–220 to +300	200–1000	High duty 'super-packing'	1–14
PTFE – Silk	Braided PTFE – silk yarns with added lubricant	5	280	200	Water, steam, acids, alkalis, oils, oxidizing agents, dye stuffs, etc.	0–14
Graphite yarn	Woven or braided graphite yarn treated with graphite powder	–	–200 to +600	–	High temperature services	0–14
Carbon fibre	Amorphous carbon yarns treated with graphite powder	–	–200 to +600	–	High temperature services	0–14
Graphite/ asbestos	50/50 wet spun, with added inert lubricant	25	500	80	For water, air, steam, hydrocarbons	3–12
Graphite/ asbestos	50/50 wet spun, inconel wire, reinforced yarn	–	650	900	Steam, gases,	2–14
Expanded graphite	Flexible plait or tape form	35	–200 to +600	300	High temperature services, foodstuffs, etc	0–14
Glass fibre	Braided glass fibre yarns with added lubricant	–	–	–	Suitable for use with acid and corrosive fluids (except caustic solutions)	–
Alumina silicate	Plaited with inconel wire	–	1200	–	Extremely high temperature services. Limited to static seals or slow movements	–

*Depending on actual construction, type of lubricant used, etc.

TABLE 5 — TYPICAL EXAMPLES OF MODERN PACKINGS

Proprietary Name*	Construction	Performance Parameter and Limits				Applications
		Rubbing Speed m/sec	Temp. °C	Pressure bar	pH	
LATTYflon 4788	Aramid yarn (patented process), braided and impregnated with inert lubricant and stabilized PTFE	25	−220 to +300	200	1−13	Sealing all fluids as high duty rotary shaft packing, non-toxic, non-contaminating, suitable for all food, drink and pharmaceutical processing
LATTYflon 4789	Similar to Lattyflon 4788	20	+275	100	2−12	Lower cost general purpose shaft packing, similar to above
LADDYflon 4757	Special plaited packing featuring Aramid corners and graphitised PTFE fibre on the faces	10	−220 to +300	1000	1−14	Specifically developed for reciprocating applications — particularly suitable for piston sealing in high pressure chemical pumps, etc.
LATTYflon 3206	Pure PTFE fibres treated with a dispersion of PTFE and inert lubricant	10	−200 to +300	100	0−14	Sealing all media, particularly suited for vacuum and gas seals
LATTYflon 3206B	Without lubricant		−200 to +300	100	0−14	Oxygen services
LATTYflon 4207	Superplaited PTFE fibre with graphite		−200 to +300	100	0−14	Aggressive chemicals and high speed pumps
LATTYflon 3210	Extruded from fibrillated PTFE and graphite with inert lubricant	5	−100 to +250	50	0−14	Seals all media except molten alkali metals, some fluorinated compounds and oxygen
LATTYflon 3215	Extruded from fibrillated PTFE with inert lubricant	5	−100 to +200	100	0−14	Food use, etc; Also general purpose seal for maintenance work, etc.
LATTYtex 861	Plaited homogeneous asbestos graphite	29	+500	80	3−14	Water, air, steam, gases (except oxygen), hydrocarbons, chemicals
LATTYgraf	Pure expanded graphite in flexible plait, tape or die-formed rings	35	−200 to +600	300	0−14	High duty packing for heat transfer fluids, aggressive media, demineralized water, food products, nuclear reactors, etc.
LATTYflon 2790	High tenacity polyacrylic packing, PTFE treated, impregnated and food quality lubricated	15	260	100	1−13	For rotary pumps, moderately abrasive and aggressive media
LATTYflon 2775	High tenacity polyacrylic, PTFE treated, diagonally plaited and re-impregnated with PTFE	10	260	300	1−113	For valves and static sealing

Aramid fibre also shows up favourably as regards resilience or the ability to react to varying pressure and adapt to the gland space available. Graphite fibre also has good resilience, but PTFE fibre or PTFE/asbestos packings generally show lack of resilience. These comparisons are drawn from empirical tests made on packings of the same plaited construction but different materials.

Graphite packings

Graphite fibre packings produced from graphite filament yarns have the potential advantage of being suitable for temperatures up to more than 400°C, with similar chemical resistance to PTFE (with the exception of strong oxidizing agents). They also have low friction with excellent heat dissipation properties. When they first appeared they did not live up to their original expectations, largely because they needed extreme care in fitting and also high surface finishes on both housing and shaft. They have been further developed to meet the requirements for operating at extremes of temperature, pressure and speed and the most aggressive chemicals but remain a high cost material though this is not necessarily reflected in the time cost effectiveness of a packing (see *Seal Selection Guides,* Section 8).

Flexible graphite foil is another second-generation packing material. A special production technique is used to produce an exfoliated graphite that is calendered into a foil without the use of a *separate binder* – a process that ensures purity and consistent density in service, there being no volatile substance that could be driven out at high temperatures and no additives for the media to leach out.

Gland packings in exfoliated graphite are preformed from tape into rings before fitting. Care must be taken to prevent extrusion from the gland. In high pressure valves, one solution giving near homogenity with the required anti-extrusion protection is illustrated in Figure 11.

Certain limitations exist however even with this near perfect material. For rotary work, the limitation is the more rapid thermal expansion of the exfoliated graphite packing than that of the iron or steel housing in which it is contained. This causes a rapid build up of pressure within the housing and against the shaft, leading to rapid increases in friction and of temperature, which are self-sustaining. To overcome this a 'cushion' to

Figure 11
Packings pre-formed in exfoliated graphite

absorb the expansion of the graphite, is needed and generally provided by end rings of graphite or carbon fibre packing, one against the neck bush and one under the gland follower.

The more common use for expanded graphite as a packing material is in valve spindle sealing and here the limitation is extrusion around the clearances at the bottom of the gland and around the follower.

At this stage it is useful to study the method by which a packing 'ring' of exfoliated graphite is made. From experience, better results are obtained by preforming rings to the required sizes before fitting. Simple three part but accurately sized tools are all that is needed, (Figure 12).

To charge the tool, the tape is wound as tightly as possible around the former to the required o.d. (less than 0.5 mm) before inserting in tool. If a tape of specific gravity

COMPRESSION TOOL FOR
LATTYGRAF® E1 AND LATTYTEX® 117

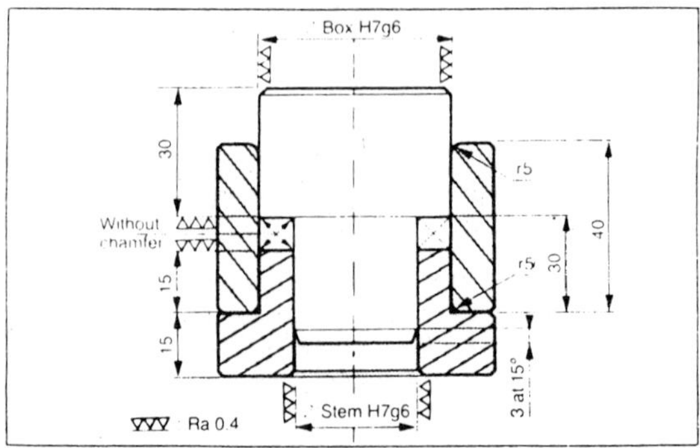

Figure 12

0.7 is being used and a ring of specific gravity 1.4 is required, compression should be such to reduce the ring height to 50% of that of the original tape plus an allowance for driving out the air left between the turns during winding. A good rule of thumb is to compress to 40% of the original tape height, so that a 25 mm tape would give a ring of 10 mm in height.

The advantage of starting with a tape of relatively low density is that the resultant ring will have good mechanical cohesion while still retaining a good resilience and recovery factor (see Figure 13).

Returning now to the use of expanded graphite in valve sealing, the danger of extrusion if a higher than original ring forming pressure is applied became clear. Graphite or carbon fibre packings are of little help as end rings in this case because of the fragility and low shear strength of the fibres when subjected to pressure. The answer for high pressure, high temperature valves including those on the primary circuits of nuclear reactors, has been found in an inconel wire reinforced packing plaited from a 50/50 graphite/asbestos chemically pure, dust free, wet spun yarn. (Figure 14).

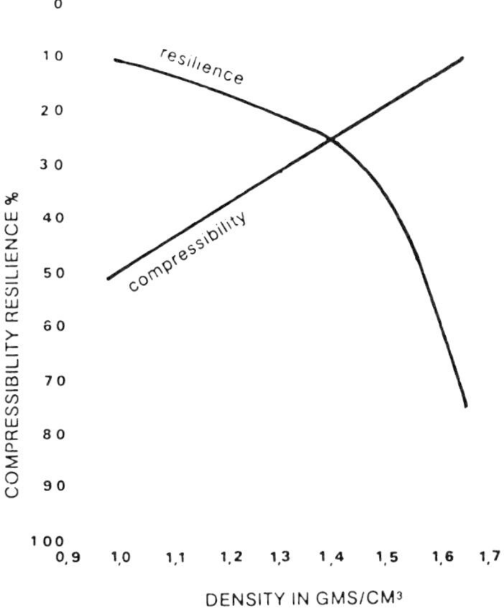

Figure 13

Figure 14
Plug of fibrous self-lubricating glazed sealing compound showing anti-extrusion ring gasket.

Glass and ceramic fibres

Relatively limited use is made of glass and ceramic fibres as a possible alternative to asbestos, although they have similar or better temperature and chemical resistance properties. A major drawback, as yet, is that the yarns made from these fibres do not lend themselves to the construction of traditional patterns of plaited/braided packings. *Glass fibre,* too, is held by some authorities to have something of the health hazard of asbestos.

Polyacrylic fibres

Plait into good, general service, PTFE-threaded packings, handling temperatures up to 260°C and fluids of pH value 1 to 13. White, clean to handle, at a lower first cost than pure PTFE. (LATTY flon 2788,2775, 2790. For pumps, valves and static seals.

Soft and hard packings

The majority of packings are quite resilient, ranging from soft or semi-plastic to moderately hard, to suit the service required. There is a further class of *hard* packings normally based on a metal outer covering and intended primarily for rotary applications, often at high temperatures. Wedging action packings form an intermediate group, for although presenting a hard face to the rubbing surface this is backed by a resilient section which provides the wedging 'squeeze' when compressed. *Wedging* rings, on the other hand, may be of hard construction throughout and virtually incompressible.

In some cases glands may be packed with combination packings to advantage. Thus where the service demands the use of a soft or semi-plastic packing to seal a fluid of low viscosity, a combination of this packing with harder end rings can be employed to minimize the risk of extrusion and support the soft rings properly. Similarly, hard and soft packing rings may be employed alternately to give rigidity to a relatively soft packing or to provide resilience in a basically hard packing (*eg* to absorb side loads).

Metallic packings

Metallic packings are fabricated from specially treated, lubricated metal foil, either in a complete wrap or around a resilient core (*eg* asbestos or sometimes flax). Metal foils commonly used are aluminium, copper and lead (see also Table 6).

Spiral-wrap packings are made from metallic foil ribbon wound in layer-upon-layer construction. Each layer is normally lubricated. This produces rigid sections (*ie* not resilient), but with a certain degree of flexibility.

Folded/twisted packings are made from spiral wrapped foil ribbons, folded and twisted to final size. Again each layer is lubricated and the final section has voids which act as a lubricant reservoir.

With *cored* packings the metal wrap may completely enclose the resilient core, or be open on one side, *ie* the sealing face.

Loose packings

Loose packings have a particular application for the sealing of valves or pump glands and have the advantage of being particularly easy to fit. They have only to be tapped home lightly to pack the gland. They become dense and homogeneous during operation but keep a high degree of permeability. As a general rule, it is advisable with loose packings to employ end rings of solid section or split rings of solid packing in order to prevent extrusion of the loose packing, although this may not be necessary if the gland and neck bushes are in first class condition.

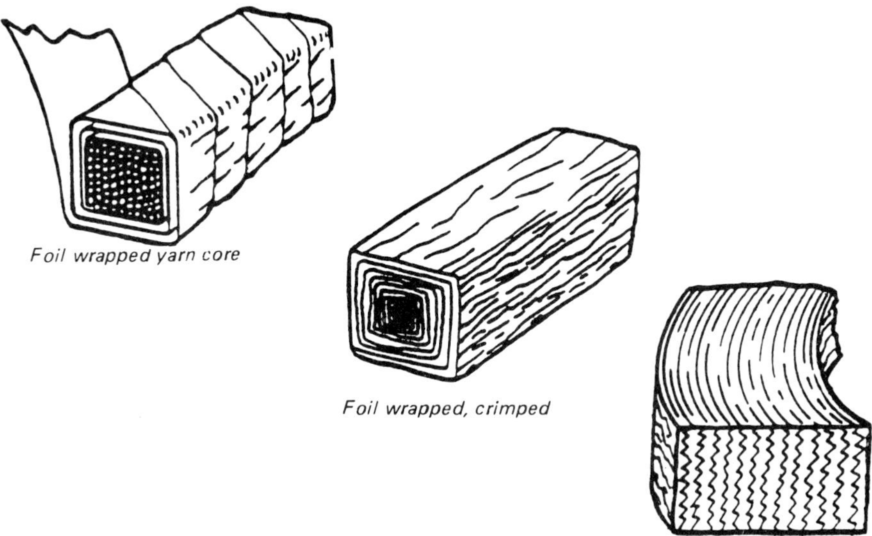

Foil wrapped yarn core

Foil wrapped, crimped

Foil wound, pressed

TABLE 6 – EXAMPLES OF METALLIC FOIL PACKINGS

Basic Type	Construction	Max Service Temp °C	Max Service Temp °F	Motion and Speeds	Application
All metal	All aluminium foil	280	1000	rotary to 5 m/sec (1000 ft/min)	High temperature services
	Braided copper wire	340	1200	rotary to 5 m/sec (1000 ft/min)	High pressure steam services
Foil	Aluminium with asbestos core	280	1000	rotary to 5 m/sec (1000 ft/min)	High pressure/high temperature steam
	Aluminium, soft core	*	*	rotary or reciprocating	Oils, organic chemicals and most aqueous solutions (not alkalis)
	Copper, with asbestos core	340	1200	rotary up to 5 m/sec (1000 ft/min)	Temperatures above 280°C (1000°F)
	Copper, braided	340	1200	rotary or reciprocating	Organic chemicals, sulphur-free oils, brines, high pressure steam services
	Lead with asbestos core	230	450	rotary up to 7.5 m/sec (1500 ft/min)†	High speed services
	Lead	230	450	rotary or reciprocating†	Water, salt solutions, dilute acids, etc
	Whitemetal foils with asbestos core	200	400	rotary or reciprocating†	Mildly corrosive acid and alkaline solutions
	Whitemetal foils with cotton, flax or hemp core	120	250	rotary or reciprocating†	Aqueous solutions
Tubular metal	Whitemetal, graphite lubricant	240	500	rotary or reciprocating	Air, water, steam, oils, etc.
	Copper, lubricated asbestos	340	1200	rotary or reciprocating	Air, water, steam, oils, etc.

*Depends on core material †Low friction packings

Valve packings

PTFE cord provides a simple, effective material for making *valve* packings. Several turns of cord are wrapped around the valve stem to fill the packing space and the gland follower adjusted as necessary.

There are various other types of packing materials used for simple *valve spindle* seals, such as graphited asbestos yarn or asbestos yarn coated with PTFE dispersion; pure PTFE fibre in woven ribbon form or *moulded* rings; *moulded graphite* rings, *etc.*

Nuclear Valve Stem Seals

The very highest integrity is an obvious requirement of valve seals in nuclear plant, demanding satisfactory performance in the following areas:

(i) *Mechanical* – deformation, resistance and recovery.

(ii) *Dynamic* – behaviour under spindle movement, under pressure and at elevated temperatures.

(iii) *Corrosion* – elimination of valve steam corrosion.

Main fluids to be handled are demineralized water, saturated steam (BWR plants) and borated water (PWR plants); but with possibly 1500 pieces of equipment in each division of a nuclear power plant, the multiplicity of problems involved can be immense.

The gland material now generally considered the most suitable in application is nuclear purity expanded graphite, with incorporated corrosion inhibition and anodic protection to eliminate electrolytic attack on stem surface. This is readily formed in mechanically sound rings from tape to provide the necessary resilience and deformability to behave as an efficient seal. To accommodate extrusion of the packing, expanded graphite rings are normally combined with plaited rings at the top and bottom of the gland – a logical choice here for high temperature working being pure graphite/asbestos braided packing which can also incorporate a corrosion inhibitor and a suitable proportion of anodic material to act as a sacrificial anode. An alternating arrangement of graphite fibre and expanded graphitie rings does not produce as satisfactory a seal.

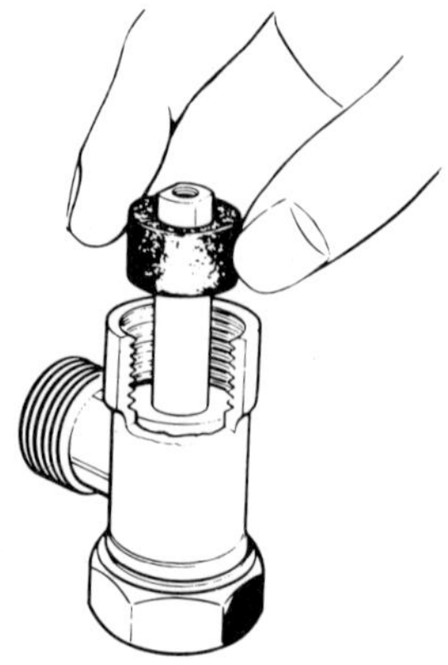

Figure 15a
One-piece valve packing in asbestos/graphite or asbestos/PTFE.

The braided packing rings serve to eliminate the risk of extrusion of the expanded graphite where radial play at the bottom of the box exceeds 0.5 mm and 0.3 mm around the gland follower.

This packing arrangement, LATTYGRAF E1 with LATTYTEX 117, after extensive laboratory testing by EDF, has been used successfully in French nuclear power plants for many years and is rapidly being extended through all primary circuits.

Should it be necessary to avoid asbestos products, rendering even wet spun, dust free product unacceptable, a graphite fibre packing can be employed for anti-extrusion purposes. This would however introduce problems of fragility and of potential corrosion risk (the stem alloys having to be chosen with extreme care).

Gland dimensions

The dimensional relationship between stem, box and packing rings is of prime importance. Interference between ring and stem with play between ring and box is to be avoided as leading to high stem torque and poor sealing. It is preferable to begin with a tight fit to the box and a small clearance to the stem (for example 0.1 mm).

Surface finish is important, particularly on the valve stem to realize minimum packing wear and low operating torque. Recommended values are $R_a = 1.6\ \mu m$ for the gland surface.

Live loaded packing

The maintenance of a leak free seal is directly dependent on the maintenance of an adequate loading on the gland packing. Disc springs (Belleville type washers) can compensate for loss of loading due to:

- relaxation of the packing (very slight for expanded graphite, of the order of 4% at 350 bar).
- wear
- differential expansion
- temperature variations

The introduction of disc springs also assists the precision with which gland loadings can be determined (*ie* by height reduction of disc). The cost of suitable disc springs is small in relation to the advantages they bring.

Gland geometry

The depth of a valve gland should be 1.5 to 2 times the stem diameter. A greater depth serves no purpose. Many glands are too deep. Should the number of packing rings exceed 6-7, transmission of gland loading becomes very uneven and stem torque increased disproportionately. Tests made by the EDF showed that an increased number of rings/depth of gland could result in increased leakage.

Recommendations for stem diameter/ring size are:

Stem diameter (mm)	Ring square section (mm)
10	4
20	6
30	8
40	8–10
50	10
60	12

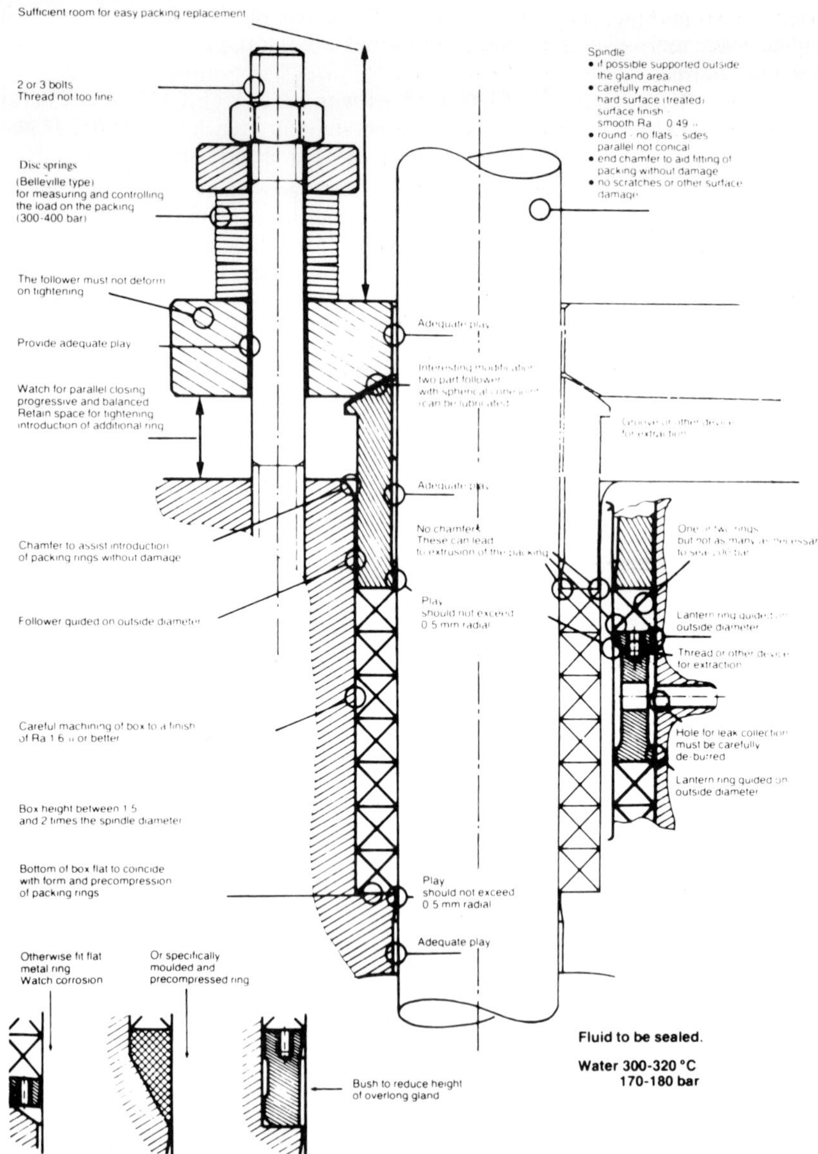

Desirable features of a nuclear valve gland. (Electricite de France)

Causes of leakage

The initial cause of leakage developing is not always apparent, particularly as one fault can lead to another. Experience has indicated that in order of seriousness, likely causes are:

(i) The use of braided packings that lose volume too readily and harden in use.

(ii) Damage to packing rings during fitting.
(iii) Bad meeting of ring ends where cut rings are used.
(iv) Incorrect disposition of ring joints, forming leak path.
(v) Insufficient gland loading.
(vi) Poor support to stem.
(vii) Poor ability of packing to withstand thermal shock.
(viii) Reduction in gland loading due to packing relaxation, packing wear/volume loss.
(ix) Incorrect dimensional tolerances between stem/packing/box.
(x) Gland too deep.
(xi) Stem surface finish of low order.
(xii) Play at bottom and top of box too great.
(xiii) Corrosion of stem and abrasion of packing during stem operation.
(xiv) Too many rings above lantern.

Corrective action

The following are points to observe, not necessarily in order of significance. Any one is important in achieving satisfactory gland performance.

(i) Correct gland design.
(ii) Entry to facilitate fitting of packing rings.
(iii) Optimum gland depth.
(iv) Correct surface finishes.
(v) Adequate capacity for gland loading/adjustment.
(vi) Disc springs to compensate for wear.
(vii) Good stem support.
(viii) Use of expanded graphite sealing rings particularly recommended (expanded graphite is permanent, resilient, has better relaxation, maintains its volume and withstands thermal shock), with Lattytex 117 as the second choice for steam and hot fluid valves.
(ix) Correct dimensional tolerances between packings and gland.
(x) Correct fitting and loading of gland before service operation.

Gland packings – installation and maintenance

Gland packings are supplied either as a continuous length in the form of a spool coil or spiral of packing material or as a pre-shaped split ring, when the exact size required must be specified when ordering.

The recommended method of cutting rings from a continuous length of packing is as follows:

(i) Place the packing round the shaft, or round a mandrel of the specified diameter (Figure 16a). The bore of metallic and extruded packing spirals should conform to this diameter.
(ii) To assist in cutting rings, two guide lines parallel to the shaft axis and separated by a distance equal to the packing section may be drawn on the spiral (Figure 16b).
(iii) Cut the rings from the spiral at an angle of 45°, diagonally across the guide lines so that no gap is left between the ends (Figure 16c).
(iv) Metallic and extruded packing rings are spirally open ready for fitting by pulling the ends axially apart. (Figure 16d).

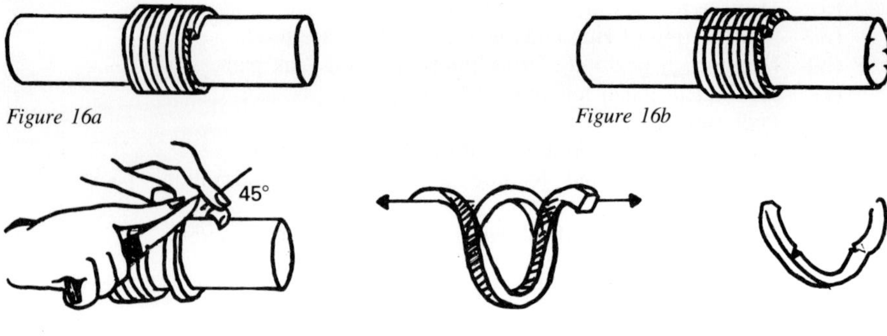

Figure 16a

Figure 16b

Figure 16c

Figure 16d

Figure 16e

(v) When the rings cannot be opened without risk of damage to the packing section, they are hinge-cut, cutting diagonally about two-thirds through the packing (Figure 16e).

(vi) Check the first ring to ensure a correct fit in the stuffing box, before cutting further rings in the same way.

Removing old packings

The system should be depressurized where appropriate and/or drained. The nuts holding the gland follower are then released, allowing the follower to be withdrawn along the shaft and clear of the stuffing box (Figure 17a). The old packing rings should then be withdrawn carefully. The best way is to use a pair of extractor tools inserted on diametrically opposite sides of the shaft (Figure 17b). After removal of the old rings, make sure that the inside of the stuffing box is clean. Check also that the shaft is in good condition, *ie* not scored, pitted, ridged or corroded. New packings will only have a short life rubbing on a damaged or rough shaft.

The following checks should be carried out:

(i) Check the shaft for concentricity with the stuffing box bore.

(ii) Check the shaft to ensure run-out does not exceed 0.025 mm (0.001 in) TIR.

(iii) Examine the gland follower for general condition and fit. The inner radial clearance should be 0.4 mm (0.015 in) maximum, and the outer radial clearance should be 0.25 mm (0.010 in) maximum, to prevent risk of cocking or touching on the shaft (Figure 17c).

(iv) Check the clearance between the neck bush and the shaft (Figure 17d). If this is greater than 0.25 mm (0.010 in) radially, it may be advantageous to employ a thin, close clearance spacer ring in the bottom of the stuffing box, to prevent risk of packing extrusion.

The number of rings required will be obvious from the number of old rings removed. However, if a different packing is to be used, measure the depth of the stuffing box and deduct a suitable allowance for the entry of the gland follower to arrive at the total depth of packing rings required.

Fitting new packings

Having checked that the shaft turns freely and is concentric with the housing bore, rings can be fitted up individually. A tamper can be used to ensure that each ring is pushed

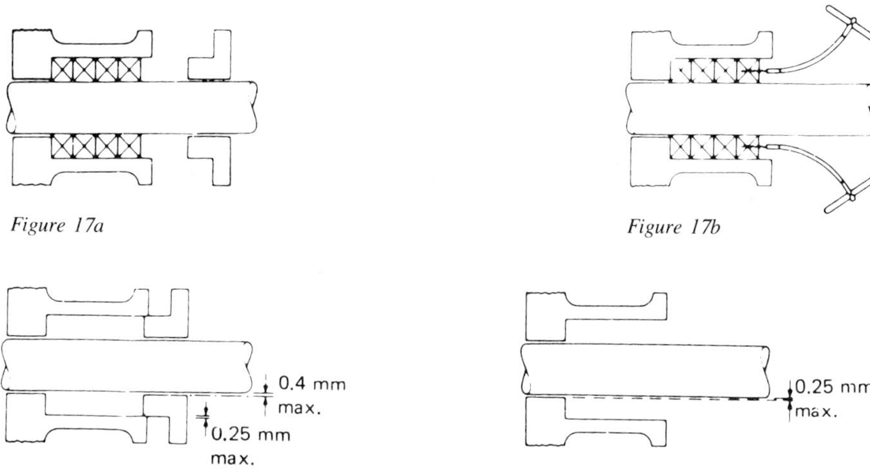

Figure 17a

Figure 17b

Figure 17c

Figure 17d

home true and square, but the use of a split sleeve in conjunction with the gland follower is better (Figure 18a). *There are exceptions to this practice.* Gland rings of PTFE multifilament yarns or graphite should not be tamped or driven home. They are only to be slid in place, otherwise they may suffer permanent damage.

Joints in succeeding rings should be staggered by 120° (Figure 18b). After fitting each ring, check that the shaft can be turned.

The actual position of each ring after installation can be checked by measuring the free depth remaining, *ie* this should be equal to the gland depth less the depth of the ring(s) installed. This is a particularly useful check in the case of a gland which incorporates a lantern ring to ensure that the lantern ring is positioned below its inlet. Note that the initial position of the lantern ring will be biased towards the 'open' end of the gland (Figure 18c) to allow for the compression of the bottom rings when the gland is tightened.

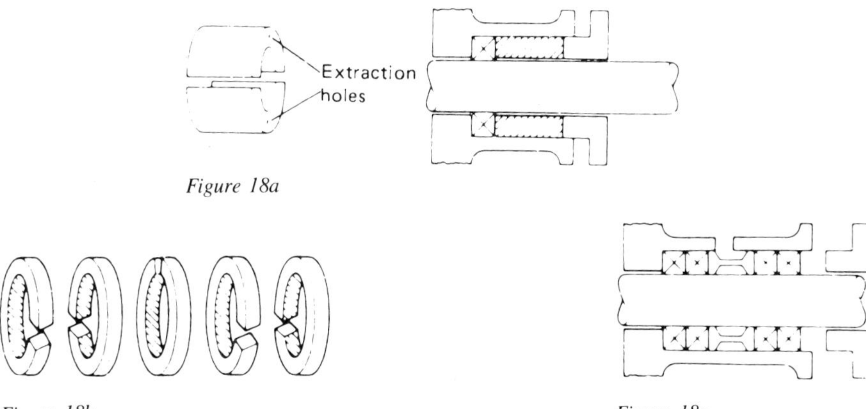

Figure 18a

Figure 18b

Figure 18c

After fitting the final ring the gland follower is brought up into position and tightened evenly with finger pressure on the nuts. Check that the shaft turns freely. The system can then be started and pressurized, which should produce appreciable steady leakage past the gland. After running for some time in this condition (*eg* five to ten minutes) the gland nuts should then be spanner-tightened in rotation about one-sixth of a turn at a time until leakage is reduced to an acceptable level.

This adjustment should be repeated at ten to fifteen minute intervals until leakage is reduced to drip leakage at an acceptable level. This level depends very much on the application, *eg* with careful running-in and light operating conditions, leakage may virtually be eliminated; with heavier operating conditions, a fair rate of drip leakage may be necessary to ensure that the gland does not overheat.

After the running-in period, further adjustments should only be necessary after prolonged intervals. All such adjustments should be made with the system operating at normal temperature and pressure, *ie* not under static conditions.

TABLE 7

	Observations	Possible causes/actions
At start-up	No leakage	1. Incorrect installation – negative stuffing box pressure requiring lantern ring. 2. Slight overtightening – follow stop/start procedure.
	Excessive leakage	1. Wrong section packing. 2. Incorrect installation procedure. 3. Shaft run-out.
Packing set removal	Packing section reduced under horizontal shaft	1. Shaft misaligned with stuffing box. 2. Worn bearings – packing acting as bearing.
	Packing section reduced above horizontal shaft	1. Shaft misaligned with stuffing box. 2. Worn bearings.
	Whole or part of an end ring missing	Excessive clearance between shaft and stuffing box neck or gland follower allowing extrusion.
	Wear on the outside of one or more rings	Rings rotating with shaft, packing section too small.
	Rings next to gland follower worn, other all right.	Incorrect installation procedure, gland follower overtightened.
	Ring i.d.'s burnt, dried or charred, remaining material all right.	1. Incorrect packing selection re : temperature limit/shaft speed. 2. Inadequate lubrication.
	Innermost ring deteriorated	Incorrect packing selection re : compatibility with product.
	Packing seizure on shaft after equipment shutdown	Product crystallizing or solidifying in packing bore through: 1. Inadequate lubrication. 2. Lack of heating or cooling.

Selection

There are so many reasonable alternatives available to satisfy each application, that it is only possible to set out general guidelines for selection.

In any event, it is essential that the following information be supplied, before a recommendation as to the type of packing and installation can be made.

A. Application
- (a) A pump – is it rotary or reciprocating?
- (b) A mixing vessel – top, side or bottom entry?
- (c) A blower or fan?
- (d) A valve or expansion joint?
- (e) A stern gland?

B. Product *ie* liquid being sealed.
- (a) Product name.
- (b) Does it contact the packing? If not, what does?
- (c) Is it corrosive – acid or caustic? Give pH value.
- (d) It is abrasive?
- (e) Will it crystalise or give off noxious vapours in contact with air?.
- (f) Can leakage to atmosphere be tolerated?
- (g) What is its vapour temperature and pressure?
- (h) Is a clean compatible liquid available?

C. Temperature
What is the temperature at the stuffing box?
- (b) Is it constant? If not, give details.

D. Pumped product pressure
- (a) What is the pump product pressure at the stuffing box?
- (b) What are the suction and discharge pressures?
- (c) Is the pressure constant? If not, give details.

E. Shaft details
- (a) What is the shaft diameter
- (b) What is the shaft speed?
- (c) What is the shaft material and hardness?

F. Other considerations
- (a) Does the packing have to comply with any standards or regulations? If so, give details.
- (b) Is the pumped product contamination by the packing materials acceptable.

Given this information, the appropriate installation arrangement may be determined, and the acceptability of the various packing types and forms available considered.

The data given in Tables 8 and 9 provide a good starting point for the selection of the types of standard asbestos and non-asbestos based ranges most appropriate to a particular application.

Self-lubricating gland sealing compounds

Self-lubricating gland sealing compounds are a non-compression type of seal using anti-extrusion discs at each end of the stuffing box, the space between the discs being filled with a compound. The compounds consist of a colloidal mixture of lubricants, amalgamated with non-asbestos fibres and then pressure treated to ensure uniform dispersion, as shown in Figure 19. It is intended for use in both valves and pumps, but the method of application varies slightly.

When used in valves, the lantern ring must be removed and then connections plugged. A new ring of appropriate packing is then placed at the bottom of the stuffing box. For maximum effect this should be graphite braided. The stuffing box is then filled with

TABLE 8 — COMPRESSION PACKINGS

MAIN CONSTRUCTION MATERIALS	ROTARY	RECIPROCATING	VALVE	ACIDS/ALKALIS pH	WATER	SIEAM	AMMONIA	AIR	OXYGEN	PETROLEUM OILS	SYNTHETIC OILS	GASES	SOLVENTS	NUCLEAR	SI URRIES
Asbestos, PTFE Impregnation	✓	✓	✓	2-14	✓	✓	✓	✓		✓	✓	✓	✓		
Asbestos, PTFE, Wax Impregnation					✓	✓	✓	✓		✓	✓	✓			
Wire reinforced Asbestos, Binders, Graphite					✓	✓	✓	✓		✓	✓		✓		
Asbestos, MoS_2, Wax				4-10	✓	✓	✓	✓		✓	✓	✓			
Wire reinforced Asbestos, Binders, Graphite, Inhibitor	—	—	✓	2-11	✓	✓	✓	✓		✓	✓	✓	✓		
Asbestos, Binders, Graphite	✓	✓	✓		✓	✓	✓	✓		✓	✓	✓			
Aluminium Foil, Asbestos				4-11				✓		✓	✓				✓
Soft anti-friction Foil, Asbestos, Oil				3-12	✓	✓	✓	✓		✓		✓			✓
Aramid, PTFE Impregnation, Lubricant				2-12	✓	✓	✓	✓		✓	✓		✓		✓
PTFE, Oil				0-14	✓	✓	✓	✓		✓	✓		✓		
PTFE, PTFE Impregnation					✓	✓	✓	✓	✓	✓	✓	✓	✓		
PTFE, PTFE Impregnation, Oil					✓	✓	✓	✓		✓	✓	✓	✓		
PTFE, Graphite, Oil					✓	✓	✓	✓	✓	✓	✓	✓	✓		✓
Graphite, PTFE Impregnation					✓	✓	✓	✓		✓	✓	✓	✓		
Graphite, Graphite Impregnation					✓	✓	✓	✓		✓	✓	✓	✓	✓	
Expanded Graphite					✓	✓	✓	✓	✓	✓	✓	✓	✓	✓	

TABLE 9 — COMPRESSION PACKINGS

MAIN CONSTRUCTION MATERIALS	SHAFT SPEED m/s			PRESSURE bar			TEMPERATURE °C
Asbestos, PTFE Impregnation	10	2	2	20	40	70	−75 to +260
Asbestos, PTFE, Wax Impregnation	12			15	40	70	−75 to +260
Wire reinforced Asbestos, Binders, Graphite				20	100	200	−40 to +540
Asbestos, MoS_2, Wax				15	40	70	−40 to +150
Wire reinforced asbestos, Binders, Graphite, Inhibitor	—			—	—	200	−75 to +650
Asbestos, Binders, Graphite	15			20	40	70	−40 to +450
Aluminium Foil, Asbestos				20	70	200	−40 to +540
Soft anti-friction Foil, Asbestos, Oil				20	50	70	−40 to +230
Aramid, PTFE Impregnation, Lubricant				15	100	200	−75 to +260
PTFE, Oil	10			15	40	70	−40 to +260
PTFE, PTFE Impregnation				20	200	370	
PTFE, PTFE Impregnation, Lubricant	12			15	40	70	
PTFE, Graphite, Lubricant	18			20	200	350	−75 to +260
Graphite, PTFE Impregnation	25			20	50	350	−200 to +500
Graphite, Graphite Impregnation	30			20	50	350	−200 to +2000
Expanded Graphite				20	50	350	−200 to +2000

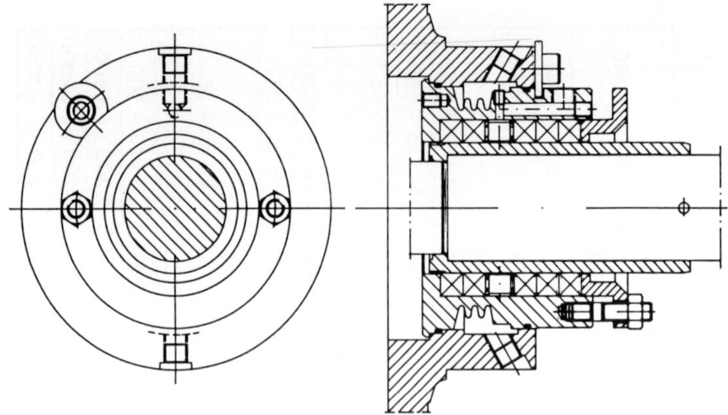

Figure 19
"Modular gland with braided compression packing".

the compound, using the gland follower to compact the compound and to remove any air pockets. The gland nuts are then firmly tightened. Since the compound is self-lubricating, the valve is easy to turn. In addition, the compound does not harden, crack or dry out.

Parameters.

Pressure	75 bar (1050 psi)
Temperature	−40 to 250°C (−40 to 600°F)
pH	2 to 12

For use in pumps or other dynamic applications, any old packing should be removed, as well as the lantern ring and the connections plugged. Two gaskets, each 3 mm (0.125 in) thick and having a 45° slit cut in them, are fitted one at each of the stuffing box.

The space between the gaskets can now be filled with the compound, using a special pressurised dispenser. The gland nuts are now moderately tightened, the pump started, and the nuts slowly tightened until zero leakage is reached.

Parameters.

Pressure	15 bar
Temperature	−40 to 315°C (−40 to 600°F)
Speed	3600 r/min
pH	2 to 12

Shaft run-out not to exceed 0.025 mm (0.001 in) total indicator reading.

Over a period of time it will be found necessary to top up the compound, but this is normally only a five minute job, and the interval between top ups usually longer than that for gland packing replacement.

The compound is not recommended for strong solutions of Black Liquor, heavy concentrations of lime or methyl ethyl ketone and similar very strong solvents. If there is any doubt as to the chemical compatibility of the compound, a simple test should be made. This consists of rolling two balls of the compound, and keeping one of them as the control sample. The other should be immersed in a jar of test liquid for 24 hours, and if it has not dissolved, or is only slightly affected, it will perform satisfactorily.

SECTION 5

Mechanical Seals

Mechanical Face Seals

ALL MECHANICAL face seals, generally known as mechanical seals, consist essentially of three parts. These are a) A rotating component, variously referred to as a rotary seal ring or face; b) A stationary component, known as a seat or stationary seal ring; c) A spring or springs, sometimes comprising as many as twenty-four. Where no springs are present, such as in the case of metal bellows seals, the convolutions of the bellows themselves, act as springs, producing a face load, (Figure 1).

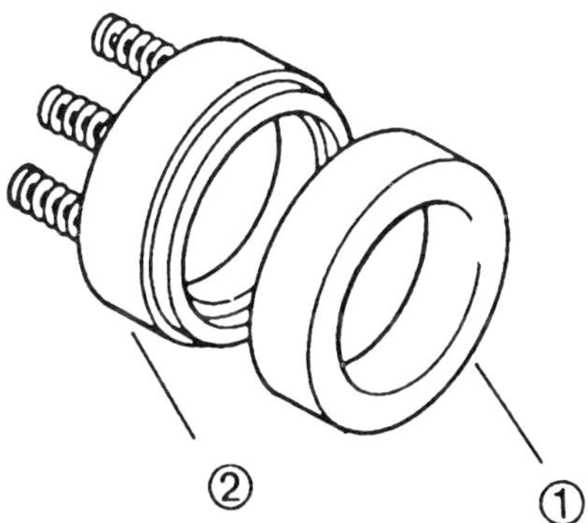

Figure 1

The *surface finish* of the face and seat must be of a very high order, and in most cases is to 2 light bands (1 light band is 11.6 millionths of an inch, or 295 nanometers). This surface finish is achieved by lapping, and can only be verified by the use of optical flats and a monochromatic light source. Hence the term *light bands* to denote the degree of flatness obtained by the method.

Successful operation of the seal is dependent on achieving the right conditions at the seal face. The surfaces of the opposing rotary and stationary components are lubricated by a thin hydrodynamic film, some 3 to 5 microns thick, formed from the liquid being sealed. As the liquid passes through this very small gap, there is a pressure gradient and the pressure falls from that surrounding the seal to atmospheric pressure, (Figure 2).

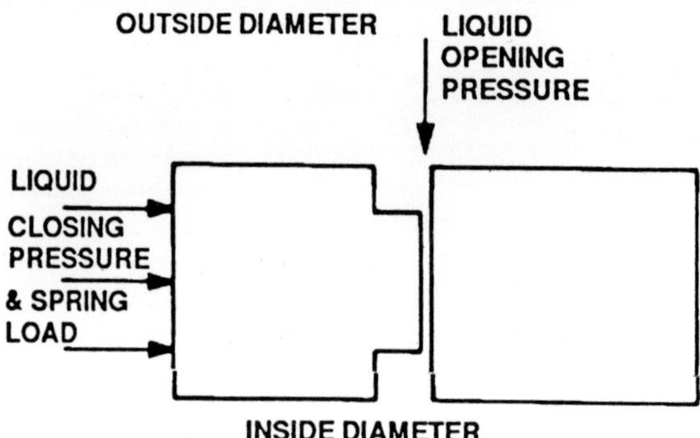

Figure 2
Mechanical face seal operation

A mechanical seal is a remarkable piece of engineering equipment. It can contain:-

Liquids of all descriptions.

Pressures from full vacuum to 80 bar.

Temperatures from −100°C to +400°C.

Shaft speeds up to 3,600rpm (and higher speeds with special designs).

All this is achieved with a face width of 1 to 5 mm, depending on size and seal design, (Figure 3).

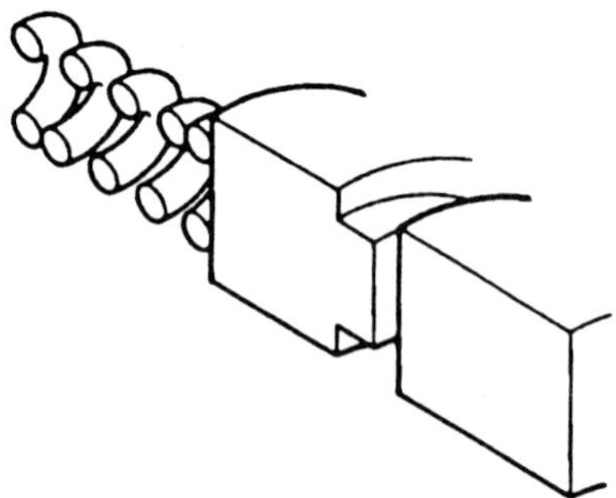

Figure 3
Example of mechanical face width

Of course, seals require more than the components so far referred to. They require some means of preventing leakage between the seal and the shaft (or sleeve) on which they are mounted. This area is known as the secondary seal, and can take the form of O-rings, wedges, chevrons or rubber bellows (Figure 4).

Another area which must be considered is where the seat sits in the cover plate (or gland plate). This is often referred to as the tertiary sealing area. The seat is fitted with either an O-ring, a square section piece of PTFE, or other high temperature resistant material. An L section piece of rubber is used in the case of the low cost types of mechanical seal, (Figure 5).

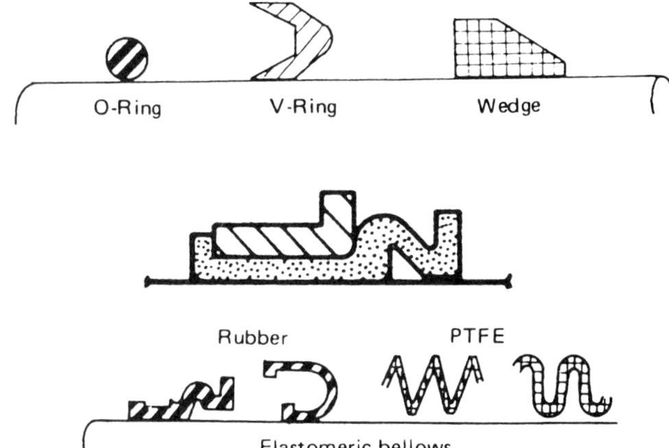

Figure 4
Examples of secondary seals

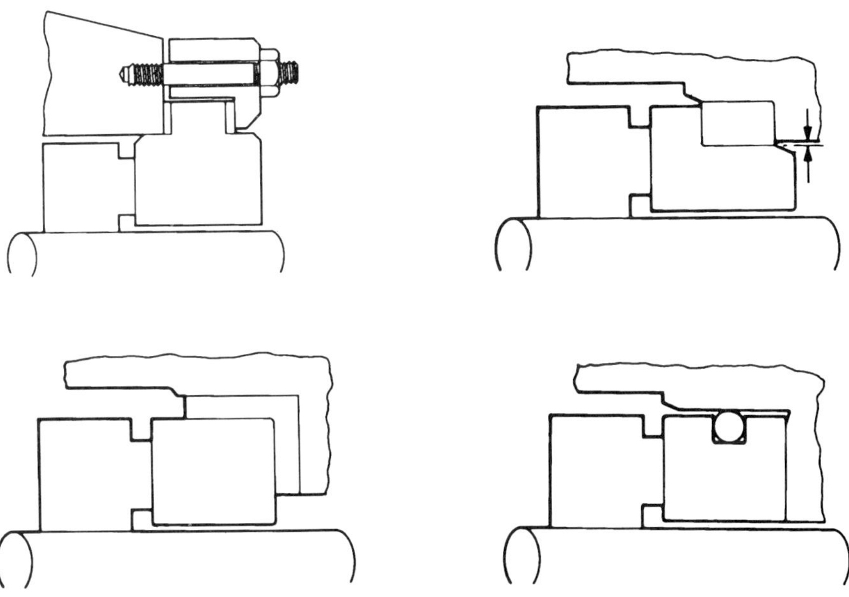

Figure 5
Examples of tertiary sealing.

Shaft of sleeve surface finish

This is a most important contributory factor to the efficient working of the mechanical seal. Different types of seal require different surface finishes if they are to function properly. To illustrate this, consider those seals which have to move along the shaft (pusher seals) as wear takes place at the seal faces. To do this, a surface finish of 100

TABLE 1 — SECONDARY AND TERTIARY SEAL TEMPERATURE LIMITS

Seal material	Temperature °C	
	Minimum	Maximum
High nitrile rubber	−30	120
Ethylene propylene	−50	150
Cloropren	−40	150
Fluorocarbon	−5	200 (1)
Ailixonw	−50	200 (2)
Fluorosilicone	−60	250 (2)
Perfluoroelastomer	−20	260 (2)
PTFE (Pure)	−40	230
PTFE (Glass loaded)	−100	280
High temperature polymers	−100	280
Compressed asbestos fibre	−100	300
Pure graphite materials	−200	500 (3)

N.B. (1) For use on water max. temperature +100°C.
(2) Unsuitable for use as bellows.
(3) In non-oxidising atmosphere max. 2000$C.

to 250 nanometers is desirable, and equates to a ground and polished finish. Mechananical seals in this category are of the O-ring, wedge or chevron type.

By contrast, rubber bellows seals have to grip the shaft, since the drive is transmitted to the shaft through the rubber bellows, then via the drive ring and drive sleeve and finally to the face. It is imperative, therefore, that a shaft surface finish is provided which will effectively allow the seal to bond itself to the shaft.

A surface finish of 800 to 1200 nanometers is recommended. This is equivalent to normal fine machine turning.

For other categories of seal, such as metal bellows, which contain O-rings, a surface finish of about 400 to 600 nanometers gives the best result.

In this case, when wear takes place at the seal face, the secondary seal does not have to slide along the shaft. Wear is taken up by the convolutions of the bellows. It does however have to slide along the shaft during fitting. Any shaft surface imperfections may damage it, so a good surface finish is essential, (Figure 6).

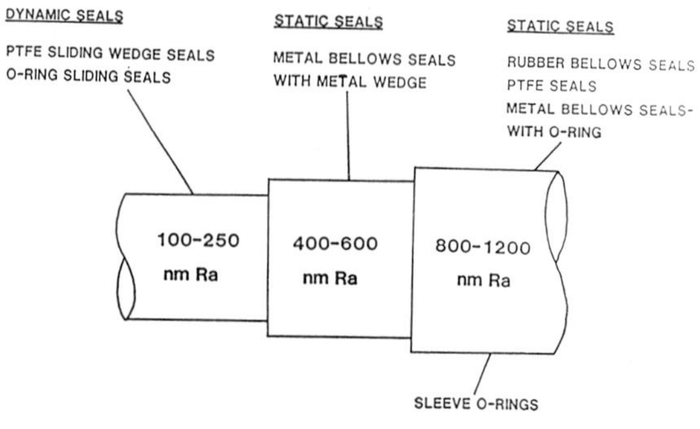

Figure 6
Nanometers roughness average.

TABLE 2 – SURFACE FINISH AND HARDNESS REQUIREMENTS OF SECONDARY SEALING DEVICES

Secondary Seal	Material of Construction	Sliding	Stationary	Surface Finish Nanometers (nm)	Hardness Brinell Minimum
'O' Rings	Rubber coated PTFE or Coated	✓		100 – 250	125
Wedge	Pure PTFE	✓		100 – 250	125
Wedge	Glass/Carbon Loaded	✓		100 – 250	500
Bellows	Rubber		✓	800 – 1200	125
Bellows	PTFE		✓	800 – 1200	125
Bellows and 'O' Rings	Metal		✓	800 – 1200	125
Bellows and Metal Wedge	Metal		✓	400 – 600	125

Springs – single versus multiple

Single springs normally have a much greater cross-sectional area than multiple ones. They are also ground on both ends in order to present as flat a surface as possible. The ends can never be absolutely parallel to each other, however, and a moment is produced diagonally throughout the length of the spring.

The advantages of the single spring are that it is better able to resist corrosion, and very good when handling slurries. The centrifugal action of the rotating spring tends to throw any particles away from the seal and so prevent clogging.

The disadvantages of single spring seals are that they can be difficult to fit, since on the larger sizes, the force required to compress them is quite considerable. They also suffer from a need for a much greater working length than multi-spring seals. This may pose problems if the stuffing box area of a pump, say, is limited due to the proximity of the bearings.

Finally, it has to be accepted that it is extremely difficult to achieve an even face loading around the complete seal face circumference.

Multiple springs, by contrast, give a very even face loading because the moment of one spring is cancelled out by that of another. They have the further advantage that, due to their much shorter length, they can be fitted into a relatively small area.

Their disadvantages are, that they are not so well suited to handling corrosive liquids because of their small cross-sectional area. Futhermore, they are prone to clogging when handling slurries.

When selecting a seal, therefore, the use of single and multiple springs is a factor which must be considered.

Shaft sizes and PV

The Pressure x Velocity (PV) value determines in some measure the suitability of material combinations for the seal faces, in particular the amount of heat generated at the face.

Large diameter multi-spring seal.

TABLE 3 – ADVANTAGES AND DISADVANTAGES OF SPRING TYPES

Spring type	Advantages	Disadvantages
Multiple	Smaller seals. Even face load. Higher rotational speeds.	High stress levels. Low corrosion allowance. Prone to blockage.
Single	Low stress levels. Greater corrosion allowance. Readily coatable in *eg* PTFE.	Uneven loading. Larger and longer seals. Speed restricted.
Rat-trap Wavy washer Belleville washer	Space-saving.	High rate.

For a given rotational speed, an increase in shaft diameter will increase the PV.

From this, it follows that for a given shaft size, an increase in rotational speed will also increase the PV.

This is because:

PV = pressure x π x seal diameter x speed

In imperial units; pressure in psig.
seal diameter in ft
speed in r/min

Note that the seal diameter is the mean face diameter of the contact area.

In metric units;
pressure in bar.
seal diameter in m.
speed in rev/s.

Refer to Figure 7 for conversions.

The formula shown refers only to unbalanced seals. For balanced seals, the PV value obtained must be multiplied by a figure equivalent to the degree of balance designed into the seal. In most cases this figure will be about 0.75.

Note that the pressure value used is the actual pressure on the seal. In the majority of instances this will be somewhere between suction and discharge pressures because of the use of product recirculation.

All seal manufacturers publish figures for PV, and typical material combinations and PV values are shown in Table 4. N.B. One way of reducing PV is to use a balanced seal.

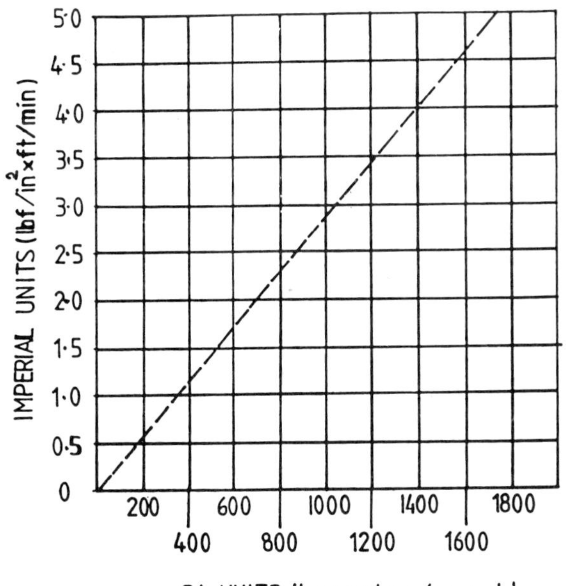

Figure 7
P.V. Conversions.

TABLE 4 — PV (PRESSURE X VELOCITY) LIMITS

Face Materials RSR/SSR	FACE MATERIAL PV LIMITS BAR M/S†			
	Water and Aqueous Solutions		Other Fluids	
	Unbalanced Seals	Balanced Seals	Unbalanced Seals	Balanced Seals
Stainless steel/carbon	5.5	–	30	–
Lead bronze/carbon	23	–	36	–
Stellite/carbon	49	85	52	580
Chrome oxide/carbon	70	440	–	–
Alumina ceramic/carbon	36	250	88	420
Tungsten carbide/tungsten carbide	44	500	71	420
Tungsten carbide/carbon	70	700	88	1225

†1 bar m/s = 2855 lb/in^2 ft/min

Single and double seals

Single seals have one set of sealing faces and are generally suitable for sealing fluids which provide adequate lubrication at the seals faces. In the great majority of cases, product recirculation from the pump discharge to the seal will be provided, and this will give some degree of cooling to the seal. Even so, there will be an increase in temperature at the seal faces. This can be as high at 20°C.

Single seals can also be used with non-lubricating fluids by introducing a separate flushing fluid into the seal chamber. This will cool and lubricate the seal faces. However, it has to be accepted there will be a problem controlling the flushing liquid. It will, for example, dilute the product being pumped, even when a close fitting neck bush is fitted in the pump stuffing box. In addition there is the loss of the flushing fluid itself to consider, with its attendant cost, (Figure 8).

Where such losses cannot be tolerated, double seals are used.

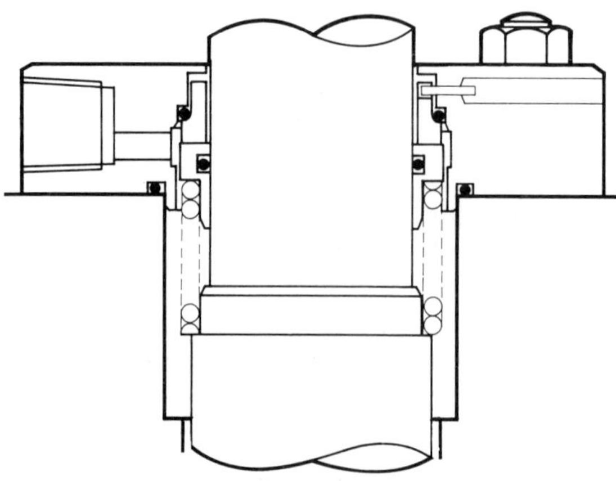

Unbalanced single spring seal.

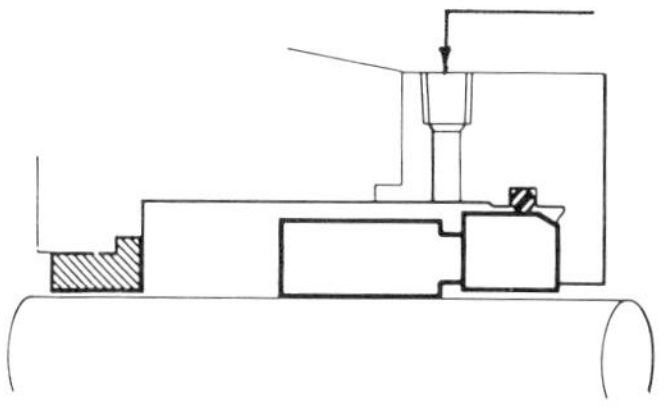

Figure 8
Single seal with flushing fluid.

Double seals.
These are of three types:-

(a) Back-to-back
(b) Tandem
(c) Concentric – See section on *Vessel Seals*.

Back-to-back seals are mounted with one seal facing inboard towards the pumped liquid, and the other facing outboard towards atmosphere.

In all cases it is necessary to circulate between the two seals a separate liquid which is compatible with the pumped product. It must also be at a pressure of at least 1 bar greater than the pressure of the pumped product. In many cases it is preferable that this differential pressure be even more than 1 bar. In all pumping systems, transients (sudden surges of pressure) occur for no very obvious reason. When this happens, a higher differential pressure gives a greater degree of protection to the seals. This limiting factor on the value of the differential pressure is, in fact, the maximum pressure which the outboard seal can accept.

The separate liquid is known as the sealant, and in theory any leakage at the inboard seal should be of this sealant across the seal faces into the pumped product. In practice, however, there is an interchange of product and sealant. If the sealant is contained in a tank, this can be detected, and a decision made to change the inboard seal. This will entail shutting down the process. Should the loss of sealant or pumped product not be excessive, then this may be accepted, and the process allowed to continue until a more convenient time for plant shut-down.

Double back-to-back seals are used because the product either crystallizes, oxidises on meeting atmosphere, or alternatively, is abrasive. In addition, there may be a need to use double seals for safety reasons, (Figure 9).

Tandem seals are mounted one behind the other, both seals facing in the same direction, that is to say towards the atmosphere side.

There are two ways in which these seals may be used, the first one being as a safety device. When used in this mode, the inboard seal is immersed in the pumped liquid, and product recirculation is provided from the discharge flange of the pump to the seal, adjacent to the seal faces.

The outboard seal has a separate low pressure sealant injected into the space between it and the inboard seal. A barrier has thus been formed by the sealant between the pumped liquid and atmosphere. Should the inboard seal fail, and product start to leak through the seal faces and into the space occupied by the sealant, this can be detected, and a decision made as to whether to shut down the system or not. Remembering that the two seals are identical, it will be realised that the outboard is capable of accepting the same

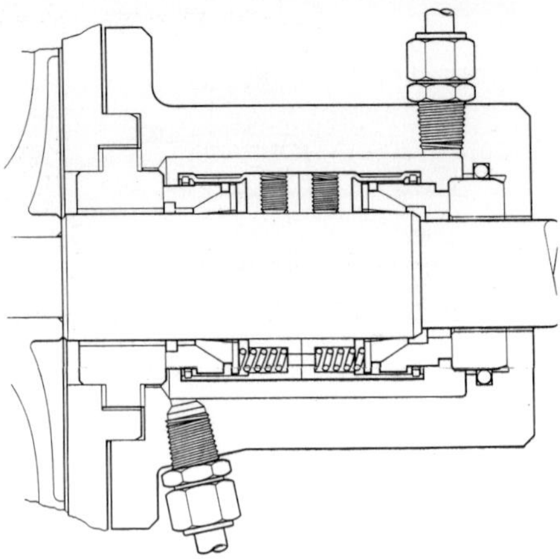

Figure 9
Double Seal.

pressure as the inboard one. This being so, it may not be thought necessary to shut down until the leakage becomes unacceptable, (Figure 10).

In the event that the outboard seal only should fail, this will be easily observed and the requisite action taken. The leakage in this case will be of sealant, which is normally water or oil, and therefore not harmful to plant operators.

Tandem seals in this configuration are used to seal light hydrocarbons such as butane, propane, ethane and methane, as well as other products which may be flammable, toxic or explosive. The other way in which tandem seals may be used is as pressure break-down seals.

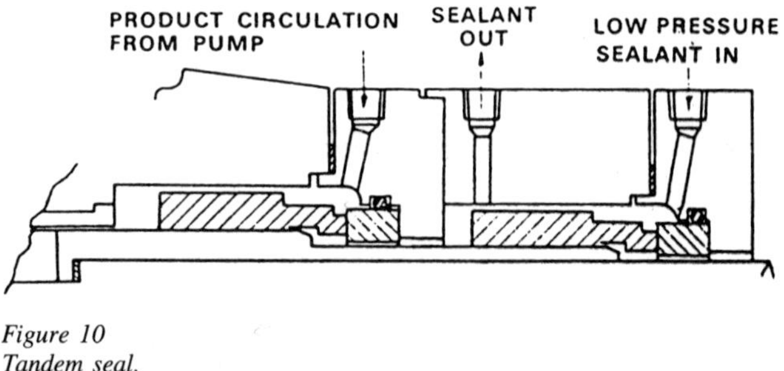

Figure 10
Tandem seal.

When the pumped product pressure is too high for a single seal alone to handle, two and sometimes more seals are mounted in tandem. The inboard seal is subject to the full product pressure, and product recirculation provided in the usual way. The outboard seal is supplied with sealant at half the product pressure, and coincidentally the inboard seal is also subject to this sealant pressure. In this case, however, on the atmosphere side of the inboard seal. As a result, the inboard seal is now on differential pressure, *ie*, the difference between product and sealant pressures, (Figure 11).

Possibly the best known use for this method is in the nuclear power industry.

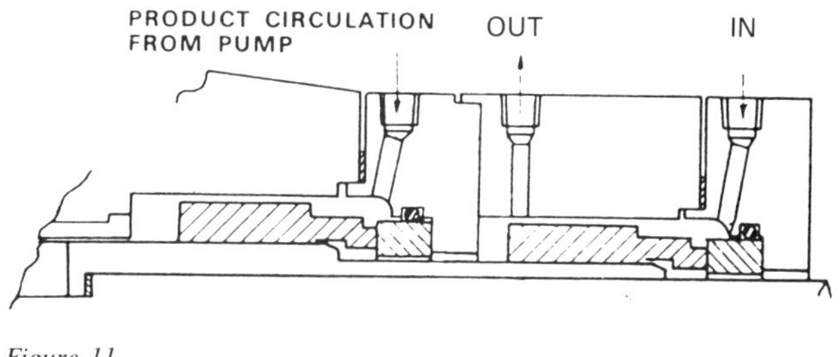

Figure 11
Tandem seal.

Rotary versus stationary carbon faces

The great majority of seals use one of the 'contact' faces in carbon, its main advantage being that it has excellent self-lubricating properties. This is a particular advantage on start-up. It also has very good heat transfer, which helps keep the seal cool by passing generated heat into the pumped product. Various theories have been advanced as to whether it is better to rotate the carbon face and stationary mount the seat, or to rotate the metal/ceramic face and stationary mount the carbon seat.

There is no doubt that both methods have proved successful, so the argument is largely academic.

It has to be said, however, that fitting the carbon into the clamp plate requires a good deal of care, since this can otherwise result in fracture due to the fact that carbon is mechanically weak against distortion.

Conversely, it is common practice to stationary mount the carbon and rotate the seal by those manufacturers who normally do the opposite. This is because the carbon is driven by dimples, and when handling viscous products, the seal is trying to rotate a carbon which is stuck to the seat by the product following shut-down.

Of course the position can be alleviated somewhat by applying a steam quench to the bore of the seat/face contact area, and maintaining this heat even when the pump is stoppen the pump is stopped.

Most manufacturers rotate the metallic face, and press the carbon into the clamp plate. Of those who do not, the best known are *Sealtec* and *John Crane International*. (An amalgamation of Crane Packing Ltd and John Crane USA).

Most manufacturers rotate the metallic face, and press the carbon into the clamp plate. Of those who do not, the best known are *SEALTEC* and *JOHN CRANE INTERNATIONAL*. (An amalgamation of Crane Packing Ltd and John Crane USA).

Standards for mechanical seals

The development of European national standards has produced the German DIN 24960 and ISO 3069 seal installation standards; and also in the UK the BS 5257 : 1975 pump standard which effectively incorporates ISO 3069. A number of other national standards generally require seals conforming to the dimensional sizes specified in DIN 24960. Nomenclature and material coding given in this standard are not necesssarily followed by all manufacturers, however.

The (German) VDMA nomenclature for single internal mechanical seals with rotating seal head is as follows (see also Table 5):

GLRD (specifying a mechanical seal) VDMA
3 symbols (specifying type and size) letter code (specifying materials)

In the first group of letter symbols:

1st letter – U = unbalanced or B = balanced.

2nd figure gives nominal diameter in mm.

3rd letter gives direction of rotation or direction of winding of spring, *viz:*

R – right hand, with single spring.
L – left hand, with single spring.
S – rotation independent of single spring.
A – direction independent of multiple springs.

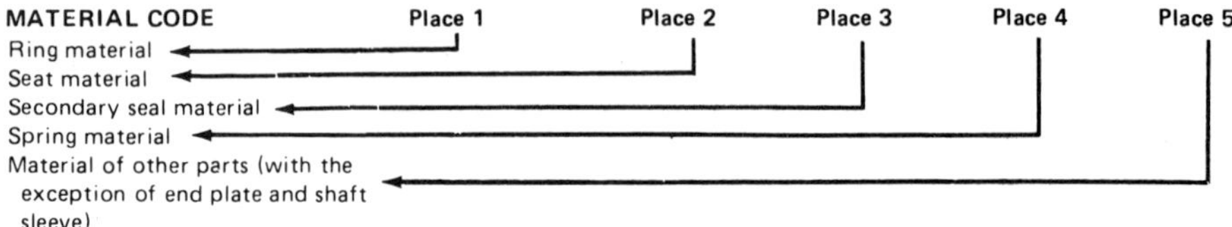

TABLE 5 – MATERIAL CODE

Place 1/2	Place 3	Place 4/5
Material of ring or seat	**Material of secondary seal**	**Material of springs and other metallic parts (with the exception of end plate and shaft sleeve)**
Mechanical Carbon	**Elastomers**	
A = Hard carbon impregnated	P = Nitrile rubber	
B = Hard carbon resin impregnated	N = Chloroprene	
	B = Butyl rubber	
C = Other carbon	E = EP rubber	
	S = Silicone rubber	
	V = Fluor rubber	D = C-steel
Metal	M = PTFE-encased	E = Cr-steel
D = C-steel	X = Misc. elastomers	F = CrN1-steel
E = Cr-steel		G = CrN1Mo-steel
F = CrN1-steel		
G = CrN1Mo-steel		
H = CrN1-steel, stellited	**Non-elastomers**	
K = CrN1Mo-steel, stellited	T = PTFE	
L = Stellite	A = Asbestos impregnated	M = High nickel alloy
M = High nickel alloy	F = IT seal	N = Bronze
N = Bronze	Y = Misc non-elastomers	
P = Cast iron		
R = Alloyed cast iron		
S = Cast chrome	**Special case**	
T = Misc metals	U = Differences in secondary sealing materials	T = Misc metal
Metal Carbide		
U = Metal carbide		
Metal Oxide (ceramic)		
V = Al-oxide		
W = Cr-oxide		
X = Misc metal oxides		
Plastics		
Y = PTFE, strengthened		
Z = Misc plastics		

Type classifications

Mechanical seals can be classified in several ways, *eg:*

(i) Inside or outside seals (Figure 12).

(ii) Unbalanced or balanced seals.

(iii) Single or double seals.

Mechanical seals are described as *inside* if the hydraulic pressure being contained acts in a 'closing' direction, when any leakage is from the o.d. of the seal. With an *outside* seal hydraulic pressure acts in an 'opening' direction, with any consequent leakage from the i.d. of the seal. Actual seal mounting position on the shaft (*ie* externally or internally) does *not* define whether the seal is an internal or external type.

The internal seal is a more stable type; the external seal tends to lose face 'contact' with a sudden surge and quite often it is impossible for it to remake normal face 'contact'. Excessive leakage will then occur with the only remedy usually being to shut the unit down and restart. This phenomenon is far less likely with internal seals. Secondly, the cooling of an internal seal is usually better than the external seal, with the pumped fluid having better contact to remove heat. Even with a cooling injection, the injection positioning for heat removal is usually easier to place in the optimum position on an

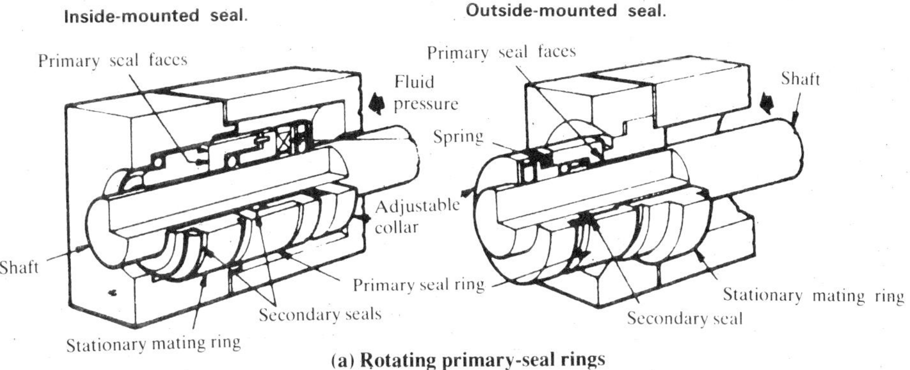

(a) Rotating primary-seal rings

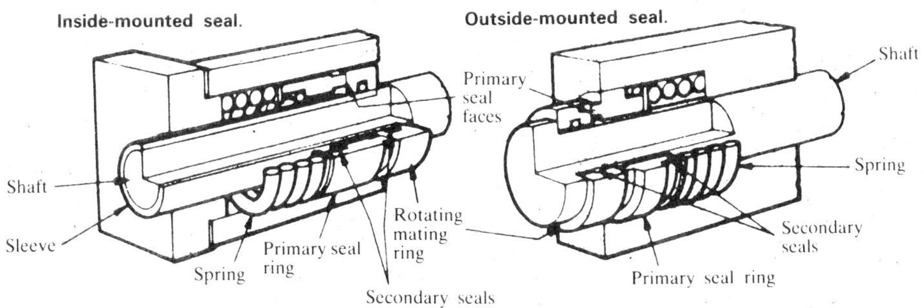

(b) Stationary primary-seal rings

Figure 12
Inside-mounted and outside-mounted mechanical face seals incorporating rotating and stationary primary-seal rings.

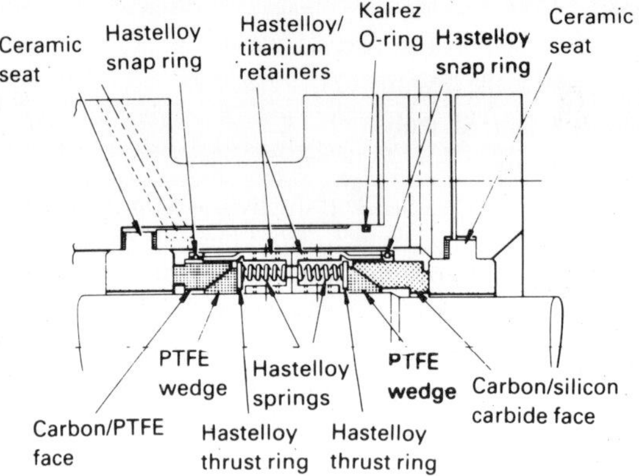

Detail of double mechanical seal with V-seats.

internal seal installation. This cooling problem makes thermal distortion (and, therefore, leakage failure) a greater problem on the external seal. A basic limitation of the external seal is that solid particles in the fluid which is being sealed can readily collect by centrifugal force against the sealing faces, and cause physical damage by chipping the edges and then entering between the faces to accelerate the process.

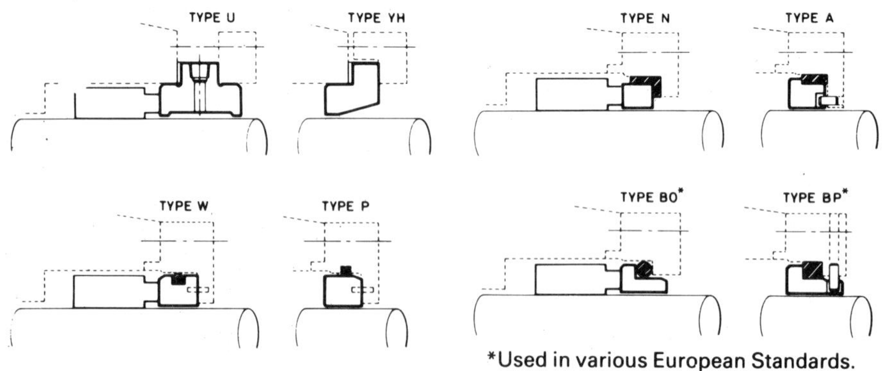

Examples of stationary seals.

Hydraulic balance

Hydraulic balance refers to the relationship between the pressure of the fluid being sealed and the contact pressure between the seal faces. In the case of an *unbalanced seal,* hydraulic pressure is unrelieved by the face geometry so that the face contact pressure is equal to or greater than the fluid pressure (Figure 13). With a *balanced seal* the design proportions the radially disposed areas so that the effective contact pressure is always less than the fluid pressure (Figure 14). In practice, the lowest face pressure that will give effective sealing is normally selected.

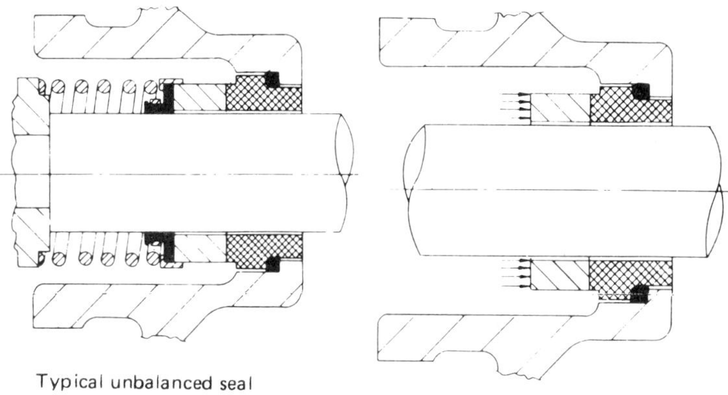

Figure 13

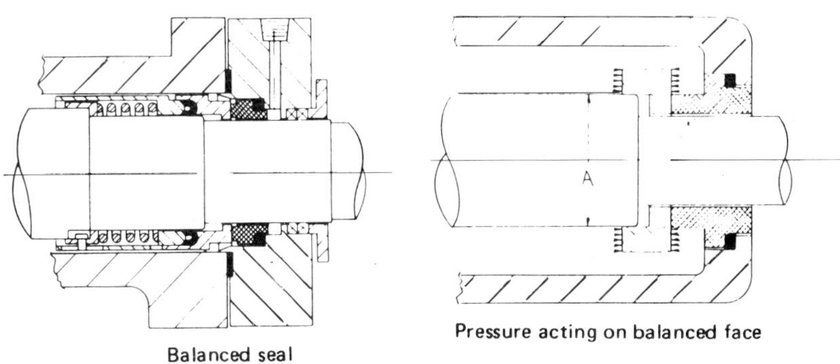

Figure 14

The immediate advantage of a balanced seal is that friction is reduced at the seal faces (because of the lower face contact pressure), with less heat generated and reduction in rubbing wear. Also such a seal can accommodate higher pressures than its unbalanced counterpart. The disadvantage is that construction is more complicated and costly, and commonly calls for a *stepped shaft* (but not in all designs). Thus unbalanced seals are more widely used for pressures up to 10 bar (150 lb/in^2). API 610 specification calls for balanced seals to be used for pressures above 75 lb/in^2, and specifically balanced seals only for all process duties, while balanced seals extend this application of mechanical seals up to about 85 bar (1250 lb/in^2).

Variations of seal balance are illustrated in Figures 15, 16 and 17, showing various face arrangements. Here 'S', the diameter of the seal sleeve shoulder, represents the effective sealing diameter which may also be referred to as the *hydraulic* or *sliding diameter* because all thrust loads are computed about this surface. 'A' represents the hydraulic piston area of the *sliding seal ring* and 'B' represents the contact face area of the sealing faces. The spring force is ignored in all cases.

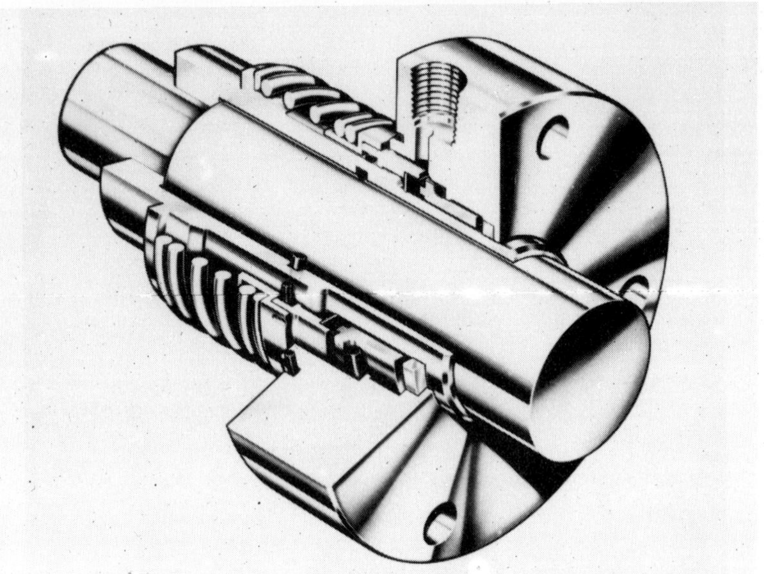

Advanced balanced seal for API610 applications.

In Figure 15 all the contact face area B is disposed outside the effective sealing diameter S and the hydraulic piston area A is equal to the contact face area B. This diagram, therefore, represents a condition of 100% out-of-balance which also indicates that the average contact face pressure will be exactly 100% of the hydraulic pressure sealed.

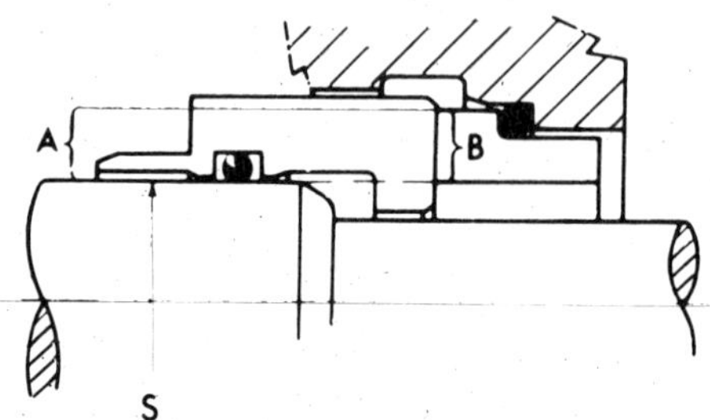

Figure 15

In Figure 16 all the contact face area B is disposed outside the effective sealing diameter S and the hydraulic piston area A is greater than the contact face area B and the loading will actually be higher by an equivalent percentage than the hydraulic pressure being sealed. This is the condition existing in most unbalanced seals.

Figure 17 shows the relationship of most balanced seals. Here part of the contact face area B, designated as B_1, is disposed outside the effective sealing diameter S. Area B_2 is, therefore, equal to the hydraulic piston area A. Because the rest of the seal face area

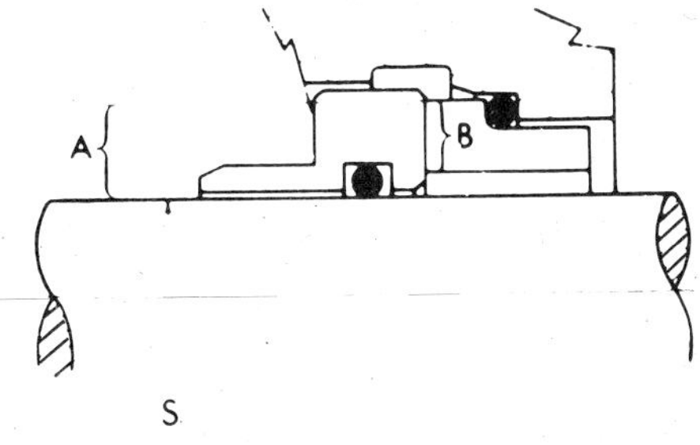

Figure 16

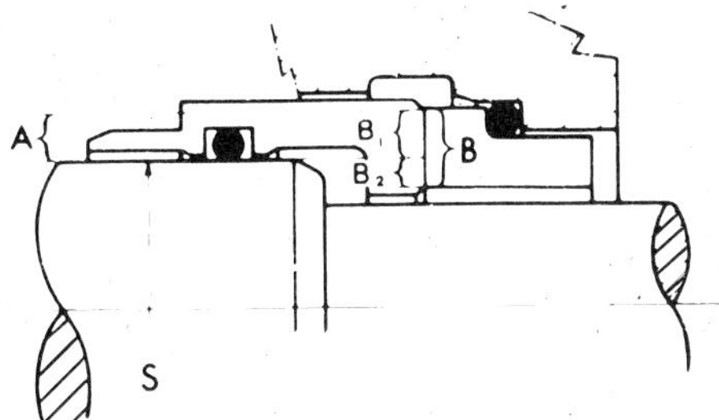

Figure 17

B_2 is located inside the effective sealing diameter S, the total seal face area B will equal the sum of $B_1 + B_2$ and the unit face loading will be less than the pressure sealed in accordance with the ratio of the two areas, *ie*

This value, expressed as a percentage, indicates by how much the seal is out of balance.

Face temperature

To maintain an adequate lubricating film between the seal faces it is essential to limit the temperature at the interface to below that at which the film tends to vaporize or boil away.

The amount of heat generated will depend on the face contact pressure related to stuffing box pressure. Figure 18 shows the amount of heat generated by a typical seal at different pressures.

The heat generated at the seal faces has to be conducted through the seal rings to the liquid in the stuffing box. Thus the product temperature, or, strictly speaking, the difference between the temperature at the seal faces and the temperature of the product in the box, has an influence on seal stability. Similarly, the thermal properties of the seal ring material play an important part.

As a consequence of this heat dissipation the temperature within the stuffing box will increase and some means must be provided to remove it, *eg* by jacket cooling of the stuffing box or more generally by the circulation of the relatively cool product *via* a tapping from the discharge nozzle to the seal faces.

It is possible to determine stability limits for a seal by running it at various speeds and pressures and increasing the temperature until the seal becomes unstable. Figure

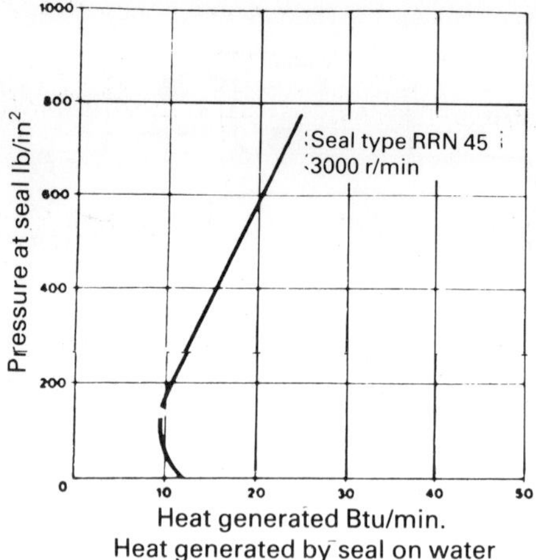

Figure 18

19 shows limits produced in this way. The curve represents the maximum allowable temperature in the stuffing box in order for the liquid film to remain stable.

The most common cause of premature seal failure arises from the loss of the liquid component of the fluid film between the seal faces. Just before this occurs, the seal emits a puff of vapour every few seconds and there is local boiling-off of the liquid film, causing the seal faces to open and tilt momentarily. This allows more liquid to enter between the seal faces giving a temporary cooling effect. When the frictional heat generated at the seal faces has built up sufficiently to vaporize more liquid, another puff of vapour is emitted and the cycle is repeated. At this stage the only bad effect suffered by the seal is a chipping of the edges of the carbon seal face caused by the tilting of the rotating seal ring.

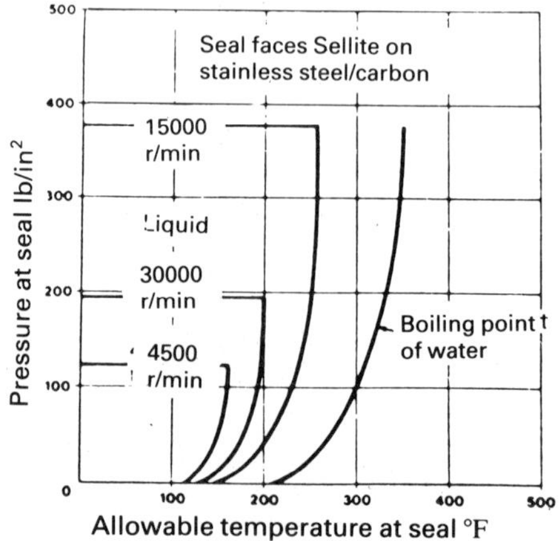

Pressure temperature limitations at various speeds RRN 45 on water

Figure 19

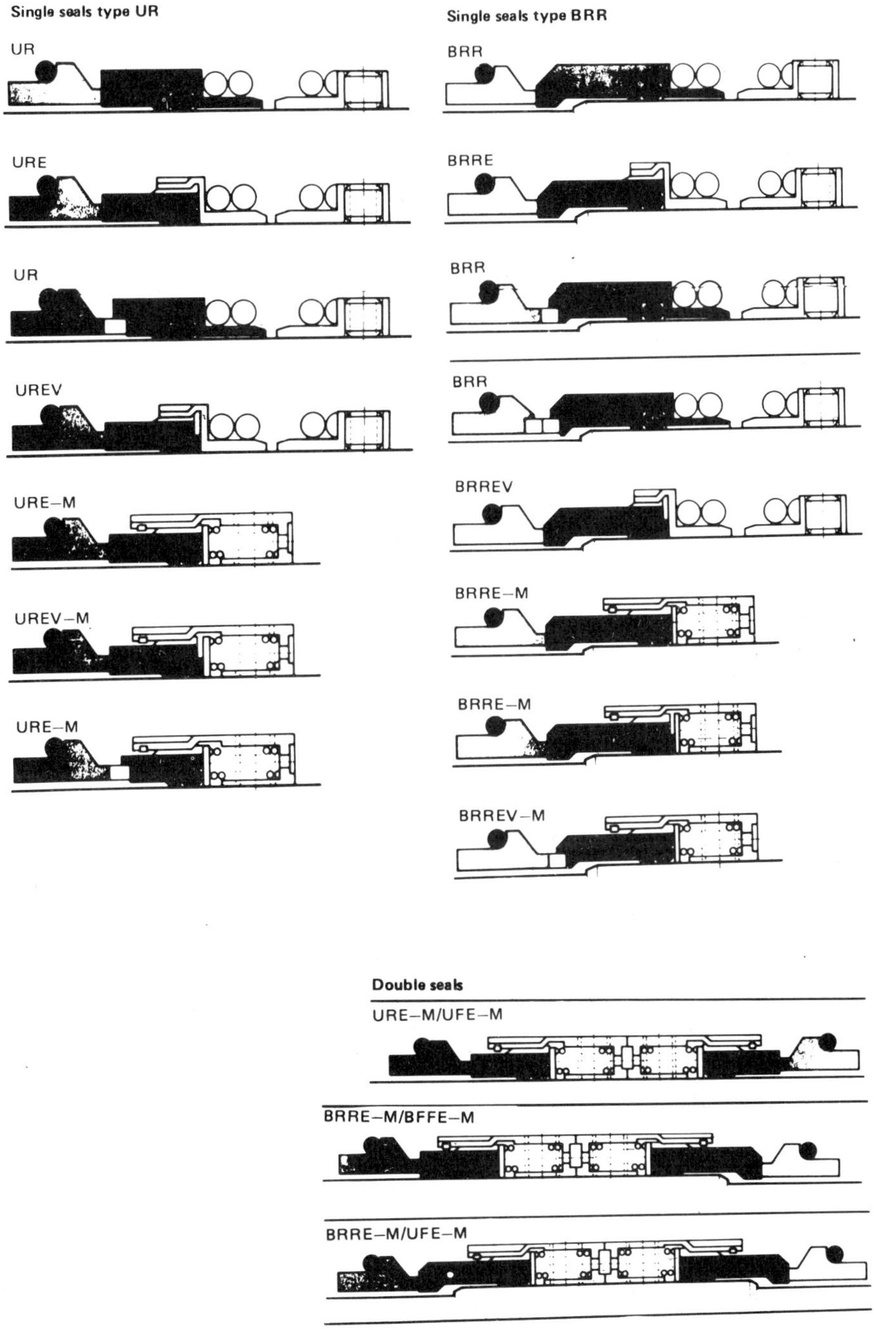

Single seals type UR
UR
URE
UR
UREV
URE–M
UREV–M
URE–M
Single seals type BRR
BRR
BRRE
BRR
BRR
BRREV
BRRE–M
BRRE–M
BRREV–M
Double seals
URE–M/UFE–M
BRRE–M/BFFE–M
BRRE–M/UFE–M

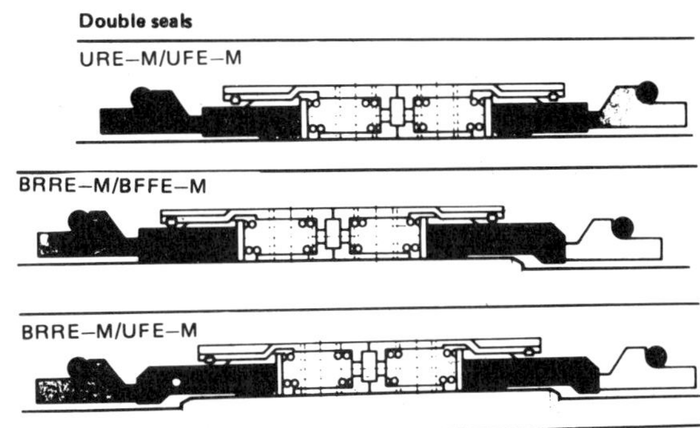

Schematic outline drawing of a range of proprietary mechanical seals

EXAMPLES OF PROPRIETARY MECHANICAL SEAL RANGE
MECHANICAL SEALS – STANDARD SERIES

TYPE B1100
SERVICE CONDITIONS
Diameter: 10 to 80 mm
Pressure: 25 bars
Temperature: 20 to + 180 C
Speed: 20 m/sec

MAIN APPLICATIONS
Water pumps. Refinery and chemical industry pumps. **Aggressive, slightly abrasive products.**

VARIANTS: Types B1110 (with pin), B1120, B1122

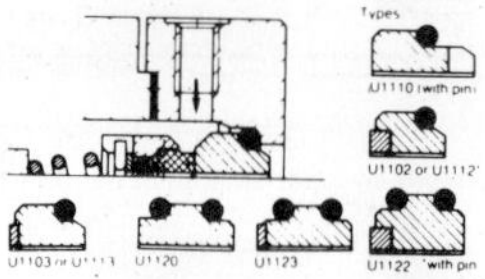

TYPE B1102
SERVICE CONDITIONS
Diameter: 10 to 80 mm
Pressure: 25 bars
Temperature: 20 to + 180 C
~~**Speed**: 20 m/sec~~

MAIN APPLICATIONS
Hot water and chemical industry pumps
Clear, non-abrasive liquids.

VARIANTS: Types B1112 (with pin), B1152

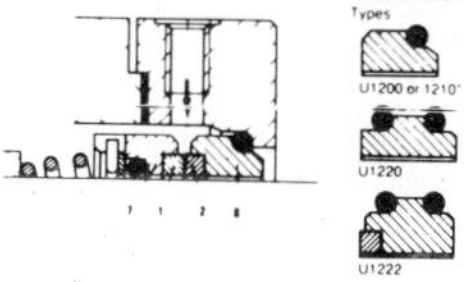

TYPE B1202
SERVICE CONDITIONS
Diameter: 10 to 80 mm
Pressure: 25 bars
Temperature: 20 to + 120 C
Speed: 10 m/sec

MAIN APPLICATIONS
Chemical industry and refinery pumps
Paper-mill and sugar-refinery pumps
Abrasive, slightly aggressive liquids.

VARIANTS: Types B1212 (with pin), B1220, B1222

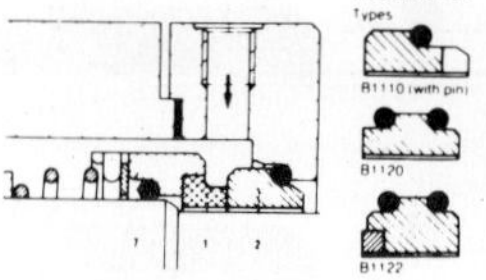

TYPE U1100
SERVICE CONDITIONS
Diameter: 10 to 80 mm
Pressure: 10 bars
Temperature: 20 to + 180 C
Speed: 20 m/sec

MAIN APPLICATIONS
General purpose pumps for refineries and chemical industry. **Corrosive and slightly abrasive products.**

For dry starting and occasional dry running fit a stellited Cr Ni Mo steel stationary ring (K).

VARIANTS: Types U1110 (with pin), U1102 or U1112*, U1103 or U1113, U1120, U1123, U1122 *with pin

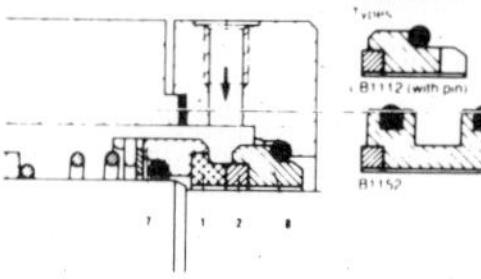

TYPE U1202
SERVICE CONDITIONS
Diameter: 10 to 80 mm
Pressure: 10 bars
Temperature: 20 to + 120 C
~~**Speed**: 10 m/sec~~

MAIN APPLICATIONS
Pumps in refineries, in the chemical and paper-making industries. **Highly abrasive and corrosive products.**

VARIANTS: Types U1200 or 1210*, U1220, U1222

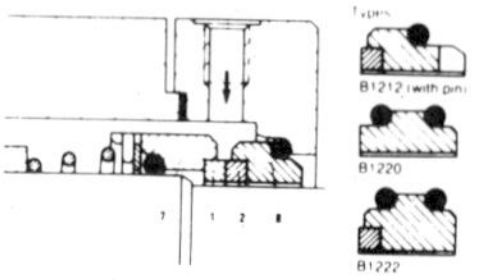

TYPE U1303
SERVICE CONDITIONS
Diameter: 10 to 80 mm
Pressure: 10 bars
Temperature: 20 to + 120 C
Speed: 10 m/sec

MAIN APPLICATIONS
Submersible pumps, waste water pumps. **Highly abrasive non-aggressive products.**

VARIANTS: Types U1300 or 1310*, U1320, U1323

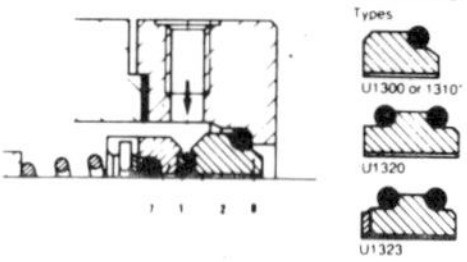

TYPE B2100
SERVICE CONDITIONS
Diameter: 80 to 200 mm
Pressure: 25 bars
Temperature: 20 to + 180 C
Speed: 20 m/sec

MAIN APPLICATIONS
Water pumps. Refinery and chemical industry pumps. **Aggressive and slightly abrasive products.**

VARIANTS: Types B2110 (with pin), B2120

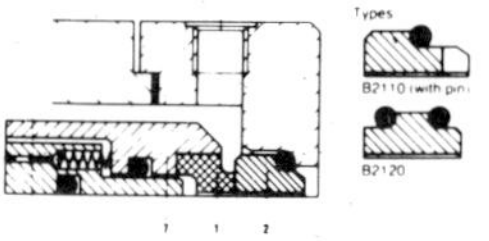

TYPE U2100
SERVICE CONDITIONS
Diameter: 20 to 200 mm
Pressure: 10 bars
Temperature: 20 to + 180 C
Speed: 20 m/sec

MAIN APPLICATIONS
General purpose pumps, refinery and chemical industry pumps. **Corrosive and slightly abrasive products.**

VARIANTS: Types U2110 (with pin), U2102 or U2112*, U2123, U2120, U2103 or U2113 *with pin

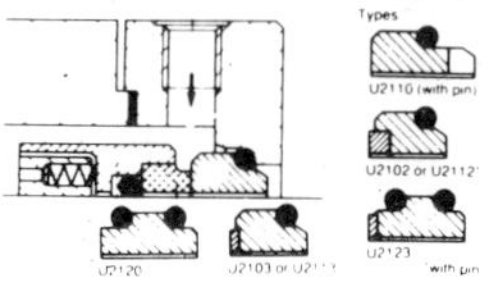

TYPE B6610
SERVICE CONDITIONS
Diameter: 20–80 mm
Pressure: 25 bar
Temperature: −20 to +180 °C
Speed: 20 m/sec

MAIN APPLICATIONS
Chemical industry and refinery pumps
Aggressive and slightly abrasive products

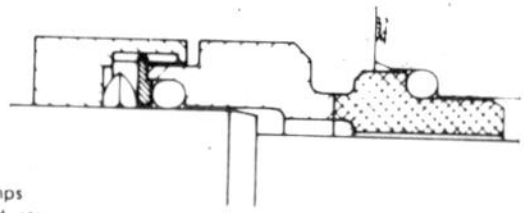

TYPE U6610
SERVICE CONDITIONS
Diameter: 20–80 mm
Pressure: 10 bar
Temperature: −20 to +180 °C
Speed: 20 m/sec

MAIN APPLICATIONS
Chemical industry
Aggressive, slightly abrasive products

TYPE B7110
SERVICE CONDITIONS
Diameter: 20–100 mm
Pressure: 25 bar
Temperature: −20 to +180 °C
Speed: 20 m/sec

MAIN APPLICATIONS
General purpose and chemical industry pumps.
Slightly corrosive and slightly abrasive products.

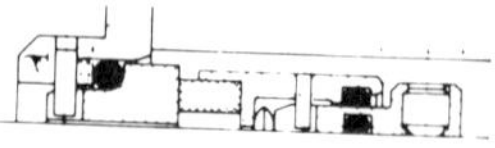

TYPE B7210
SERVICE CONDITIONS
Diameter: 20–100 mm
Pressure: 25 bar
Temperature: −20 to +180 °C
Speed: 10 m/sec

MAIN APPLICATIONS
Pumps in refineries, chemical and paper making industries.
Highly abrasive and corrosive products.

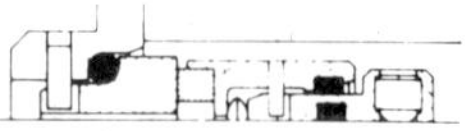

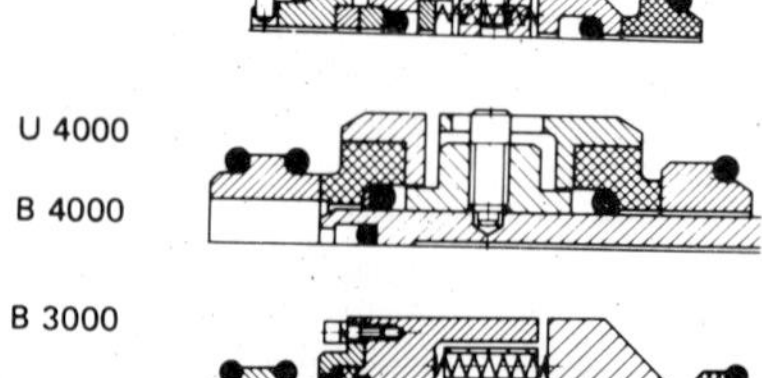

U 3000  — Sizes 20 – 200 mm; P max 10 bar; V max 15 m/s; Temperature range −60 to +220 °C; Unbalanced double seal, independent of direction of rotation

U 4000 — Unbalanced seal for agitators PV max 300

B 4000 — Balanced seal for agitators PV max 500

B 3000 — Sizes 40 – 300 mm; P max 40 bar; V max 10 m/s; Temperature range −60 to +220 °C; Balanced seal, independent of direction of rotation

MECHANICAL SEALS – SPECIAL SERIES

Type	Geometry	PV Maximum	Remarks
U 5000		50	Sizes 20 – 100 mm; P max 10 bar; V max 20 m/s; Temperature range –20 to +370 °C; Unbalanced, independent of direction of rotation.
B 9000	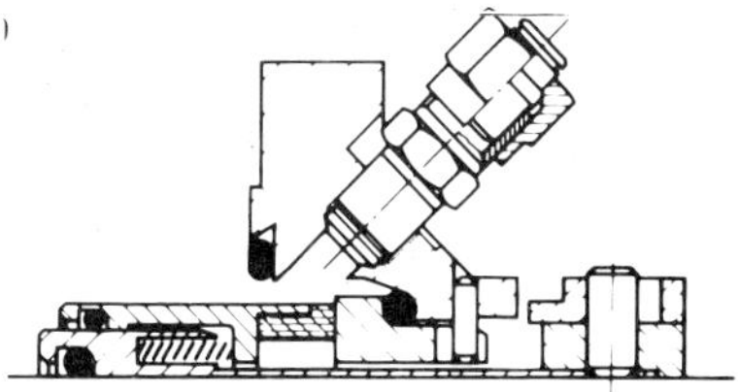	500	Sizes 38–200 mm; P max 35 bar; V max 10 m/s; Temperature range –20 to +220 °C; Balanced seal, cartridge type, independent of direction of rotation.
U 10000	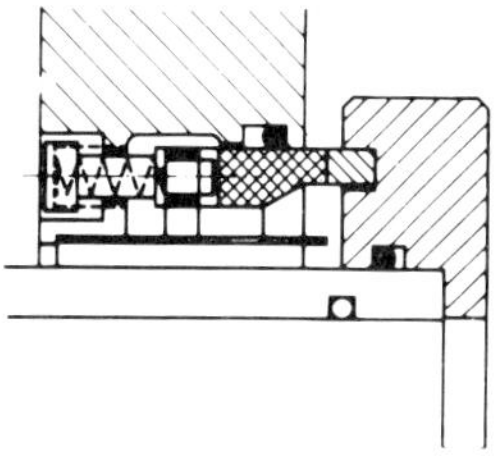	50	Unbalanced seal; P max 15 bar; V max 15 m/s; Spring/face assembly stationary. Designed for hygienic operation in the food industry, or as double seal for mixers.
B 11000	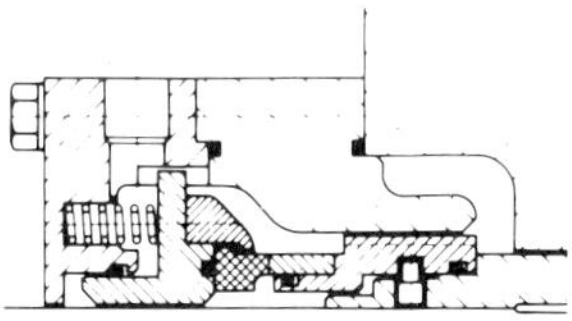	4000	Balanced seal; P max 50 bar; V max 60 m/s; T max 250 °C; (with coding); High performance seal designed for PWR nuclear plant main charging pumps. Self aligning, floating sealing faces. Pumping ring incorporated. High security, low maintenance design.
U 16000		20	V max 20 m/sec; T max 180 °C; External seal for chemical applications with faces ceramic/loaded PTFE.
U 20000	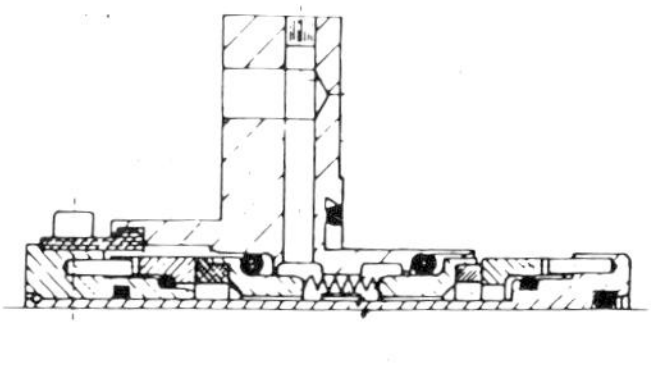	500	V max 35 m/sec. T 200 °C; Double cartridge seal. Factory engineered and preset.

If the temperature of the product surrounding the seal is increased slightly, the rate of emission of puffs of vapour also increases until there is heavy leakage of liquid. This is caused by the eruption of carbon from a small crater in the face of the seal when product which has permeated below the surface of the material is vaporized suddenly. The carbon debris from the crater is ground between the seal faces leaving a 'comet trail' pattern on the carbon face.

This is the prelude to complete loss of the liquid component between the seal faces and 'dry running' with resultant heavy face wear occurring. Both carbon and metal rotary seal ring faces are then heavily grooved – somewhat like a gramophone record.

A dry-running seal may also produce shaft wear under the secondary seal. As the liquid at the face gasses off, it pushes the secondary seal along the shaft. When the gas has escaped into the atmosphere, the box pressure closes the faces. This phenomenon is known as 'popping' and can occur with great rapidity. The result is shaft wear, known as 'O-ring drag' or 'wedge-etching'.

The cure is to stabilize conditions at the seal face by cooling; hardening the shaft under the secondary seal also helps.

Cooling

The heat generated by the rubbing faces of a mechanical seal must be removed if the seal is to work with minimum leakage and wear. Equally, to be most effective, the coolant should be directed at the point where the most effective heat transfer occurs. This applies whether cooling is achieved by recirculation of the pumped product, or by an external source of coolant independent of, and not in direct contact with, the rubbing faces. Examples of good and bad practice in inlet connection location are summarized in Table 6.

TABLE 6 – INLET CONNECTION PRACTICE

Ideal	Troublesome
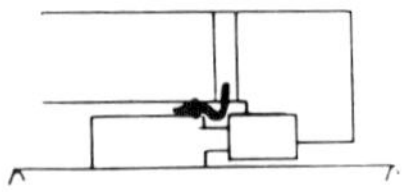 Vapour ring broken down, maximum cooling effect.	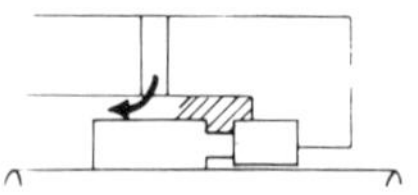 Avoid dead pockets – overheating will result.
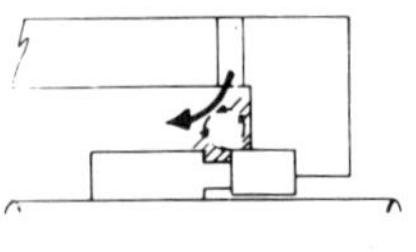	Clearance around seal too large – vortex prevents liquid flow.
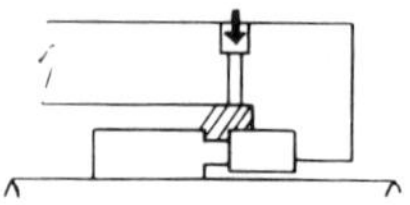	Avoid restrictor in clamp plate – change of velocity on entering seal chamber may cause vaporization.
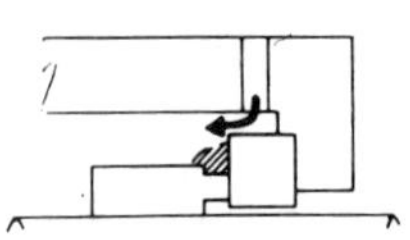	Avoid axial flow with entry remote from seal face – vaporization may occur in vicinity of rubbing face.

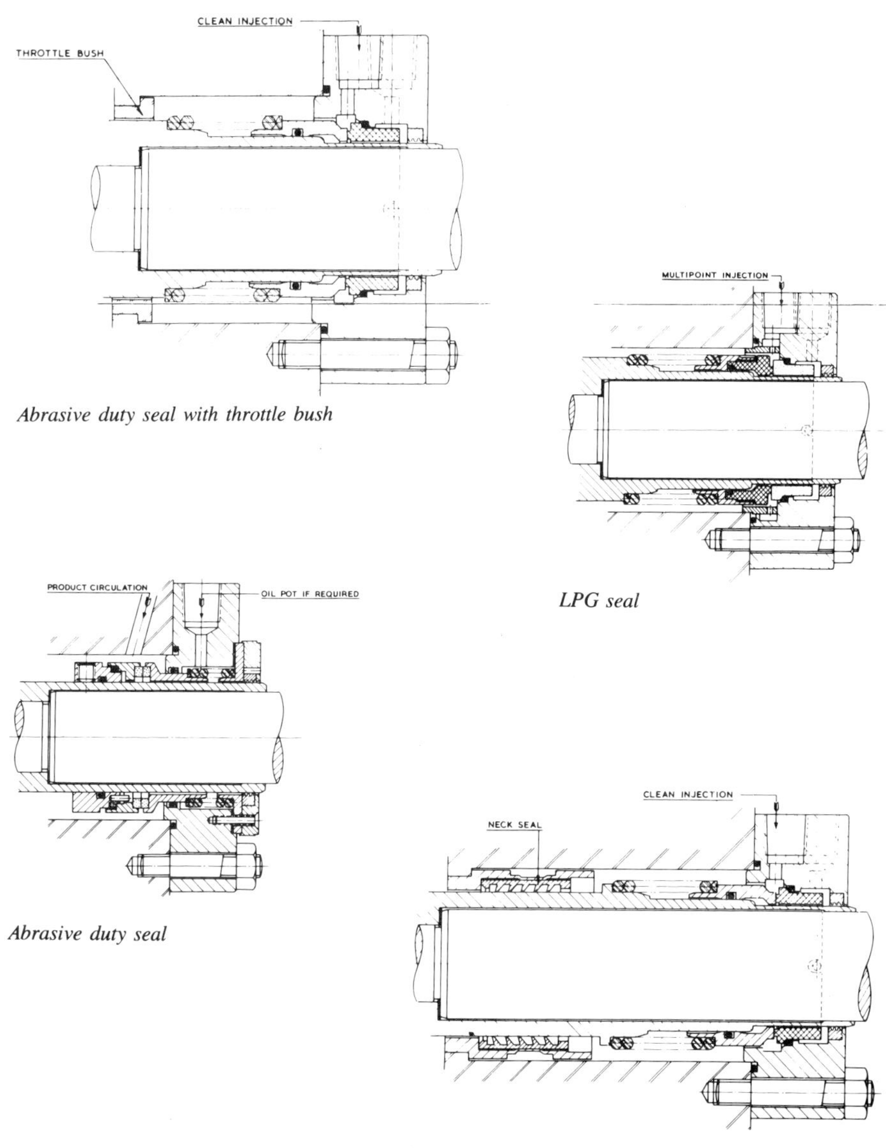

Abrasive duty seal with throttle bush

LPG seal

Abrasive duty seal

Abrasive duty seal with neck seal

Examples of special duty seals

Leakage

Because mechanical seals are designed to operate with a lubricating film between the contact faces there is always the possibility of some leakage, if only minimal. Thus, in practice, all mechanical seals tend to leak although this may be of such a low order that the fluid evaporates or vaporizes and is not apparent as liquid leakage, or is perhaps observed as occasional drips. Any more generous leakage, however, is likely to be an indication of wear on the seal faces, bearings, wear rings, *etc.* (See Figure 20).

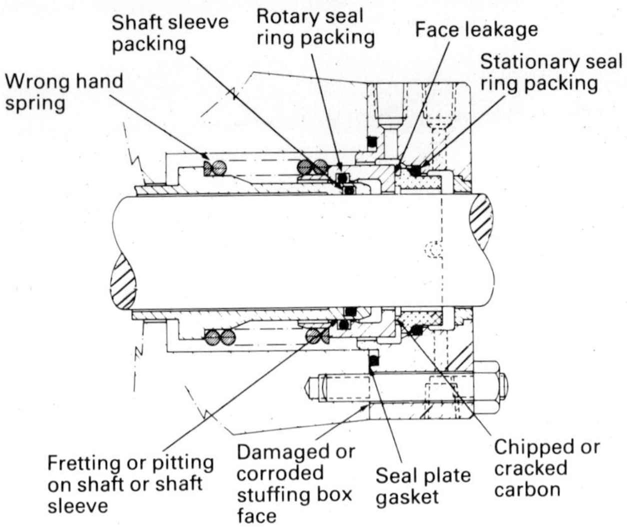

Figure 20
Possible leakage paths from a standard balance seal.

Factors which have an overriding influence upon the leakage experienced are:

(i) Accuracy of fitting and installation.
(ii) Plant mal-operation and plant upset.
(iii) Integrity of the secondary seal.
(iv) Characteristics of the liquid handled.
(v) Adequacy of cooling.
(vi) Correct selection of the seal and materials.
(vii) Vibration and stability of the equipment.

Wear

Manufacturers establish wear rates by testing different face material combinations on several products until a satisfactory level of wear rate is established from which they are able to specify an average seal design life. Figure 21 is an example of such data expressed as operating limits for a proprietary range of unbalanced and balanced seals covering product, temperature, pressure, shaft size and speed. Also indicated is the symbol (cT), a parameter used to denote the temperature difference between product temperature at the seal and vapour or boiling point. The curve on the graph indicates the maximum stability conditions for this size seal with different body and face materials. Therefore, a 45 mm size seal with stainless steel body and tungsten carbide face will not give a satisfactory life beyond this curve. Built into the manufacturer's seal selection guide is a minimum cT to indicate this curve for seals working under different conditions. The condition shown on the graph could be a boiler feed duty at 150 lb/in^2 and 170°C failing at point 'A' – an unstable region. The nearest curve to the left of point A selects a seal with a copper alloy rotary and ceramic face, this being the only single seal that will operate under these conditions. Even then cooling would have to be applied to the seal

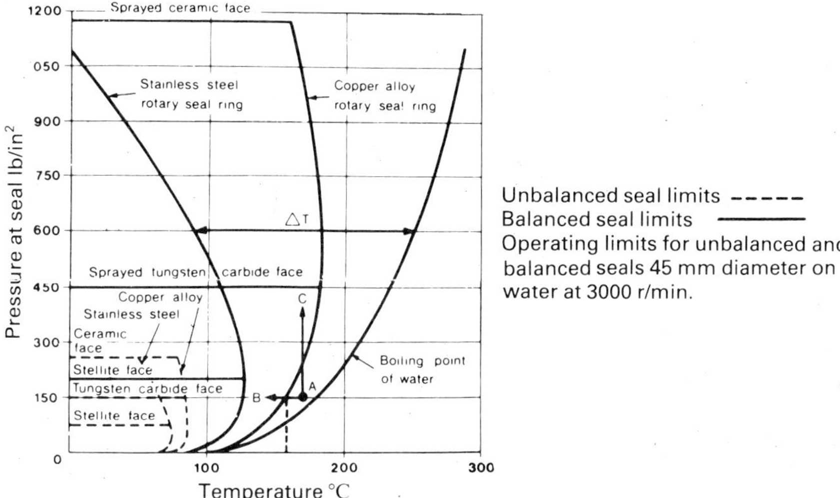

Figure 21
Low profile mechanical face seal.

area to bring the temperature down to where it intersects our curve at approximately 160°C.

Seals for abrasive services

The most common sources of abrasive content in a fluid being handled are crystallization solidification of the product occurring as a result of temperature changes or chemical reaction(s), or the normal presence of abrasive solids in suspension. Crystallization occurs mostly as a result of the product passing across the seal faces, with an accompanying pressure drop to atmospheric pressure. The liquid is lost in the form of vapour, leaving behind a deposit of crystals. These crystals can build up sufficiently to disturb the stationary face in the gland plate. This can be overcome by fitting a lip seal, close clearance bush or an auxiliary packed gland in the outer portion of the gland plate, with a quench to the space between it and the seal faces. This creates a liquid barrier between the product and atmosphere, and any crystals that form and pass across the faces are washed away by the quench medium.Liquids with abrasive solids in suspension can be more difficult to deal with, particularly as the solids may vary widely in size, density and concentration. Also they can produce wear both by impingement erosion and abrasion of the seal face(s). Problems associated with impingement erosion can be minimized by arranging a suitable flow pattern to carry abrasives away from the seal face. Thus centrifugal pumps with bell housings can have a vortex breaker fitted in the bell housing to prevent solids being drawn to the seal face and held there by the flow pattern. Restrictions which could trap solids should be eliminated and circulation pipes (where fitted) should be of generous diameter and direct re-entry of fluid into a region where solids do not accumulate. In more severe cases, the addition of a cyclone separator will remove the bulk of large solids and reduce the possibility of erosion and blockage of the seal mechanism.

It has to be remembered, however, that cyclone separators will only remove particles down to 5 ;m in size, and anything smaller will reach the seal faces. Since the gap between the faces is of the order 3 to 5 ;m, these smaller particles can still damage the faces.

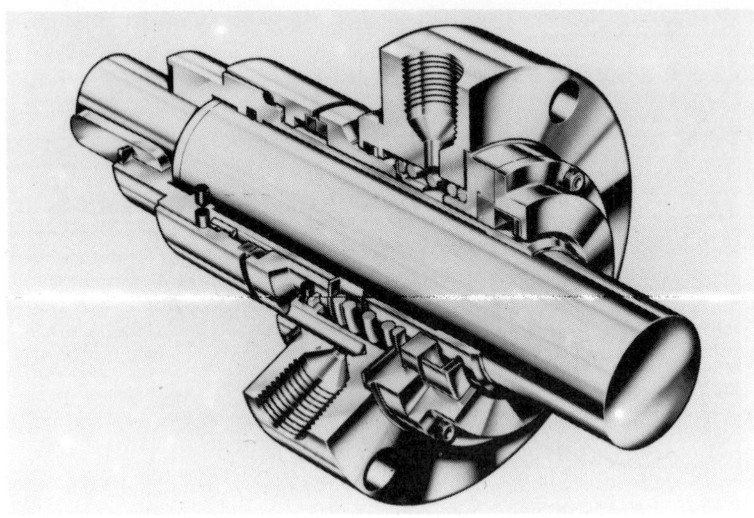

External stationary spring seal for abrasive and high viscosities.

Normal practice is to make one or both faces in hard materials such as tungsten carbide or silicon carbide. Another way of overcoming the problem is to make the product recirculation connections semi-tangential, rather than diametric, so that the abrasive particles are centrifuged away from the seal.

In the most severe cases, unless such precautions are taken, the abrasive particles can 'bombard' the seal and abraid the face and seat. This phenomenon is known as high velocity impingement. Table 7 shows various methods of dealing with abrasive fluids.

TABLE 7 — SEALS FOR ABRASIVE FLUIDS

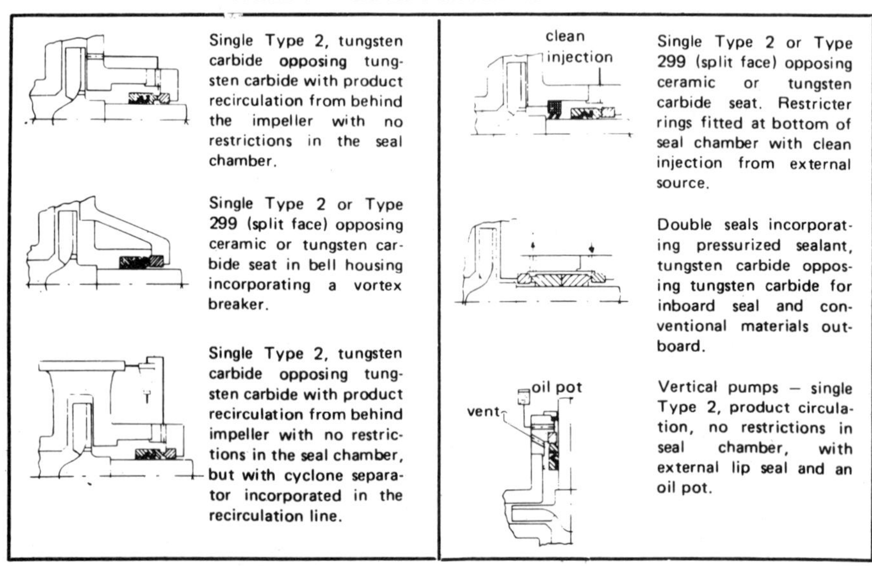

Single Type 2, tungsten carbide opposing tungsten carbide with product recirculation from behind the impeller with no restrictions in the seal chamber.	Single Type 2 or Type 299 (split face) opposing ceramic or tungsten carbide seat. Restricter rings fitted at bottom of seal chamber with clean injection from external source.
Single Type 2 or Type 299 (split face) opposing ceramic or tungsten carbide seat in bell housing incorporating a vortex breaker.	Double seals incorporating pressurized sealant, tungsten carbide opposing tungsten carbide for inboard seal and conventional materials outboard.
Single Type 2, tungsten carbide opposing tungsten carbide with product recirculation from behind impeller with no restrictions in the seal chamber, but with cyclone separator incorporated in the recirculation line.	Vertical pumps — single Type 2, product circulation, no restrictions in seal chamber, with external lip seal and an oil pot.

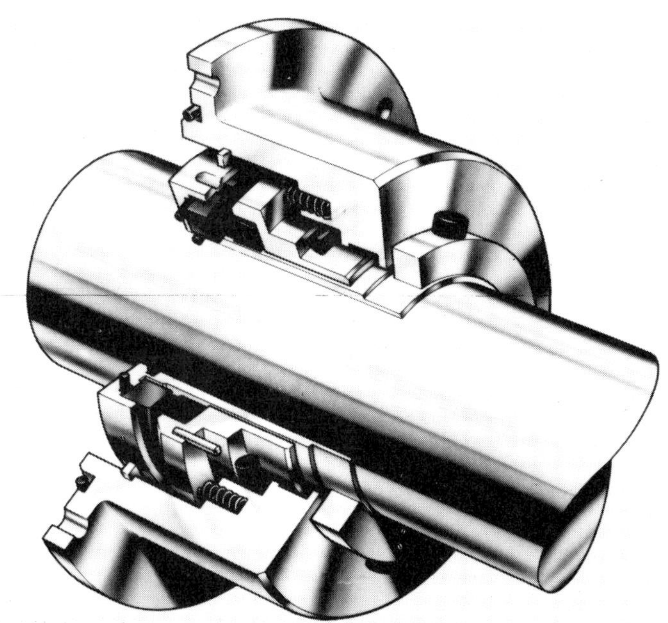

Dry-running standby seal.

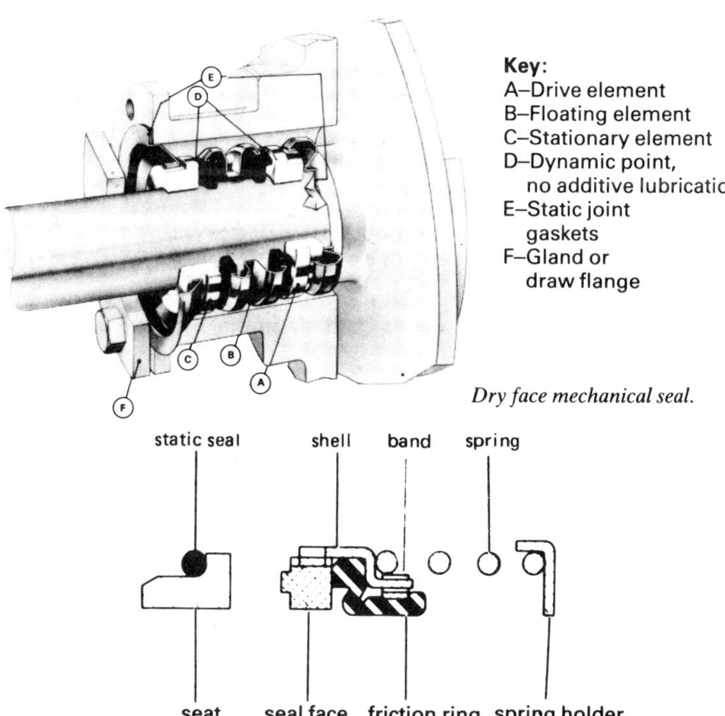

Dry face mechanical seal.

In the case of fluids which crystallize on meeting atmosphere, the best way of dealing with this is to provide a barrier fluid on the atmosphere side of the seal. This means that should the pumped product leak it will be washed away by the barrier fluid and be prevented from reaching atmosphere. Of course, some means of containing the barrier fluid will be required, and Figure 22 shows five of these.

If a metallic bush is used, the cross-sectional area between the bush and the shaft should be less than that of the bottom return connection for the barrier fluid. This will ensure the minimum loss of this fluid and at the same time, the bush can act as a steady bearing when there is radial movement of the shaft due to worn bearings.

Non-metallic bushes do not require this clearance.

Both gland packing and lip seals work better if the area of the shaft immediately under them is hard coated. This will eliminate shaft fretting.

Lastly there is the split curled bearing. This is made from a special PTFE which has been cut to a fish tail shape at one end and an arrow head at the other so that the two ends mate when the PTFE is rolled into a ring. To fit this component a small recess is machined into the bore of the clamp plate.

Auxiliary systems

The efficiency of mechanical seals depends on creating the right environment for them to work in.

If the liquid is clean, always present at the faces to lubricate and cool, has a 'condition' which is well below the vapour pressure, does not require heating prior to start-up to reduce viscosity and free the faces, has sufficient heat transfer to remove the generated heat, does not require a quench gland because of very dangerous or toxic vapours, and does not require extra pressurisation because of very low *flash-point*, then consideration of auxiliary devices does not arise.

Unfortunately, however, this ideal situation is not the norm. Because of this, systems have been developed by seal/pump manufacturers, and in particular by institues such as API (The American Petroleum Institute). The latter produces a frequently up-dated document, usually referred to as API610 and currently in its seventh edition.

As its name implies, it has been produced specifically for the refining industry. In fact, however, its recommendations are much more widely used and applied in other industries.

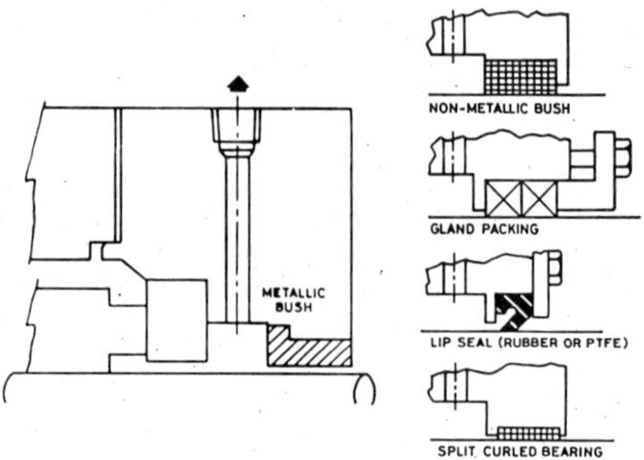

Figure 22
Sealants used to contain barrier fluid.

Its recommendations include among other things the pressures related to shaft size and speed at which balanced seals should be used. In practice, this results in balanced seals being used on all continuous refinery duties, and only the off-site duties (*eg* loading pumps) using unbalanced seals.

API610 publishes a series of figures which enable pump manufacturers to quote the figure reference denoting the auxiliary system to be used. This is usually in accordance with the pump makers experience. Examples of commonly used plans are shown in Figures 23 and 24, respectively cooling water-piping for between-bearing pumps and for overhung impeller pumps. In the same way, API610 publishes a series of figures relating to mechanical seals. These are shown in Figures 25 and 26.

When recommending seals where cooling of the pumped product is necessary, seal manufacturers tend to ignore any effect that cooling of the pump may have, and regard the latter's effect as a bonus.

There are several reasons for this. First of all it is difficult to calculate accurately the contribution made by cooling the pump. Secondly, and possibly more importantly, the source of the cooling water supply needs careful consideration. In hard water areas, or where iron salts are present, this can result in deposits forming in the cooling jackets of pumps and drastically reducing the cooling effect.

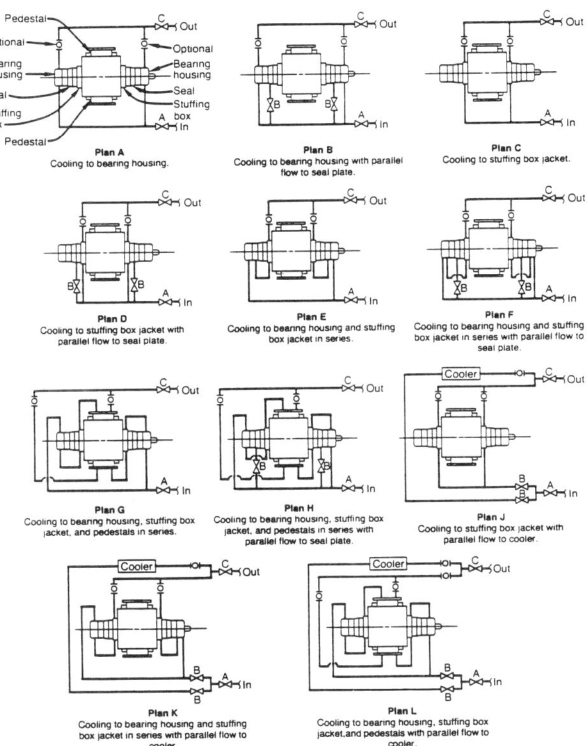

Figure 23
Cooling-water piping for between-bearing pumps.

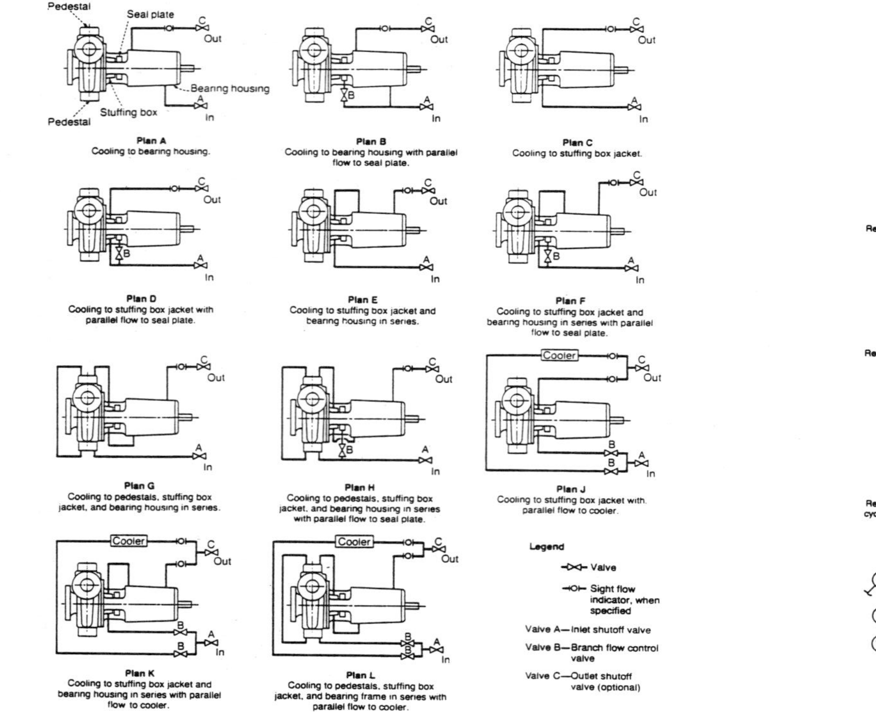

Figure 24
Cooling-water piping for overhung impeller pumps.

Clean Pumpage

Plan 1
Integral (internal) recirculation from pump discharge to seal.

Plan 2
Dead-ended seal box with no circulation of flush fluid. Water-cooled box jacket and throat bushing required unless otherwise specified.

Plugged connections for possible future circulating fluid

Plan 11
Recirculation from pump case through orifice to seal.

Plan 12
Recirculation from pump case through strainer and orifice to seal.

Plan 13
Recirculation from seal chamber through orifice and back to pump suction.

When specified

Plan 21
Recirculation from pump case through orifice and cooler to seal.

Plan 22
Recirculation from pump case through strainer, orifice, and cooler to seal.

Plan 23
Recirculation from seal with pumping ring through cooler and back to seal.

Dirty or Special Pumpage

When specified
By vendor
Recommended by purchaser

Plan 31
Recirculation from pump case through cyclone separator delivering clean fluid to seal and fluid with solids back to pump suction.

Plan 32
Injection to seal from external source of clean fluid (see Note 2).

Plan 41
Recirculation from pump case through cyclone separator delivering clean fluid through cooler to seal and fluid with solids back to pump suction.

Legend

Cooler
(PI) Pressure gage with block valve
(TI) Dial thermometer
(PS) Pressure switch with block valve
Cyclone separator
(FI) Flow indicator
Y-type strainer
Flow-regulating valve
Block valve
Check valve
Orifice

NOTES:
1. These plans represent commonly used systems. Other variations and systems are available and should be specified in detail by the purchaser or mutually agreed upon by the purchaser and the vendor.
2. For Plan 32, the purchaser shall specify the fluid characteristics, and the vendor shall specify the volume (gallons per minute) and pressure (pounds per square inch gage) required.

Figure 25
Piping for primary seals.

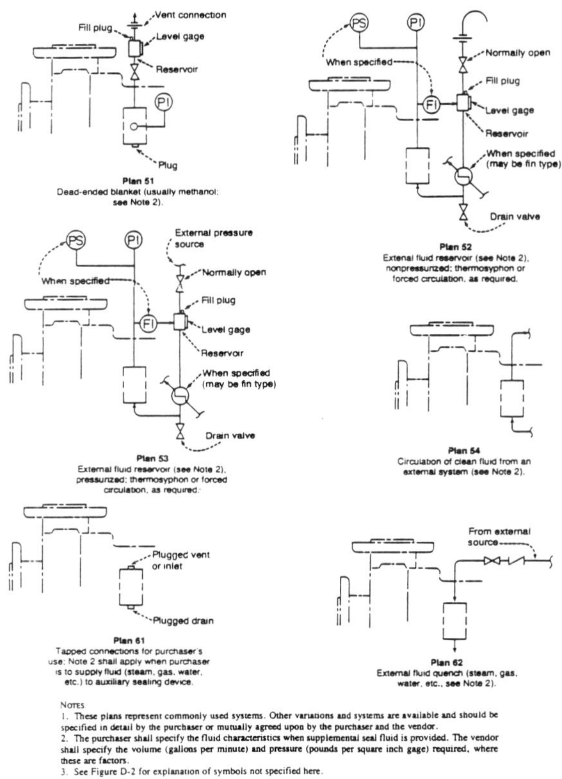

Figure 26
Piping for throttle bushing, auxiliary seal device, tandem seals, or double seals.

Vortex or Cyclone separators

A vortex separator is used when the liquid handled is dirty; its function is to clean the liquid which is being supplied to the mechanical seal for lubricating/cooling purposes, thus increasing the seal life and reducing maintenance down-time. The separator operates as follows: the dirty liquid enters the separator, being piped from the pump discharge; and enters tangentially into the side/top of a conical chamber, the tangential flow creates a vortex so that the liquid swirls around in the conical chamber, thus forming a 'coarse' centrifuge. The rotation throws the solid particles to the wall of the cone by centrifugal action and then dirt moves gradually downwards to the outlet from the cone apex (bottom) and back into the main stream of the liquid at a low pressure point. The clean liquid is extracted from the top of the cyclone separator and led to the mechanical seal faces (Figure 27).

In general the following conditions must be observed for the unit to operate successfully:

(i) The solid particles must not be too large, in general less than 1 mm (very large cyclone separators can be manufactured for larger particles, but the flow rates would not be economical). Minimum size depends on

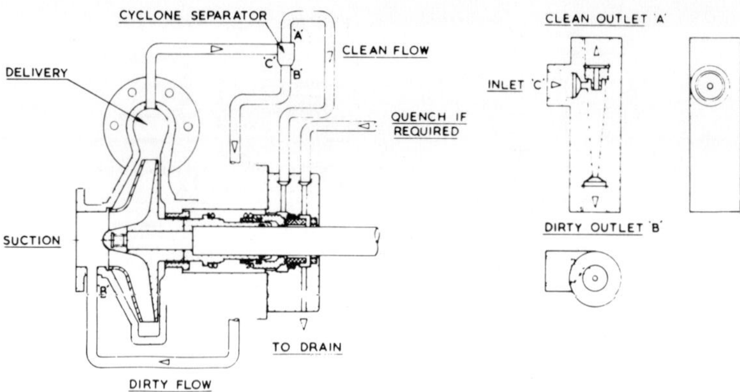

Figure 27

specific gravity of the solid and the liquid viscosity, but is in the order of 5 micron.

(ii) The solids should not exceed 10% by weight, but 5 to 7% is the normal maximum.

(iii) Because the operation is by centrifugal force the solid particles must have a higher specific gravity than the liquid.

(iv) If the liquid is viscous the centrifugal action is so retarded that it is commercially non-operative. Maximum viscosity should therefore not normally exceed 20 centistokes.

(v) There must be a pressure difference between the inlet and outlet of the cyclone to achieve the necessary velocity of flow for separation purposes.

Since the advent of metal bellows seals, and in particular the asymmetric type, many duties can now be handled without the use of cyclones. This is especially true of slurry duties.

Filters

Filters are used when the solids content has specific gravities which are close to or even lower than the liquid media (otherwise a vortex separator is a preferred choice) or in a closed circuit (say through a heat exchanger) where the mechanical seal is separated from the main body of dirty liquid by a neck bush. The greatest problem with filters is that they choke with remarkable rapidity and therefore require frequent attention.

Leakage detectors

Leakage detectors are normally only used where the contained liquid is very dangerous (explosive or toxic) The detectors are fairly simple in operation provided the contained liquid does not solidify.

A typical leakage detector is shown in Figure 28. The operation is extremely simple in that normally valve 'A' is open and valve 'B' is closed so that any leakage will build up in the chamber and trigger the flow alarm. Generally only an alarm is actuated, but the signal can also be used to shut down the unit if this is deemed necessary.

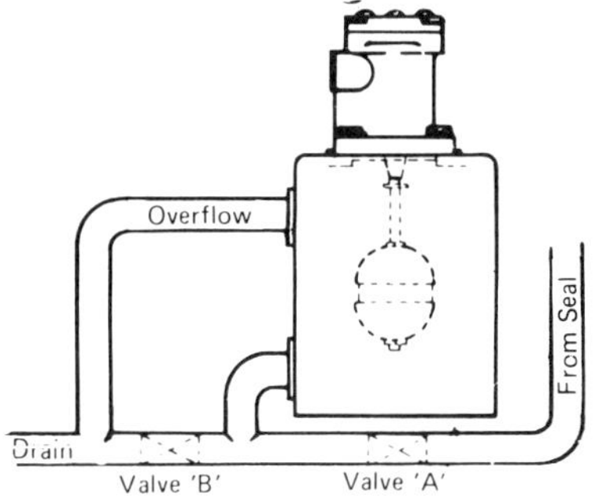

Figure 28

Sealant systems

Sealant systems are a means of ensuring that a sealing liquid which is safe, mobile and has good lubricating properties is provided to a double seal arrangement. The seal arrangement may be back-to-back, tandem or concentric. The system can control the environment, be used for safety reasons and will extend seal life. A typical arrangement is shown in Figure 29.

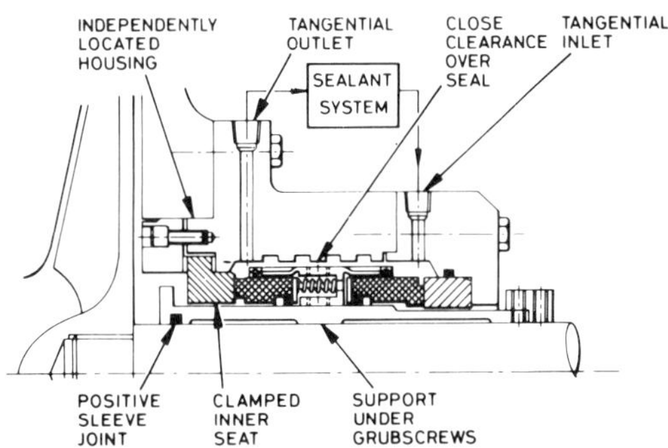

Figure 29

Vessel sealing

The term vessel is often used to denote mixers, agitators, and even autoclaves.

Vessels may conveniently be divided into three catagories. Top entry, side entry and bottom entry, and additionally there are double ended vessels which may also be classified as side entry (Figure 30).

Side entry mixers often utilise single seals since the seal is always immersed in the product. Where necessary, an auxiliary seal such as gland packing, a lip seal or close clearance bushing can be provided to contain a separate supply of cooling liquid or to contain a barrier liquid in the event that the product in the vessel is toxic, flammable or crystalline.

MODES, CAUSES AND CORRECTIONS OF MECHANICAL FACE SEAL FAILURES

Failure mode	Cause	Correction
PRIMARY SEAL FACES		
Overall corrosion	Improper materials.	Upgrade materials.
Intergranular corrosion	Improper materials.	Upgrade materials.
	Improper processing.	Correct processing, such as welding and heat treating.
Stress-corrosion cracking	Improper material.	Upgrade material.
	Improper processing.	Correct processing, such as stress relief.
Leaching corrosion	Improper material.	Upgrade material.
Seal-face distortion (concavity, convexity, waviness or non-uniform contact)	Excessive fluid pressure on seal.	Lower fluid pressure; upgrade material, upgrade design.
	Excessive swell of confined secondary seal.	Change secondary-seal materials or design.
	Improper seal assembly – excessive or non-uniform clamping or bolting stresses, or weight or misalignment of suction piping on end-suction over-hung-type pumps.	Correct assembly.
	Internal stress.	Stress relieve and refinish.
	Improper finishing.	Refinish.
	Foreign material trapped between faces.	Remove foreign material.
	Excessive PV value of seal operation.	Reduce PV value or upgrade design or materials.
	Improper equipment operation resulting in adverse seal environment (for example, insufficient seal lubrication).	Correct environment, such as cooling of fluid flow rate; upgrade materials or design.
Fracture	Improper handling.	Handle parts with care.
	Improper assembly (jamming).	Correct assembly.
	Stress-corrosion cracking.	See above, at Stress Corrosion Cracking.
	Excessive thermal stress by: Improper operation such as insufficient seal lubrication.	Improve equipment operation or environment; upgrade design or materials.
	Excessive PV value.	Reduce PV value.
	Excessive fluid pressure on seal.	Reduce fluid pressure or upgrade design or materials.
	Excessive swell of secondary seal.	Upgrade materials; change fluid or lower temperature.

cont...

Failure mode	Cause	Correction
PRIMARY SEAL FACES contd...		
Edge chipping	Excessive fluid pressure on seal.	Reduce fluid pressure; upgrade design or materials.
	Excessive shaft run-out.	Reduce run-out or upgrade seal design.
	Excessive shaft deflection.	Reduce deflection or upgrade seal design.
	Excessive shaft whip.	Reduce shaft whip or upgrade seal design.
	Seal faces out-of-square.	Square faces to shaft axis.
	Seal face vibration.	Reduce fluid temperature to 28 °C below boiling point.
Severe, uniform adhesive wear	PV value too high.	Reduce PV value; upgrade materials; improve seal balance or lubrication.
	Deposition of dissolved solids.	Reduce fluid temperature to 28 °C below boiling point; reduce dissolved solids; upgrade face materials.
	Incorrect assembly (jamming).	Correct assembly.
	Fine abrasives in sealed fluid.	Remove abrasives from fluid; upgrade materials.
	Failure of axial holding hardware (jamming).	Correct for axial slippage of set screws or collar.
	Excessive shaft end play.	Correct end play.
	Poor environment, such as insufficient seal lubrication or loss of cooling.	Correct equipment operation; upgrade face materials or design.
Heavy, non-uniform wear, galling and grooving	Abrasive contaminants.	Remove contaminants or upgrade materials.
	PV value too high with one or both faces of metal.	Reduce PV value; upgrade materials or design.
Erosion	Abrasive flow into seal face.	Remove abrasive from sealed fluid; shroud seal faces or direct flow away from seal.
Blistering of carbon-graphite faces	Improper materials.	Upgrade seal-face material.
	PV value excessive or cyclic.	Reduce PV value and cycling.
	Inadequate seal cooling.	Improve seal cooling.

cont...

MODES, CAUSES AND CORRECTIONS OF MECHANICAL FACE SEAL FAILURES(contd.)

Failure mode	Cause	Correction
SECONDARY SEALS		
Extrusion	Excessive pressure.	Reduce pressure or upgrade design rr or material.
	Excessive temperature.	Reduce temperature or upgrade design or materials.
	Excessive swell.	Upgrade materials or change seal fluid.
Chemical attack	Improper material.	Select proper material.
Cracking	Excessive temperature.	Reduce temperature (by cooling); upgrade materials; reduce PV value.
	Chemical attack.	Select proper material.
	Ozone attack.	Upgrade materials; reduce ozone concentration; reduce stress.
Cuts, tears and splits	Improper handling.	Correct handling.
	Poor dispersion.	Correct manufacturing.
	Poor knit.	Correct manufacturing.
	Inclusion.	Correct manufacturing.
	Material overstressed.	Lower stress or upgrade materials.
Corrosion of interface	Crevice corrosion.	Eliminate crevice; apply corrosion-resistant coating.
	Fretting corrosion.	Eliminate vibration caused by out -of-square faces or excessive deflection; end-play or run-out of the shaft.; apply corrosion-resistant coatings. Lubricate interface. Upgrade resistance to corrosion and wear.
HARDWARE, GENERAL		
Overall corrosion	Improper material.	Upgrade material.
Stress-corrosion cracking	Improper material.	Upgrade material.
	Improper processing.	Correct processing, such as stress relief.
Intergranular corrosion	Improper material.	Upgrade material.
	Improper processing.	Correct processing, such as heat treating and welding.
Hydrogen embrittlement	Improper processing.	Correct processing.
Fatigue	Excessive stress or vibration.	Reduce stress or vibration.
	Seal-face vibration.	Reduce fluid temperature to 28 °C below boiling point.

cont...

MODES, CAUSES AND CORRECTIONS OF MECHANICAL FACE SEAL FAILURES (contd.)

Failure mode	Cause	Correction
MECHANICAL DRIVE		
Torsional shear	Excessive torque due to: Improper lubrication.	Correct for stoppage of lubricant flow to seal faces; clean out or install (if needed) bypass flushing line.
	Failure of axial holding device.	Check set screws or other means of securing collar for slippage or jamming.
	Excessive fluid pressure.	Reduce fluid pressure or re-design (for example, seal balance).
Axial shear	Excessive shaft end play.	Correct end play.
	Improper assembly (jammed).	Correct assembly.
	Excessive pressure loading.	Reduce pressure or upgrade design.
Wear	Excessive torque.	See Torsional shear.
	Out-of-square seal faces.	Square faces to shaft axis.
	Excessive shaft run-out.	Reduce run-out.
	Excessive shaft deflection.	Reduce deflection.
	Excessive shaft end play.	Reduce end play.
	Seal face vibration.	Reduce fluid temperature to 28 °C below boiling point.
Seal hang-up	Deposition of dissolved solids, or products of corrosion, seal-fluid oxidation or decomposition.	Change to non-pusher seal; upgrade corrosion resistance of material; lower temperature of fluid; quench seal.

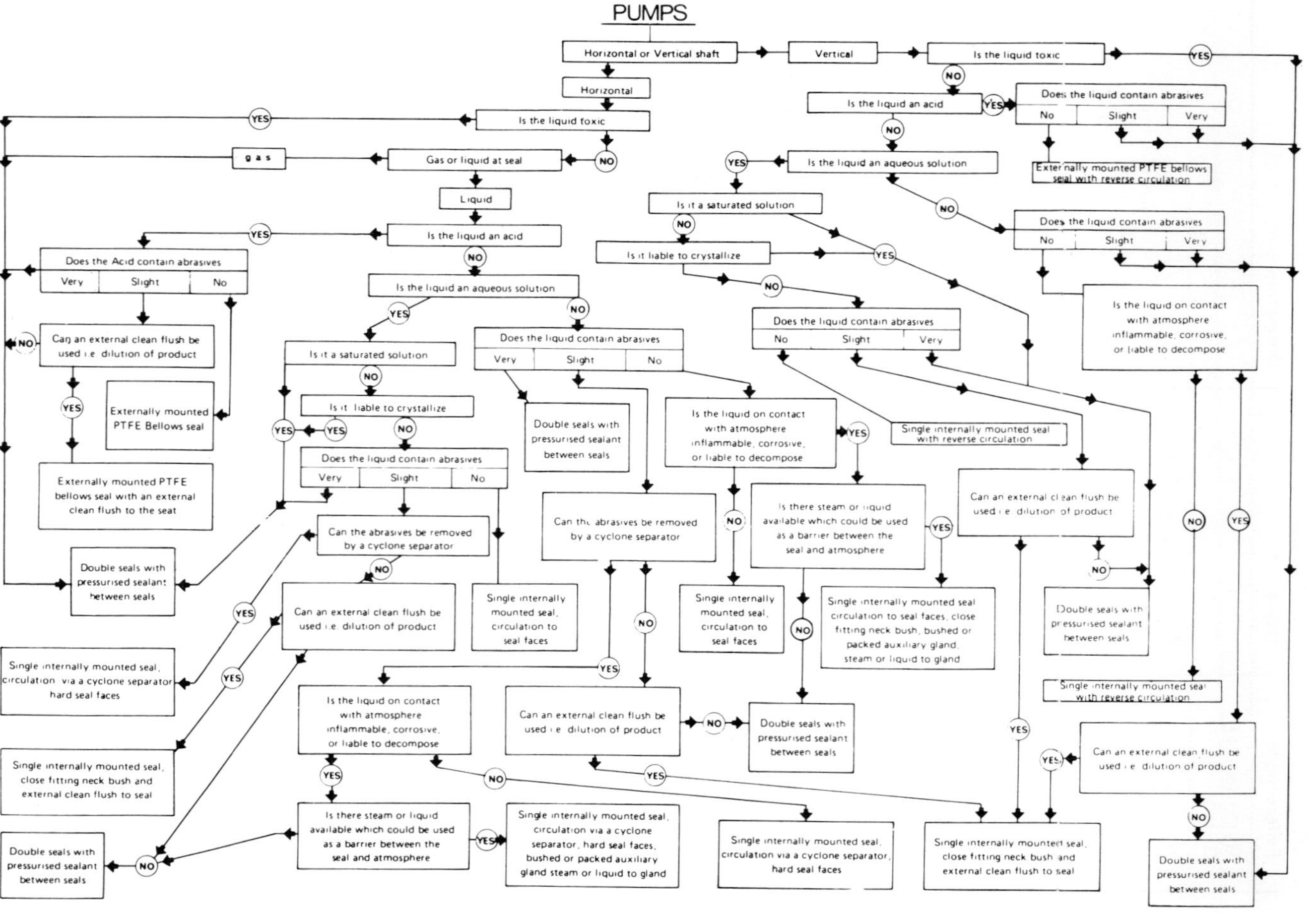
MECHANICAL SEALS SELECTION GUIDE
PUMPS
Horizontal or Vertical shaft
Vertical
Is the liquid toxic
Horizontal
Is the liquid toxic
Is the liquid an acid
Does the liquid contain abrasives
No
Slight
Very
Externally mounted PTFE bellows seal with reverse circulation
Gas or liquid at seal
gas
Is the liquid an aqueous solution
Is it a saturated solution
Liquid
Does the liquid contain abrasives
No
Slight
Very
Is the liquid an acid
Is it liable to crystallize
Does the Acid contain abrasives
Very
Slight
No
Is the liquid an aqueous solution
Does the liquid contain abrasives
No
Slight
Very
Is the liquid on contact with atmosphere inflammable, corrosive, or liable to decompose
Can an external clean flush be used i.e. dilution of product
Is it a saturated solution
Does the liquid contain abrasives
Very
Slight
No
Is it liable to crystallize
Externally mounted PTFE Bellows seal
Double seals with pressurised sealant between seals
Is the liquid on contact with atmosphere inflammable, corrosive, or liable to decompose
Single internally mounted seal with reverse circulation
Does the liquid contain abrasives
Very
Slight
No
Externally mounted PTFE bellows seal with an external clean flush to the seat
Can the abrasives be removed by a cyclone separator
Is there steam or liquid available which could be used as a barrier between the seal and atmosphere
Can an external clean flush be used i.e. dilution of product
Double seals with pressurised sealant between seals
Can an external clean flush be used i.e. dilution of product
Single internally mounted seal, circulation to seal faces
Single internally mounted seal, circulation to seal faces
Single internally mounted seal circulation to seal faces, close fitting neck bush, bushed or packed auxiliary gland, steam or liquid to gland
Double seals with pressurised sealant between seals
Single internally mounted seal, circulation via a cyclone separator hard seal faces
Is the liquid on contact with atmosphere inflammable, corrosive, or liable to decompose
Can an external clean flush be used i.e. dilution of product
Double seals with pressurised sealant between seals
Single internally mounted seal with reverse circulation
Can an external clean flush be used i.e. dilution of product
Single internally mounted seal, close fitting neck bush and external clean flush to seal
Is there steam or liquid available which could be used as a barrier between the seal and atmosphere
Single internally mounted seal, circulation via a cyclone separator, hard seal faces, bushed or packed auxiliary gland steam or liquid to gland
Single internally mounted seal, circulation via a cyclone separator, hard seal faces
Single internally mounted seal, close fitting neck bush and external clean flush to seal
Double seals with pressurised sealant between seals
Double seals with pressurised sealant between seals
Can the abrasives be removed by a cyclone separator

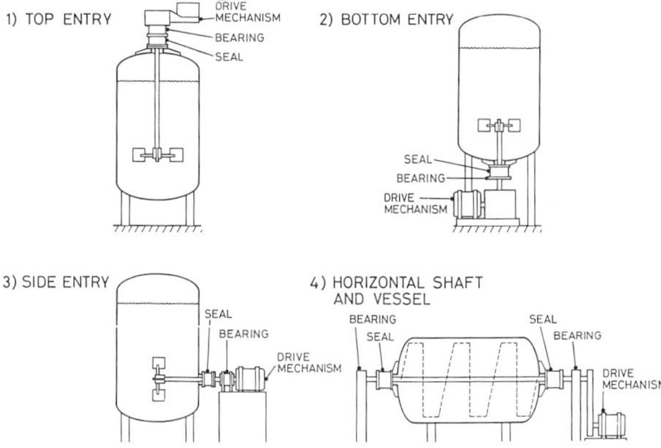

Figure 30
Examples of vessels using seals.

Top entry vessels pose a different problem in that there is normally a gas or air pocket above the product to allow for expansion of the product and the increase in pressure that would otherwise ensue.

This means that the seal would be sited above the product, and so be dry running.

To overcome this condition, double seals are used, and are provided with a separate source of liquid to lubricate and cool them. Back-to-back seals are commonly used, with a sealant between them at a somewhat higher pressure than that in the vessel. It is also possible to use face-to-face seals where pressures allow, with a low pressure injection of coolant, (Figure 31).

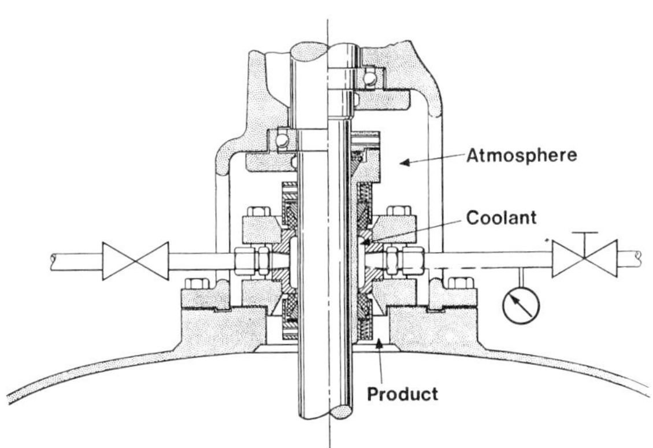

Figure 31
Face-to-face seal with coolant.

In the case of larger vessels with long shafts, problems can be met with due to shaft deflections (Figure 32). Conventional back-to-back or face-to-face seals are secured to

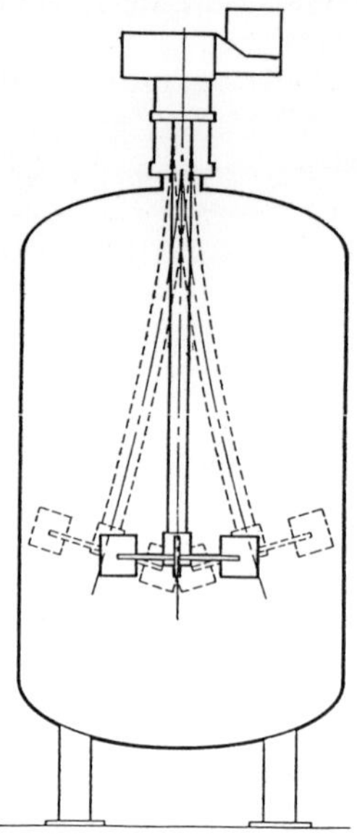

Figure 32
Diagram showing shaft deflections in large vessel.

the shaft and, therefore, any deflection of the shaft is reflected in movement of the seal relative to the stationary face. When this happens there is an ingress of sealant into the product. The result is either contamination or dilution of the product, and since the product is normally of a very high value, this cannot be tolerated.

A seal which overcomes this difficulty is shown in Figure 33.

In this type there are two seals mounted concentrically and running against a common rotating seat, usually of metal, chrome oxide deposited metal or ceramic. The seals are

Vessel sealing

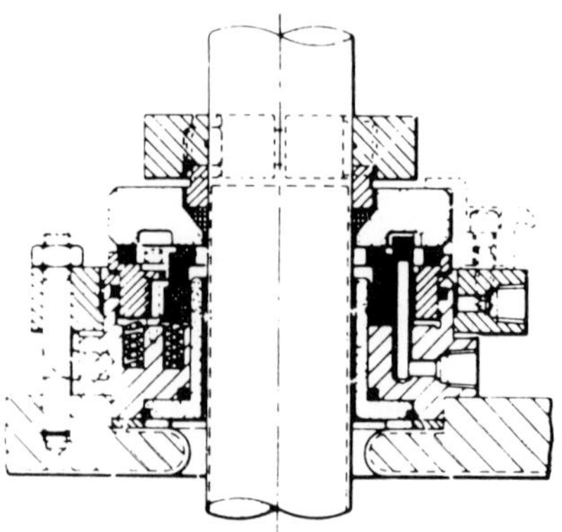

Figure 33

contained in a housing which is bolted to the top of the vessel, and there is a relatively large gap between them and the shaft. The only rotating parts are the seat, which is sealed to the shaft by a PTFE wedge, and the split clamp ring which imparts motion to the seat by means of drive pins.

In some cases, vessels are not only glass lined, but the shafts are glass coated as well. This is to provide resistance to the corrosive effects of certain products. In this case, special seals using ceramic and PTFE components are specified.

Bottom entry vessels.

The nature of the product will, as in the case of both top and side entry mixers, determine the type of seal and the materials from which it is made. There is, however, a further consideration to be taken into account when selecting a seal for bottom entry vessels. This is the tendency for the product to fall to the bottom of the vessel, especially when there are particles in suspension. These can have a serious effect on the seal by way of clogging or abrasion. To alleviate the problem, special designs have been evolved where the seal is totally enclosed in a rubber bellows-like shroud, or deflectors are fitted to prevent settling of the product in and around the seal, (Figures 34 and 35).

As with all double seals, a separate sealant is required, and where this is not readily available a sealant system may be used.

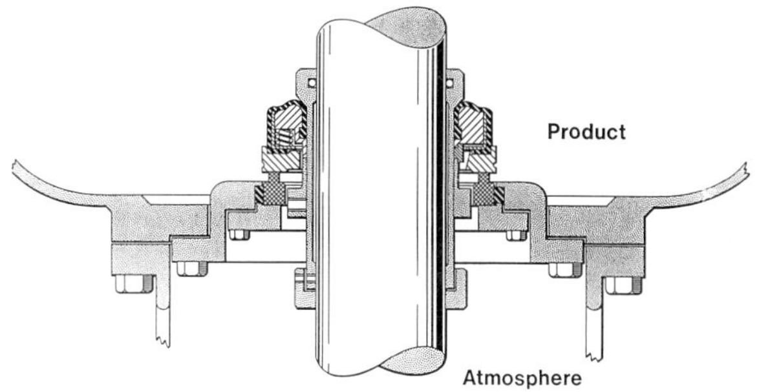

Figure 34
Seal with rubber bellows cover to prevent settling.

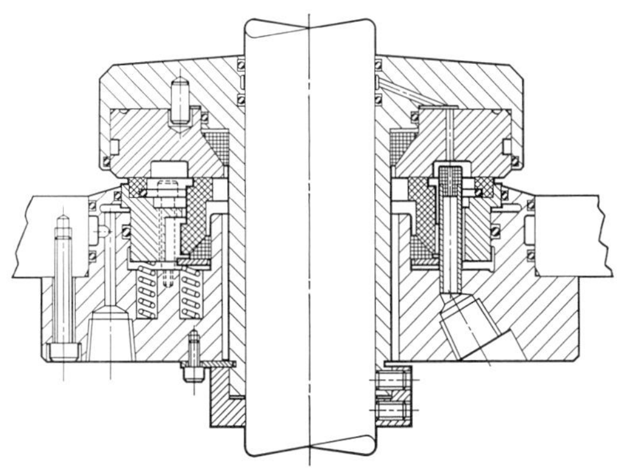

Figure 35
Seal with deflectors to prevent settling.

Food industry

Most of the pumps used in the food industry are either the progressive cavity type screw pump, such as those produced by *Mono Pumps*, or are rotary lobe double shaft gear driven pumps produced by manufacturers such as *Stainless Steel Pumps, Ibex* and *Howard Pneumatic.*

In the case of Mono Pumps, most of the duties are satisfied by use of a rubber bellows. The large spring helps prevent clogging, and sterilisation is easily carried out when the prevention or elimination of bacteria is necessary.

This question of prevention of bacteria requires serious consideration, and is the subject of numerous standards. Probably the best known of these, and one to which most food industry equipment manufacturers adhere, is the American Food and Drug Administration standard. Conformity to this standard has resulted in seals made predominantly of stainless steel, and designed to eliminate awkward crevices in which bacteria may form. Typical of these seals is the one shown in Figure 36.

This seal can be supplied both in single and double configuration, and where necessary have hardened seats and faces of silicon or tungsten carbide.

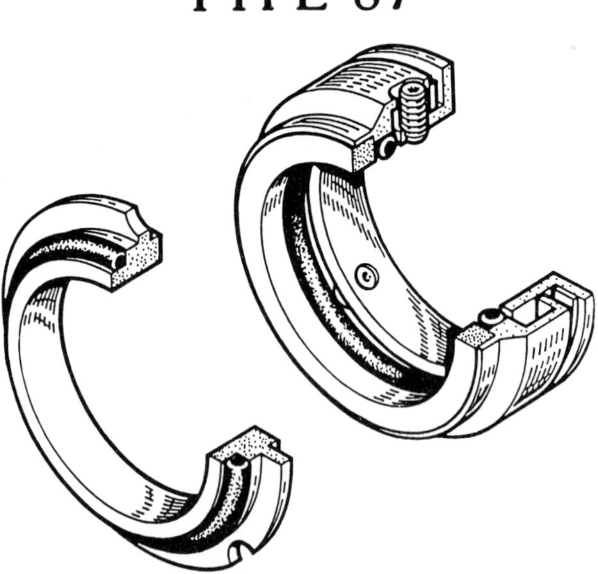

Figure 36
A typical stainless steel seal for the food industry.

Earth moving equipment

On this type of equipment, the track rollers and bearings are subjected to sand, mud, rain and gravel, *etc*, and obviously a seal that can withstand this is essential. The type 27 shown in Figure 37 has stationary and rotary faces made of solid stellite. Face loading is achieved by compression of the rubber sealing members, and the seal is simply pressed into the bore as a single unit.

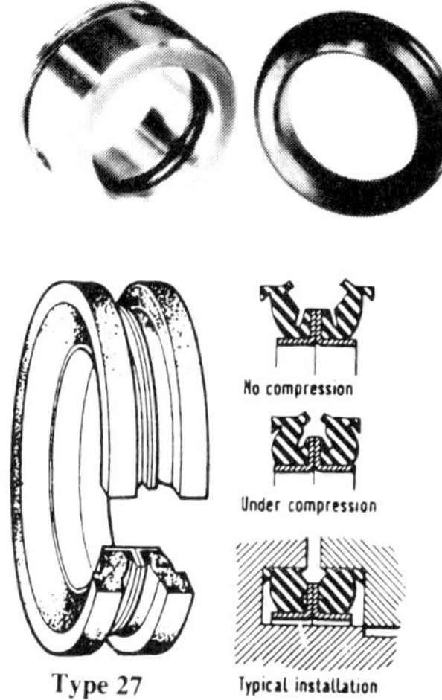

Type 27

Figure 37

Refining industry

Temperatures in the industry vary widely, from as low as −100°C to as high as 400°C. In consequence, the range of seals to meet these parameters is equally wide. Starting at the top of the temperature range, metal bellows seals of the edge-welded variety give the best results, their only requirement being a steam quench when handling products like bitumen, heavy crude or bottoms. These tend to solidify when the pump is stopped, and without a steam quench left on during standstill would give problems on re-starting. The steam should enter the clamp plate via top connection, and as it condenses, leave by way of the drain at the bottom.

In the middle temperature range, say from −20 to 200°C, PTFE wedge, O-ring or metal bellows of both edge-welded and the asymmetric type are commonly used. In most cases product recirculation is sufficient.

There are occasions, however, where special considerations apply. For example, when handling products having low s.g., and where the vapour and suction pressures are close to each other, generated heat must rapidly be dissipated to prevent the product vaporising. This entails using tungsten carbide seats having metal-to-metal contact with the clamp plate, and an auxiliary gland with cooling in conformity with API 610.

Very low temperatures can sometimes be sealed using a single seal and an auxiliary gland. Usually, however, double seals are employed. In the past, methanol has been used as the sealant, but modern practice has been to regard methanol as dangerous, and to use low pour point oils instead.

When a seal manufacturer is asked to make a seal selection, it is necessary to know the product being handled, the suction discharge and sometimes vapour pressures, the temperature, speed of rotation, shaft diameter, stuffing box bore and depth, the pH value, and, for some products, the material from which the pump is made. If the wrong materials are chosen for the seals, it is possible to produce electrolytic action which corrodes the seal components. An overall guide for taking all the operating conditions into account is given diagrammatically in the *Mechanical Seal Selection Guide* chart, but see also Table 8.

TABLE 8 — SEAL TYPES RECOMMENDED FOR DIFFERENT HAZARD RATINGS AND DEGREE OF RISK

Nature of hazard	
Very high risk — fire and explosion.	
Extremely hazardous toxicity rating 3	
Hazardous toxicity rating 2	
Hazardous toxicity rating 1	
Hazardous — result of temperature of change of liquid characteristics	
No hazard — awareness of leakage desirable	
No hazard	

Degree of risk →

TABLE 9

PUMP PRODUCT	USES	HAZARDS
ACRYLONITRILE	— RUBBER	— *AFFECTS RESPIRATORY SYSTEM CAUSES NAUSEA AND VOMITING*
AMMONIA	— FERTILIZERS	— *CAUSES CONJUNCTIVITUS—IRRITATES NOSE AND THROAT—VOMITING*
BENZENE	— INSECTICIDES	— *IS A CARCINOGEN OF THE BLOOD FORMING TISSUES*
FURFURAL	— WEED KILLER	— *AFFECTS THE CENTRAL NERVOUS SYSTEM*
PHENOL	— REFINING LUBE OIL	— *AFFECTS LIVER AND KIDNEYS DEATH FROM SKIN ABSORBTION RAPID*
VINYL CHLORIDE	— PLASTICS	— *A CARCINOGEN—SKIN BURNS BY RAPID EVAPORATION & CONSEQUENT FREEZING*

TABLE 10 — AUXILIARY GLAND QUENCH REQUIREMENT

Liquids	Requirement	Quench medium	Reason
Sub zero hydrocarbons	Start up	Low pour point oil	Prevent icing.
Normal hydrocarbons	Emergency	Steam	Usually readily available.
High temperature stable hydrocarbons	Continuous	Steam	Decomposed products flushed.
High temperature unstable hydrocarbons	Emergency	Steam	Usually readily available.
Crystallizing solutions	Continuous	Water or oil	Dissolves crystals.
Toxic liquids	Emergency	Steam or water	Leakage dilution.
Heat transfer media (chemical)	Continuous	Oil or other suitable liquid	Moisture avoidance.
Expensive liquids	No quench	None	Leakage collection.
Acids	Emergency	None	Contain leakage.

Sealing water turbine main shafts

In a typical hydro-electric power station a turbine is driven from a higher water source such as a lake or river, and discharges to a lower reservoir or water course. The shaft of this machine drives a generator, and the arrangement may be such that the shaft is either vertical, horizontal or inclined. In all cases, the water must be prevented from leaking along the shaft in sufficient quantities to cause unreasonable costs in pumping out or, in extreme cases, flooding the plant.

Sealing a hydro-electric water turbine main shaft makes greater demands on the design than seemingly related other sealing duties. It is, in its simplest form, analogous to sealing the propeller stern tube of a ship, but with the added problem of much higher speeds and pressures. Furthermore, costs which would be acceptable for a stern tube seal cannot be justified in a public utility industry.

A lower cost solution had to be found, and this meant consideration of traditional methods, despite their deficiencies, and the fact that these deficiencies may be exacerbated in arduous environments. Examples of such environments are, where the water quality is not good, and where the installation is used for pump storage applications, *ie*, where the same machine generates and pumps. This necessitates bi-directional rotation of the shaft, accompanied by changes in pressure, and often by considerable vibration when changing mode. *Pump turbine* seals often meet further problems because the pressures sealed are usually higher than ordinary turbines. These machines are frequently run in the 'blow-down' condition. This involves suppressing the level of water by injecting compressed air into the machine so that the turbine runner can spin without contacting the water. This creates two problems for the seal:-

(1) The very high pressure surges that accompany air injection.

(2) The need to run, sometimes for hours, with the turbine water level well below that of the seal.

Use of a labyrinth or close tolerance bush type device will result in a very high water leakage, and unacceptable air leakage in the 'blow-down' condition. Such devices are usually incorporated in the turbine design as a back up to the main seal. A family of mechanical seals for water turbine and pump turbine applications has been developed by Bestobell Seals, their success depending on carefully developed design and choice of sealing face materials. The seals have split faces, enabling them to be assembled round

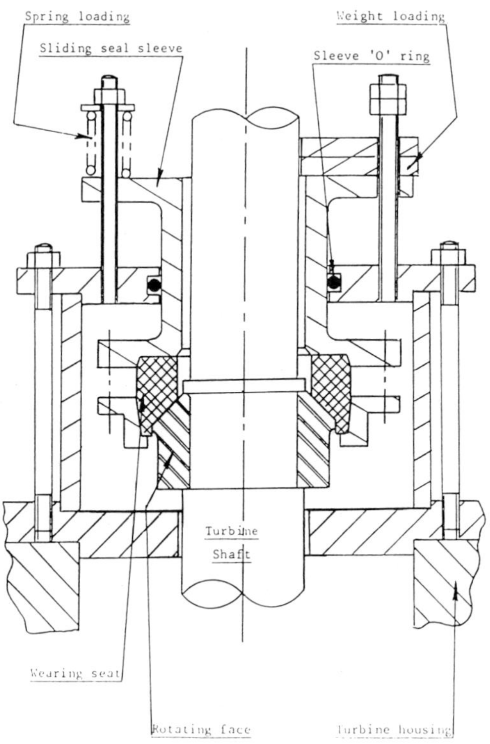

Figure 38
Conical facet type water turbine seal.

the shaft in-situ, and are tolerant of abrasive particles in the water. The faces do not have to be lapped, and in fact 'bed' themselves in. There are two basic diagrams, the first having a split stainless steel cone fixed to the shaft, and rotating within the water, (Figure 38).

To ensure adequate cooling, a flow of about 4 l/min per 100 mm of shaft diameter is required through the seal area. The water need not necessarily be clean, untreated penstock water normally being used. To prevent dry running of the seal, arrangements can be made for this cooling water to be trapped in the seal area during 'blow-down' conditions. Further lubrication of the faces can be supplied through hydrostatic feed holes drilled in the stationary seat, but is normally only necessary when vacuum conditions obtain.

In vertical applications, an alternative is to fill the annulus between the seal faces and the shaft with water. In most applications, cooling and lubrication is effected by the leakage of water across the faces. Spring pressure is set to give a small positive leakage. This is acceptable, as pumps are always installed on site to keep the sump of the power house from flooding. Such settings ensure cool running and a low wear rate, typically 1 mm/1000 hours, although higher where water quality is poor. A wear path of 30 and 65 mm can be designed into the softer seal face, giving a very acceptable seal life. In the radial face design, loading is achieved by hydraulic pressure behind one seal face, the face being free to slide axially, (Figure 39).

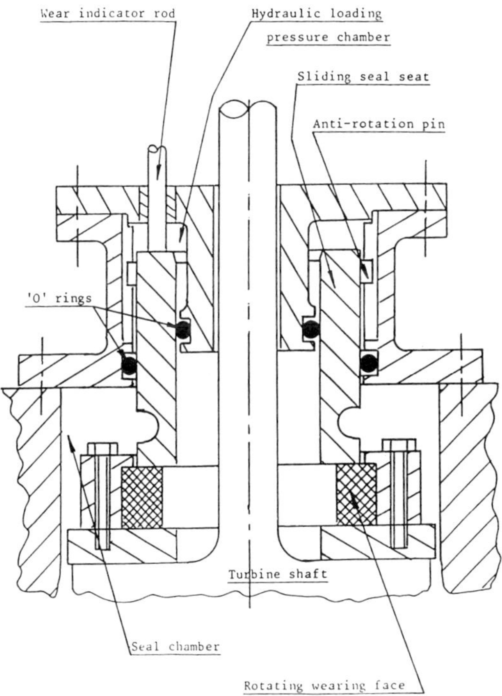

Figure 39
Radial face, water pressure regulated type turbine seal.

Cutaway of Mechanical Seal on shaft showing quench arrangement

In the simplest installations, this can be provided from an external source, such as a regulated water main or a header tank. This has the advantage that it can be adjusted remotely. The load on the face can be made to respond to changes such as pressure surges by linking the hydraulic load chamber to the sealed water pressure. This can be done via a control panel, using orifices to maintain the loading pressure at a fixed ratio of the sealed pressure. Together with the hydraulic balance built into the seal, this enables it to adjust to a much wider range of pressures, including situations where, for instance, a high pressure is present when the machinery is stationary.

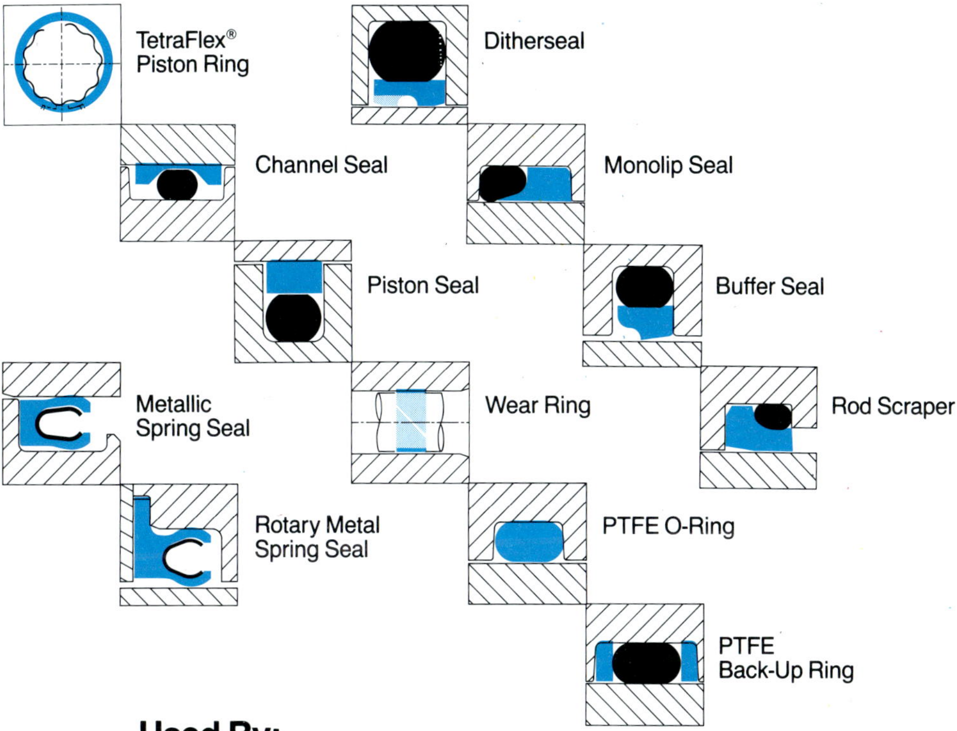

Bellows Seals

BELLOWS ARE essentially exclusion devices, normally capable of accommodating both axial and radial movements. They may be used in the form of gaiters specifically to exclude dust, dirt *etc*, from a machine; as secondary sealing components in mechanical face seals; and as primary seals replacing packed glands or conventional pusher type mechanical seals on such machines as agitators, pumps, compressors, mixers, *etc*.

Bellows may be made of elastomeric or polyamid materials, or from metal, and are used in a very wide range of sealing duties – from automobile water pumps and domestic appliances through high temperature refinery duties and high speed turbocompressors to the sealing element in a submarine propeller (stern) shaft seal.

In all cases, the bellows seal eliminates the need for the sliding elastomer which is necessary for secondary sealing of conventional pusher type seals.

Elastomeric bellows

Elastomeric bellows are normally formed by moulding. Deep, narrow convolutions are best for low pressures and rounded, wide-spaced convolutions for higher pressures. For high pressure either a substantial increase in thickness or laminated construction is necessary, which has the effect of stiffening the bellows.

Most elastomeric bellows seals normally also incorporate a metal spring to provide correct face loading, to ensure faces remain closed in static (no pressure) condition, and to offset fatigue. They have been designed for economical production and are manufactured in high volume.

They are mostly supplied to the OEM, including pumps for hot and cold water, mild chemicals and solvents, oil, hydraulics, slurries and sludges. They are also used extensively for oil lubricated refrigeration compressors.

There are a number of alternative designs supplied by various manufacturers,and three examples are shown over the page.

The design principle is similar in each case – a rubber diaphragm is compressed on to the shaft and in turn transmits torque to the seal face component by a mechanical drive. The advantages of the rubber bellows seal (compared to the conventional pusher seal) are the low cost, and elimination of the sliding packing (hang up hysteresis, shaft wear). Disadvantages are considered to be:

(i) Less robust.

(ii) Lower pressure/temperature/speed capability.

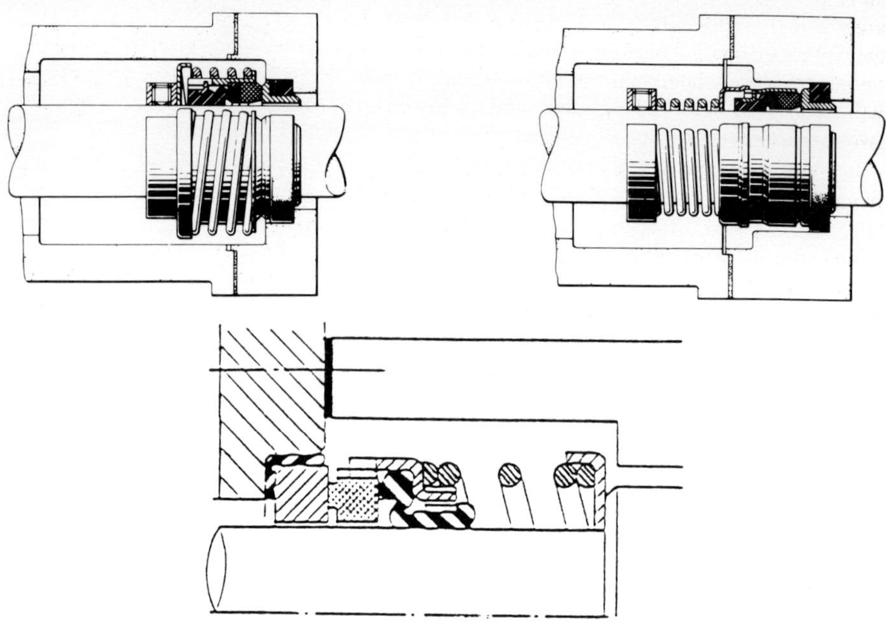

Figure 1
Examples of rubber bellows seals.

(iii) Rubber bellows require specifically designed components in a variety of materials to cope with different media.

A radical new design of elastomeric bellows seal is shown below, designed for the industrial and domestic pump market. It is a simple four part seal with the bellows acting not only as a sealing and frictional drive element, but also provides an evenly distributed spring load.

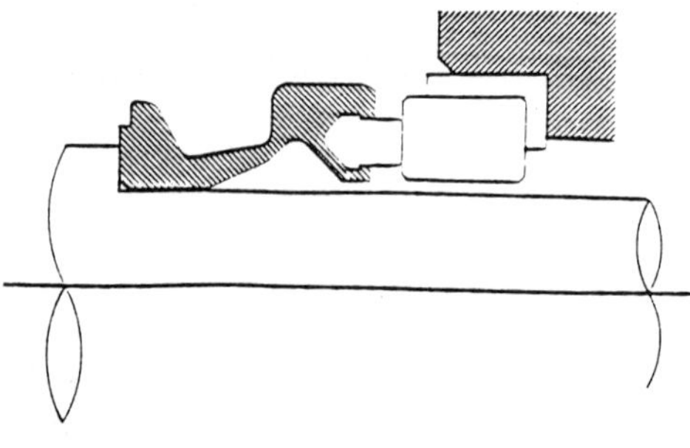

Figure 2

Metal bellows

Metal bellows have inherent spring action and may be of convoluted (formed from tube by rolling) or welded construction. (Other methods of construction include hydro-forming and electro-deposition). Edge welded bellows represent the more costly construction but are stronger, can be given more precisely controlled operating characteristics and have the highest operating temperature range. They are also generally less subject to vibrational fatigue than convoluted bellows. Further improvement in performance for high pressure working can be achieved by using laminated welded construction instead of single ply (welded) construction.

Edge welded metal bellows are fabricated from a series of metal discs or plates normally from 0.10 to 0.17 mm (0.004 to 0.007 in) thick in stainless steel or corrosion-resistant alloy in the form of a nesting ripple (Figure 3). This configuration produces a bellows with a long stroke capability without overstressing the material. They are by far the most common form of metal bellows seal. Figure 4 shows a hydroformed bellows with an assymetric convolution form. This construction gives more evenly distributed stresses than the conventional symmetric formed bellows.

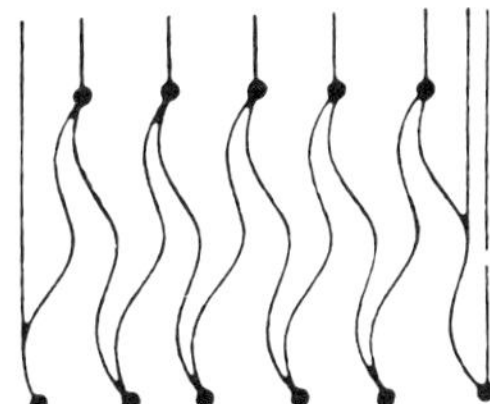

Figure 3
Bellows formed by welded discs.

High temperature metal bellows seal.

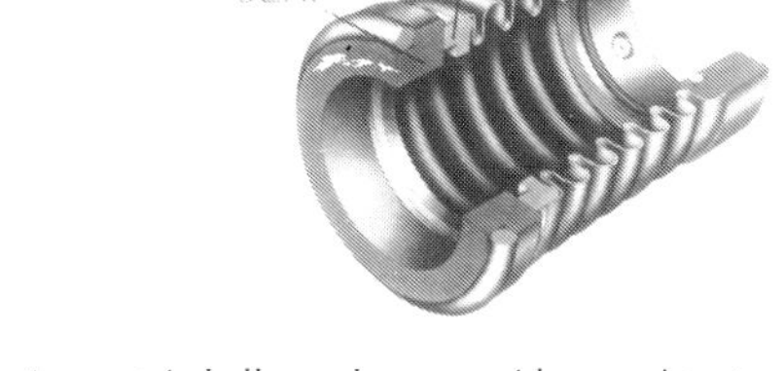

Figure 4 *Assymetric bellows shape provides consistent and even distribution of stresses.*

General arrangement of typical bellows (face) seal is shown in Figure 5. The bellows itself acts as a spring to keep the faces together, provides a dynamic seal, and transmits the torque from the shaft to the face. One end of the bellows is welded to an adaptor which is usually clamped to the shaft or sleeve by set screws and which houses the static secondary seal. The other end of the bellows is welded to the shell which retains the seal face, both of these being carried clear of the shaft.

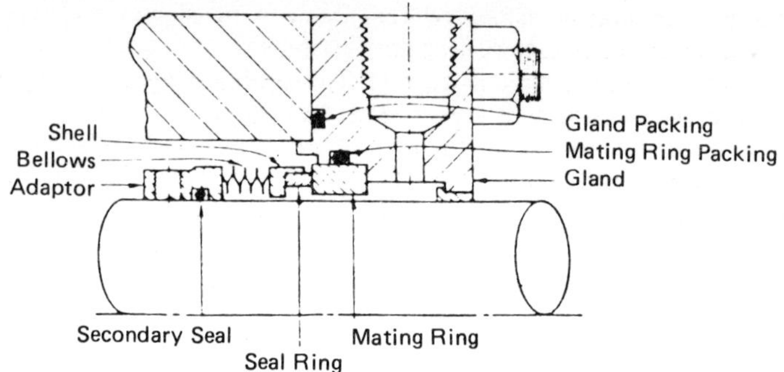

Figure 5
General arrangements of typical bellows seal.

Alternative arrangement using hydroformed bellows.

With the product being contained at the o.d. of the seal, the centrifugal action throws any particles away from the bellows and so prevents clogging, hence this design of seal is self-cleaning. As the bellows seal has no sliding elastomer there is no drag or hang-up on the shaft to make the faces separate in the event of shaft misalignment, probably the biggest cause of leakage with face type seals. This feature, which also eliminates shaft or sleeve wear, another common fault or face type seals, has been a major attribute in rapid popularity of bellows seals. This design of seal can be mounted either directly on to the shaft, on to a hook sleeve, or to a cartridge sleeve in the case of inaccessible or difficult arrangements. Cartridge sleeve arrangements allow the seal to be assembled and set before fitting to the machine, which can be a valuable asset.

Temperature limitations

A major limitation of the face type mechanical seal when replacing soft gland packing is the temperature range that can be accommodated. The limiting factor in high temperature sealing is the safe operating range of elastomers, which is generally 175 to 200°C (350 to 400°F). Besides external cooling of the media, which is usually undesirable and expensive both in the initial equipment purchase and in the constant maintenance required to retain effectiveness, the only remaining approach to high temperature sealing is the use of metal bellows designs. The absence of sliding elastomers in bellows seals eliminates the temperature restrictions imposed by them.

Coking is another problem experienced with face type seals when sealing high temperature viscous products such as oils and asphalts, because solidification around the sliding elastomer causes it to 'freeze' to the shaft. Coking and the resulting hang-up and leakage can be eliminated by the use of metal bellows seals, as none of its moving parts is in contact with the shaft. It is normal to introduce a steam quench to the atmospheric side of the seal to cleanse, preclude oxygen and cool the area around the seal faces to prevent the formation of coking.

In general, temperature extremes are catered for as follows:

Cryogenic:	−200°C to −40°C. Austenitic stainless steel or Inconel bellows. Expanded graphite packing/metallic gasket/metal wedge secondary seal.
Medium Temperature:	–40°C to +200°C. Almost any available metal bellows material (weldable strip). Mainly Alloy 20, AM 350, Hastelloy C. Elastomeric O-ring secondary seals – mainly Viton, but also Nitrile, EP, Neoprene, Kalrez.
High Temperature:	+200°C to 450°C. Hardenable and weldable metal strip. Usually AM350 or Inconel. Expanded graphite packings/metallic gasket/metal wedge secondary seal.

A rotating bellows seal able to handle temperature ranges from -40° to +450° C.

High performance seal for extreme temperatures.

Two seal arrangements are shown below, one with a rotating seal (Sprung member) and one with a stationary seal. Whilst the stationary seal is generally favoured for high temperature duties since there will be less coking and the steam quench can be better directed to the faces, the rotating seal is favoured where the product contains solids, or if the torque needs to be transmitted by drive lugs (thermo-setting fluids *etc*).

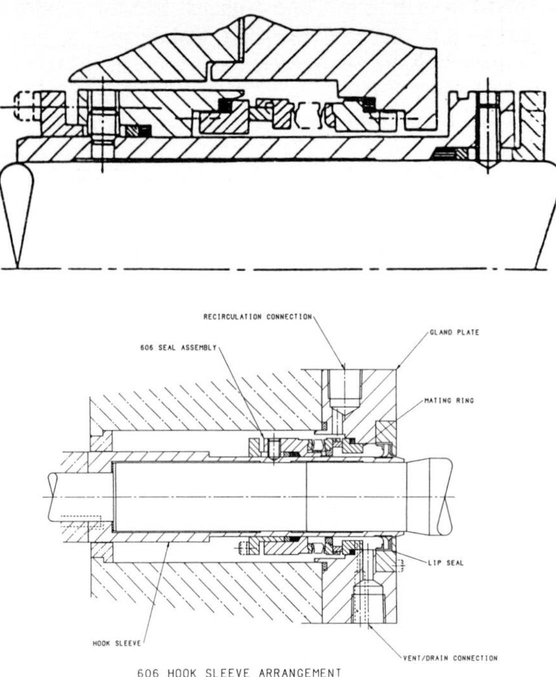

Figure 6
High temperature metallic bellows seal.

In high temperature applications transient expansion is a major problem in machines (*eg* pumps) because the shaft expands at a faster rate than the body of the pump, due to its smaller mass. The advantage of the bellows design is that the flexibility of the bellows will accommodate this growth, quite often with specially designed long bellows cores, without the risk of "bottoming" or hang up experienced with conventional pusher type mechanical seals.

Speed limitations

Due to the elimination of the secondary packing and resulting hysteresis, the metal bellows seal will cope with much higher shaft speeds. For rotating seals, the limiting factor is normally the dynamic limitation of squareness of the stationary counterface to the shaft. Figure 7 shows the comparison between bellows seals and pusher seals.

For stationary seals, there is virtually no dynamic limitation of metal bellows seal. The limitation is almost always associated with friction, fluid dynamics and materials of the rubbing surfaces. Stationary bellows seals are used for sealing oil lubricated turbo-compressor seals to 20000 r/min, 20 bars and with diameters to 200 mm plus.

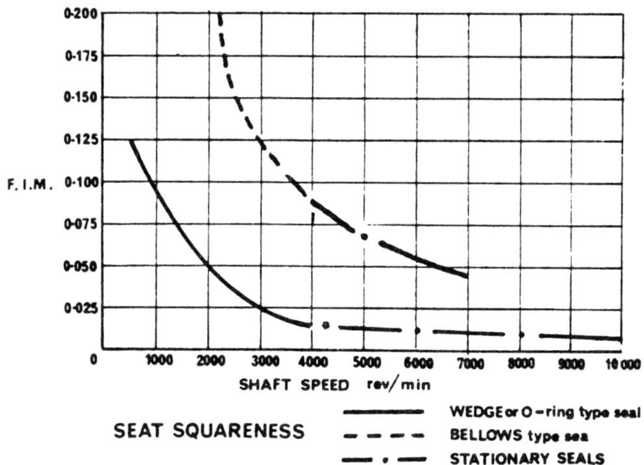

Figure 7

In the aerospace and defence industries, bellows seals cope with much higher speeds and combinations of speed, pressure and temperature. Some examples are given below:-

	Size (mm)	Speed r/min)	Pres. (bar)	Temp. (°C)
Weapon steam turbine	16	140000	–	180
Aircraft fuel pump	35	30000	20	150
Aircraft gas turbine starter	20	50000	2	400
Aircraft refrigerant compressor	16	48000	9	−54 to 180

Balanced bellows seals

Another particular advantage of a bellows seal is that its construction is inherently pressure balanced without the need for a stepped shaft. In the bellows design balancing is related to the effective diameter (see Figure 8). Fluid in the stuffing box goes between the convolutions and exerts pressures against the bellows plates, creating an axial force in each direction. The amount of the force depends upon the plate area and the pressure exerted on it.

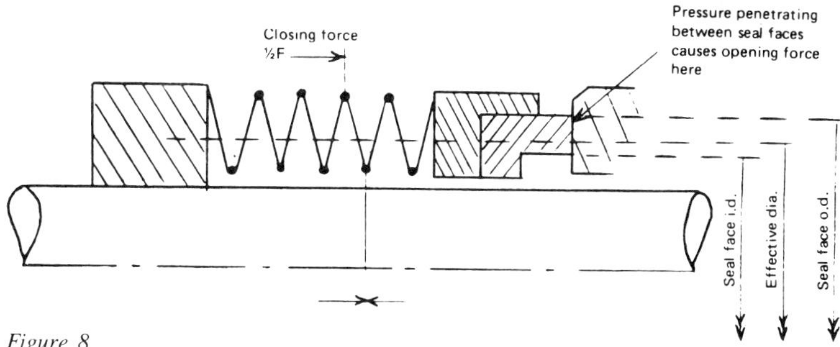

Figure 8
Pressure balancing of bellows seal.

Each bellows plate will be in equilibrium, but an axial force will be transmitted at each end to the end fittings. At the rear (fixed) end, this force will be reacted by the mechanical attachment to the shaft (normally a setscrew). At the front end this force will be reacted between the seal faces, normally by a self generated squeeze film. The magnitude of this end load can be adjusted by configuring the nose profile of the seal face to increase the closing force (by increasing the face diameters) or to decrease the force. Typically, the effective diameter of a metal bellows is approximately the mid diameter of the bellows span (although this can vary slightly with plate shape, pitch and applied pressure). For a 70% balance ratio (typical).

$$\frac{(\text{Seal Face od})^2 - (\text{Effective Diameter})^2}{(\text{Seal Face od})^2 - (\text{Seal Face id})^2} = 0.7$$

The diameters are arranged to always ensure that the faces are held together during all operating conditions without generating excessive frictional heat.

High pressure media

Single ply welded metallic bellows may be designed to accommodate internal (bursting) pressures up to 500 bar (7500 lb/in^2) depending on pitch. Further improvement in pressure rating is possible by using laminated bellows as shown in Figure 9 increasing the maximum pressure rating to 700 bar (10000 lb/in^2) while retaining the same nesting capabilities and low spring rate.

It should be emphasised that metal bellows seals are not specified to their limiting burst pressures. Most performance limitations are based on infinite life model prediction taking into account cyclic pressure impulse, axial extension and torsion caused by marginal face lubrication (stick-slip). Pressure limitations are generally listed as being 15 to 100 bars depending on size, strip gauge and whether single or double ply lamination.

Handling abrasive media

When sealing abrasive fluids, the entrained solids become trapped, either between the packing and the shaft when using soft packed glands or between the elastomer and the shaft with face type mechanical seals. This causes badly worn shafts and elastomers, which allows the product to leak out and incurs high replacement costs. This is eliminated with the bellows seal design as there are no sliding elastomers, *etc* running against the shaft. With fluids which are heavily entrained with solids the bellows can be mounted internally so that the media only come into contact with the outside of the bellows. Any particles will be thrown away from the bellows convolutions by the centrifugal action and so clogging is prevented, a common fault with the small springs used in face type seals.

Quite often single bellows seals have replaced double conventional pusher seals which were using a high pressure clean liquid barrier system for face lubrication. In other cases single bellows seals replacing single conventional seals work with drastically reduced clean flush flow, sometimes with either unfiltered product or dead-ended seal housings. For sealing abrasive fluids, single seals are normally recommended with hard faces (Silicon Carbide v Silicon Carbide) and very low clean filtered flush. An external quench will often extend seal life.

Handling corrosive media

The limiting factor is generally the elastomer, but care must be exercised in selecting the correct bellows and face materials. Alloy 20 is now the most widely used bellows material and is considered to be very cost effective for a very wide range of compatibility.

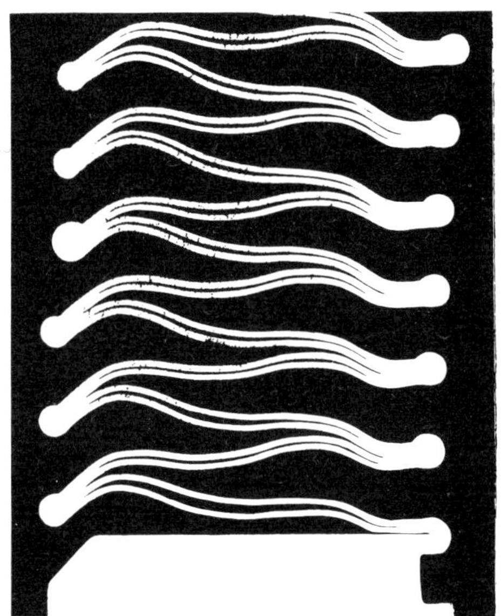

20x macrography of laminated bellows in cross section. Note the long compression stroke and almost total nesting capability which the plate and weld bead configuration afford.

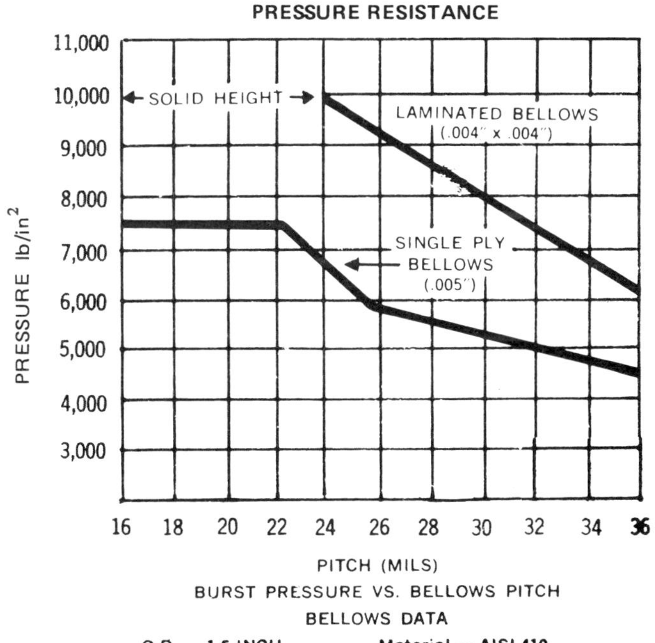

Figure 9

Its corrosion resistance is superior to both austenitic (300 series) and precipitation hardening (AM 350) martensitic stainless steels. For temperatures between −40°C and 250°C, Hastelloy C will be used for duties requiring improved corrosion resistance. Hastelloy C will resist all but the most aggressive media. For extremes of temperature, coupled with corrosion resistance and strength, Inconel will normally be used.

The following table provides brief details of the most commonly used bellows materials.

TABLE 1 – METALLIC BELLOWS MATERIALS

Materials	Temperature °C	Description
Alloy 20	−40 +250	Superior stainless steel with 20% Chromium and 32.5% minimum Nickel. Wide use in chemical industry. Readily weldable and very stable in cold worked condition. Most commonly used material for bellows seal for the process industry.
AM350	−40 +450	Martensitic stainless steel. Toughest and most commonly used material for high temperature seals. Very high strength after heat treatment. Corrosion resistance equivalent to 303 (free machining 18/8 stainless steel) before heat treatment and as 400 series after heat treatment.
300 Series	−250 +450	Austenitic stainless steel. Unheat treated. Can be used for cryogenics, but weak at high temperatures. Has tendency to take permanent set, so undesirable when spring rate is important. Useful as bellows device in corrosive conditions.
Inconel	−250 +750	Good strength at high and low temperatures. Improved corrosion resistance over stainless steel but marginal with sulphur duties. Difficult to weld. Higher cost raw material and expensive heat treatment.
Hastelloy	−250 +250	Best corrosion resistance. Unheated treated. Limited pressure capability. Good strength. Higher raw material cost than Inconel.

Face materials

Table 2 shows an extensive list of material pairs and their mechanical PV limitations. Face materials for metal bellows seals are normally specified as follows:-

General Purpose/High Volume –	Carbon v Aluminia
Best life performance –	Carbon vs Silicon Carbide
Temperatures above 300°C –	Special or Tungsten Carbide v Silicon Carbide
Abrasive –	Tungsten Carbide v Silicon Carbide
Abrasive and Corrosive –	Silicon Carbide v Silicon Carbide.

General comments

Metal bellows have been developed from designs used for the aircraft and defence industries. Their penetration into the process and industrial pump market was established

TABLE 2 – FACE COMBINATION PV LIMITS

Seal face	Mating face	Maximum PV	
		lb/in^2 ft/min	bar m/s
Carbon	Stainless steel	150000	50
	Stellite	700000	240
	Chrome oxide coating	450000	150
	Tungsten carbide coating	800000	280
	Tungsten carbide	1000000	350
	Sealide	5000000	1750
	Ceramic	500000	170
	Lead bronze	700000	245
	Ni-resist	850000	300
Tungsten carbide	Tungsten carbide	150000	50
	Sealide	500000	170
Stellite	Sealide	250000	90
Sealide	Sealide	400000	140

The above PV values are calculated using mechanical force only.
For values using higher pressures as would generally be found in pumping applications, the above figures may be increased by as much as up to x 2.

by solving difficult seal duties – particularly *high temperature, high speed, abrasive* duties where pusher seals were known to give limited performance. However, there is now a wide range of products available to meet most market requirements, where they now compete on cost as well as on performance. They are extensively used on both industrial pumps and high volume industrial (mainly refrigeration) compressors. They can be incorporated into cartridge (modular bench assembled) units and into complex environmentally controlled seal systems using multi-seals (tandem/double with lubrication systems).

Their advantages over pusher seals are:-

(i) Elimination of sliding packing, hang-up hysteresis, sleeve wear.
(ii) Can be used at higher temperatures and speeds.
(iii) Inherently balanced without stepping shaft/sleeve.
(iv) More compact (particularly larger shaft sizes).
(v) More easily fitted.

Their disadvantages are considered to be:-

(i) Lower pressure capability.
(ii) Less robust.
(iii) Fatigue failure can occur in marginal lubricating conditions.

Teflon bellows seals

Highly corrosive acids can only be handled by seals made from Teflon* and mounted externally, so that the metal half-clamps securing the seal to the shaft are outside the product.

* *Registered Trademark*

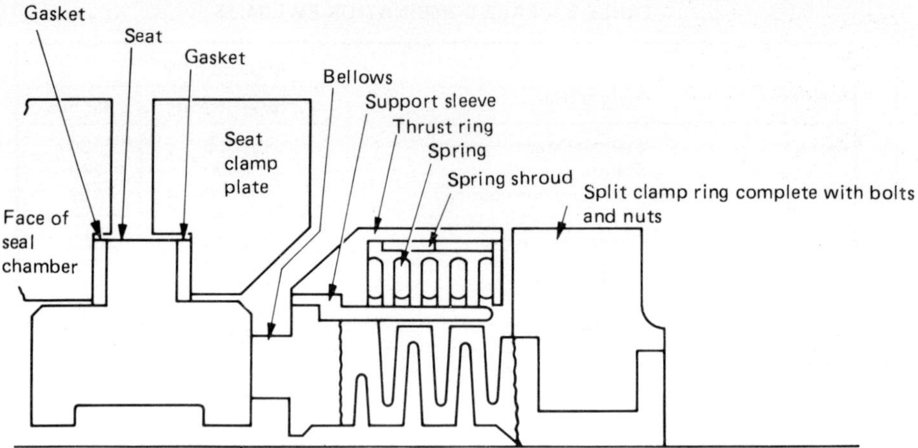

Details of a sophisticated PTFE bellows seals. The support sleeve limits bellows expansion under pressure surges, thus inhibiting over-stressing which could other wise lead to failure.

Teflon cannot be moulded into complicated shapes and is available only in solid bars or tubes. Teflon bellows, therefore, have to be machined, as have all the other seal parts in this material.

In almost all cases, the clamped stationary seal ring, or seat, is made of ceramic, which has a high aluminium oxide content.

The opposing rotating face may be of Rulon, tungsten or silicon carbide, or more generally, glass-loaded Teflon.

Pure Teflon is slightly porous and would allow penetration by the pumped liquid, which could result in an uneven film thickness.

Dirty/abrasive acid duties

Abrasive particles in the product can fret the inside of the bellows, allowing the product to escape to atmosphere. This can be dangerous to plant operatives.

Injections of the same product from a separate, clean source, at a slightly higher pressure than stuffing box pressure, will solve the problem. Its drawbacks, however, are that the clean injections are lost into the dirty product.

A remedy is to introduce the clean liquid into the seal area via a connection in the bottom of the seat, and to return it to its source via a similar connection in the top of the seat.

To prevent loss of the clean liquid into the product, a Teflon lip seal, or close-clearance neck bush, should be fitted in the bottom of the stuffing box.

Cartridge Seals

MECHANICAL SEALS have traditionally been mounted either directly onto the shaft or onto a sleeve. In recent years, however, the use of cartridge mounted seals has steadily gained favour. There are many reasons for this, amongst them being the following:

(1) Ease of fitting (no measuring required)

(2) No damage to the shaft resulting from burrs caused by grub-screws.

(3) Down time reduced.

(4) No special skills required.

(5) Reduction in seal failures.

Cartridge seals consist of a conventional mechanical seal, mounted on its own sleeve, and attached to its own clamp plate. In use the whole assembly is pushed onto the shaft and the clamp plate bolts tightened. Spacers between the stuffing box face and the clamp plate are then swung away and the seal assumes the correct compression setting. An alternative to this method is to have a thin PTFE ring interposed between the stuffing box face and the clamp plate, and to finger tighten the clamp plate bolts. The external grub screws holding the cartridge to the shaft are then tightened. The clamp plate bolts can then be pulled up tight using a spanner. In use the PTFE spacer will gradually wear away, but the need for loose spacers is eliminated, and PTFE spacers are cheap.

Prior to fitting the seal, a visual and dimensional check of the shaft is all that is required.

Incorrect assembly of the seal is eliminated since they are all factory assembled, and where appropriate pressure tested. Most of them will be internally mounted, although there are some exceptions, and are available in many design variants. They can be balanced or unbalanced, single- or multi-spring, bellows type, be pusher or non-pusher and use as the secondary seal wedges, O-rings, chevrons or U-rings. In short, have all the attributes of conventional seals without their drawbacks.

In addition they have benefits deriving from the seal manufacturers control of the design of the clamp plate. This permits positioning of the circulation connections in the most favourable place, and the inclusion of a variety of auxiliary devices to assist seal performance.

Standards

There are currently no complete standards covering dimensions, and in consequence seal manufacturers tend to design seals to operate within the cavity dimensions relating to non-cartridge seal standards.

A recent addition to DIN 24960 does, however, include cartridge seals to fit 30, 40, 50 and 60 mm shaft sizes only. Outside these sizes, manufacturers tend to work to non-cartridge seal standards (*eg* ISO 3069 and DIN 24960).

A further feature of many of these seals is the use of slotted clamp plates which enable the seal to accept a wide range of clamp plate bolt pitch circle diameters, (Figures 1 and 2).

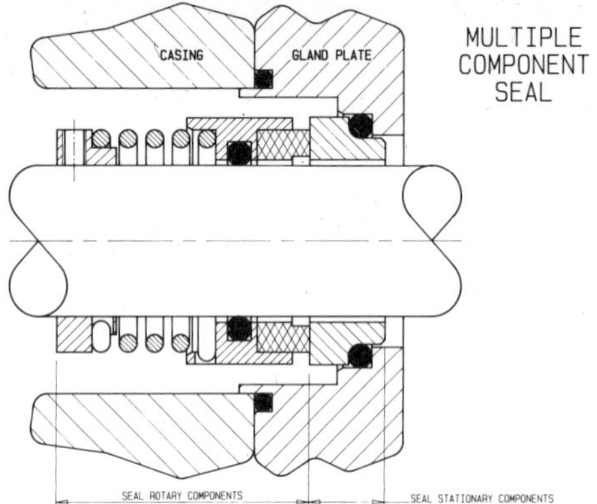

Figure 1
Cartridge seal suitable for use in ISO 3069 or DIN 24960 seal cavities.

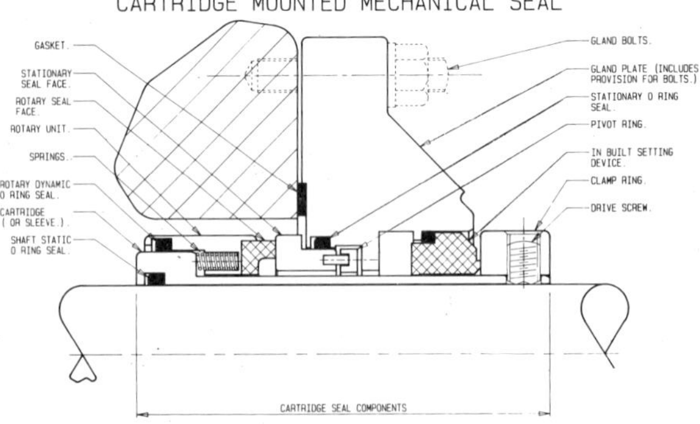

Figure 2
Cartridge seal with short external working length.

To offer even more versatility, some designs are made with short external working lengths. This allows the seals to be fitted to close-coupled process pumps where shaft space between the stuffing box face and bearings is minimal.

Due to the ease of fitting, cartridge seals are popular with maintenance engineers. Back pull-out pumps permit fitting on site, with considerable savings in equipment/process down time. Packed glands may be easily converted to mechanical seals since sleeve designs hava a parallel through-bores.

Bearings may also be incorporated in the seal design; a useful feature when an overhung shaft has a poor slenderness ratio. Figures 3 to 11 show a variety of cartridge seal designs.

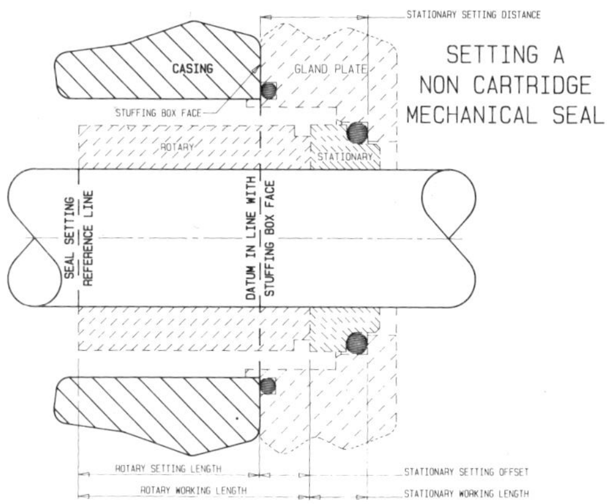

Figure 3
Multiple component seal.

CARTRIDGE SEAL FACE CONTACT

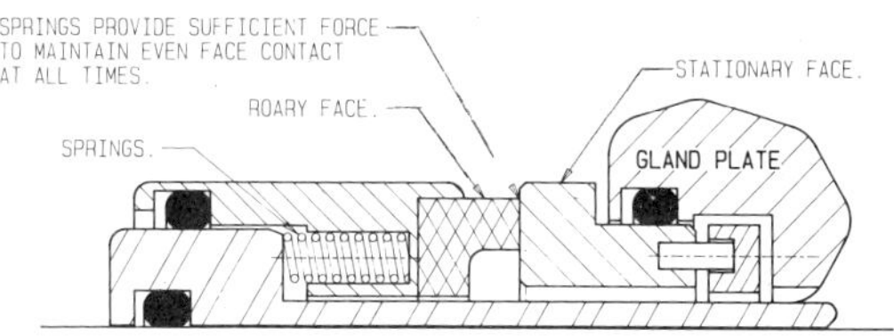

Figure 4
Cartridge mounted mechanical seal.

INTERNAL BALANCED CARTRIDGE SEAL
WITH FLUSH CONNECTION

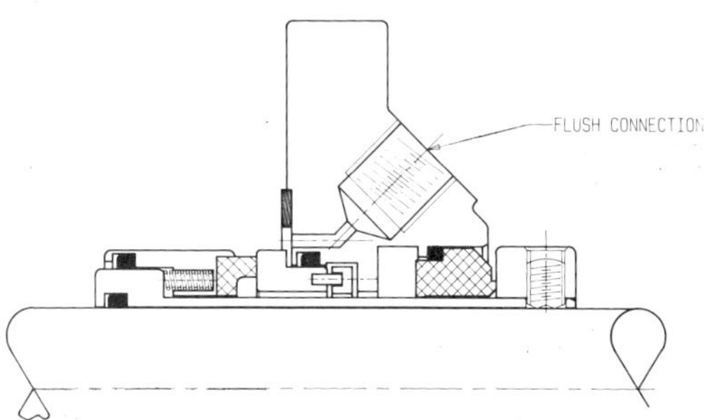

Figure 5
Cartridge seal face contact.

INTERNAL BALANCED CARTRIDGE SEAL
WITH CLOSE FITTING BUSH AND QUENCH CONNECTION

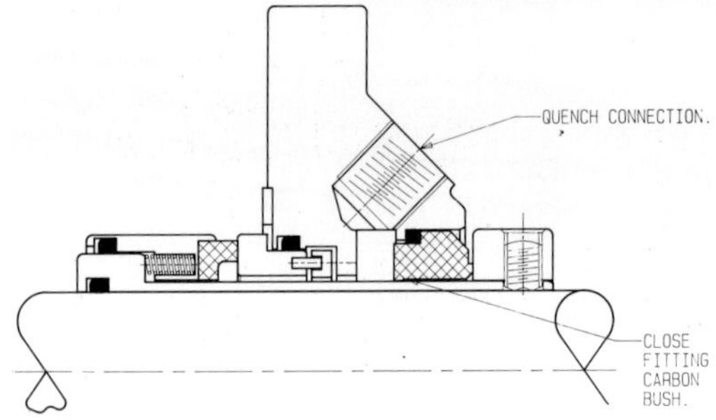

Figure 6
Internal balanced cartridge seal
with flush connectior

INTERNAL BALANCED CARTRIDGE SEAL
AUXILIARY ROTARY LIP SEAL

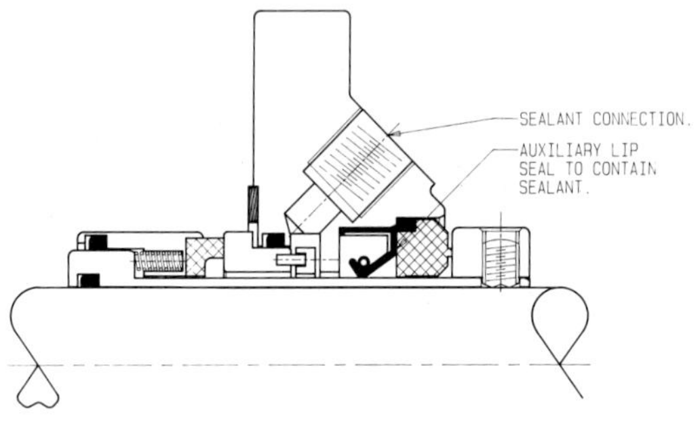

Figure 7
Internal balanced cartridge seal
with close fitting bush and quench connection.

CARTRIDGE MOUNTED DOUBLE SEAL

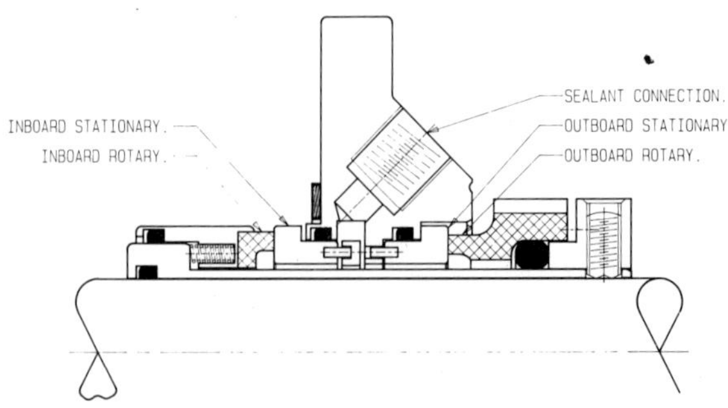

Figure 8
Internal balanced cartridge seal
auxiliary rotary lip seal.

CARTRIDGE SEAL SUITABLE FOR USE IN
ISO 3069 OR DIN 24960 SEAL CAVITIES

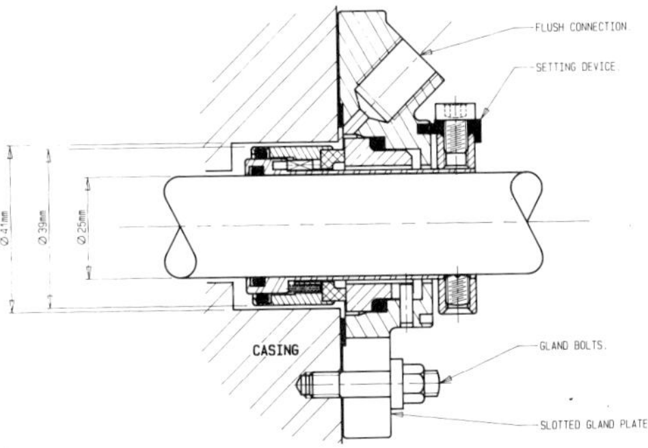

Figure 9
Cartridge mounted double seal.

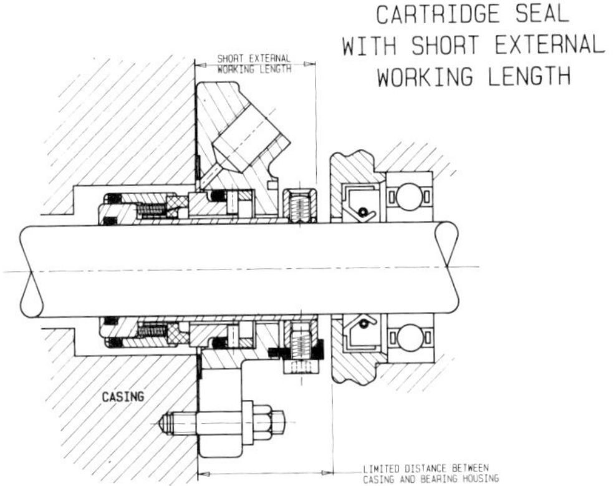

Figure 10
Cartridge seal for corrosive duty.

CARTRIDGE SEAL FOR CORROSIVE DUTY

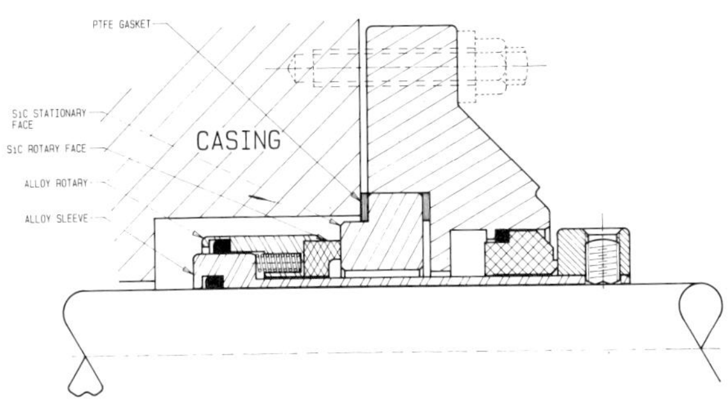

Figure 11
Cartridge seal with bearing.

Top entry mixer with FFET cartridge seal.

Split cartridge seal for top entry (left hand pic).

Split cartridge seal (right hand pic).

General purpose cartridge seal.

Marine Stern-Tube Seals

MARINE STERN-TUBE (propeller shaft) seals represent a specialized requirement in having to provide adequate sealing with low friction with the external component(s) in contact with a corrosive medium (salt water). The seal may also be required to have self-aligning properties.

Marine seals are normally located around the propeller shafts, where stabilizers operate through the hull, in way of rotating shafts passing through watertight bulkhead structure, where rudder stocks enter the stern aperture and instances where manoeuvring 'thrusters' are employed.

It is possible for a ship (*eg* a cross-channel ferry) to require all five types which could mean the employment of a total of 18 seals.

Methods of propulsion

The type of seal required depends upon the propulsion to be used.

A *fixed pitch* propeller replaced the sail and paddle wheel and is now the most commonly used especially on the larger, less flexible ships. Such vessels are helped by tugs when required to move in restricted waters.

Thrust units enable a ship to move transversely, *ie* from side to side. Their flexibility is such that it can also be used as the chief means of propulsion. They are particularly useful on semi-submersible platforms where they can be coupled to computerized positioning equipment to maintain station regardless of tide or wind drift during rig-drilling operations. Coasters and tugs also find thrusters to be of particular service.

Water jets are similar in principle to the manner in which a jet engine moves an aircraft through the air. These are found on fast cruise ships and patrol boats and, being a elatively new innovation, they are likely to be extended for use on larger warships.

A simple but very effective type of stern-tube seal suitable for small craft with high shaft speeds is shown in Figure 1. A simple moulded bushing gland is a generous clearance fit on the shaft and accommodates a number of packing rings. It is mounted clear of the stern-tube on a length of rubber hose. A spigotted gland nut is tightened to compress the packing rings against the shaft so that the complete seal rides on these rings.

With suitable choice of packing, such a gland seal can provide long service with no attention, other than periodic tightening of the gland nut to maintain minimum leakage. Initial adjustment is to provide slight drip leakage, re-adjusted after a 'bedding-in' period, and then at longer intervals as necessary. Such a gland should not be adjusted to run

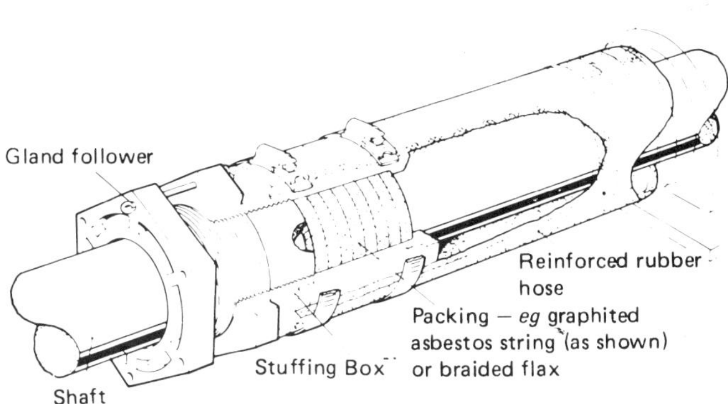

Figure 1
Typical form of stern-tube gland for small craft.

dry as this will increase friction and heat resulting in degradation of the pre-lubricated packing rings and possible scoring of the shaft.

Commonly used gland materials are graphited asbestos and braided flax (grease-impregnated). The life of such a gland, regularly and properly adjusted, can be of the order of 300 to 400 hours. Modern packings (*ie* PTFE/aramid fibre packings) are an attractive alternative and should have a much longer life as well as generating lower friction and running cooler.

Other simple glands in this category may be of rigid construction with backing rings or O-ring seals, relying largely on the presence of adequate lubricant (waterproof grease). This is normally provided by a grease cap, tightened a quarter-turn or so at regular intervals to show fresh grease just emerging from the end of the gland. Similar forms may be used as floating glands (*ie* mounted on the end of a rubber hose as in Figure 1). Some types are also water cooled, tapping the engine coolant system for supply to the water jacket.

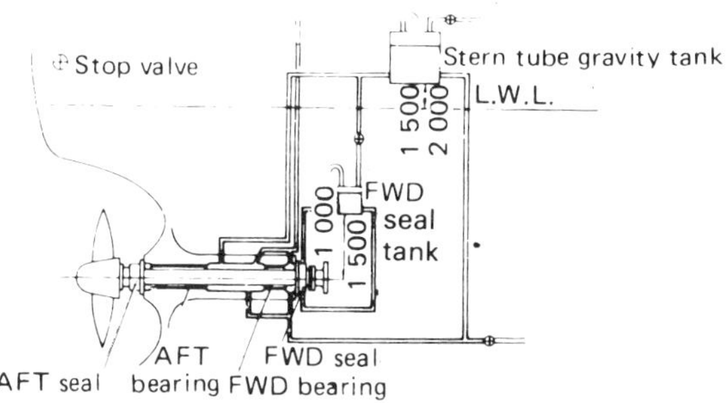

Oil lubrication system for stern gland seal.

Mechanical face seals are also used, an example can be seen in Figure 2. A protective sleeve (ST6) attached to the propeller boss, transmits the drive to the rotating cup (ST3)

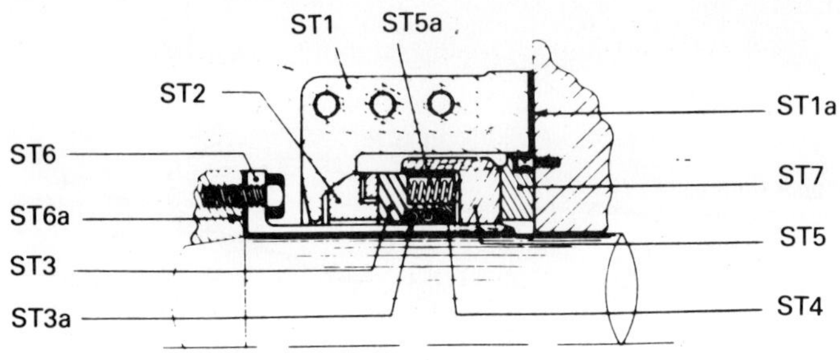

Figure 2

via a driving ring (ST5). The cup (ST3) is 'sealed' on the sleeve by resilient synthetic rubber packing rings (ST3a), and is urged towards a stationary ball joint (ST2) under the action of corrosion-resistant springs (ST5a) to a thrust plate (ST7) mounted against the aft end of the stern-tube bush.

The ball joint (ST2) is positioned in a counterseat machined in the robust case (ST1), securely fixed to the stern-tube nut, which protects the internals of the packing from damage.

The wearing faces are lubricated by oil from the stern-tube, thus ensuring long life.

The resilience of the cup (ST3) and the freedom of the ball joint (ST2) will accommodate vibration of a reasonable amplitude and frequency and/or axial shaft movement without the packing leaking.

On larger craft the traditional packed gland remains a primary choice. Its performance has been enhanced by the appearance of modern packings based on PTFE-impregnated aramid fibres with very low friction and considerably extended life compared with most other packing materials. For new installations, however, modern designs of stern-tube sealing devices are often preferred as being simpler to fit and maintain.

Figure 3 is an example of a seal designed especially for converting existing ships with water-lubricated stern-tubes to a mechanical seal (as well as being suitable for first fitting). Particular features are:

(i) It fits inboard of the existing stuffing box, replacing the existing gland, and uses a previously unworn part of the shaft.

(ii) It uses the existing gland studs.

(iii) It is fully split and can be assembled without disturbing the shaft. The rubber seal elements are bonded in place to form continuous rings.

(iv) The inflatable seal permits maintenance afloat, and the use of a running spare permits rapid return to service.

(iv) Further shaft wear is eliminated and the seal will accommodate radial movement and vibration.

This seal also provides freedom for radial or axial shaft movement (Figure 4). In the event of large movements the garter springs permit the shaft to slip axially through the seal rings while maintaining sufficient grip to provide a positive drive to the rings.

A particular feature of this seal assembly is the *inflatable* seal fitted outboard of the seal cavity (Figure 5). This is inflated by air pressure to grip the shaft and provide a seal with the shaft stationary, permitting maintenance work to be carried out on the seal without the need to dock the ship or trim down by the bow.

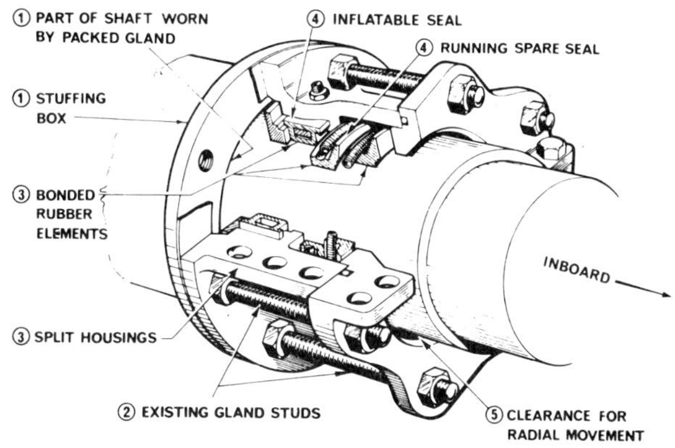

Figure 3
Seal for water lubricated stern-tubes.

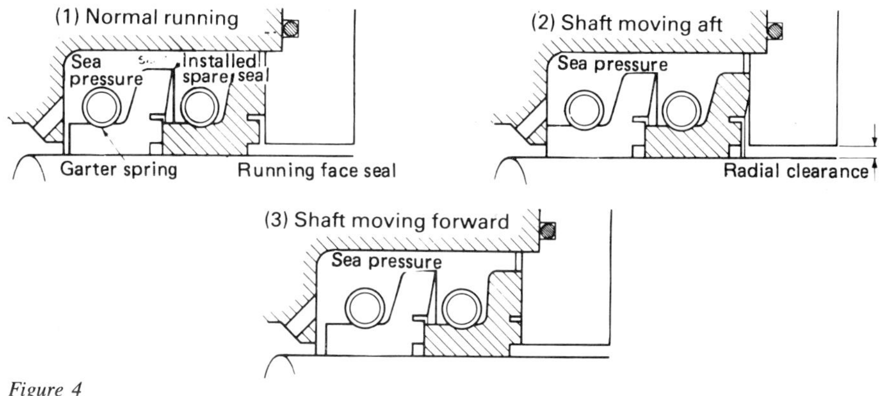

Figure 4
Water lubricated stern-tube seal.

Oil-lubricated stern-tube seal.

Figure 5
Stern-tube seal with added inflatable seal.

The Safeguard Sterngear system

The system incorporates a drained and vented space (interspace) at atmospheric pressure between the water excluding sealing lips and the bearing oil retaining sealing lips.

An emergency sea water seal is incorporated in the inboard side of the oil retaining seal to ensure that any sea water passing into the interspace is retained without contaminating the bearing lubrication system.

The interspace is drained and vented inboard by means of special pipes housed within the oil circulating channels of the stern-tube bush, thus providing positive continuous monitoring of the aft seal conditions.

The system is designed to be fail-safe so that under normal operating conditions neither wear nor failure of any seal element will allow oil to leak outboard, avoiding pollution, nor will sea water enter the oil system, ensuring reliability.

The seal components are of standard design, well proven in service, and are arranged to operate under the most favourable conditions in order to achieve extended life in service.

The bearings are of standard form, cast iron castings lined with a high integrity whitemetal lining, using pre-treatments which ensure a high quality metallurgical bond. A requirement of the manufacturer's quality assurance programme is that bond strength tests are carried out on all stern-tube bushes.

In order to avoid the problems with split supplier responsibility this system is offered as a complete package, including seals, bearings and internal pipes.

The lubricating oil system is simple and economical, using low pressure (0.2 bar) from a single low level head tank, resulting in minimal wear on the lip seals and liners by virtue of the low PV factor.

There is provision for continuous or periodic monitoring of the condition of the aft seal system from inboard, as required.

The system does not require any change in the propeller boss diameter, propeller overhang or aft lines of the vessel from the conventional.

It can be fitted to fixed or controllable pitch propellers and to all reputable sterngear systems, including withdrawal types such as the Glacier-Herbert Sterngear System.

The Sternguard system

Initially developed for deep draft tankers and high powered container vessels, the 4SC seal is now becoming acceptable for use on almost any type of vessel, regardless of seal size, because of its high reliability.

Lip seals and shaft liners of the Safeguard Sterngear system.

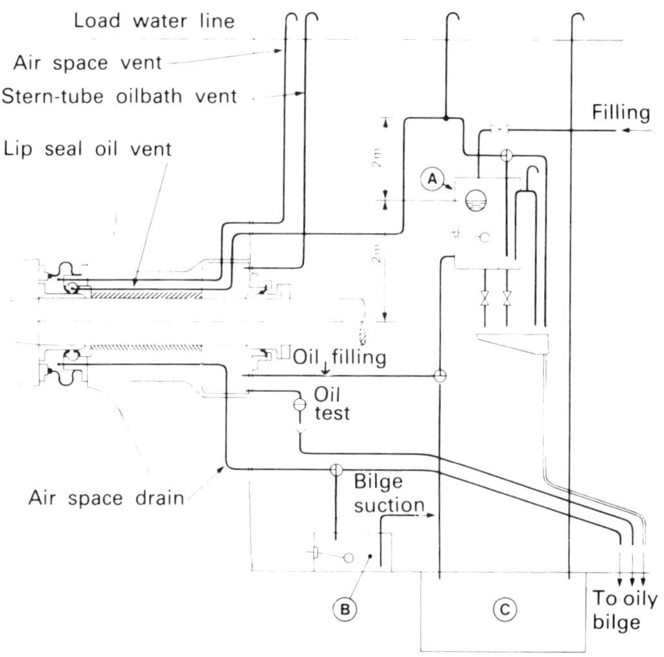

Key:
A–Stern-tube bearing oil tank.
B–Suitable air space drain tank.
C–Stern-tube oil drain tank.
Minimum capacity to be stern-tube capacity plus 300 litres.

(Tank A can be supplied at ship builder option.
Tanks B and C and all pipework outside the stern-tube to be supplied by the ship builder.)

The Coastguard Sterngear system Mark 1.

The 4SC seal has been designed to withstand pressure fluctuations and has a built-in safeguard against oil pollution.

Special stern-tube lubrication/circulation systems allow for constant monitoring of seal performance and enable countermeasures to keep the vessel operating without dry-docking between major surveys.

Most important is the 'drain concept' which, in the case of damage to seal rings, prevents oil leakage and/or water ingress, therefore eliminating oil pollution and bearing damage.

Marine stern-tube

In an age where extended periods for surveys are in vogue, the reliability and freedom from trouble offered by the product makes a strong appeal to the hard-pressed commercial ship-owner. The seals are able to meet exacting operating conditions when sustaining high speeds as is frequently required by naval craft manoeuvring in surface attack, rapid sprinting, sudden reversing and absorbing flexing and shock from typhoons. With controlled inflation the inflatable seal may be used dynamically, thus enabling a damaged ship to return to port under its own power.

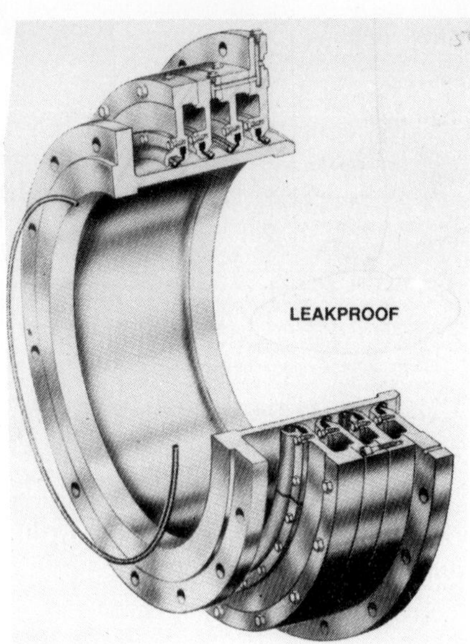

The Sternguard system.

HMS Brilliant

A type 22 Broadsword class frigate with primarily an anti-submarine role; 3500 tonnes displacement at 30 knots with controllable pitch propellers and stabilizers. Fitted with Deep Sea Seals Limited's type MA split stern shaft assembly for water lubricated stern-tubes.

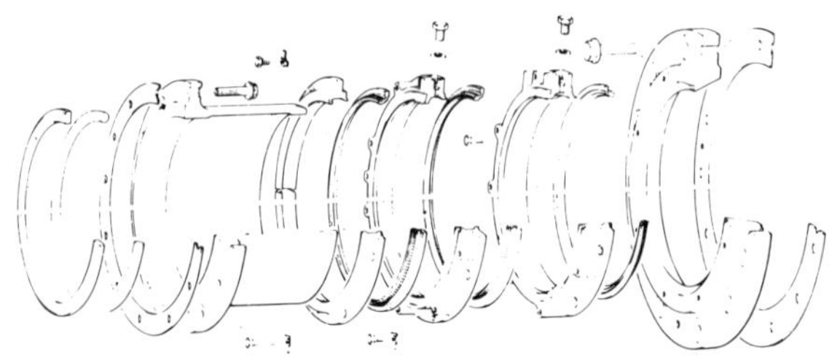

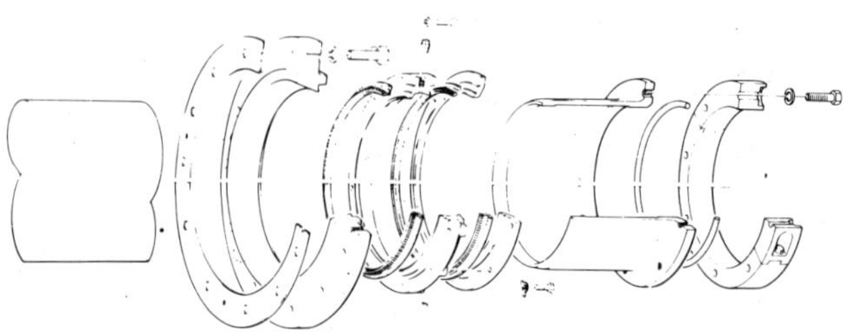

Stern-tube seal for large ships.

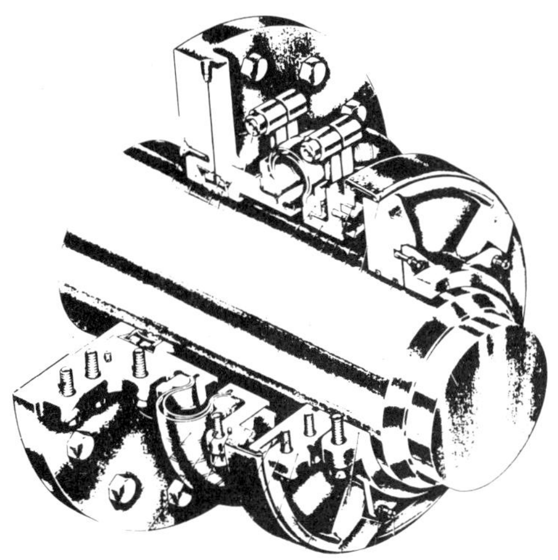

The MX9 stern-shaft seal.

The ManeSeal line:
MA–Fully split.
MD–Partially split.
MX9–Fully split, incorporates auxiliary packing provision. (US Navy approved).

The seal designs of the ManeSeal offer watertight integrity to critical propulsion system components. Corrosion problems with the line shaft bearing and other components in the immediate area of the seal are virtually eliminated.

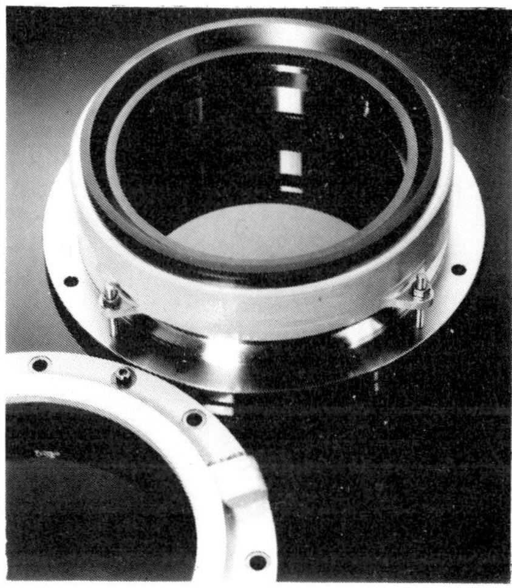

The ManeDive A8 stem shaft seal.

'M' series stern-tube seals

These fall into basic categories: (a) the *ManeSeal* for water applications – 160 to 1040 mm diameter; (b) the *ManeBrace* for oil applications – 280 to 1500 mm diameter.

These can be used inboard (forward) and outboard (aft). Both are derived from the same basic design and offer the owner the facility to replace components without disturbing other items of machinery.

The illustration shows a typical set-up for use with oil-lubricated bearings *(ManeBrace)* which protects the stern-tube bearings from the ingress of sea water and prevents oil pollution of the environment.

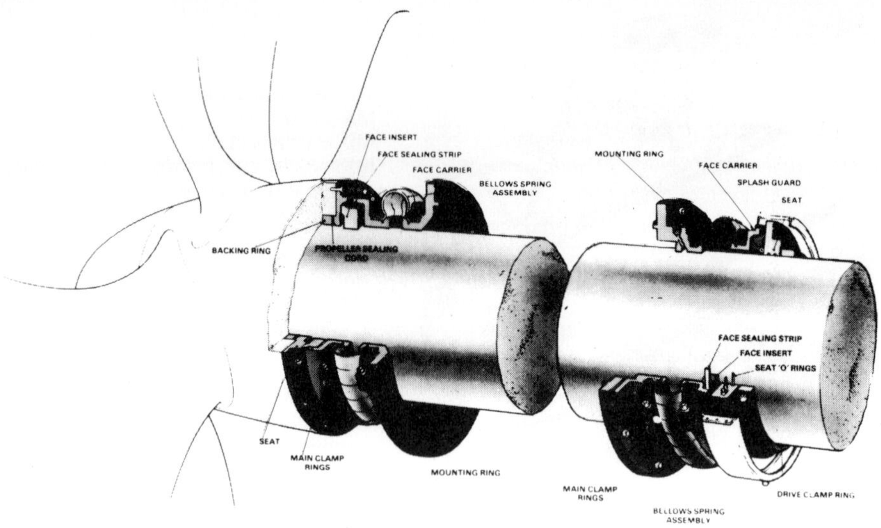

The ManeBrace seal which is used on the largest installation at sea – 1500 mm diameter (58 in diameter).

Application of 'M' series seals

'M' series seals can be fitted to more or less any kind of ship, although in general they are normally fitted to those with shafts ranging from 285 to 1500 mm.

The types of vessel range from small coastal ships up to VLCC's and cruise liners. They are also used extensively on warships, and have been fitted on ice-breakers and other vessels subject to arduous conditions.

They are particularly suitable for retro-fits, on the occasions where owners wish to change from their existing methods of sealing. Over recent years this has been happening more frequently, as owners are looking for longer periods between dockings and reliability between survey periods.

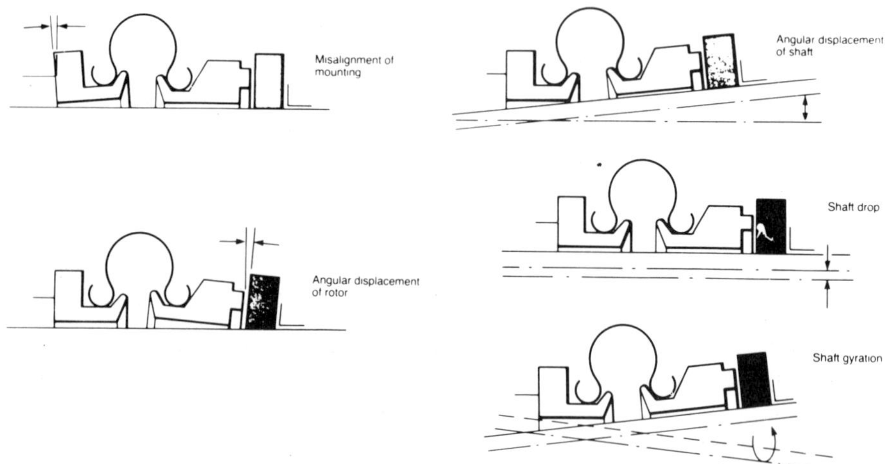

Advantages of 'M' series seals

(i) Ability to accept mounting and operational errors.
(ii) All wearing components visible.
(iii) No shaft or liner wear.
(iv) Maintenance possible without removal of the propeller or tailshaft.
(v) Oil circulation systems not required in the majority of installations.
(vi) No bonding vulcanizing or special adhesives required.
(vii) Fully-split and semi-split designs available.
(viii) Suitable for ten-year shaft surveys, now offered by major classification societies.
(ix) Only one header tank required.
(x) Low wear rates due to hydraulic balance.
(xi) Interchangeable with other seal designs.

Application of 'A' series seals

The following is an indication of the various applications where *'A' series* seals are involved: submarines, semi-submersible drill rigs and support vessels, thrust units, water turbine installations.

While naval applications are an extremely important part of the 'A' series market, the units are now extensively used by major propeller and thrust unit manufacturers, who are continuously extending their own ranges. This tends to dictate the size range, and currently, a range up to 500 mm is being supplied.

The units are particularly suitable for multiple installations on drill rigs, as the design only requires a single head oil tank per unit.

'A' series seals are also one of the options available for water turbine installations.

Advantages of 'A' series seals

(i) Balanced seal design automatically allows for variation in pressures.
(ii) Low wear rates due to balanced design.
(iii) Service can be carried out (if required) without removing the propeller or its shaft.
(iv) Fulfils the requirements of ten-year survey periods.
(v) Designed to suit the customers' requirements.
(vi) The seal will accommodate axial movement without wear on the shaft.
(vii) Major wearing components are visible.
(viii) Split components do not require special bonding or vulcanizing equipment.
(ix) Complicated header tank or balancing systems are not required.

'N' series bulkhead seals

The *'N' series* seals fall into two basic designs: the *ManeSafe type ND* or diaphragm bulkhead seal and the *ManeSafe type NB* or inflatable bulkhead seal.

Both are designed to seal off watertight bulkheads in the event of an emergency. They also meet the requirements of both naval and commercial markets.

The ManeSafe seal was developed to enable naval architects and ship designers to incorporate an efficient type of bulkhead sealing other than the packed gland.

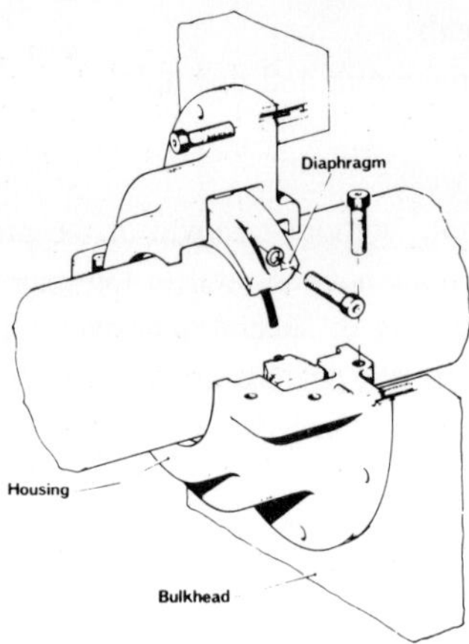

The ManeSafe type ND.

The seal is intended for use under emergency conditions and, for a number of years, a purpose-built version of a standard inflatable seal unit, in the form of the NB, has been and is still being supplied for warship applications. However, there can be a requirement for the seal units to be permanently in operation. To accommodate this the diaphragm type ND seal was developed.

Its development was originally for use in the fast patrol boat market, which has proved to be extremely successful, and supplied extensively to builders both in the UK and abroad. In addition to its intended market, there are a number of other areas for investigation:

(i) *Ferries* – Modern ferry design still utilizes the mid-ship engine. This means that there are long lengths of transmission shaft, passing through water-tight compartments. Most ferries have at least two shaft lines plus auxiliaries.

(ii) *Cruise liners* – As with ferry designs, this type of vessel tends to have the engine room mid-ships, and at least two shaft lines.

(iii) *Tugs* – These days some tug owners are moving towards rotating thrust units, driven by centrally mounted engines or motors. Often these require flexible shafts or couplings that pass through watertight bulkheads.

Advantages of 'N' series bulkhead seals

(i) No shaft wear.

(ii) Inspection and replacement of sealing elements with shaft *in situ.*

(iii) Ability to seal the bulkhead automatically in an emergency.

(iv) Dynamic operation without special lubrication system.

(v) The seal will accommodate both radial and axial movements, under normal or flooded conditions.

(vi) The seal will accept shock loadings (*ie* for naval applications) without detriment to its sealing ability.

'E' series or ManeBar seals

In its present form, the *ManeBar* seal is a development of the successful *Beavic* design. This was introduced in 1947, and several hundred of these earlier seals supplied throughout the world, gave sufficient in-service data to enable the advanced *'E' series* to be established.

The original concept of the 'E' series was for an extremely basic seal, that was cost effective to produce and market, as well as being sound and reliable in service. It was also aimed at the smaller end of the commercial ship market, for vessels such as coastal tankers and supply boats, *etc,* which tend to use oil-lubricated stern-tubes.

However, over recent years the seals have been supplied for a considerable range of ships and marine propulsion equipment, to the extent that they are now used on vessels ranging from suction dredgers to fast patrol craft. In addition, they are also supplied to manufacturers of propellers and thrust units, who have also carried out extensive test programmes, prior to standardizing on the ManeBar equipment.

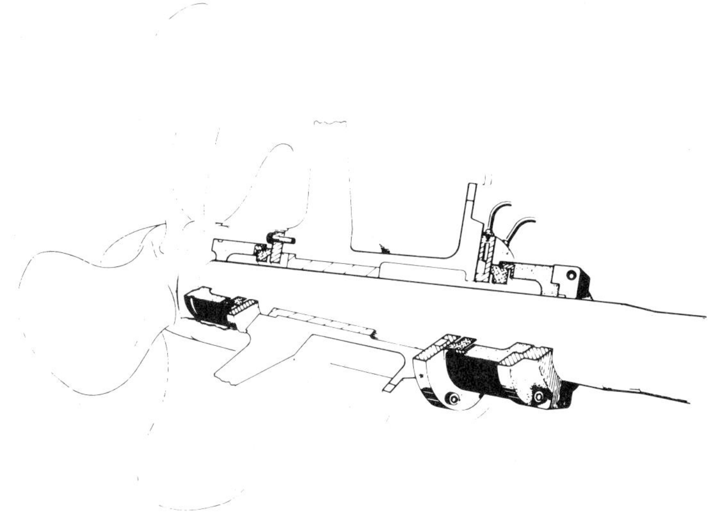

ManeBar seals.

Design and materials

The design is, as with the ManeSeal, a face type of seal, but with this design, a moulded elastomer body replaced the omega-shaped bellows. Also, unlike its predecessor, the seal's body of neoprene is attached to, and rotates with, the revolving shaft, using pre-set compression to maintain contact between the rotating seal face and a simple cast iron or *Ni-resist* seat that is mounted on to either the bulkhead or the end of the stern-tube. On oil-filled systems, an outboard seal, compressed between the propeller and the necessary stern-tube equipment, is also fitted.

As the original concept was for a simple, easy-to-use seal, the basic *EJ* and *EK* units are of a moulded elastomer construction, with an integral face insert, (Figure 6). However, the success of this design was quite phenomenal and, following requests from a number of sources, a semi-split *EL* version of the unit was introduced. This enables work on the inboard wear faces to be carried out with the shaft in position.

Figure 6

Later, an outboard *EN* version completed the range, so that now the ManeBar series can be fitted as a simple and inexpensive non-split *EJ/EK system* or, alternatively, in the semi-split form of the *EL/EN* (Figure 7). In a number of instances, the combination of EL inboard and EK outboard has proved to be a very workable solution where ease of servicing, combined with economy, has been of prime importance.

Where warships are concerned, water-lubricated bearing systems tend to be used and, under these conditions, only the inboard EJ or EL seals are necessary, and are fitted normally in conjunction with a standard inflatable seal.

Figure 7

Applications

(i) Vessels ranging from work boats to small coastal vessels.

(ii) Bow thruster or outboard propulsion units.

(iii) Water turbines (in specific cases).

(iv) Rudderstocks.

(v) Small standard propeller and stern-tube packages, supplied by equipment manufacturers.

Advantages

(i) Its simple design, means that installation by trained personnel is not essential.

(ii) Eccentricity, bearing wear and shaft whirl, are of no consequence as far as the seals are concerned, because as they are fitted to and rotate with the shaft, they automatically follow whatever the shaft does.

(iii) Once fitted, unlike the packed gland, the inboard seal needs no further adjustment, as the flexible body of the unit automatically adjusts for wear.

(iv) With the *EL,* the seal faces may be changed within one hour, without having to vulcanize or use special adhesive. This can be carried out with the vessel in the water and without disturbance to either the shaft or coupling. This can also be carried out on a water-lubricated system, where an inflatable seal has also been fitted.

(v) No shaft or liner wear in way of the seal, because the seal rotates with the shaft.

(vi) Inexpensive – comparative costs by a number of prominent yards and equipment suppliers, have shown that the use of the equipment has reduced their overall costs significantly.

(vii) Shaft salvage – in a number of instances owners have been able to salvage otherwise unrepairable shafts, because of wear in way of the original packed gland. With the consent of the classification society it can normally be arranged for the seal to fit over the worn area of the shaft or liner.

(viii) Proven success under abrasive conditions – ManeBar seals have been proved in service on dredgers where, as replacement units, they have doubled the time between seal services and have reduced the quantity of oil used to an insignificant amount.

(ix) Shock resistant to *NATO* standards.

EM ManeCraft propeller shaft seal

Type *EM ManeCraft propeller shaft* seal is a development orientated towards the smaller craft and leisure market. The seal replaces traditional orthodox packing. It is compressed against a metal seat fitted to the end of the stern-tube and rotates with the shaft, thus eliminating shaft wear. Being self-adjusting, the seal does not leak.

Efficiency is not impaired by shaft whip or engine misalignment.

Installation of this inexpensive seal is said to be easily within the capacity of the do-it-yourself boat-owner and can be as easily fitted to existing worn shafting.

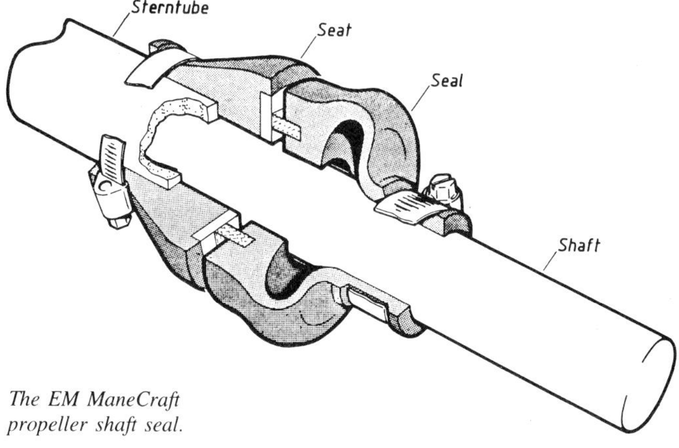

The EM ManeCraft propeller shaft seal.

A further example of a stern-tube shaft seal assembly incorporating a mechanical face seal and an inflatable seal is shown in Figure 8. One housing is provided for the face seal and one housing for the inflatable seal, with all parts of the seal assembly split to facilitate installation around the shaft. Provision is made for flushing the rotating parts of the face seal; also to ensure an adequate flow of water through the base ring to the stern-tube bearing. The seal housing contains an air-bleed valve outlet for venting the seal chamber.

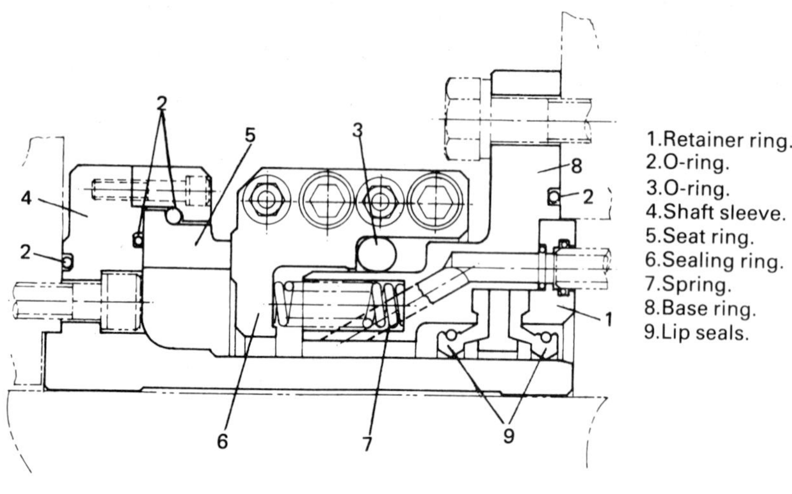

Figure 8
Mechanical stern-tube seal

The *Tyton* seal for oil lubricated stern-tubes (Figure 9) is designed to eliminate the need for a pressurized oil system and provide easy monitoring of performance and function of both the sea water side face seal and the oil side lip seal. The aft (outboard) seal comprises a face type sea water seal and a double-lip oil seal. The inboard seal comprises a single-lip (oil) seal. Major items of the seal are of split design to allow for replacement of wear parts without pulling the shaft.

Rudder stock seals

Rudder stock seals need to accommodate stock deflections and dynamic bearing loads under all operating conditions while preventing sea water leakage. Ideally, too, they should provide capability of renewing the seals without removing the rudder, crosshead, tiller or rudder stock. For smaller craft these requirements can usually be met most economically with a packed gland. Special seal designs are normally employed on larger vessels.

An example of a modern rudder stock seal is shown in Figure 10. This comprises a *lower bearing* seal to prevent ingress of sea water into the bearing housing and a *hull* seal to prevent ingress of water into the hull. The sealing rings are stabilized by a carrier ring and cover, and are assembled in a split base ring. The sealing ring areas are grease-filled. No other lubrication is required.

1. Base ring (split.
2. Cover ring (split).
3. Retainer ring (split).
4. Inflatable seal.
5. Sealing ring (split).
6. Pressure ring (split).
7. Compression spring.
8. Clamp ring (split).
9. Drive ring (split).
10. P-rings.
11. Stern-tube bearing lubricant inlet.
12. Seal flush water inlet.

Figure 9
Shipshaft seal for oil-lubricated stern-tubes

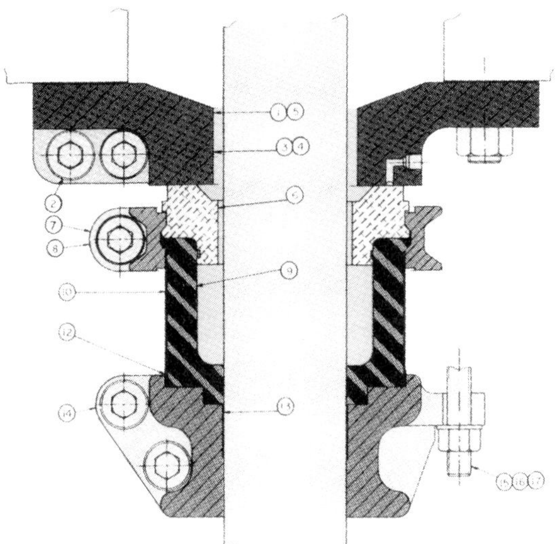

ITEM	DESCRIPTION
1	SEAT
2	BUTT BOLTS
3	ALIGNMENT PIN
4	LOCKING PIN
5	STUDDING PLUG
6	FACE
7	CLAMP RING
8	BUTT BOLT
9	BODY

ITEM	DESCRIPTION
10	WRAP
11	WRAP ADHESIVE
12	BUTT BRIDGE PIECE
13	DRIVE CLAMP RING
14	BUTT BOLT
15	STUDDING
16	NUT
17	WASHER

Figure 10
Rudder stock seal.

Upstream Pumping Seals

THIS IS a new high technology mechanical seal concept, developed as an alternative to conventional double seal arrangements. Irrespective of design, mechanical seals have been used to prevent leakage from a pressurised source to atmosphere. In most cases, single seals have been fitted, but where safety or other considerations dictate, double seals of the back-to-back or tandem configurations have been put forward.

Back-to-back seals require a sealant at greater than pump pressure between the seals, so that they are running on a fluid film derived from the sealant. Their disadvantages are that there can be an interchange of pumped liquid and the sealant; the size and cost of the sealant system needed to maintain the sealant pressure above pump pressure; the difficulty of guarding against sudden pressure surges, or transients, which can push the inboard seal seat out of its housing.

Tandem seals, although not suffering from these defects, do have the problem of single seals in that there will inevitably be leakage across the inboard seal faces. This means that the sealant surrounding the outboard seal can become contaminated with the pump product. In addition, if the pump product is a light hydrocarbon, *eg*, butane or propane, provision must be made to direct it to the flare stack in the event of seal failure, and to bring about shut-down of the whole system. Double seals of both configurations, despite their limitations, have proved their worth. It must be accepted, however, that their development was largely due to the lack of high performance materials to combat abrasion, and the inadequate lubrication they possessed.

The new unit is a cartridge design, which is stationary mounted, with a rotating seat, the face of which has a series of recessed spirals to form a groove pattern. These grooves extend only some ⅔ of the radial distance across the face, the remaining ⅓ forming a sealing dam for sealing under static conditions. This is the 'business' part of the seal, the other components being springs and O-rings. So much for the inboard seal.

The outboard seal is of conventional design. Both seals are constructed from 316 stainless steel for the metal parts, and Hastelloy for the springs. The inboard seal has a carbon face opposing a rotating seat of tungsten or silicon carbide. The outboard seal has a carbon face running against a silicon carbide seat or, where this is unsuitable, a material which is compatible with the barrier fluid, (Figure 1). In use, the space between the two seals is filled with a sealant that is compatible with the pump fluid, the sealant being supplied from a manually replenished header tank.

The effect of the spirally grooved inboard seal face is akin to that of the inboard seal

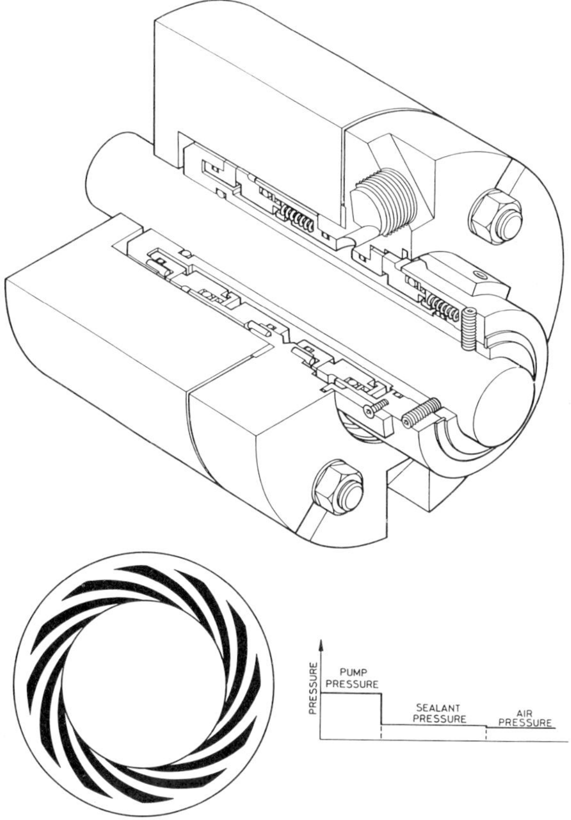

Figure 1
Outboard seal.

of a conventional back-to-back seal. In this case, however, with a liquid pressure in the bore which is less than sealed.

Concept

The seal has been developed on the principle of balancing the hydrodynamic/hydrostatic opening and closing forces. The same principle, in fact, on which the dry-running gas seal works. In this, however, the groove characteristics have been modified to suit liquids. On rotation, the grooved seat act as a pressure generating system. This causes the faces to separate, and a controlled gap of less than 0.1mm is formed. Within the gap, a calculable film stiffness is induced because the gap is automatically set by the pressure the grooves can maintain. If there is a change in the conditions, and the gap is reduced, the hydrodynamic pressure increases and the faces revert to their original position. The converse is true in the event of the face separation increasing, (Figure 2).

From this, it will be seen that the performance of the seal is dependent on pressure, speed and size. Other factors which will influence the efficiency of the seal are the temperature differential between the sealant and the product, as well as the properties

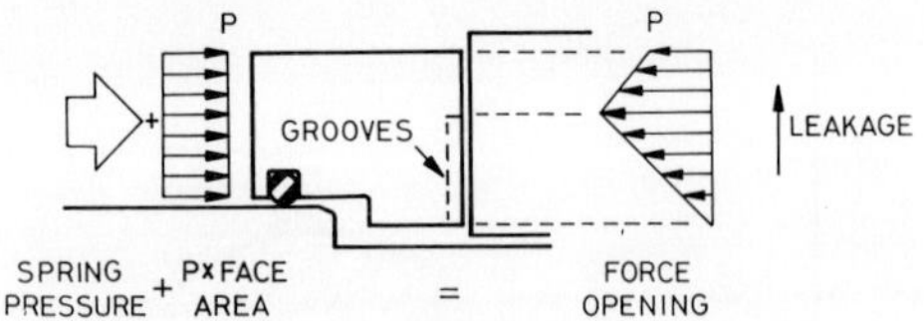

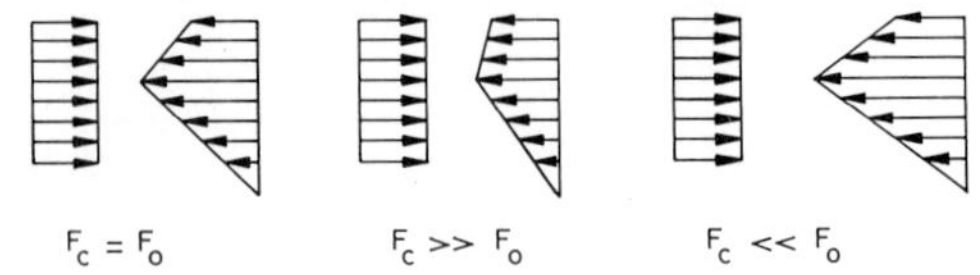

Figure 2
Diagram showing the concept of an upstream pumping seal in action.

of the sealant, *eg*, its viscosity and vapour pressure. When all the operating conditions fall within the parameters of the upstream-pumping seal, a pumping rate of from about 5 to 9 litres per day (1 to 2 gallons per day) is realisable.

The seal is not intended as a replacement for double seals, but rather to complement them. All applications for the seal must be considered by the manufacturer, but as a guide, a minimum speed of about 800 r/min and a maximum of about 3600 should be maintained. Maximum differential pressure is 3 bar for the smallest size, rising to 18 bar for the larger sizes. The barrier liquid, as well as being compatible with the product, should have a viscosity of between 0.5 and 10 cP.

Spiral Groove Seals

TRADITIONALLY, the sealing of rotating shafts on gas compressors has been achieved by the use of *carbon ring* seals, labyrinth seals and double contacting end face seals with a pressurized liquid barrier to provide lubrication and cooling.

These installations which need complicated, space-consuming seal oil systems are expensive, inefficient and require continual maintenance. Ever increasing shaft sizes, shaft speeds and pressures generate considerable friction losses in sealing oil films which represents lost energy. In addition, there is the problem of contamination of the gas by the seal oil, and of the seal oil by the gas. This requires very expensive separation systems, and on off-shore gas rigs is virtually impossible to achieve. As a result the oil has to be off-loaded and shipped to land based installations for processing. The dry running, non-contacting gas seal, however, operates by establishing a very thin gas film between the rotating faces, thus requiring no auxiliary seal oil systems, absorbing very little power and reducing wear rates to near zero values.

A dry running gas seal resembles an ordinary stationary mechanical face seal, having the same basic components: face ring, seat ring, retainer, coil springs, secondary seal, sleeve, *etc.* The main differences are: wider seal faces, specially shaped face ring and a spiral groove, Rayleigh pad or other pattern on one of the sealing faces. A typical example is shown in Figure 1.

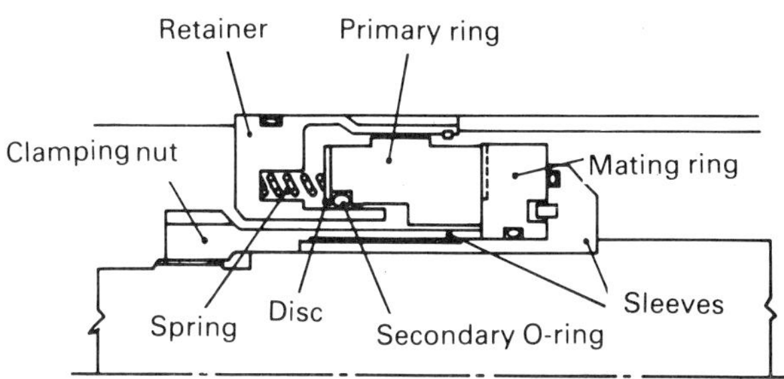

Figure 1
Basic components of a gas seal.

The rotating seat ring, usually a very hard, stiff, wear-resistant material is axially fixed, whereas the stationary face ring, made from a relatively softer, more flexible self-lubricating material like carbon, is free to move in the axial direction to establish a dynamic equilibrium sealing gap between the two rings.

Wider faces are desirable because the hydrostatic and hydrodynamic forces that keep the faces separated are much smaller with gases than with liquids.

Spiral groove seal *versus* Shrouded pad seal

Both spiral groove face pattern and shrouded pad face pattern are shown in Figure 2. Both of these designs generate pressure between the faces by means of their mutual rotation.

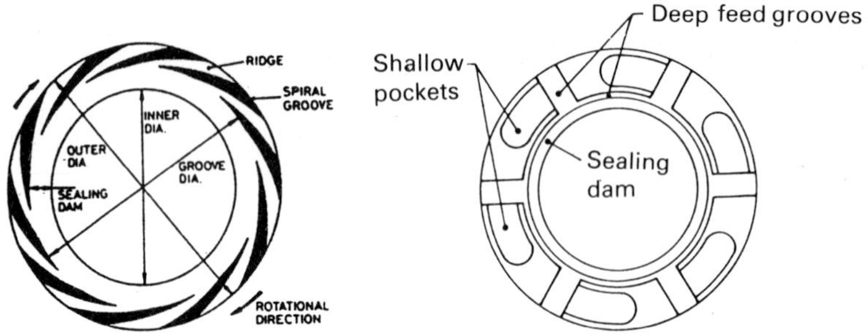

Figure 2
Features of the spiral groove and Rayleight pad sealing faces.

Pressure generating systems on both faces utilize a 'pump-restrictor' principle. The spiral groove pattern pumps the gas inward, while the dam at the i.d. restricts its flow. Rayleigh pockets scoop up the gas, raising its pressure, while the land around the pockets restricts the gas outflow. The pressure generated at certain speed, therefore, depends on the efficiency of the pumping pattern and flow restrictor.

If the pumping patterns of both designs are examined, it can be seen that, while the spiral groove pattern utilizes the entire outer portion of the face, the pad pattern needs radial grooves and an annual groove to feed the gas to the pockets and to equalize the pressure around the pads. The area of the groove does not contribute to the lift and, therefore limits the area available for pressure generation.

Flow restrictors on both faces also show significant difference between both patterns. On a spiral groove design, pressurized gas escapes inward over the dam towards the i.d. of the faces, whereas on a shrouded pad design the gas escapes in three directions over all three dams surrounding the pockets. Unless the dams are very wide, they provide much less restriction than a single width dam on the spiral groove pattern. If the dams are wide, they occupy too much area and less is available for pressure generating pads.

There is another difference between both patterns, especially important for narrow gap running. Spiral grooves as well as shrouded pads generate some areas of higher pressure and other areas with lower pressure.

Figure 3 shows that high-pressure areas on the shrouded pad design are wide apart while, on the spiral-groove design, they overlap each other. This difference becomes significant when one realizes that one of the faces is made of carbon with a low modulus of elasticity.

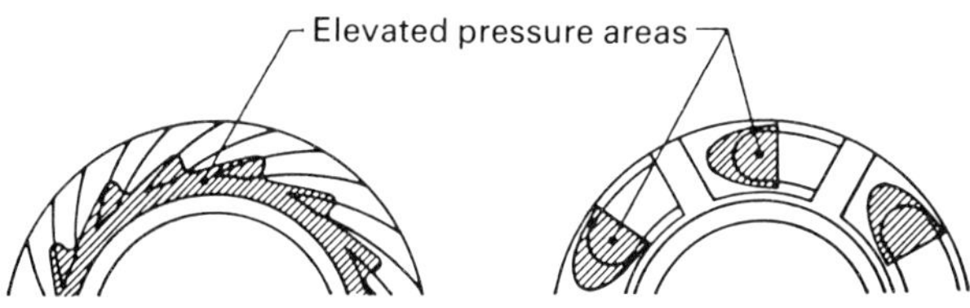

Figure 3
Areas of higher pressure on optimized spiral groove pattern and Rayleigh pattern.

With high-pressure areas wide apart, these will cause waviness on the carbon ring. As the gap decreases, pressure intensifies and waviness increases, eventually leading to face contact. With the spiral groove design, this waviness can be almost eliminated by increasing the number of grooves. The shrouded-pad design requires elongated pads and a waviness problem is hard to avoid. For the aforementioned reasons, the spiral groove pattern is especially suitable for thin film performance.

The spiral groove seal
The principle of the spiral groove face seal operation is that of a hydrostatic and hydrodynamic force balance. When pressure is applied the forces exerted on the seal are hydrostatic and are present both when the seat is stationary or rotating. Hydrodynamic forces are generated only upon rotation. The spiral groove pattern plays a vital role, since upon rotation, it provides the means of achieving a converging sealing gap. The groove pattern is a series of spirals recessed into the seat to a depth of between 0.0025 to 0.010 mm. The spiral groove pattern as shown in Figure 4, rotates in a clockwise direction. As gas enters the groove it is induced towards the centre where it is compressed and meets the resistance of the sealing dam. Pressure is increased causing the flexibly mounted face to 'lift off' thus setting the sealing gap.

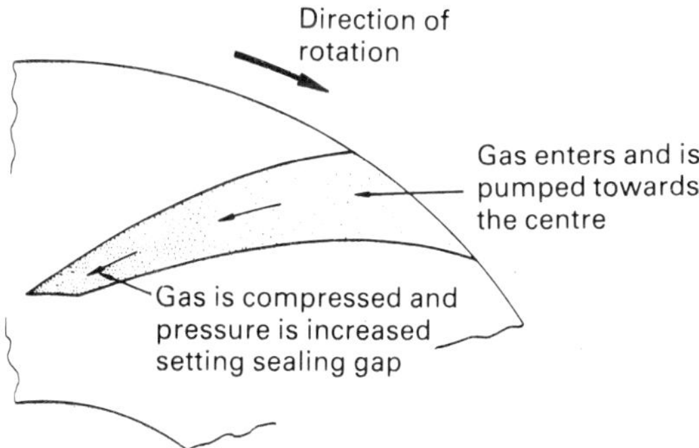

Figure 4
Principle of operation.

The forces which govern the seal operation are axial, shown graphically in Figure 5. Opening force Fo is the system pressure plus the pressure drop across the

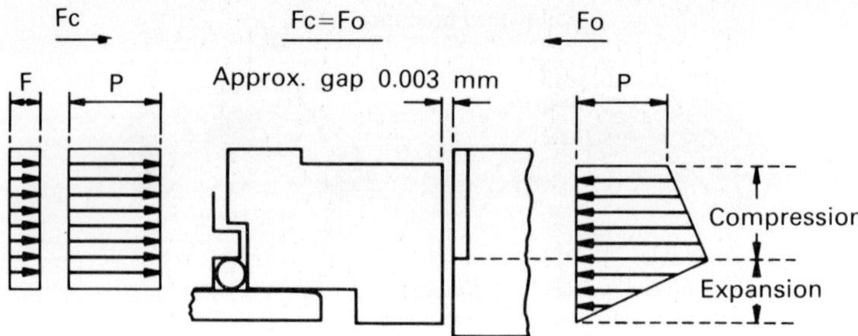

Figure 5
Pressure distribution.

face and the closing force Fc is the system pressure acting behind the face plus a small spring force. Equilibrium in operation with a designed sealing gap is achieved when Fc=Fo.

The grooves and the diameter of the sealing dam give the balance of the seal, adjustment of which determine whether the seal will tend to 'lift off' under static pressure or slow speed, a consideration necessary if the seal is to survive the crucial period of start up or shut down.

Under ideal conditions the seal will establish its operating gap creating a pressure distribution as represented by Figure 6a. If as a result of some disturbance there is a decrease in the sealing gap the forces within the film considerably increase, as indicated by Figure 6b. Similarly, if the sealing gap increases there is a reduction of the forces within the film, as indicated in Figure 6c, and in both cases the original sealing gap is quickly restored.

The force resisting this change is the film stiffness and this changes significantly as a result of a relatively small change in sealing gap. In consequence the spiral groove seal is insensitive to pressure fluctuations and other mechanical disturbances, such as axial movements, without there being direct contact between the face and seat.

Unlike the purely hydrostatic type of seal which produces the sealing gap by pressure only and is independent of speed, the spiral groove seal is both hydrostatic and hydrodynamic, thus making it speed dependent and independent of pressure.

Leakage

Leakage in any seal is undesirable but in the spiral groove seal it is a function of design, being roughly proportional to the cube of the sealing gap. The parameters that affect

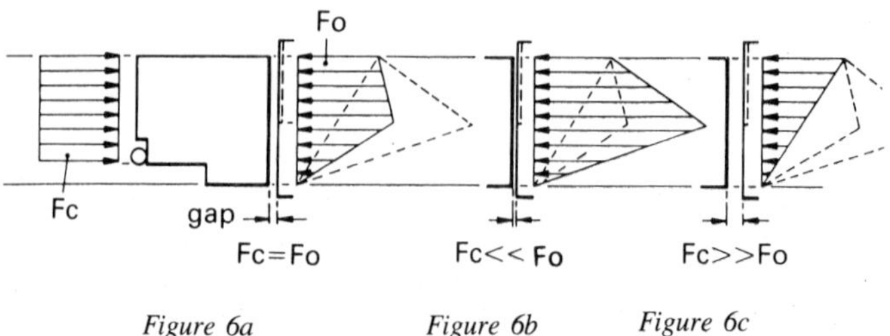

Figure 6a *Figure 6b* *Figure 6c*

Figure 7

leakage are those which affect the sealing gap. The effects of speed, pressure, viscosity and temperature are all interrelated to the seal geometry so as to provide the minimum sealing gap and thus minimum leakage. Figure 8 indicates the leakage that may be anticipated from various sized spiral groove seals at the conditions shown.

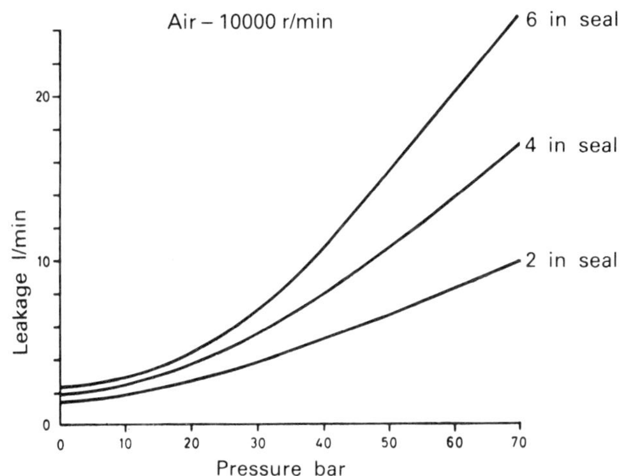

Figure 8
Theoretical leakage rates.

In some process applications, because of gas toxicity, flammability or other considerations, the leakage inherent in the operation of a single gas seal is unacceptable. In such cases inert buffer gas injection, double and tandem seal installations are used.

A single seal incorporating a labyrinth with a clean, compatible buffer gas injected, Figure 9, can be applied for hazardous fluids. The labyrinth acts as a buffer zone with

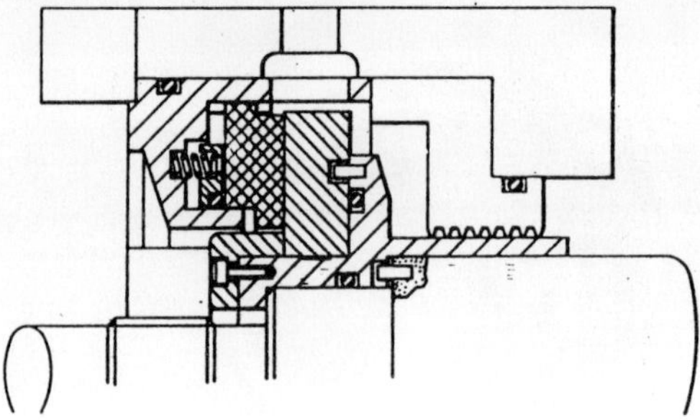

Figure 9
Single seal arrangement with labyrinth (injection).

the inert buffer gas flowing past into the process gas stream and also across the seal face. This installation can also be used for hot process gas applications where the buffer gas acts as the coolant.

For a greater degree of safety when handling hazardous fluids, the double face-to-face seal, Figure 10, is used. An advantage of this installation is that it offers a greater control of the buffer gas leakage into the process gas stream.

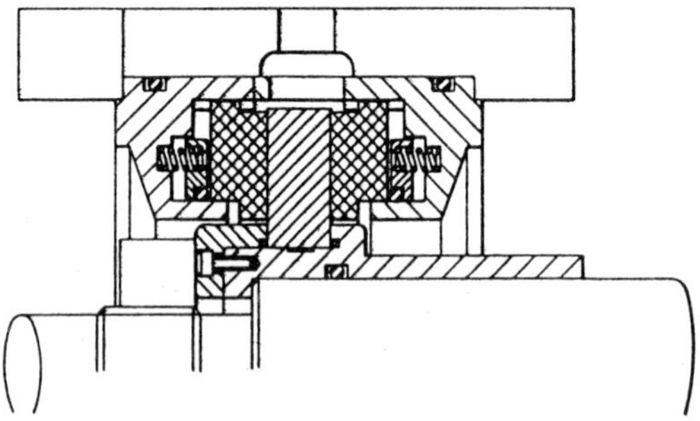

Figure 10
Double seal arrangement.

In applications where pressures are high a tandem seal, Figure 11, can be considered. The pressure differential across each seal will be approximately 50% of the product pressure, thereby enabling this installation to be used when product pressures are above that which can be accepted by a single seal. Tandem seals are particularly suitable for pipe line compressors and are usually supplied in cartridge assemblies incorporating auxiliary devices to prevent final gas emission into the atmosphere.

Safety

The dry gas seal requires no seal oil, hence there is no contamination by the process gas. Where explosive gases are present there is no need for degassing equipment, and there is no contaminated low flash point oil to be handled.

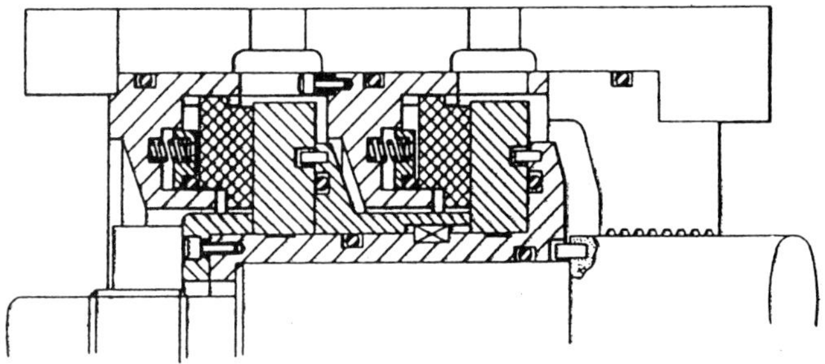

Figure 11
Tandem seal arrangement.

Maintenance
Operating experience indicates that the seal is capable of running maintenance-free for up to four years. At the end of this period, the faces may need polishing, and the O-rings replacing.

Unit design
The seal is supplied as a cartridge unit, further reducing costs, and simplifying fitting.

Gas losses when venting compressors casing
The seal is virtually gas tight when the compressor is stopped. As a result the compressor casing does not need to be vented as each shut down.

Benefits
The dry running non-contacting gas seal has many advantages over traditional gas sealing techniques. Five of the most significant are:

(i) Energy conservation – operational saving.
(ii) No wear in operation – maintenance saving.
(iii) No seal oil system required – capital saving.
(iv) No PV limit – constraint saving.
(v) No oil/gas contamination – contamination saving.

Energy saving
Consider a 5 inch double liquid cooled seal handling gas at 50 bar, 10000 r/min, the power absorbed by the seal unit will be about 20/25 kW. A gas seal installation handling the same condition would be about 1 kW, a ratio of about 20:1.

No wear in operation
Unlike the liquid cooled seal the gas seal is a controlled leakage clearance seal. The two running faces are not in contact and the heat generation which leads to wear on the gas seal is negligible as reflected in the power.

No seal oil system required
Oil lubricating systems can often be complex and expensive with the inevitable risk of failure. Liquid cooled seals are reliant upon the integrity of the seal oil system. The gas seal avoids all of these complications having the ability to operate on filtered process gas or at the most an independent buffer gas source.

No PV limit

The maximum pressure/velocity capability of a gas seal is limited by the strength of materials and its abililty to survive the centrifugal forces, whereas in the liquid cooled seals the material's ability to conduct heat is the limiting factor. In the gas seal this advantage is manifest in the low power absorbed.

No oil/gas contamination

Whenever oil is used to cool and lubricate the seal there is a risk of contamination, either by oil getting into the gas stream, giving rise to problems in the compressor and associated pipelines, or from gas contamination of the bearing oil, necessitating periodic maintenance. With the gas seal the risk of contamination is much reduced, even eliminated altogether.

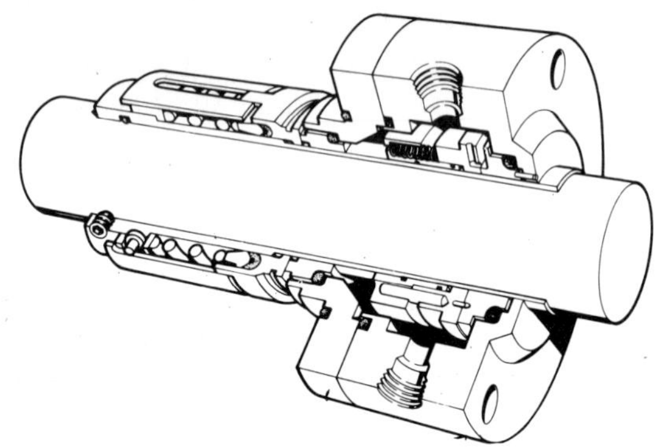

Borg-Warner type GS gas seal.

Oil Seal Types

Development of rotary shaft seals
THE FIRST Rotary Shaft Seal units were made from leather – the hat shaped lip being clamped between metal pressings to form the outer casing.

Not only were these heavy and expensive to make, but they were true 'brute force' seals relying on heavy, wide contact between lip and shaft backed up by a powerful coiled garter spring to prevent oil leakage. If it was not for the fact that leather allowed a certain amount of lubricant to find its way to the interface, the lips would have burned up at much lower speeds than they did. Even so, the peripheral speeds which could be handled were relatively low and the heavy loadings resulted in high shaft wear.

The introduction of synthetic rubber in 1938 did not improve matters much since the seal design followed the principles of leather.

Seal manufacturers began to research the function of oil seals and it was shown that a seal acted like a bearing – an oil film being generated between lip and shaft. It was demonstrated that so long as a certain oil film thickness was not exceeded the meniscus which formed at the air side of the interface would not break and leakage was avoided. Relatively light loads were required to achieve this situation and so friction was greatly reduced with lower running temperatures and much longer seal life.

For the first time, seals could be designed which would work successfully without trial and error. The heavy metal cased construction has long since been replaced by a bonded seal using a single metal pressing with an outer layer of rubber to give a better fluid-tight joint with the machine housing, Figure 1.

It will be appreciated that any damage to the shaft where the seal runs will cause leakage because the optimum oil film thickness will be exceeded locally. Even a slight scratch can cause leakage so it is of the utmost importance to provide the right finish on the shaft.

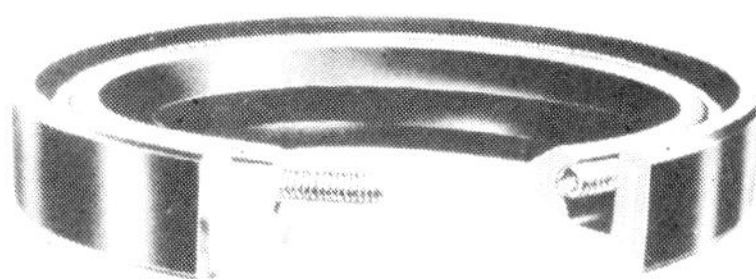

Figure 1

To maintain the oil film thickness within its required limits means that this seal lip must follow any shaft movement. This becomes difficult when the shaft is subject to eccentric running or vibration at high speeds. An application where such conditions exist is the rear end of an automotive crankshaft.

Some years ago, fluid seal manufacturers developed what was termed hydrodynamic acids or positive action features to help the seal cope with such conditions. These features consisted of raised ribs on the air side of the sealing lip which 'bedded-in' to form tiny projections to catch any oil leaking from the seal and return it to the oil film under the lip. Such features introduced great reliability to rotary shaft seals for difficult automotive applications hitherto impossible to achieve – Figure 2.

Although initially the positive action features were uni-directional, a fact not of any consequence in many applications, duo-positive features soon became available making it possible to offer a standard range of seals of this type, which could be offered for any application without knowledge of the direction of shaft rotation.

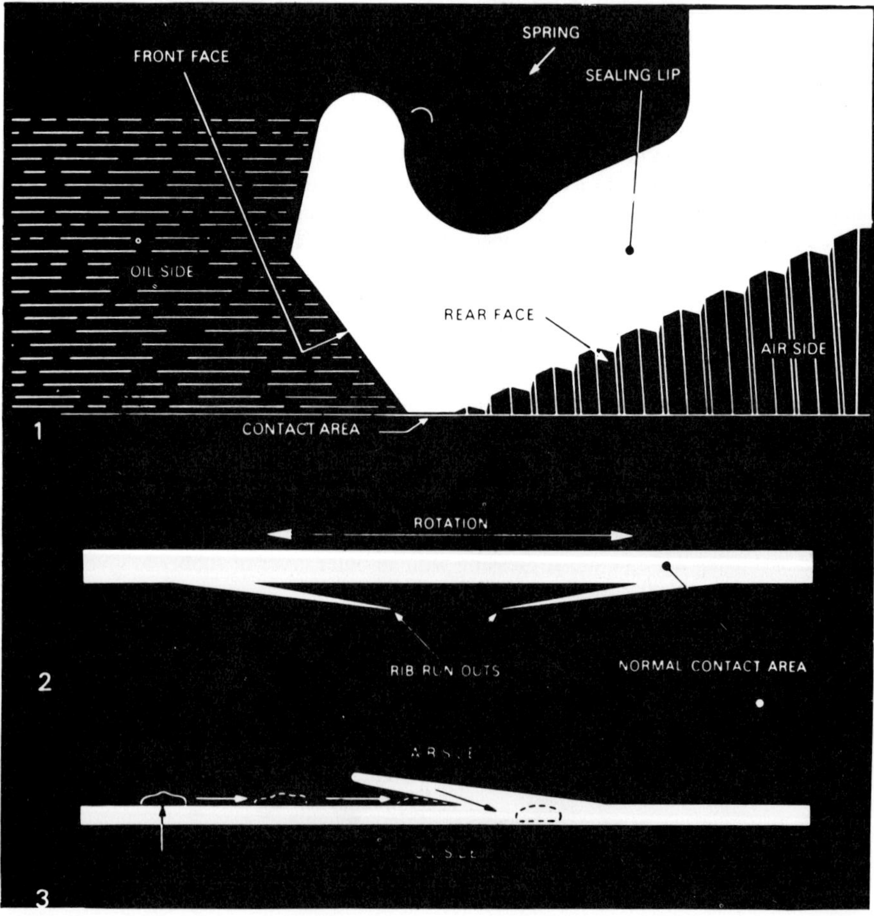

Figure 2

Seals can incorporate additional features such as auxiliary dust lips to protect the main sealing lips from damage. These dust lips can be either within the seal width or protrude beyond the seal base.

Materials from which the sealing lip is made have also developed over the years and a whole range of materials with differing properties can be specified. The chart shows the most popular grades and their uses.

Material	Oil resistance	Temperature range °C	Price indicator
Nitrile	Good	−20 to +100	1
Polyacrylic	Good	−10 to +130	1½ to 2
Silicone	Fair	−40 to +180	3
Fluorocarbon	Excellent	−30 to +200	10

It will be seen that generally the higher the temperature resistance, the higher the price. It is therefore important not to overspecify. However, it must be borne in mind that the normal effect of temperature on synthetic rubber is to cause hardening and eventual cracking. The effect is cumulative so for long life it may be advisable to specify a more expensive high temperature grade. Seal manufacturers are in the best position to advise on choice of material and should be consulted.

A more or less standard form of lip seal has been developed for *shaft oil seals,* intended for sealing low pressures of the order of 0.3 bar (5 lb/in^2). Further developed forms are available for sealing higher pressures, and some types may also be used for reciprocating seals on rods or movements having a very short stroke.

The basic form of an elastomeric oil seal is shown in Figure 4, providing a narrow contact area at the sealing point. Sealing performance depends largely on a suitable unit load being maintained at the seal-shaft interface. This may be achieved by pre-load when

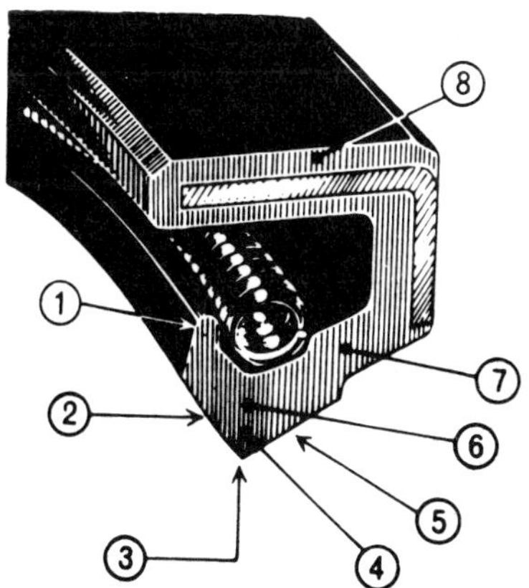

1. Deep rim affords perfect spring retention.
2. Large face angle brings sealing edge well below spring.
3. Knife-edge contact at sealing point.
4. Sealing point stiffened against local deformation by large included angle.
5. Shaft angle gives adequate clearance.
6. This section stiffened to prevent deformation under load.
7. Flexible web.
8. Gaco skin affords better fluid-tight fit in housing.

Figure 4
Typical form of oil seal and nomenclature.

assembled (*ie* the free inner diameter of the seal is an 'interference fit'); or more usually by a certain degree of pre-load together with mechanical pressure applied by a garter spring. The complete section is also normally associated with a metal stiffening member, either in the form of a metal shell casing by which the flexible member is encased (or to which it is bonded) or combining a metal insert moulded integral with the elastomeric ring – see also Table 1.

TABLE I – TYPICAL TYPES OF ROTARY SHAFT OIL SEALS

Type	Typical Configuration	Remarks
Lip seal without metal reinforcement		Energized by garter spring. Mostly used for larger shafts (*ie* 75 mm (3 in) diameter upwards)
Lip seal with inset metal reinforcement		More or less standard form for general application
Lip seal with metal case		This and similar encased configuration used where metal-to-metal housing to seal contact is preferred.
Heavy duty lip seal		This and similar configurations for arduous operating conditions. Some designs may incorporate additional dust sealing lip.
Hydrodynamic lip seal		Various designs with positive or 'windback' sealing action. Superior to plain lip seals for high speeds and accommodating vibrations or eccentricities
Oil seal Dust seal		Positive action seal with auxiliary lip to work as a dust seal

Requirements of the section shape are fairly well standardized, resulting in a general similarity of form to single lip oil seals. The actual contact point or lip approximates to, or may actually be, a knife-edge. A generous face angle is normally employed on the open side of the section to bring the sealing edge well below the spring, with a somewhat smaller angle on the other side of the sealing edge providing adequate clearance. A generous included angle between these two faces is desirable to provide stiffening of the section against local deformation under load. The forward bottom section is then generally relieved to impart flexibility.

Sealing is provided by the surface tension of the hydrodynamic oil film between the seal lip and shaft (Figure 5). Optimum oil film thickness is of the order of 0.25 μm (10 μin). Any greater film thickness tends to promote leakage; any less film thickness increases friction and wear. The continued presence of a consistent oil film is thus most important in providing lubrication for the seal. It is equally important that to maintain its effectiveness as a seal its consistent thickness must be maintained if the meniscus retaining it is not to be broken.

In practice, this means that the surface finish of the shaft at the nominal contact strip must be of a high order. Scratches or other surface imperfections can result in local thickening and breakdown of the oil film, and consequent leakage from the seal. Surface

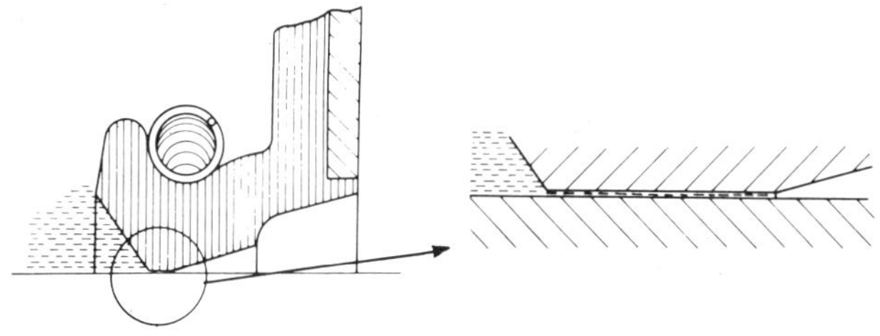

Figure 5

roughness can result in peaks extending through the surface of the extremely thin oil film to make dry contact with the seal lip itself, and consequently produce high wear. On the other hand *too* smooth a shaft finish may inhibit wetting and retention of the oil film on the shaft. General recommendation is that the shaft surface finish should be of the order of 0.25 to 0.50 μm (10 to 20 μin), with a surface pattern that promotes retention of oil. (See also later under the heading *Wear*).

Similar effects can result from the presence of grit, dirt or other abrasive material present in the bulk of the oil (or entering from the outside under certain circumstances). Basically, therefore, the ideal performance of a simple lip seal, *ie* negligible wear and positive sealing, relies on the oil being clean. Depending on the circumstances, this may be achieved by filtering, or frequent oil changes.

The same sort of trouble can arise with clean oil, the particles being produced by deterioration of the oil itself. This can take place if the seal friction is excessively high, causing over-heating of the oil film and its breakdown with the formation of carbonaceous particles; excessive heating of the oil in the system has the same effect. In the latter case it is important to note that the size of the degradation particles which can cause trouble at the oil seal are generally too small to be trapped by conventional filters with paper or felt elements.

Arduous duty seal

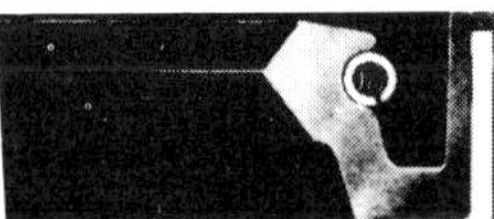

Basic standard design

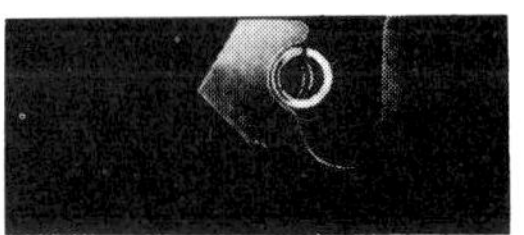

Split seal
(without reinforcement)

Positive action seal
with auxiliary lip
dust seal.

Seal with 'windback'
pumping action

Examples of variations in oil seal geometry for different services.

Almost all seals of this type depend on a suitable garter spring pressure to provide adequate radial loading on the sealing edge. The spring normally employed is a simple *closed-coil tension type,* specifically proportioned to provide the required or design load when extended to its normal working length, *ie* its actual length when the seal is assembled. Spring design and manufacture can be a critical factor because spring stiffness or rate is proportional to (wire diameter)^5 and inversely proportional to (coil diameter)^3 so that small variations in either dimension can result in considerable differences in spring rate. Also, of course, *garter springs* with different characteristics offer a method of varying the lip opening prerssure and the radial load.

The normal method of installation of oil seals is with the open or exposed spring side facing the volume to be sealed, although there are exceptions. When initially assembled with a diametral interference the sealing edge will be in intimate contact with the shaft and possibly slightly distorted if a true knife-edge. Initial rotation of the shaft will therefore produce a period of wear until the sealing edge beds down to a definite contact strip. Bedding down will normally be accomplished in quite a short period, during which friction falls rapidly from an initial fairly high value to a lower constant value, generally of the order of 50% of the initial friction with conventional seals, *eg* Figure 6.

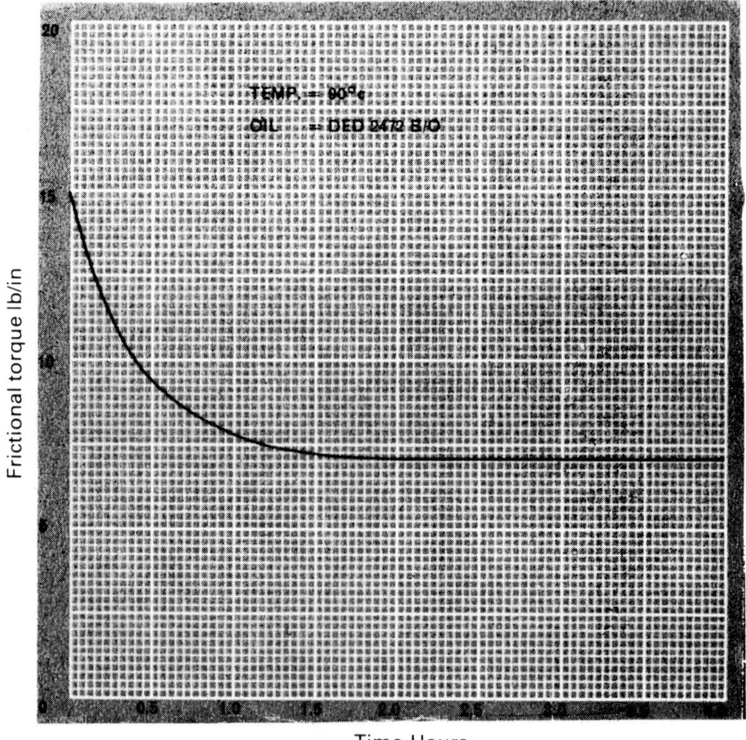

Figure 6
Frictional torque of typical standard oil seals for 75 mm dia shaft at 1000 r/min in material grade SE70.

Friction and speed

Friction is also obviously dependent on rubbing speed, although the increase in friction with increasing speed is normally more marked with large diameter seals than smaller sizes, (Figure 7).

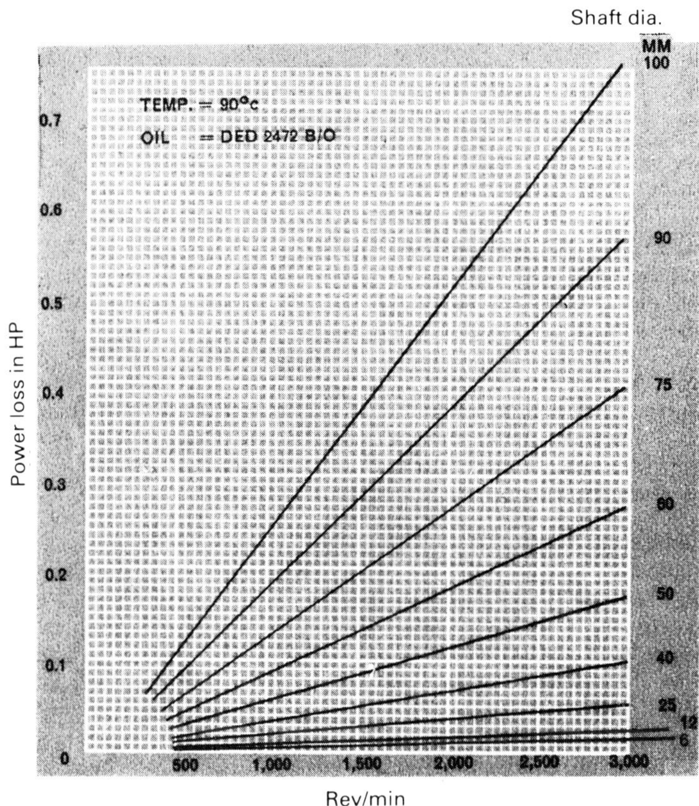

Figure 7
Effect of shaft r/min on frictional power loss of standard MI seals in grade SE70.

Limits are normally given for the peripheral speed of a particular type or series of seal and any rubbing speed up to these limits is acceptable. Typically, this is of the order of 18 m/s (3500 ft/min). In this respect it should be appreciated that with smaller sizes of shafts the actual r/min should be considered as well as peripheral speed and it is normally undesirable to exceed the following r/min figures with standard oil seals:

Shafts under 30 mm (1¼ in) diameter – 7000 to 8000 r/min
Shaft diameter 60 to 75 mm (2½ to 3 in) – 3000 to 4000 r/min

Nominal limits for intermediate values of shaft diameter can be estimated by interpolation. For shafts above 75 mm (3 in) diameter the r/min limit can be estimated purely on peripheral speed. Figure 8 provides rapid solutions for rubbing speed for various shaft diameters and r/min.

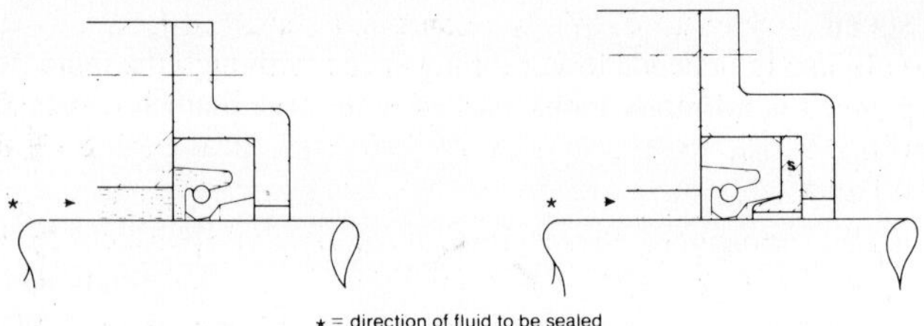

Oil seal mounting for nominal pressures (left); and with support ring for pressures about 0.2 bar (right).

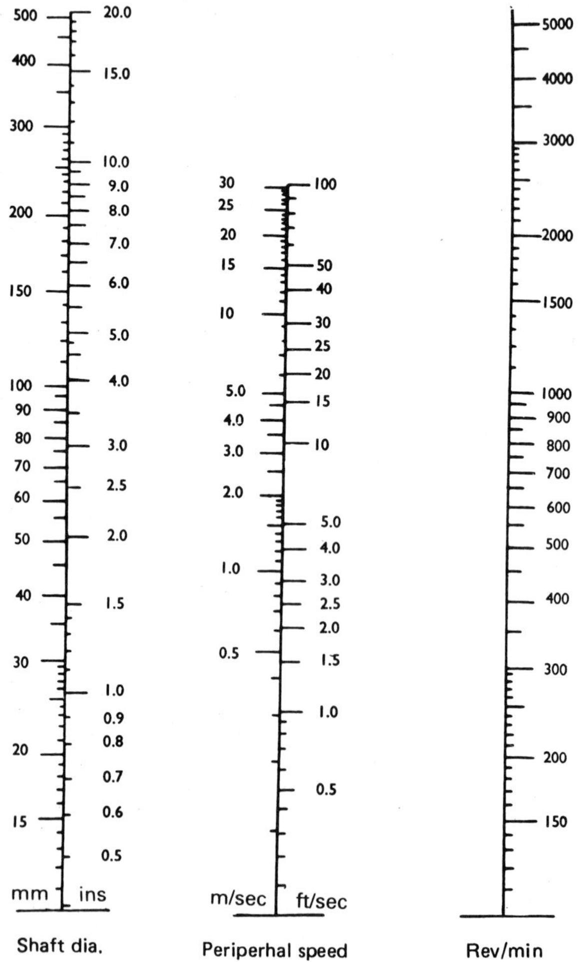

Figure 8
Peripheral speed nomogram.

For slow and medium speed applications a somewhat modified form of seal may be preferred with a broader lip. This will have a lower maximum permissible peripheral speed. A similar lower speed limit will also normally apply to lip seals designed to resist higher internal pressures than the 0.3 to 0.75 bar (5 to 10 lb/in^2) nominal characteristic of conventional oil seals.

It should be noted that while both surface finish and peripheral speed are critical factors affecting the wear and life of a sealing ring there is no direct relationship between the two. That is to say, freedom from wear depends on a specific requirement for a smooth surface finish, regardless of peripheral speed. A reduction in peripheral speed will not compensate for surface roughness and wear will be inevitable, with an elastomeric ring, if the surface finish is rougher than 0.5 μm (20 μin) at all speeds. The only practical way of accommodating surface roughness of this order or greater is to employ a material, such as leather, which is more abrasion-resistant than the normal elastomers.

Where the seal is to be used with a reciprocating instead of a rotary motion, a modified type of seal is normally required with a thicker lip. In such cases surface speed limits will normally be specified by the seal manufacturer; 0.2 m/s (40 ft/min) is a typical figure.

Double-lip seals

Some oil seal sections are designed with double lips in various configurations. These range from the use of the second lip as a secondary dust seal to back-to-back sections or *duplex* seals to provide equal sealing efficiency in either axial direction. The advantage of a single seal in duplex form for this duty is that it can be made more compact than two single-lip seals mounted back-to-back.

Other arrangements include the use of *shrouds,* scraper rings, *resilient washers* and similar devices, combined with, or incorporated in, an otherwise conventional oil seal section. Of these the *flinger* ring is about the most effective. This is mounted on the shaft and thus rotates with the shaft. Centrifugal force then tends to throw any dirt reaching this ring outwards away from the seal itself, further protection being provided by turning the outer edge of the flinger ring backwards over the seal housing.

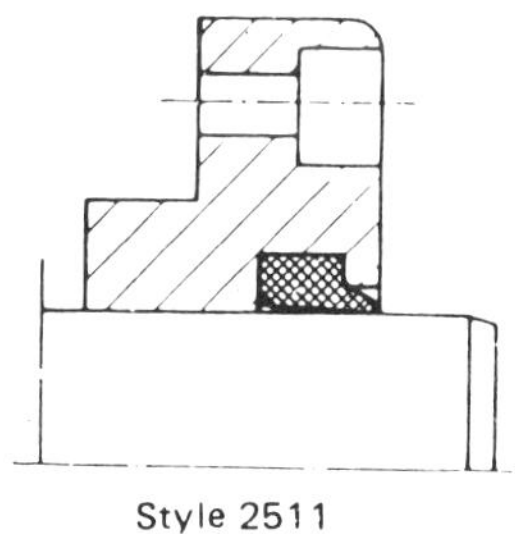

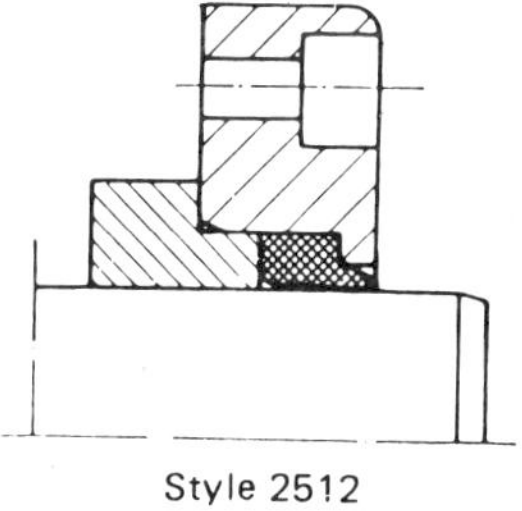

Wiper/scraper seals

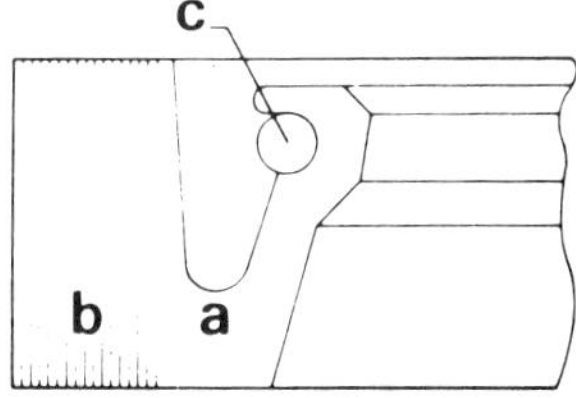

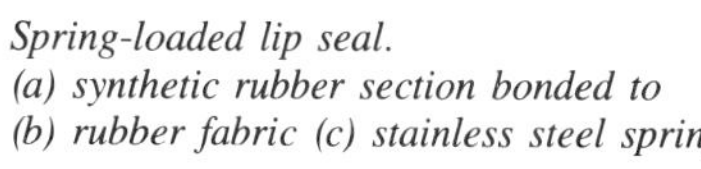

Spring-loaded lip seal.
(a) synthetic rubber section bonded to
(b) rubber fabric (c) stainless steel spring

Under certain circumstances a flinger ring may be employed to advantage mounted on the shaft adjacent to the inside of the seal when a large flow of oil is involved. This serves the purpose of deflecting most of the oil flow away from the seal lip relieving the seal of excess fluid. Lubrication of the seal will remain perfectly adequate under these conditions as this will be maintained by the hydrodynamic oil film, reinforced by oil splash or spray. Only where a seal may definitely run dry for a period of time, for instance, on starting up and before oil circulation has built up, may it be necessary to provide auxiliary lubrication *via* a *grease pocket* or *separate oil feed;* and then not always, because residual lubricant retained on stopping may be adequate to provide lubrication on starting up again until normal circulation is restored.

For operating under arduous conditions, a *hat-shaped leather* ring, radially loaded by a garter spring, may be employed instead of a conventional moulded elastomeric section. The particular advantage offered by a leather sealing element is that it will withstand the effects of surface roughness and even abrasive particles better than elastomers. Also the lip is far less likely to be damaged by splines or keyways during assembly and so can be handled more casually.

Leather sealing lips have excellent compliance and can give complete sealing with minimum wear on surface finishes as rough as 1 μm (40 to 45 μin). Main limitations are a maximum permissible peripheral speed of the order of 7.5 m/s (1500 ft/min) and a maximum working temperature of about 80°C. The friction is substantially higher than that of conventional oil seals. It can, however, be used under abrasive conditions and may also be employed as a grit seal in tandem with a conventional oil seal. In other designs intended to operate under gritty conditions a leather washer may be incorporated in the same casing or housing as a conventional elastomeric seal.

Higher pressure seals

When a conventional oil seal is subject to higher pressures the section tends to become deformed, considerably modifying its performance as a seal. The main deformation takes place at the lip, leading to much higher contact area, and consequently high friction. Its sealing performance may also be adversely affected.

A stiffened form of the same basic type of seal can be used for pressures up to about 7 bar (100 lb/in^2). Stiffening may be provided by a rigid backing washer of suitable shape (*eg* Figure 9) or incorporating a bulk of the seal section within a rigid metal casing. For higher pressure working it is usual to select an entirely different design of seal (for example, a face seal).

Where a seal is employed with higher than usual pressures a normal press fit may be inadequate to prevent the seal being blown out of its housing. Some method of securing

Figure 9
Oil seal with back-up washer.

the seal on the low pressure side must be employed, unless the seal is mounted in a housing with the blind end on the high pressure side or is supported by a shoulder on the high pressure side.

'Positive-action' seals

New designs of oil seals continue to appear in an attempt to improve performance still further. These are difficult to classify under a single heading because they are mostly individual designs, but incorporate some 'positive-action' or automatic compensation for conditions which adversely affect the performance of conventional oil seals. Many of these are based on hydrodynamic pumping effects.

It is well known, for example, that under certain conditions even the most minute scratch or damage mark on the surface of the shaft in the area of the seal track can produce dynamic leakage. A diagonal scratch or transverse helical grinding marks can develop into a pumping leakage or sealing condition depending on the direction of rotation of the shaft. Figure 10 illustrates the path of a particle of fluid under the influence of the rotational drag of the shaft when confronted with an obstacle such as a scratch. It is also known that the surface produced by plunge grinding is vulnerable to this type of dynamic leakage if considerable end play or float is present in the shaft bearings allowing the surface to oscillate back and forth in relation to the seal lip. It follows, therefore, that this hydrodynamic pumping effect should be deliberately designed and built into rotary lip seals to produce an inward pumping situation thereby extending the useful life and reliability of the seal.

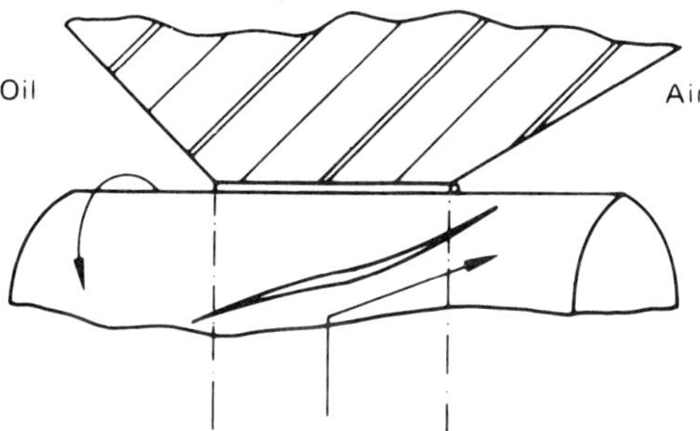

Figure 10
Direction of oil particle due to scratch on surface of shaft.

A number of designs have been produced including those with raised helical ribs or scrolls moulded on to the back face or air side of the lip so that the end of the thread or rib runs into the static sealing line, *eg* Figure 11.

Potential leakage is carried round by the rotational drag of the shaft and directed by the opposing wall of the rib back across the static sealing line to its source. These walls are for use with uni-directional shafts. Seals for use with bi-rotational shafts based on the same principle are available. These consist mainly of either a raised zig-zag rib or a series of triangular depressions on the air side of the lip (Figure 12). Any fluid finding its way past the sealing lip is directed along the opposing wall of the rib or depression by the rotational drag of the shaft back to its source.

In practical terms the inclusion of hydrodynamic pumping features provides two major benefits:

(i) Extended life and uprated performance due to improved lubrication resulting in lower running temperature and hence lower friction.

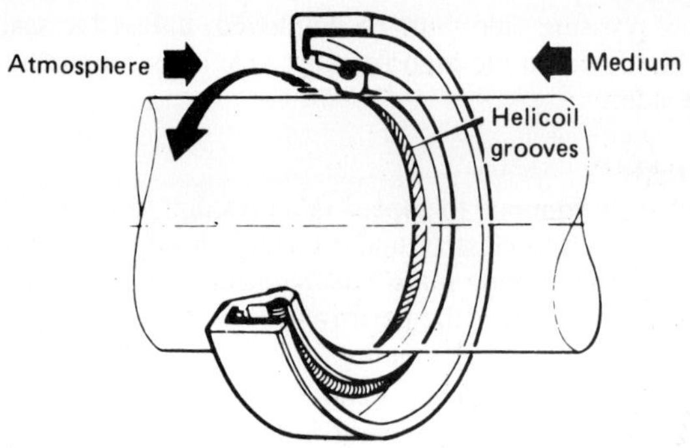

Figure 11
Helicoil oil seal.

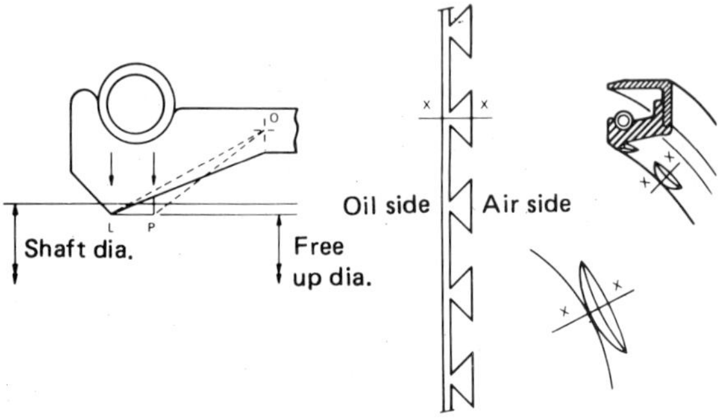

Figure 12
Hydrodynamic shaft seal.

(ii) Increased reliability, due to conformity of manufacture and controls, *ie* the completely-moulded lip eliminates trimming variations and many manufacturing quality defects normally associated with conventional rotary lip seals.

Brief details of some of these later seal types are summarized in Table 2.

Sealing against grit

The conventional oil seal has a strictly limited ability to provide protection against the ingress of dirt, grit, *etc*, and where such protection is specifically required a face seal is to be preferred. However, the simple oil seal may provide satisfactory protection under certain circumstances. If used to seal grease, for example, the grease film under the seal lip may provide an effective barrier against the entry of dirt, provided the external ambient conditions are reasonably dry. If wet, then similar protection can be provided only by employing two seals mounted back-to-back with the space between them filled with grease. Provision should be made to replace this grease regularly, preferably by fitting a grease nipple.

Back-to-back seal pairs are intended to provide a seepage of grease outwards so that recharging with grease expels any grit or other harmful particles which may have worked under the front seal with the old grease extruded. It is important, therefore, to adjust

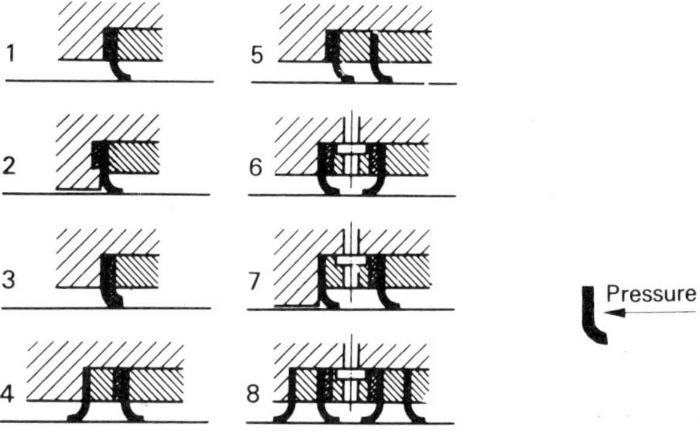

1. Pressures up to 5 bar.
2. For high pressures up to 25 bar.
3. For very low pressures, the elastomeric gasket is mounted on the pressure side and presses the seal lip like a spring against the shaft surface.
4. Double-lips for pressure and vacuum service.
5. Double-lips for abrasive service.
6. Arrangement for flushing, heating, cooling, lubrication and quenching.
7. Arrangement for pressure and vacuum and possibility for flushing, heating, cooling, lubrication and quenching.

Lip type shaft seals in improved PTFE.

the frequency of application of the grease gun to the prevailing conditions so that the grease chamber can never become heavily contaminated. The operating temperature of the seal unit is also significant, as in order to be most effective the grease must be reasonably free flowing, neither too hard nor too liquid, and the grade chosen accordingly.

Where a *single* seal is used to seal grease, sealing performance will be virtually the same whichever way round the seal is fitted. Mounting the wrong way round when oil is to be sealed will normally result in some loss of sealing efficiency and oil leakage, but again this may be an advantage under certain circumstances to wash dirt away from the outside of the seal.

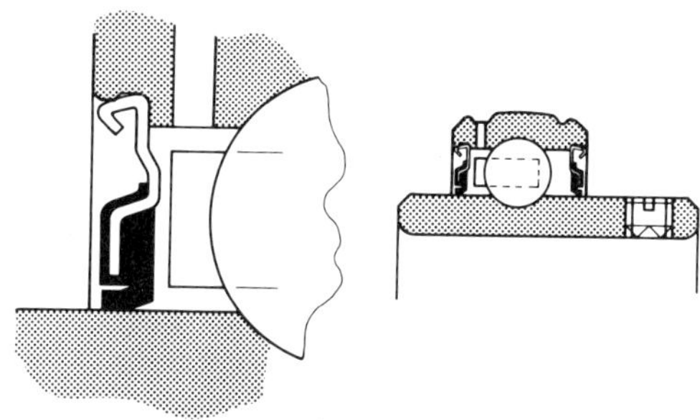

The new S-type seal for self-lube bearings makes use of the latest technology for bonding nitrile rubber to steel.

Self-lubricated bearing seal.

TABLE 2 – 'POSITIVE ACTION' LIP SEALS

Name	Configuration	Remarks	Proprietary examples		Remarks
			Name	Configuration	
Helix seal	Inclined ribs or ridges behind sealing lip to provide back-pumping action.	'Mirror image' ribs on each side of sealing lip will give bi-directional seal.	Angus Spiro-seal SG	Spiral grooves.	Particularly developed as a crankshaft main seal.
			Angus Spiro-seal DP	Inclined concentric ridges in opposite pairs.	Developed from SG seal as a bi-directional seal.
Hydro-seal	Inclined helical grooves behind sealing lip to provide back-pumping action.	Description can also apply to system where the grooves are on the shaft. Can also be a bi-directional seal.	Angus Spiro-seal TG	Inclined concentric ridges with tangentially touching grooves.	Developed from DP seal for simpler production.
Wave-seal pattern	Special lip configuration providing a sine wave contact pattern on the rotating shaft for back-pumping action.	Less subject to dust injection than the Helix-seal or Hydro-seal.	Dowty DR	Normal sealing lip with triangular or wedgeshaped 'pressure pad' pumping areas.	Helical pumping with additional bearing pad feature.

Non-metallic seals

Non-metallic oil seals normally comprise a flexible rubber section with laminated fabric back-up reinforcement, moulded as an integral section. The only metallic component involved is the garter spring. Oil seals of this type are particularly suited for heavier duties and larger sizes, but also have wide application where seal failure might have serious consequences. Fitting is usually simpler than with metal-encased seals, also there is less likelihood of seal damage through rough handling or careless assembly. A further great advantage is that seals of this type can readily be split for ease of assembly with no loss of performance, if properly housed.

PTFE oil seals

PTFE is an extremely attractive seal material because of its very low friction (and excellent chemical stability). It is particularly noteworthy for its low 'dry' friction rubbing on steel. Its main limitation as a lip seal material, however, is its poor flexibility, reducing its oil sealing performance in the presence of shaft vibrations or eccentricities.

This particular limitation can be overcome by using PTFE seals with a wide contact surface. At the same time such a surface can be formed as a positive action seal with back-pumping effect to maintain a high degree of sealing efficiency. An example of a seal of this type is shown in Figure 13. The PTFE sealing member is clamped between an inner and outer metal casing, the seal section itself incorporating a positive action feature. The same type of seal is also produced with a second (dust-excluding) lip.

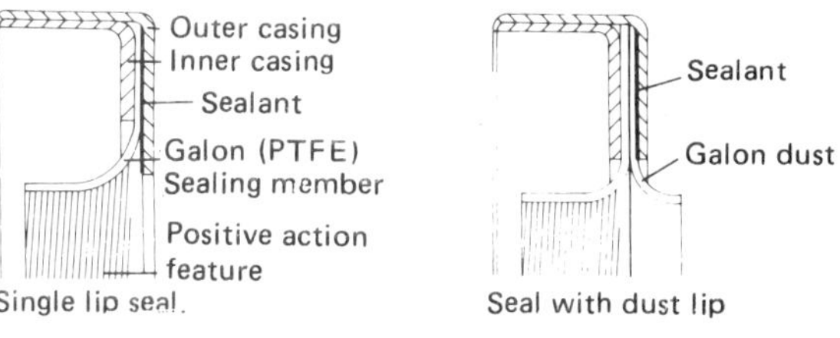

Figure 13
special duty rotary shaft seal.

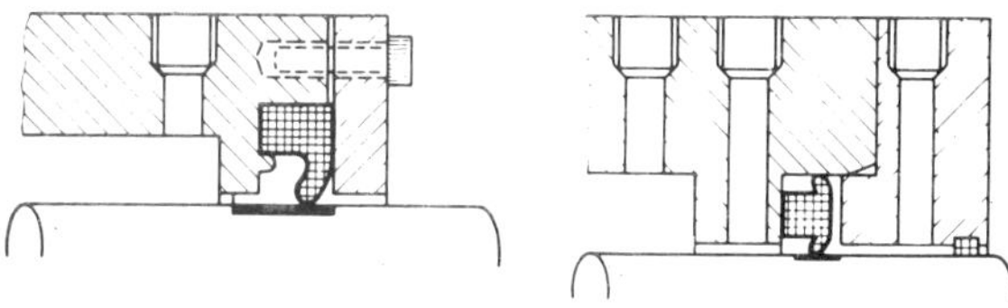

PTFE lip seals for intermediate gland sealing.

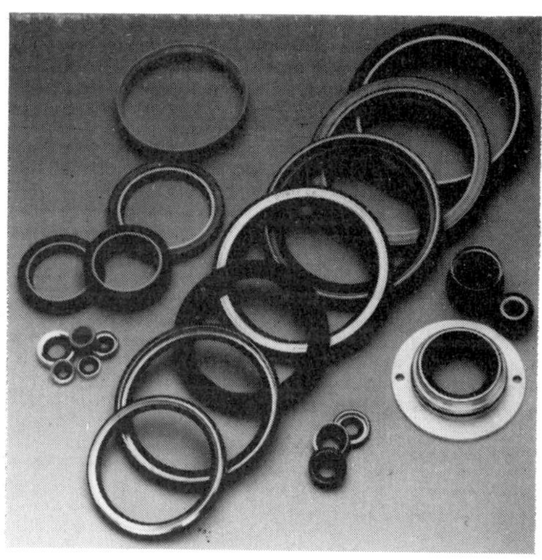

Automotive PTFE shaft seals.

An example of an extreme heavy duty oil seal is shown in Figure 14. This works on similar principles to a *mechanical shaft seal* and comprises two identical alloy steel sealing rings, lapped and mated and two opposed *Belleville* washers of specially compounded material, one of the washers having a (patented) retainer lip. The opposed washers act as springs to load the sealing rings and provide positive sealing at both the i.d. and o.d. At the same time they provide uniform loading around the entire circumference of the seal. Seals of this type are rated for pressures up to 3.5 bar (50 lb/in^2) at low speeds and for speeds up to 2 m/s (400 ft/min) at low pressures.

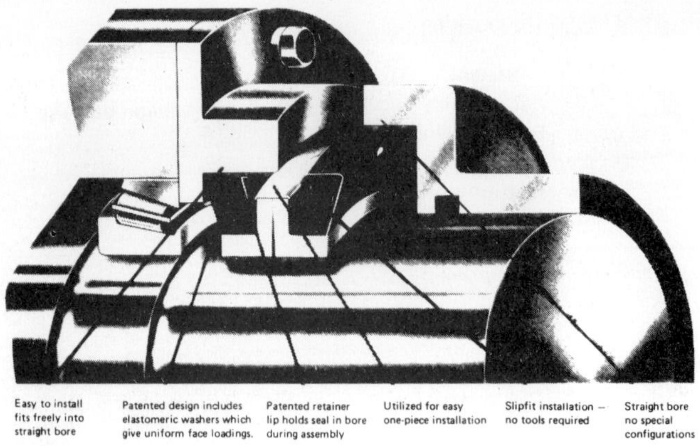

Figure 14
DF heavy duty rotary shaft seal.

Split seals

Split seals may be employed for ease of assembly where it is impractical to fit an integral ring in position. While being conventional in section and construction, these seals are specially manufactured as such and should be handled only as specified by the manufacturers. Thus some split seals are made with the metal casing separated into halves and can be opened with a hinge movement for assembly. Others need opening with a twisting movement, displacing the two cut faces in an *axial* direction only to obtain clearance for fitting over the shaft. The garter spring is then added after the seal has been positioned. This may have a hook and eye joint for assembly as a circular spring, or require twisting and the two spring ends screwing together.

Properly treated and correctly installed a split seal should have the same performance as an integral ring, provided the butt joint in the seal does not correspond with any split or joint in the retainer plate. Normally the seal should be positioned in its housing with the split on the highest point of the shaft, where applicable. The seal should then be retained by a suitable *cover plate* or *retainer plate* which supports the ring and ensures that the cut edges are correctly lined up with each other (Figure 15). Cover plates must be sufficiently rigid to the housing bore to resist dishing or distortion when tightened in place, and fixing screws or bolts should be as near to the housing bore as is practical.

Wear

Provided an adequate lubricating film is maintained under the lip of the seal and the surface finish on which it rubs is of the order of 0.37 μm (15 μin), wear should be negligible. Surface pattern can play a significant part in oil retention, and shaft material and shaft hardness are other significant parameters. Hardened and stainless steels are the most favourable shaft materials, or chrome-plated shafts in the presence of corrosive fluids. Softer metals such as aluminium, brass, bronze, *etc*, may promote higher rates of seal wear.

A ground finish is preferred, plunge grinding being better than traverse grinding, although not essential. The finished surface should be completely free from longitudinal marks or scratches, and free from helical scores or turning marks. Adequate care should also be taken to protect the finished shaft from damage during handling prior to assembly (*eg* by fitting it with a plastic sleeve for storage and transport).

Figure 15
Assembly sequence of a split oil seal with cover plate.

Hardened shafts are also preferred, even if not necessary from a strength point of view. Broadly speaking, unhardened steel shafts of Rockwell C13 to C24 hardness should be quite suitable for normal operating conditions, provided that there is no danger of the shaft being damaged or scored by handling or assembly. Recent research shows that Ford Motor Company in West Germany has found that there is less wear on the seal if the shaft is not hardened. As a protection against accidental surface damage, or for use in conditions of poor lubrication or where abrasive matter is likely to be present, a shaft with a minimum hardness of Rockwell C40 is to be preferred to minimize shaft wear. An alternative solution sometimes adopted is to chrome-plate the shaft, but for satisfactory performance this must be sound, hard-plating as poor quality plating can peel and rapidly destroy the seal lip. For shafts or shaft liners in other materials than steel it is generally advisable to consult the seal manufacturer as regards suitability and choice of seal material.

The type of elastomer used for the seal ring will also have some effect. Certain materials are more resilient, and thus more compliant than others, which means that they will have a greater tolerance towards following irregularities in the shaft contour. This may make certain elastomers more suitable for sealing on shafts which are slightly eccentric or out-of-round. Elastomers with high compliance will also need less radial load to seal on any shaft and in consequence should have less friction and generate less frictional heat.

Tolerances

Seal size is based on a specific shaft size (imperial or metric) and will have a design interference and garter spring pressure consistent with being assembled on that size of shaft.

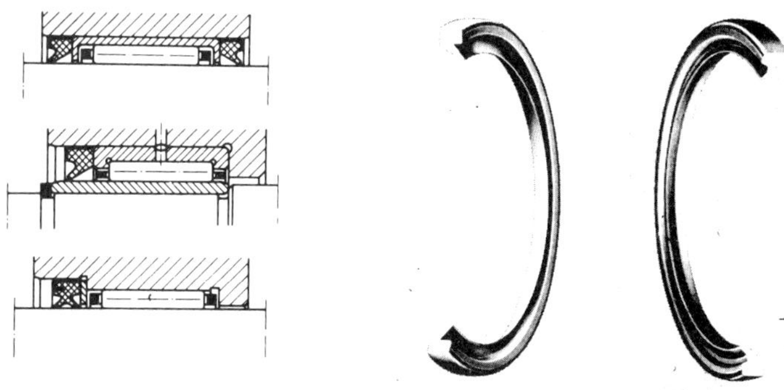

Sealing rings for needle bearings.

Nominal tolerances on shaft diameter relative to quoted seal sizes for diameter are normally of the following order:

Shafts up to 1.999 in diameter – ±0.002 in.
Shafts 2 to 3.999 in diameter – ±0.004 in.
Shafts 4 to 7.999 in diameter – ±0.006 in
Shafts 8 to 15.999 in diameter – ±0.008 in.
Shafts 16 in diameter and greater – ±0.010 in.

More specific tolerance recommendations for metric size shafts are given in Table 3.

TABLE 3 – RECOMMENDED SHAFT FINISHES FOR ELASTOMERIC OIL SEALS
(All dimensions in mm)

Shaft diameter	Tolerance h11	Shaft diameter	Tolerance h11
6 to 10	+0 –0.09	80 to 120	+0 –0.22
10 to 18	+0 –0.11	120 to 180	+0 –0.25
18 to 30	+0 –0.13	180 to 250	+0 –0.29
30 to 50	+0 –0.16	250 to 315	+0 –0.32
50 to 80	+0 –0.19	315 to 400	+0 –0.36

Housings

Each particular shaft size seal will have a corresponding bore and width dimension, consistent with the housing geometry required for satisfactory fitting, and again embody the necessary interference to provide a fluid-tight joint.

Permissible tolerance on bore diameter for the machined housing is normally of the order of ±0.025 mm (±0.001 in) for bore diameters 100 to 177 mm (4 to 7 in) and ±0.051 mm (±0.002 in) for bore diameters over 177 mm (7 in). Note, however, that with some types of seals with rigid metal casings or inserts the housing may have to be finished undersize relative to the specified bore diameter of the seal to accommodate a desired amount of press-fit interference. It is important that the bore be finished smooth and free from longitudinal scratches which could provide a leakage path; also that a generous lead-in chamfer is incorporated on the leading edge of the housing to facilitate assembly of the seal without damage (Figure 16).

To ensure correct location of the seal a flange or shoulder should be provided against which the seal can be presssed. Alternatively, the seal may be pressed and located against an accurately machined face. Recommended tolerances for machined housings are given in Table 4.

Normally an oil seal is a stationary unit, fitted into a housing and surrounding a rotating shaft. There may be requirements, however, where the seal is fitted into a member rotating

It is essential to provide lead-in chamfers to prevent damage to seal during fitting.

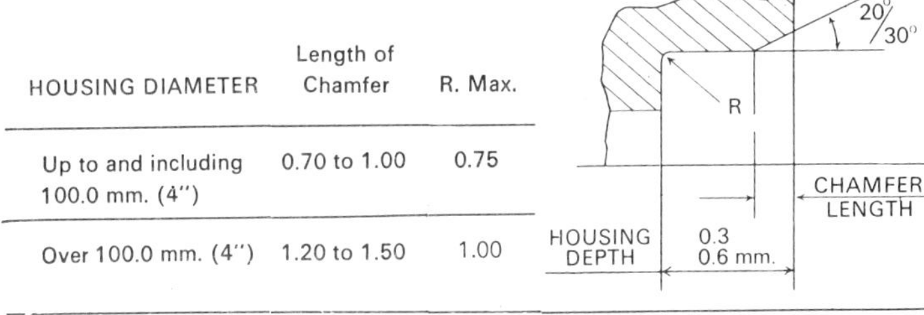

HOUSING DIAMETER	Length of Chamfer	R. Max.
Up to and including 100.0 mm. (4")	0.70 to 1.00	0.75
Over 100.0 mm. (4")	1.20 to 1.50	1.00

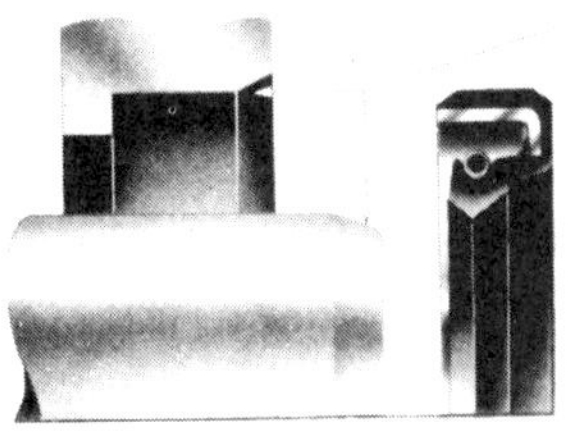

Figure 16
Recommended lead-in chamfers on housings.

TABLE 4 – RECOMMENDED TOLERANCES FOR MACHINED HOUSINGS

mm – H8				in	
Housing	**Tolerance**	**Housing**	**Tolerance**	**Housing bore diameter**	±
6 to 10	+0.022	50 to 80	+0.046	0 to 4 (inclusive)	0.001
10 to 18	+0.027	80 to 120	+0.054		
18 to 30	+0.033	120 to 180	+0.063	4 to 7 (inclusive)	0.0015
30 to 50	+0.039	180 to 250	+0.072	Over 7	0.002

Housing bore should have a surface finish better than 3.2 micrometers CLA.

about a stationary shaft, in which case a standard seal can be used, and others where the seal design may be inverted to provide an external type of seal (Figure 17). Both will normally need some modification of the garter spring design.

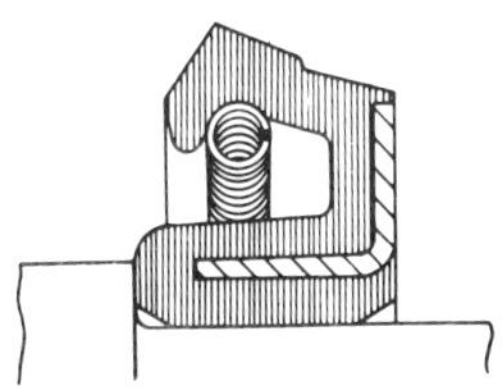

Figure 17
External oil seal.

In the case of a rotating seal the actual radial load achieved will be modified by centrifugal force and at speeds above about 1000 r/min, especially with larger seals, may need an increase in garter spring pressure to maintain adequate lip presure on the shaft. While this is an obvious solution it has the disadvantage of increasing the contact pressure at low speeds and thus the amount of frictional heat generated at such speeds. At very high speeds, this difference, and the difficulty of establishing an 'optimum' spring pressure at running speeds, may preclude the use of lip seals when the seal must rotate.

The external seal is equally critical as regards the determination of optimum garter spring pressure. It also presents a production problem in that the surface on which the lip of the seal runs must be finely finished if a high rate of seal wear is to be avoided.

To prevent or control the leakage of the quenching medium from between the outer end of the seal plate and the pump shaft, it is necessary to incorporate an additional 'P' feature in the form of a special lip seal or a labyrinth bush, or even an auxiliary packed gland. Controlling the quench medium leakage could be for safety reasons, or to aid the efficiency of the seal and to precent the leakage from entering the pump bearing housing.

Eccentricity

The performance of a simple lip seal can be adversely affected by shaft or housing eccentricity. The former produces a dynamic unbalance which can only be counteracted by the compliance of the seal, both as regards flexibility of the section and flexibility of the material. In practice this can be expressed in terms of maximum permissible eccentricity for a given design of seal and seal material, within which limit the compliance of the seal should be satisfactory. This will, of course, also depend on shaft speed, as the higher the rate of cyclic displacement the lower the eccentricity which can be tolerated by the seal. Note, however, that the performance of a particular section may be modified by (i) increasing the garter spring pressure and (ii) the use of a more flexible elastomer to accommodate greater eccentricity; or in extreme cases it may be necessary to redesign the section to provide greater flexibility.

Eccentricity of the housing is far less significant because this merely produces a static distortion of the seal in a radial direction which, if not severe, can be accommodated by the normal interference. If necessary, extra lip interference can be provided on the seal to meet cases of excessive housing eccentricity, or where actual distortion of the housing may occur.

While they will vary to some extent with the design of the seal and the elastomer used, Table 5 gives generally recommended maximum values of total (static and dynamic) eccentricity which can be tolerated by most conventional oil seals.

Seal selection

For the majority of applications where a conventional single lip seal will be suitable, seal selection is based on seven basic points, namely: shaft diameter, housing bore diameter and width, shaft speed, working temperature, fluid pressure differential and service conditions.

As a general rule, standard lip type oil seals will provide a satisfactory performance up to 0.7 bar (10 lb/in^2) or possibly 1 bar (15 lb/in^2) differential. To seal against higher fluid pressure a modified type of seal employing a thicker lip section, or a back-up washer will be required. Such seals can provide adequate performance up to about 7 bar (100 lb/in^2). For higher differentials a different type of seal should be used, such as a face seal.

TABLE 5 – MAXIMUM VALUES FOR ECCENTRICITY FOR ACCEPTABLE OIL SEAL PERFORMANCE*

Shaft Speed		Maximum Eccentricity	
m/sec	ft/min	mm	in
0.5	100	0.635	0.025
1.0	200	0.508	0.020
2.5	500	0.457	0.018
5.0	1000	0.457	0.018
7.5	1500	0.380	0.015
10.0	2000	0.254	0.010
12.5	2500	0.229	0.009
15.5	3000	0.203	0.008
20.5	4000	0.178	0.007

*Average makes,may vary widely with individual designs of oil seal.

Service conditions may introduce factors tending to degrade the performance of the seal and may modify the choice of type of seal required. If the service conditions are likely to be arduous, or a standard seal suffers premature failure, the specialist seal manufacturers should be consulted on the subject. Alternatively, it may be possible to determine the likely cause of unsatisfactory performance from the previous descriptions.

Choice of elastomer

Nitrile rubber is the most common elastomer used, being compatible with a wide range of lubricants. Special nitrile components are also available giving working temperatures up to 130°C.

Polyacrylic rubbers may be selected for applications where nitrile rubbers appear prone to hardening and cracking, but have distinct limitations, *eg* poor wear resistance, poor low temperature characteristics and incompatibility with water, acid and alkaline solutions and most solvents.

Polyurethane rubbers may be chosen for maximum wear resistance, but are not widely used for oil seals. The main special purpose rubbers are silicone and fluorocarbon – see Table 6.

Fitting seals

The most common cause of failure of oil seals is damage caused to the seal during handling, storage or, more usually, fitting. It cannot be emphasized too strongly that oil seals should always be handled with extreme care. They should be stored in a clean, dry atmosphere (preferably in air-tight containers) and never in random heaps where the sealing lips could get damaged.

Before fitting, individual seals should be examined thoroughly to ensure that the lip is not scratched or damaged, that the garter spring is correctly located and that the seal is clean and dust free. The lip can then be coated with clean grease immediately prior to assembly. The actual method of assembly will depend on whether the seal has to be assembled over the shaft first and then pressed into the housing, or pressed into the housing first and the shaft then entered. The former method is to be preferred, where possible.

TABLE 6 – MATERIALS FOR LIP-TYPE OIL SEALS

Material	Service Temperature Range °C	Properties	Remarks
Medium Nitrile	−40 to +100	Good sealing with low torque and friction. Resistant to mineral oils and greases.	Low cost materials with no production problems.
High Nitrile	−20 to +100	Improved resistance to oils and synthetic lubricants.	Some temperature limitations.
Special Nitrile	−40 to +125/135	Improved high temperature performance.	Most seal manufacturers have developed their own grades of nitrile rubber(s) with enhanced properties.
Chloroprene	−70 to +100		Little used.
Polyacrylate	−20 to +150	Good sealing with low torque. Good high temperature performance. Resistent to oils and greases, but not normally preferred to nitrile.	Lowest cost high temperature material, but poor low temperature performance and limited flexibility. Can also present production problems.
Fluorocarbon Elastomer	−20 to +200	Excellent resistance to most lubricants and greases. Chosen for high temperature services.	Best high temperature and oil resistance but highest cost. May require individual tooling to produce because of high shrinkage.
Polyurethane	−50 to +90	Excellent resistance to rubbing wear and abrasion, but compatibility limited with certain fluids.	Little used.
PTFE		Very low friction. Excellent resistance to all fluids. Different seal designs required.	Used in different design of seal. Highest cost.
Silicone	−70 to +200	Very wide working temperature range and excellent heat resistance. Often preferred for crankcase seals. Limited compatibility with more active lubricants.	High cost. Wear resistance can be a problem.
Leather	up to 160	Light wiping contact with maximum flexibility and robust construction. Works well on rougher shafts or in dirty ambient conditions.	

The use of a tapered fitting sleeve or bullet is recommended for fitting the seal over a shaft, and is virtually essential if the seal is to be passed over a splined shaft end or keyway (Figure 18). Progress of the seal will be assisted by a slight rotary motion.

For pressing into the housing a bell piece is used, slightly smaller in diameter than the seal diameter, with firm uniform pressure applied, preferably by an arbor press. The outside of the seal can be lightly greased to assist entry, facilitated also by the leading edge chamfer on the housing. The seal is then pressed home dead square and the process of fitting should be completed in one continuous operation so that the seal is not subject to any unusual loading or stressing. Correct alignment will normally be ensured by the self-guiding action of the bell piece, the bell centre positioning on the shaft or in the shaft hole as the seal is advanced up to the housing. The bell piece leading edge must

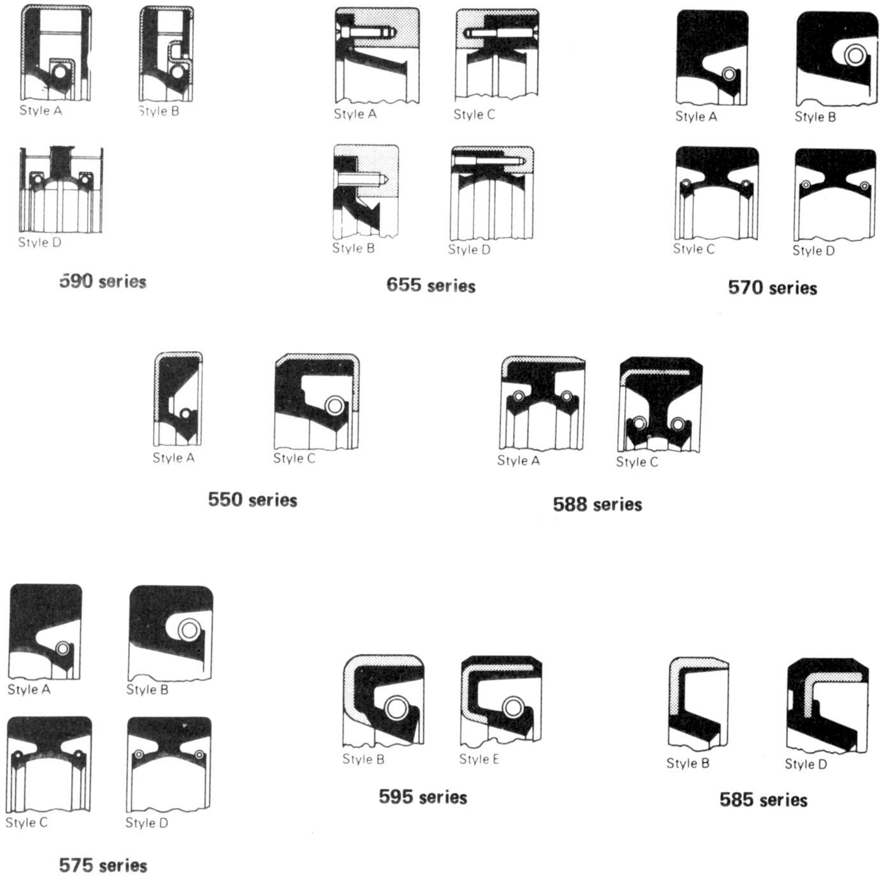

Style A
Style B
Style D
590 series
Style A
Style C
Style B
Style D
655 series
Style A
Style B
Style C
Style D
570 series
Style A
Style C
550 series
Style A
Style C
588 series
Style A
Style B
Style C
Style D
575 series
Style B
Style E
595 series
Style B
Style D
585 series

Examples of proprietary oil seals.

*
*
Housing bore
H9
R
SECTION
'B'
Housing Bore
H8
‡
Depth
'A'
SHAFT DIAMETER
h11
Depth
'A'

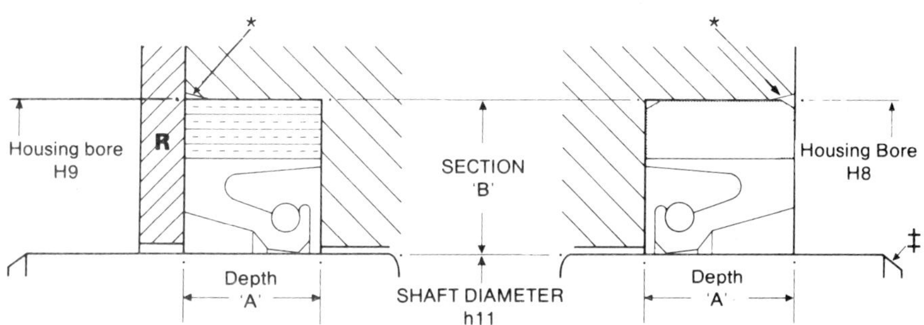

Mounting of oil seals with retaining plate (R) left; and self-retaining seals (right).

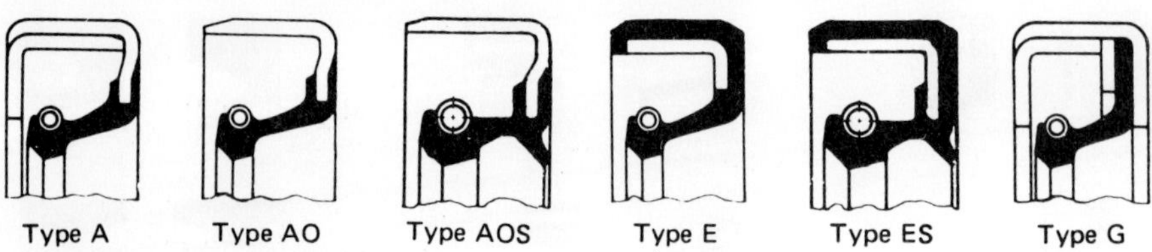

Examples of German oil seal sections.

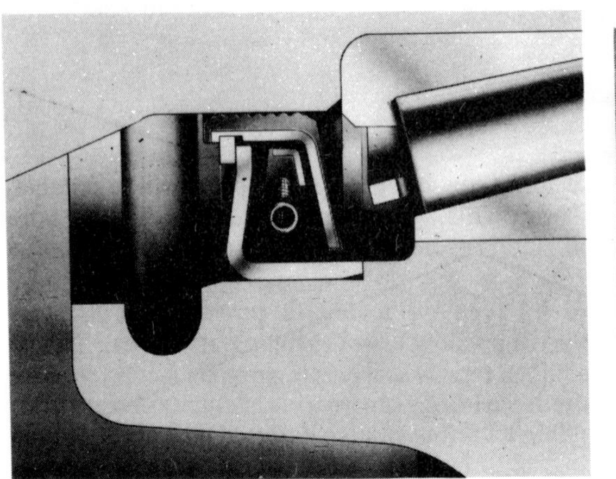

Floating elements seal.

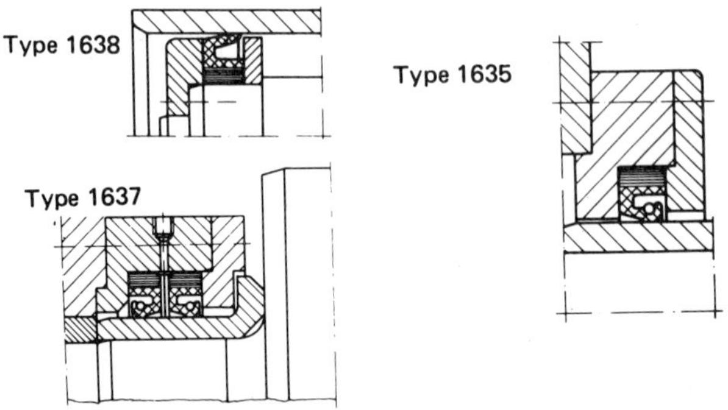

Rotary shaft seals – type 1638 is unusual in that it seals at the outer lip housing i.d.

also be of suitable shape to match the type of seal being assembled. Where the seal is being assembled back to front this bell piece may clear the garter spring and allow the seal to be assembled complete with its spring. Alternatively, it may be necessary to remove the spring for pressing the seal in place and then refit the spring when the seal is in position.

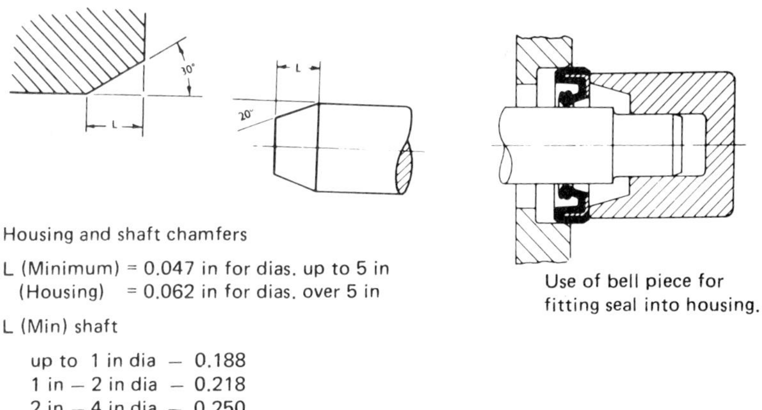

Use of bell piece for fitting seal into housing.

Figure 18

As a general rule, if a seal is removed for any reason, for example during a strip-down, it should be discarded and a new one fitted. The cost is likely to be negligible compared with the inconvenience and further stripping down necessary if the original seal is damaged during disassembly and reassembly, as is very likely.

Seal life

In the absence of conditions which could lead to high wear rates, seal life is largely dependent on the working temperature of the sealing lip. The seal material must first be chosen with regard to the working temperature of the fluid involved, when any localized increase in temperature at the seal lip will be directly caused by seal friction, governed by purely mechanical factors.

Elastomers when exposed to heat tend to harden and become brittle, the longer the exposure or the higher the temperature, the greater the ageing effects. Ultimately the material may become brittle to the point where it cracks under vibration, or has lost virtually all its compliance and no longer seals effectively. It is an interesting point that embrittlement cracking may not take place with the seal *in situ* – unless it is subject to mechanical shock, such as vibration – and a seal which has failed due to hardening may crack only when it is removed and mechanically distorted. Visual inspection of the seal *in situ*, therefore, is not a reliable guide to its condition when ageing is suspected.

Basically, for any elastomer the lower the operating temperature of the seal lip the longer its life. With low fluid temperatures, excessive seal temperature is generally due to excessive radial load, which may be due to poor design of seal, wrong garter spring, or the wrong size of seal being fitted resulting in excessive pre-load. Ideally, again, the lowest radial load consistent with satisfactory sealing performance should be employed, although some allowance is normally made in design on the side of sealing. Excessive radial load will not provide better sealing, but only excessive frictional heat and possible other troubles leading to degradation of the seal.

Temperature/life characteristics of a range of elastomers are shown in Figure 19. Characteristic life performance shows rapid ageing (short life) at higher temperatures, ultimately levelling out to prolonged life at some lower temperature. This levelling out value is representative of the maximum continuous service temperature of the material

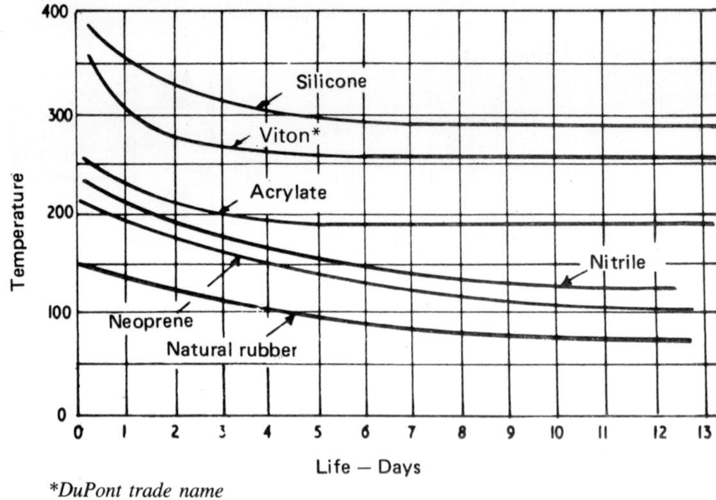

**DuPont trade name*

Figure 19

which, preferably, should be above the likely service temperature of the seal. Note that this critical temperature is the working temperature of the seal lip, which may be higher than the bulk fluid temperature.

For higher temperature working other factors are involved as well as choice of elastomer. Thus for temperatures above about 120°C stainless steel is usually the preferred material for the garter spring. At similar temperatures it may also be desirable, or necessary, to modify the metal reinforcement so that it is in the form of a casing rather than an insert – see Figure 20. This will ensure a more consistent fit on the outside diameter of the seal as assembled.

In general, for very high temperature seals silicone or fluoroelastomer is the logical choice, offering a maximum service temperature of the order of 200 to 260°C, depending on grade. Both also have excellent compliance, making them particularly suitable for high speed shaft seals as well as high temperature applications.†Fluoroelastomer material is recommended over silicone for automotive applications, particularly engine crankshaft seals as it has been found to out-perform silicone because of its improved oil and heat resistance. Failure of silicone seals due to reversion (or breakdown of the material) in automotive crankshaft applications is more prevalent today due to increased temperatures and poor oil resistance to the latest classification of engine oil (SF grade). The mode of failure for fluoroelastomer seals is lip hardening which eventually reduces shaft followability. Polyacrylic rubbers, on the other hand, have poor compliance, and thus while attractive as a high temperature material they are not widely used for shaft seals.

Footnote:

†SAE Technical Paper Series No.850332 – Performance evaluation of silicone and fluoroelastomer automotive engine crankshaft seals.

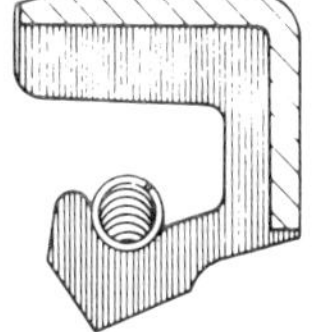

Figure 20
Reinforced oil seals.

Realistically, the performance of an oil seal can be judged by its leakage rate. Table 7 emphasizes that even small constant leakage rates can, in fact, amount to substantial oil losses over an extended period. Common causes of excessive leakage, and other oil seal faults, are summarized in Table 8.

TABLE 7 – OIL VOLUME LOST BY LEAKAGE

Leakage rate	Loss per day		Loss per week		Loss per month
	litres	gallons	litres	gallons	barrels*
1 drop every 10 sec	0.43	0.09	3.00	0.66	0.06
1 drop every 5 sec	0.86	0.19	6.00	1.32	0.125
1 drop every sec	4.75	1.05	33.25	7.32	0.7
3 drops per sec	14.30	3.15	100	22	2.09
Drips breaking into a stream	91.20	20.0	639	140	13.34

*1 barrel = 205 litres = 45 gallons

Axial shaft seals

This type of seal consists of a synthetic rubber packing ring, a metal reinforcing ring (totally enclosed within the packing ring), a sealing lip and a radial spring.

The seal is unconventional in that the sealing action takes place against the axial face of a shoulder on the shaft, or in a bearing seat. The surface against which the seal operates must be hardened and ground.

Peripheral speeds of up to 30 m/sec, with a low coefficient of friction and minimum temperature rise are possible. There are three basic types, which between them cover a wide variety of sealing applications.

Type VI has an internal lip, and is primarily for use on liquids.

Type VA is for use with grease, and has an external sealing lip.

Type DI is designed for liquids under elevated pressure conditions, and has an external sealing lip.

All the seals are produced to the approriate DIN standards, and quality assured to ISO 9000, (Figures 21 and 22).

TABLE 8 – OIL SEAL TROUBLE-SHOOTING CHECK LIST

Symptoms	Cause	Action
High leakage	Interference force too low	Check spring in correct position. Check seal size relative to shaft diameter. Check for eccentric motion to shaft. Lip may not be able to follow motion or imperfections of shaft. Increase garter spring force.
	Lip overloaded	Reduce sealed pressure. Support sealing lip with suitably shaped ring
	Seal fitted wrong way round	Install new seal in correct orientation.
	Wrong direction of rotation	Many hydrodynamic lip seals operate in one direction only.
	Rough surface	Check freedom from 'screw thread' machine marks.
	Rubber incompatible with low temperature (cryogenic)	Raise temperature. Change seal material.
	High temperature	Increase cooling. Change seal material.
	Damaged seal lips	Check installation procedure. Check seal material compatibility with sealed fluid.
	Other leakage paths	Check leakage from static side of seal. Inspect housing which may be assembled in various parts and present alternative leakage paths.
Seal damage	Excessive wear or change in seal profile.	Reduce sealed pressure. Reduce temperature. Reduce shaft speed. Increase lubrication. Reduce interference force. Employ smoother contacting surfaces. Reduce eccentricity. Reduce abrasive particles in fluid. Install suitable wiper ring if abrasive particles are present on the atmospheric side of the seal.
High friction	Stick-slip. Incomplete lubricating film	Check shaft diameter relative to seal size. Check shaft surface finish. Reduce interference force. Increase lubrication. Change seal material. Reduce temperature if high.

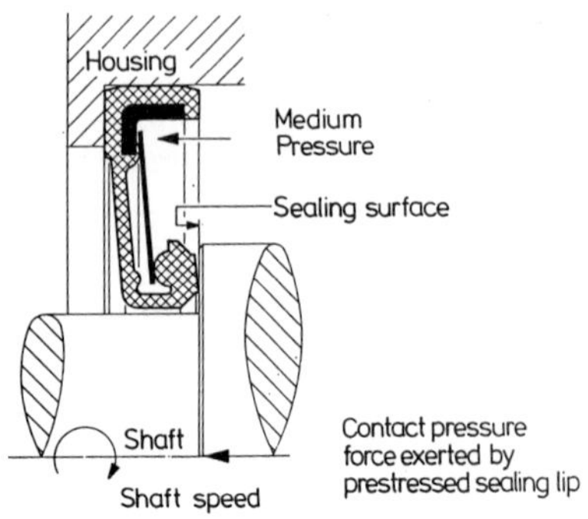

Figure 21

Sealing lip
Reinforcing ring
Axial Shaft
Packing ring
Radial spring

Figure 22

SECTION 6

Fluid Power Seals

HYDRAULIC SEALS
PNEUMATIC SEALS

Hydraulic Seals

THE HIGH efficiency of modern seals for reciprocating motions has largely resulted from the development of piston and rod seals for hydraulic and, to a lesser extent, pneumatic cylinders, capable of holding leakage to minimal amounts with high system pressures. The basic forms of cup or U-ring and V-ring seals for pistons and the U-ring for rod seals have been much improved in performance in a variety of proprietary designs.

Compact seal designs have replaced seal sets, or are available as alternatives for heavy-duty applications. Numerous designs of double-acting seals are available to replace the *back-to-back piston* seal configuration. Low friction is a characteristic of many proprietary designs of pressure-energized seals, often associated with wear rings to prolong seal life.

Industrial hydraulic systems may operate at pressures from 100 bar (1500 lb/in^2) to 140 bar (2000 lb/in^2), or even higher. Aircraft hydraulic systems may operate at 210 bar (3000 lb/in^2), or even up to 350 bar (5000 lb/in^2). Even higher pressures, *eg* up to 700 bar (10000 lb/in^2) may be utilized in heavy duty hydraulic machines (*eg* presses). The basic requirement of a *hydraulic* seal is that it must be a high pressure type. At the same time it must not have excessive friction, although there is one favourable factor here in that the fluid being sealed is normally a lubricant (*ie* hydraulic oil).

The choice of a high pressure seal for dynamic duties depends on both the size of the component and the application involved, and particularly the service conditions under which the seal will operate and continue to be operated. Thus, while a single relatively simple seal may provide adequate performance up to pressures of 350 bar (5000 lb/in^2) when fitted to a precision hydraulic cylinder, it may fail or be quite unsuitable for use at a much lower pressure on a heavy-duty industrial cylinder or ram where the working conditions may be variable and generally unfavourable. Further limits may be set by rubbing speed, surface finish (and the maintenance of adequate surface finish) operating temperature, and compatibility. Each particular case may have to be considered on its own merits.

In general, for dynamic sealing at high pressures a pressure energized seal offers maximum performance and efficiency. The higher the pressure the less flexible the seal needs to be. Equally, of course, the higher the pressure the 'stiffer' the seal section needs to be to resist excessive deformation or even extrusion.

This, in turn largely affects the materials of construction. Thus, a *homogeneous elastomeric* seal has a definite limiting pressure which it can withstand without excessive deformation or extrusion, depending on the section, clearance space and, to a lesser extent, the material hardness. Pressure rating can be improved either by reinforcement by a 'back-up' section of more rigid or fully rigid material, or by stiffening the material itself with internal reinforcement. In the former case the back-up section may be purely an anti-extrusion or supporting device, or it may form the 'working' face of the seal with the elastomeric section providing a pressure-energized 'cushion' and initial 'squeeze' pressure through an interference fit. With a 'stiffened' material, the reinforcement may be introduced into the elastomer mix and thus provide only partial reinforcement (semi-reinforcement seals). Alternatively, the base of the seal section may be reinforced material (for example fabric laminate) which is subsequently impregnated with the elastomer. The latter is the conventional material for the construction of high pressure flexible seals, although homogeneous rubber sections combined with rigid or semi-rigid back-up may also be suitable for extremely high pressure ratings.

For heavy-duty applications, especially in larger sizes of machines and where more generous tolerances may be employed, seal sets are generally to be preferred to simple ring seals. In addition to providing a more robust seal or gland, considerably greater clearances can be tolerated and the pressure rating of the gland can be increased, if necessary, by increasing the number of seal rings in the set. Further control can be provided by the use of specially-sectioned top and bottom headers, which are normally a necessary complement of seal sets. Packed glands do, however, demand considerably

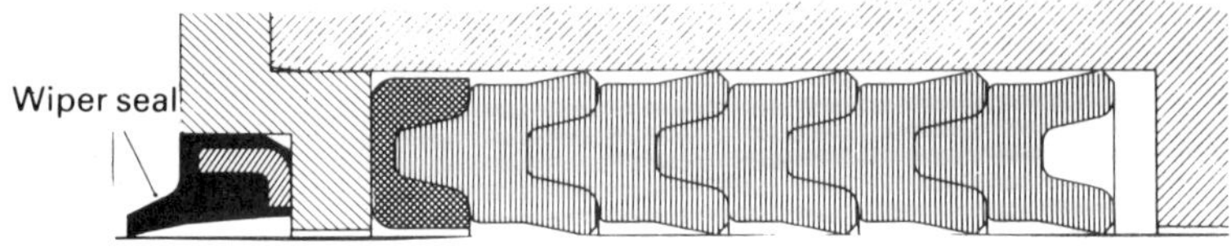

Automatic split type seal set for piston rods.

EXAMPLES OF A PROPRIETARY RANGE OF PISTON SEALS

Applications	Cross Section
Earth-moving equipment. Die-casting machinery. Injection mould machinery (up to 400 bar).	
Earth-moving equipment. Fork lift trucks. Die-casting machinery. Injection moulding machinery (up to 400 bar).	
Agricultural equipment. Ships' hydraulics (up to 250 bar).	

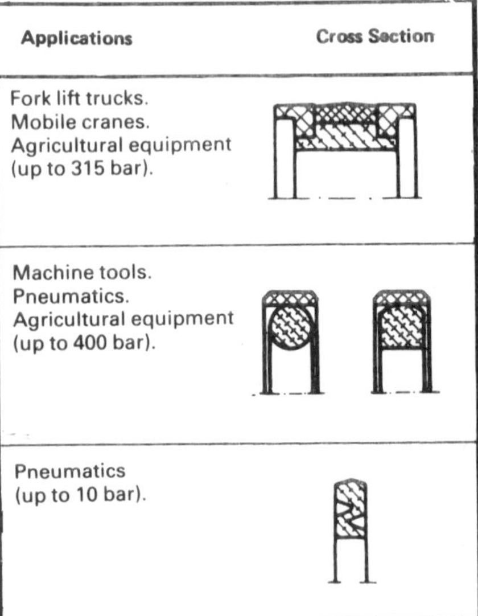

Applications	Cross Section
Fork lift trucks. Mobile cranes. Agricultural equipment (up to 315 bar).	
Machine tools. Pneumatics. Agricultural equipment (up to 400 bar).	
Pneumatics (up to 10 bar).	

more space than simple seals and the latter are normally to be preferred on smaller, compact machines and equipment. In the main the two fields of application are quite distinct, but there are instances where either type could be considered. The modern trend in such cases is to adopt the more compact solution, this having been made possible by the introduction of a large number of proprietary seal rings suitable for high pressure working.

Proprietary seal ring sets (or gland packings) have largely superseded traditional U- and V-rings for high pressure gland packings because of the superior performance offered by these special sections. All types are, however, still used although in the traditional sections the chevron or 'V' section is generally preferred to the 'U' section for higher pressures.

Friction and wear

A general effect of pressure on all high pressure dynamic seals is increased contact pressure and thus rubbing wear on the flexible lip or sealing surface, plus overall deformation of the original section and a tendency to promote extrusion into any clearance space. Friction and wear can be held to a minimum by ensuring that the contact surface is adequately lubricated. Excessive deformation which could lead to 'nibbling', cracking, or extrusion of the section can be avoided by proper fitting and support for the section – recommendations in this respect being specific to the seal section involved.

Particular attention may also have to be given to the support of rods in horizontal installations in order to avoid radial displacement which could exaggerate clearance space on one side of the seal, leading to loss of sealing or seal extrusion or nibbling. Alternatively, it may be necessary to adopt a seal type or seal set that can tolerate the degree of radial movement likely to have to be accommodated; seal ring sets of fabricated construction (or compression packings), are the most tolerant in this respect.

Surface finish

Surface finish is also an important factor, and this may in turn again dictate choice of a suitable seal material. Homogeneous rubber seals normally require a fine surface finish on which to rub (16 μin Ra or better), whereas most laminated fabric seals can tolerate surface roughness up to 32 μin Ra for extended periods. For rougher bore finishes, or where the bore surface may be subject to corrosion, leather gland packings may be selected to give a superior performance on a life basis with high pressure working. Surface finish effect on wear will, in any case, be aggravated at higher pressures and seal life under such working conditions will depend very much on the surface finish involved. Seal sets offer advantages in this respect because all the individual seal ring lips are not necessarily pressurized by the full fluid pressure, as even with vented rings there will be some pressure drop along the run of rings. With enough rings in the set, partial failure of one ring could still be accommodated by the remaining rings in the set without loss of sealing.

Piston cups

Piston cups or *cup buckets* as they are sometimes called, were originally made from leather for the hydraulic industry, and leather cups continue to be used, particularly for working on rougher bore surfaces. Greater strength and rigidity is, however, provided by laminated fabric cups. These were originally impregnated with natural rubber for water hydraulics, and later with synthetic rubber for oil hydraulics. Synthetic rubber impregnation is now more or less standard and such cups can be made suitable for water, steam, oil, and many chemical duties.

A further material combination is the use of synthetic rubber reinforced with a fibrous material rather than a fabric laminate. This results in greater flexibility and a generally smoother finish, which is particularly suitable for pneumatic duties but also employed for low pressure hydraulics. Cups of this type are generally referred to as 'semi-reinforced'. In addition homogeneous rubber cups (without any reinforcement) may also be employed for specific low pressure duties, particularly in the smaller sizes.

Cups are essentially pressure-energized seals although their relatively thin section and good flexibility enables them to be fitted with a high degree of pre-load to maintain sealing at very low pressures. They are unidirectional seals, although two cups can be mounted back-to-back, where necessary, to provide sealing in both directions without the necessity of venting the seal interspace. Satisfactory performance, however, depends on the cup being properly mounted and supported.

Mounting is normally by clamping in place between two piston plates, although on large diameter sizes this may be supplemented by through bolting or screwing. It is especially important that these plates provide the correct support and also that the clamping pressure is controlled to avoid excessive distortion of the cup base. At the same time the cup must be gripped strongly enough to resist lateral displacement which could lead to early failure of the seal. Grip in this direction can often be improved by the use of grooves (D) in the clamping plates (see Figure 1). Note also in this diagram the contouring on the supporting plate to match the rounding of the heel of the cup, but retaining an adequate clearance (C) to prevent distortion of the cup during fitting and to allow the cup to react to pressure. The other clamping plate is spigotted to control the amount of compression of the cup base.

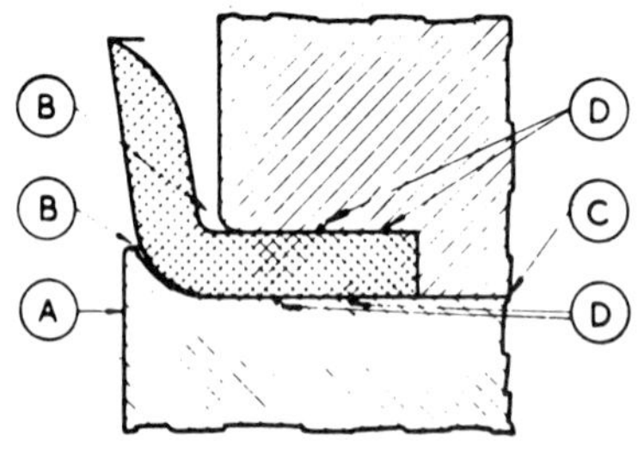

A – Clamping plate.
B – Radiused edges.
C – Controlled depth of closure to prevent distortion of cup.
D – Grooves in clamping plate.

Figure 1.

The particular advantages offered by cups are that they are relatively inexpensive compared with other types of fabricated seals, and that they can be used satisfactorily in cylinder bores which could lead to early failure with other types of seals, as, for instance, with horizontal working, or where there is an excessive clearance between piston and cylinder, or generous tolerances on tube i.d. Originally piston cups were manufactured in a wide range of section profiles to suit specific pressure ranges and working conditions, but with modern laminated fabric construction lip profiles have become more standardized.

The two main forms are the *rounded heel cup* and the *square heel* cup (or *square based* cup), see Figure 2. Both can be considered as having a similar application and performance, although the former are more common. The specific advantage offered by the square heel cup is that it can be used in conjunction with a flat supporting plate, thus simplifying machining of this component.

Examples of cup mounting for double-acting seals are shown in Figure 3. A and B

Figure 2
Modern cup sections: fabric (left) and rubber (right).

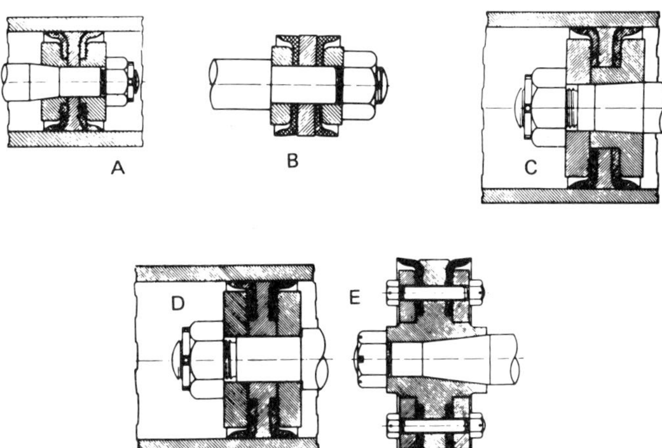

Figure 3
Examples of piston cup assemblies.

are examples for small general duty and low to moderate pressure applications, using round heel and square heel cups, respectively. C shows an alternative form of mounting for light pressures, while D and E show recommended installations for heavy-duty applications. Note that in case D joint washers stop leakage along the rod, while in E leakage along the bolts is stopped by copper washers under the head and nut, or by the use of self-locking nuts. An alternative method in this case would be to use separate clamping bolts on each side of the assembly terminating in blind holes rather than through bolts.

An important consideration in arriving at the method of mounting is the size of the centre hole. Normally a minimum size hole is specified on standard production sizes of cups which can be re-sized to a larger diameter, if required, subject to retaining a satisfactory clamping flange area. Such re-sizing should always be carried out by the cup manufacturer and cups should be ordered by diameter size and the centre hole diameter required – any suspected deficiency in clamping flange area can then be taken up at that stage. Equally, while cup specifications normally give only a minimum size of centre hole, this must not be read as a recommended size for use.

Besides adequate clamping and support, provision must also be made in the machine design to accommodate a lead-in chamfer for fitting the piston assembly. A chamfer angle of about 10° is normally recommended, with an entry diameter greater than the maximum lip diameter of the seal in the uncompressed state.

As with all types of seals it also follows that a good surface finish on the cylinder bore will result in maximum seal life, although cups are more tolerant than other types of fabricated seals in this respect. Thus, laminated fabric cups will usually be satisfactory with cylinder bore finishes up to 32 μin Ra, and leather cups will perform satisfactorily on rougher surfaces. In the case of homogeneous rubber cups or semi-reinforced cups, a minimum surface finish of 16μin Ra is called for.

Hat packings

Hat packings, also known as *collar* or *flange* packings are specifically rod seals. They can be considered as relatively low pressure seals. They would normally be chosen for small diameter rods or spindles where they can be simply supported and take up a minimum of space. They have, however, been largely replaced by modern seals for such duties and now the most common application of collars is as wiper seals.

Sections employed vary considerably, notably in the relative thickness of the base and the extent of chamfer or taper on the lip. Indeed, in some cases the lip is finished square rather than chamfered and may butt against the base of the recess into which the collar is fitted. This is to prevent elongation of the packing under the frictional drag in that direction when the seal material is quite flexible (as in the case of leather or unreinforced rubber).A bevelled lip is normally necessary for pressure-energization of the seal, with some clearance between it and the base of the recess, and essential with more rigid construction (such as laminated fabric collars). The usual forms of collar section is as in Figure 4.

U-rings

The U-ring, and its many subtle variations in section form, can be used both as a rod seal and a piston seal. While a groove-fitting seal, it can be retained by a simple assembly, *eg* Figure 5. Homogeneous rubber rings may be used for pressure up to 210 bar (3000 lb/in^2) or higher with some proprietary designs. Normal pressure limit for rubber-impregnated fabric U-rings is 350 bar (5000 lb/in^2); and for laminated rings or leather U-rings, 700 bar (10000 lb/in^2). Maximum pressure rating can, however, be influenced by the operating temperature of the system, particularly in the case of rubber rings.

A favourable feature of the U-ring piston seal is that because of the wide lip interference it will seal satisfactorily in generous tolerance cylinder tubing, thus cutting production costs.

Figure 4
Collar sections: fabric (left) and rubber (right).

Figure 5
Simple U-ring assembly on piston (left) and rod (right).

U-rings are single-acting seals and need to be mounted back-to-back in double-acting cylinders. They may also be used in sets with headers. Simple U-rings are usually supported and located by positioning rings (lantern rings), especially if they are of large diameter.

See also chapter on *Flexible Lip Seals* (Section 4).

V-rings

V-rings are always used in sets with headers. A minimum of three rings is usual, suitable for pressures up to 35 bar (500 lb/in^2). Very much higher pressure – up to 700 bar (10000 lb/in^2) can be accommodated by increasing the number of rings and/or using rubber-impregnated fabric or laminated ring construction. 'Hard' and 'soft' V-rings may also alternate in a complete set.

V-rings have less friction than the same number of U-rings, but increasing the number of rings will tend to increase both friction and wear. Careful adjustment of the gland is necessary if high friction is to be avoided.

V-ring sets are most widely used on larger sizes of cylinders designed for medium to heavy duties. They are suitable for both rod and piston seals. See also chapter on *Flexible Lip Seals* (Section 4).

Modern hydraulic seals

In addition to numerous proprietary heavy-duty seal sections and seal sets, various individual composite rings have been specifically developed for hydraulic piston seals. Many are also designed for simple assembly on a one-piece piston head, now preferred to the traditional method of assembling piston seals on a built-up piston head as being both stronger and more cost effective.

An example is the heavy-dara *multi-lip piston* seal which is double-acting and designed to spring into a one-piece piston head (Figure 6). This design is used extensively in hydraulic mining equipment under the most demanding operating conditions.

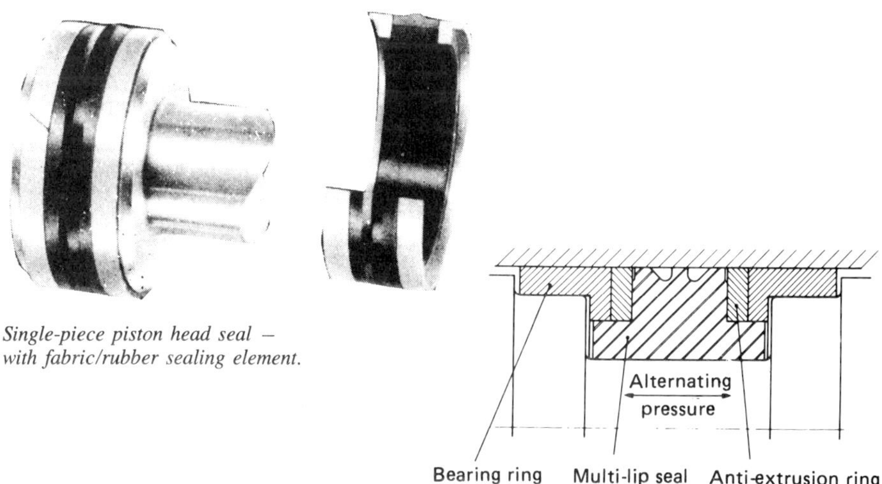

Single-piece piston head seal – with fabric/rubber sealing element.

Figure 6
Heavy-duty multi-lip piston head seal.

The complete design comprises a double-acting elastomeric multi-lip seal backed on both sides by a simple, but efficient, scarf cut anti-extrusion ring and split L-shaped bearing rings in polyacetal material. The design of the sealing tips is such that each presents a sharply defined sealing edge to the cylinder wall and provides lip flexibility to accommodate cylinder movement including tube dilation. The working pressure for this type of piston head seal is 552 bar (8000 lb/in^2) at a working temperature range of −40°C to +100°C.

Where groove space is restricted such that the space available is insufficient to use a multi-lip piston head seal then a *T-section* seal (see Figure 7) could be considered. This seal uses the same basic design as the heavy-duty type but is much smaller in section and axial length.

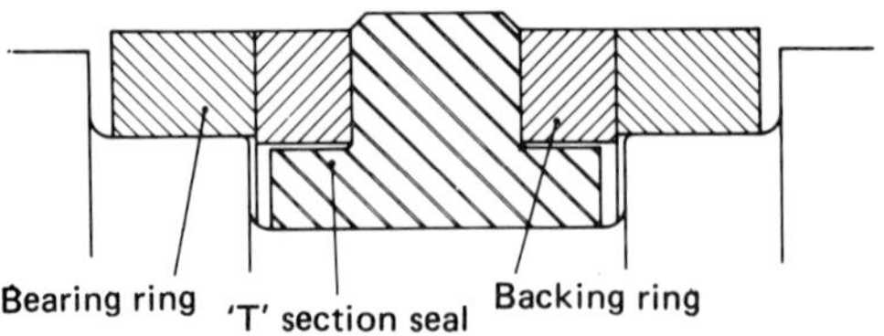

Figure 7
T-section seal for medium duty.

The complete design comprises a double-acting elastomeric T-section seal also backed on both sides by a simple, but efficient, scarf cut anti-extrusion ring, and split L-shaped or rectangular-shaped bearing rings, depending on groove width.

Chamfers are incorporated on the main elastomeric seal to give optimum sealing contact with the cylinder surface.

Similar to the multi-lip piston head seal the T-section is designed to spring into a one-piece piston head groove. The working pressure for this type of piston head seal is 345 bar (5000 lb/in^2) at a working temperature range of −40°C to +100°C.

A very compact form of high pressure hydraulic seal is shown in Figure 8. Basically a rubber-filled fabric V-ring, it is capable of sealing pressure up to 700 bar (10000 lb/in^2) and is produced in both single-acting and double-acting types. It is equally effective as a rod or piston seal, and overall performance is further improved by the addition of plastic *Delrin* wear rings.

Examples of some other simple composite rings are given in Figure 9 (but see also chapter on *Composite Seals,* Section 4).

Another type of compact piston seal is shown in Figure 10, specifically designed for

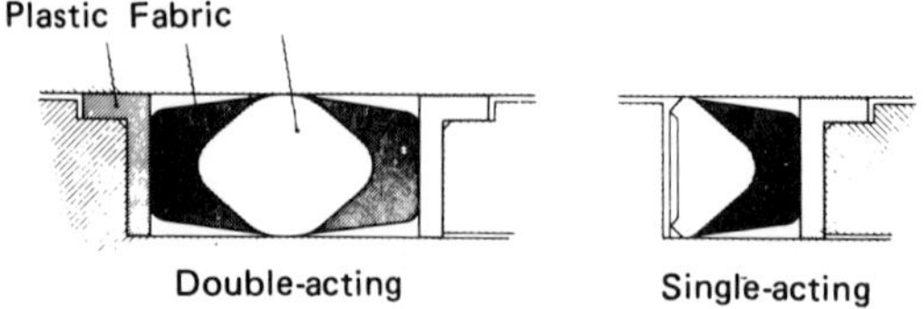

Figure 8
Compact high pressure seal used with wearing rings.

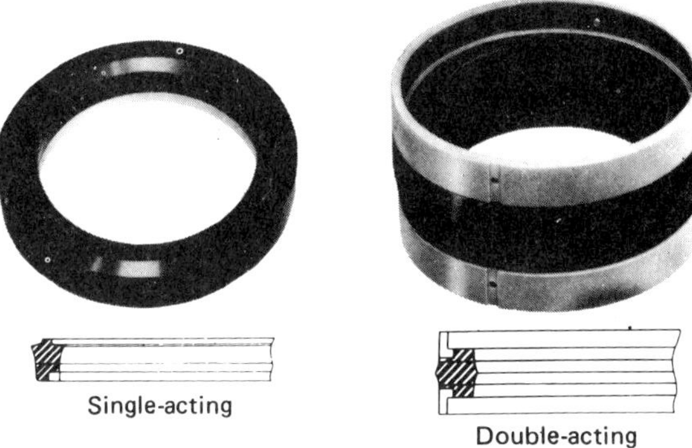

Improved Balsele

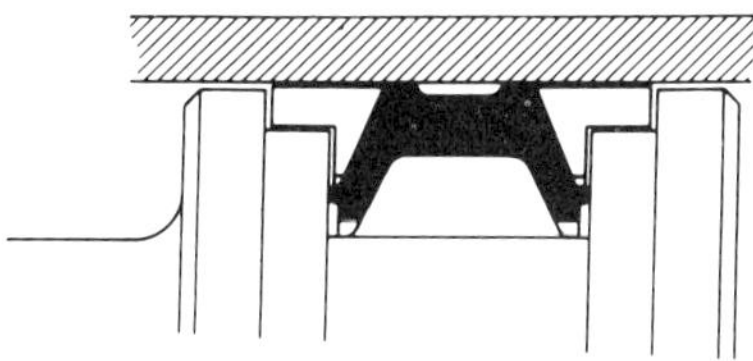

Double-acting seal for one-piece pistons.

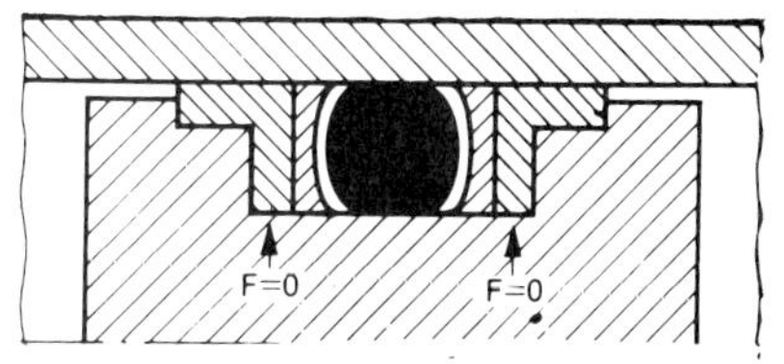

Double-acting piston seal.

Figure 9

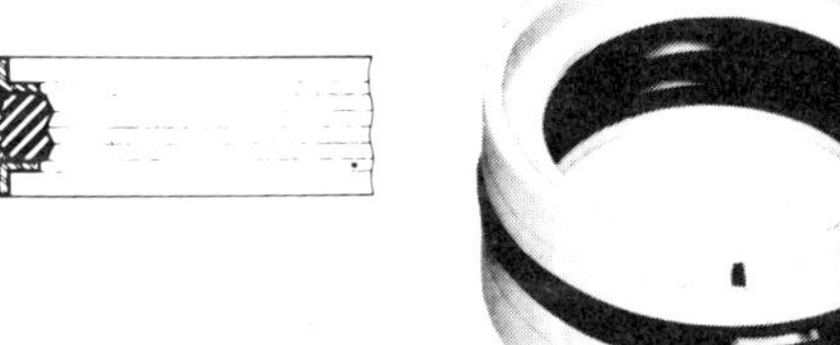

One piece piston seal.

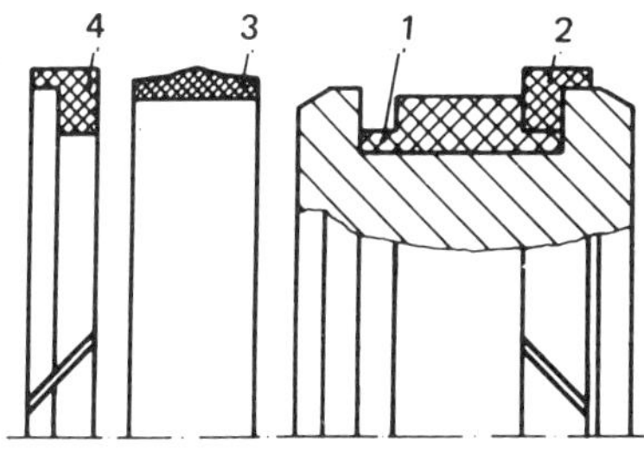

1 — elastomeric ring
2 — split plastic bush
3 — fabric ring
4 — plastic bush

Figure 10
Modern proprietary design of compact piston seal.

easy assembly on one-piece pistons. It comprises a rubber ring, a fabric ring and two angle bushes in plastic material. The fabric ring is the sealing surface, in contact with the cylinder surface. The rubber ring at the bottom of the groove serves as a *static piston* seal and provides the seal with pre-load and elasticity. The plastic rings serve both as back-up rings and piston guides. Maximum pressure rating is 315 bar (4500 lb/in^2).

Rod seals

Rod seals or *gland* seals as they are often called, exist in a wide variety of different types. The O-ring is the simplest type for smaller cylinders; and the U-ring for larger cylinders and/or heavier duty applications. The U-ring, in fact, has proved one of the most successful gland seals (as well as being suitable as a piston seal) and is very widely adopted. The best form of U-ring for a gland seal, however, can differ in detail from that of a U-ring seal. Again individual alternative designs have been developed which can offer specific advantages.

The *multi-lip gland* seal (Figure 11) is a simple, *compact rod* seal which comprises an internal *multi-lip rectangular-shaped* seal with a polyacetal anti-extrusion ring. The elastomeric part of the seal is manufactured in a high abrasion-resistant nitrile material which is suitable for a wide range of hydraulic fluids. The gland assembly can be jumped into a one-piece groove, but for small sizes the anti-extrusion ring can be scarf cut to facilitate assembly. The multi-lip gland seal was developed as a result of the excellent sealing characteristics obtained from tests on the multi-lip piston head seal. The working pressure of the multi-lip gland is 276 bar (4000 lb/in^2) at a temperature range of −40°C to +100°C.

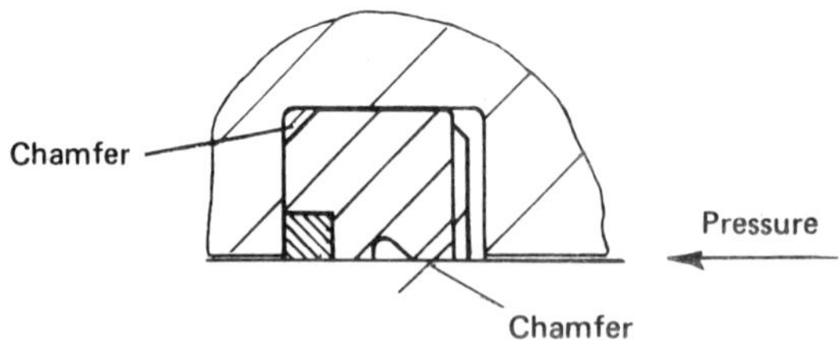

Figure 11

Another simple type of gland seal for low-duty application is shown in Figure 12. It comprises a rectangular-section seal with an integral back-up ring (but see also chapter on *O-Ring Seals*), Section 4.

The resilient *centre gland* seal (Figure 13) is used extensively in hydraulically-operated power set props in the mining industry to give a long leak-free service. The resilient centre gland seal is based on the *Dowty U-ring* configuration comprising a high-grade abrasion-resistant outer ring energized with a soft resilient rubber centre. An integral plastic anti-extrusion ring is located in a recess at the heel or base of the seal. Thus, positive contact of the inner sealing lip of the gland seal to the piston rod is maintained under all service conditions. The seal has an annular bead on the pressure side which allows fluid to flow across the seal face, ensuring fluid pressure acting on the full face of the seal at all times. The working pressure of the resilient centre gland seal is up to 207 bar (3000 lb/in^2) at a temperature range of −40°C to +100°C.

See also chapters on *O-Rings* (Section 4) and *Seal Selection Guides* (Section 8).

Examples of rod seals

Two special-purpose seals designed to withstand rotating motion under high pressure. One is suitable for operating at 200 bar while the seal with the low friction-fabric elastomer dynamic face and anti-extrusion rings is suitable for operating at up to 400 bar. Maximum continuous rotational speed is limited to approximately 0.1 m/sec and 0.2 m/sec respectively.

Rod seals

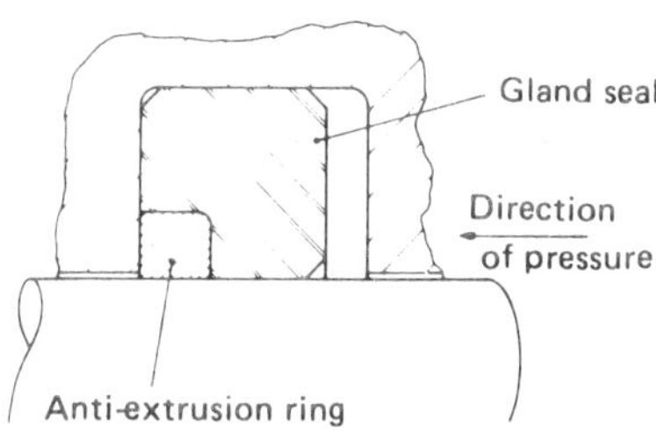

Figure 12
Rectangular sections seal with integral backing ring.

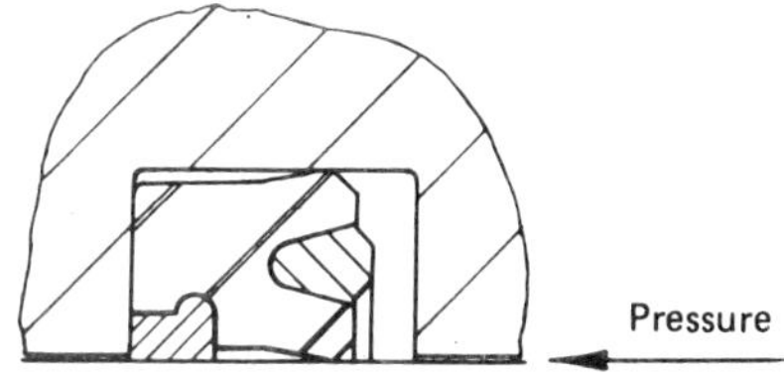

Figure 13

EXAMPLES OF A PROPRIETARY RANGE OF ROD SEALS

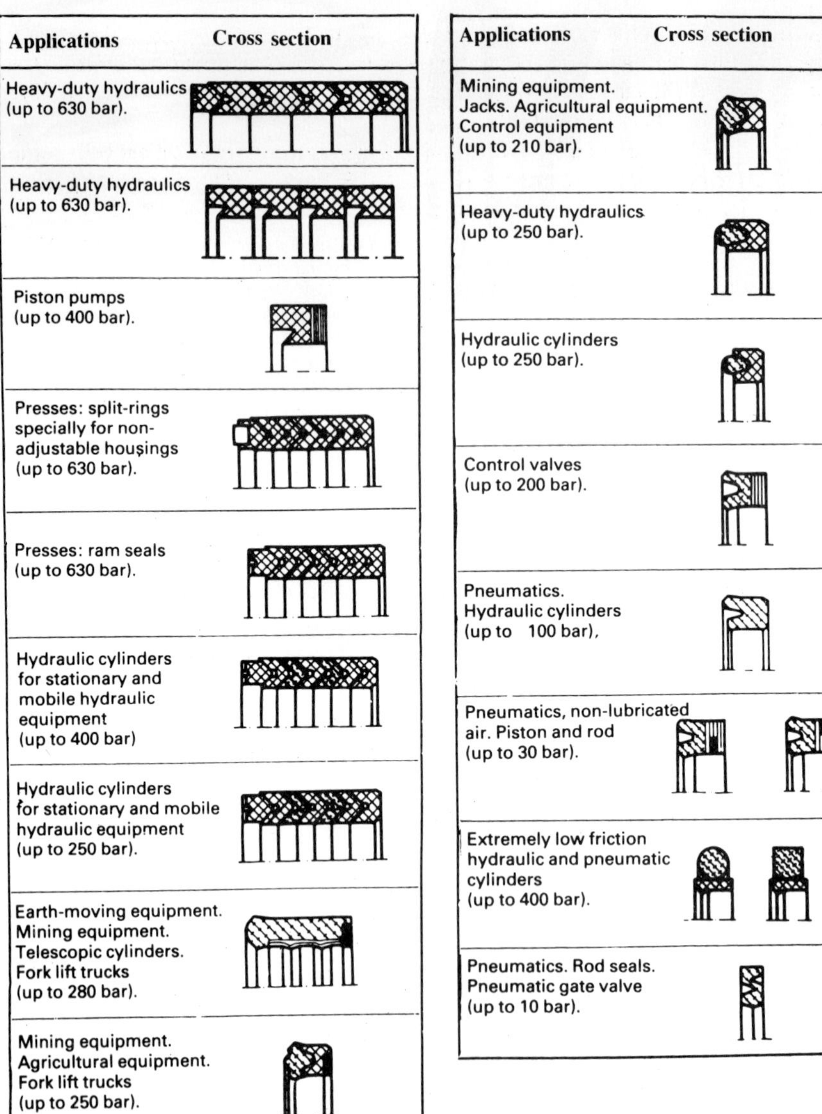

Applications	Cross section
Heavy-duty hydraulics (up to 630 bar).	
Heavy-duty hydraulics (up to 630 bar).	
Piston pumps (up to 400 bar).	
Presses: split-rings specially for non-adjustable housings (up to 630 bar).	
Presses: ram seals (up to 630 bar).	
Hydraulic cylinders for stationary and mobile hydraulic equipment (up to 400 bar)	
Hydraulic cylinders for stationary and mobile hydraulic equipment (up to 250 bar).	
Earth-moving equipment. Mining equipment. Telescopic cylinders. Fork lift trucks (up to 280 bar).	
Mining equipment. Agricultural equipment. Fork lift trucks (up to 250 bar).	

Applications	Cross section
Mining equipment. Jacks. Agricultural equipment. Control equipment (up to 210 bar).	
Heavy-duty hydraulics (up to 250 bar).	
Hydraulic cylinders (up to 250 bar).	
Control valves (up to 200 bar).	
Pneumatics. Hydraulic cylinders (up to 100 bar),	
Pneumatics, non-lubricated air. Piston and rod (up to 30 bar).	
Extremely low friction hydraulic and pneumatic cylinders (up to 400 bar).	
Pneumatics. Rod seals. Pneumatic gate valve (up to 10 bar).	

Block packings for hydraulic pumps and presses

Developments of synthetic fibres has made it possible once again to consider *block* packings for high pressure reciprocating applications. In the early days of reciprocating machinery all seals were of the block type often incorporating rubber-impregnated textile and graphited rubber surfacing. As this type of seal failed to cope with rising pressures, widening ranges of fluids and faster changes of piston direction, the lip seal in various forms, as already illustrated, began to take over until block packings dropped from favour completely.

Reversal of this trend can be seen today on hydraulic machinery from milk homogenizers and high pressure chemical pumps to pressure accumulators, due to the availability of a very tough low friction packing consisting of *Filcoat*-treated yellow fibre at the corners and PTFE on the faces.

This product is compact, contains no migratory lubricant, and has excellent volumetric stability. It is highly resistant to extrusion and can successfully face the severest mechanical conditions to pressures of 1000 bar, unsupported, and 2500 bar supported.

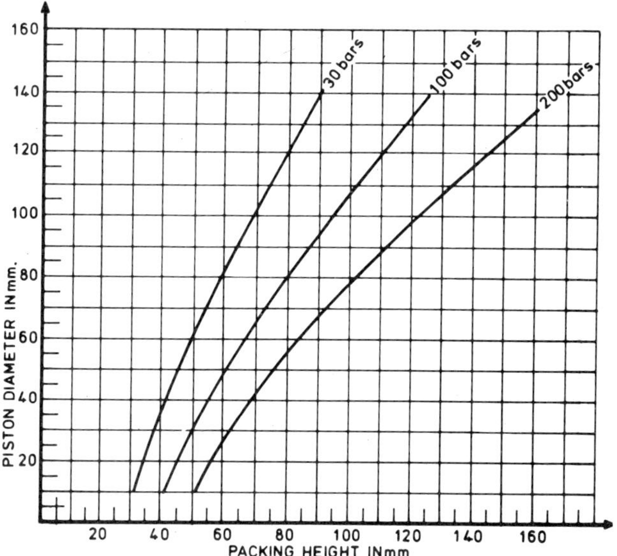

Graph 1
Recommended packing section in relation to piston diameter.

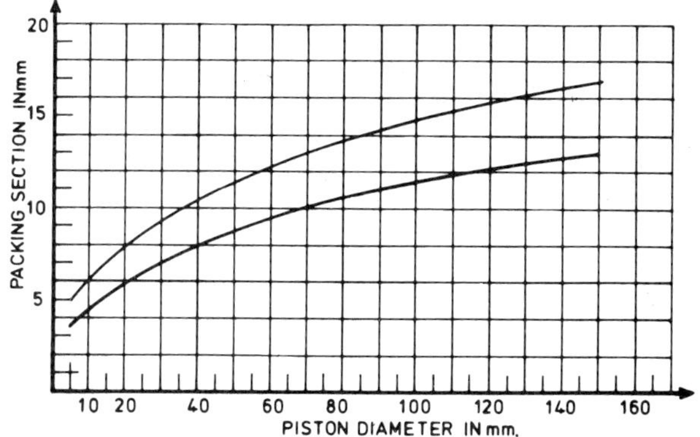

Graph 2
Recommended height of packing in relation to piston diameter and system pressure.

Hydraulic fittings and couplings

Compression type couplings are widely used in hydraulic services, relying on metal-to-metal contact and not requiring elastomeric sealing devices. A number of European manufacturers, however, produce couplings incorporating either an O-ring or profiled elastomeric ring seal, *eg* Figure 14. Most modern aircraft hydraulic systems also employ couplings with trapped O-rings.

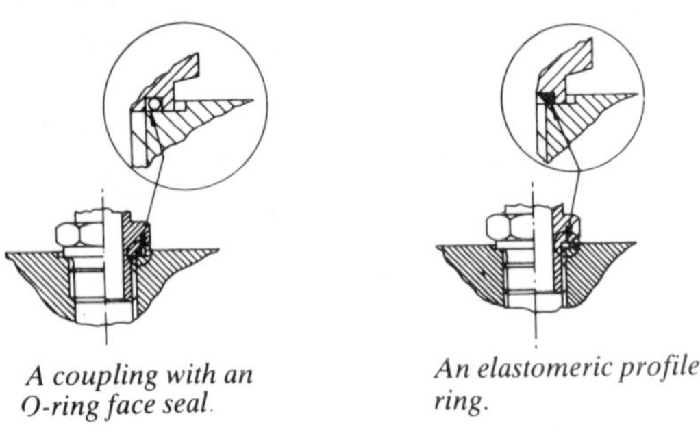

A coupling with an O-ring face seal.

An elastomeric profiled ring.

Figure 14

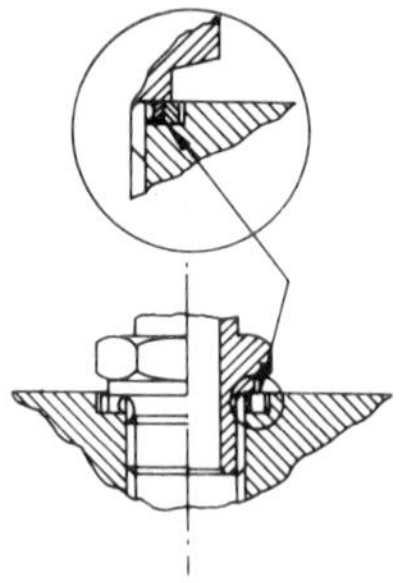

Figure 15
A coupling with a bonded washer.

One of the most widely used seals in hydraulics is the bonded washer (Figure 15). Originally designed to replace the copper washer, it is widely accepted for modern systems operating at pressures of the order of 200 to 300 bar (3000 to 4500 lb/in^2). The majority of bonded washers are used with oil fluids, the sealing element being nitrile rubber of 90 IRHD and the metal outer ring cadmium-plated steel. Composite rubber/metal washers in EPDM are used with phosphate ester fluids, but in this case are not bonded as the fluid would destroy the bond. EPDM is now the standard choice of elastomer for all seals used with phosphate ester fluids.

Two examples of hydraulic couplings incorporating O-ring seals are shown in Figures 16 and 16a. One is a welding nipple coupling (Figure 16) and the other a compression coupling (Figure 16a). Both are recommended types according to the British Steel Corporation Standard CES 25 (1978).

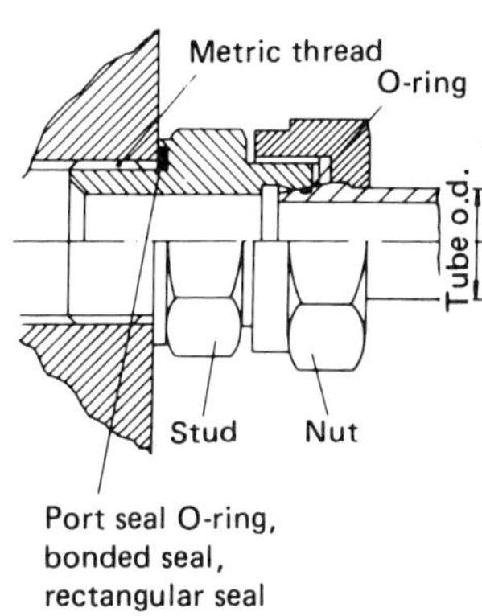

Figure 16
A welding nipple coupling.

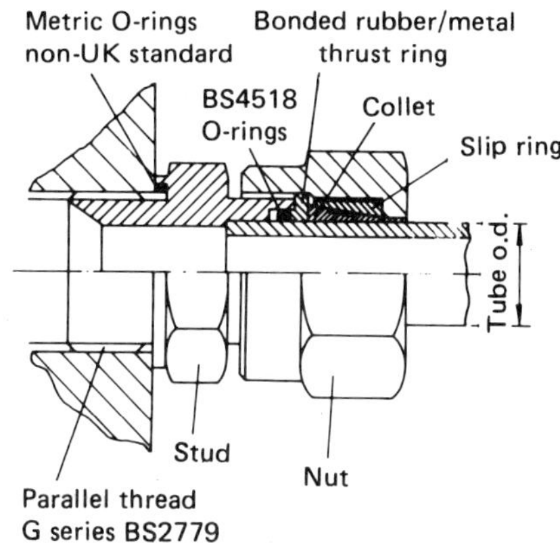

Figure 16a
A compression coupling.

Hose fittings are usually designed to provide sealing *via* cone faces. Some types, however, incorporate an O-ring, with or without anti-extrusion washers. At least one type of swivel coupling incorporates both an O-ring and a U-ring (Figure 17). In the heavier industries flanged joints are also widely used, with O-rings, V-rings, metallic seals or metallic packings.

See also *Seal Selection Guides,* Section 8.

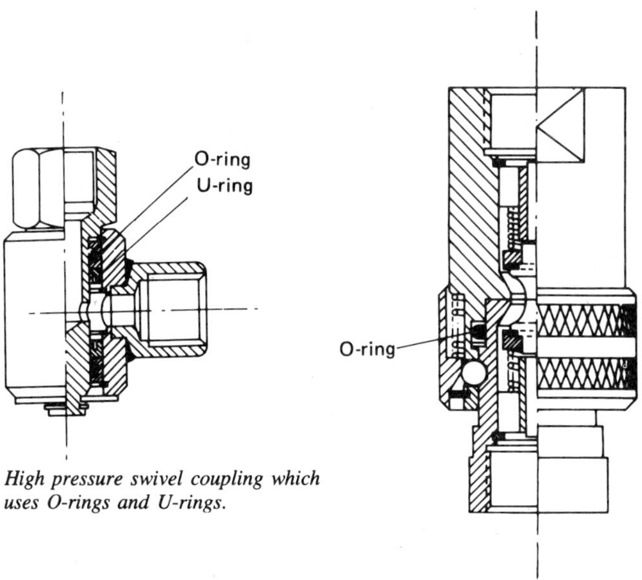

High pressure swivel coupling which uses O-rings and U-rings.

Type of high pressure flexible hose which uses an O-ring.

Figure 17

Pneumatic Seals

SEALS FOR pneumatic cylinders and valves are called upon to withstand much lower pressures than hydraulic seals, *ie* normally not more than 10 bar (150 lb/in^2). At the same time operating forces are much lower, and operating speeds much higher, calling for low friction seals. This favours the use of O-rings (or T-section or similar simple seals) for valves; and 'softer' pressure-energized seals for pistons and rods where sizes may be relatively large. Otherwise choice is similar to seals for hydraulic duties. Similar considerations apply to rod wipers fitted to exclude dirt and dust from cylinders operating in surroundings containing any form of abrasive. They should be of a form which operates with low friction. Alternatively a rod gaiter can be fitted, which will have no friction.

High conformity is required from the seal because air is more difficult to seal than fluids. Also the permeability of the seal material is important, particularly with increasing temperatures. Pneumatic systems are often used at higher temperatures than those associated with liquid sealing systems. This can affect the choice of elastomer. At normal ambient temperatures the permeability of most elastomers is low and similar, with the exception of silicone rubber. There are, however, marked differences at elevated temperatures – see Table 1. All elastomers also have definite limits as regards service temperature, above which they become hard and brittle.

Pneumatic seals may also be used to operate in 'dry' rather than 'wet' (lubricated) conditions. Here again most elastomers (with the exception of silicone rubbers) are appreciably less resistant to 'dry' heat than 'wet' heat.

The absence of lubrication can also make it difficult to keep a dynamic pneumatic seal sufficiently lubricated in the absence of separate lubrication such as oil-mist or oil-fog. Other unfavourable factors are the deteriorating effect of oxygen in air on seal materials in general and the high speed of operation normally associated with pneumatic cylinders and similar devices.

In the case of a seal operating dry, or substantially dry, the apparent permanent set of an elastomeric seal may appear much higher (typically twice as high as that of the same seal used as an oil seal). This may call for greater initial interference when assembling the seal, and subsequently higher initial friction during the first period of running when the permanent set is taken up. Selection of a suitable interference is more difficult and errors may either cause the seal to continue to bind and develop high friction and wear or run-in to a loose fit with consequent leakage.

A permanently binding seal is more likely than a 'loose' seal, particularly if the seal is proportioned as a 'dry' type and subsequently lubricated by oil-mist or oil-fog. A

TABLE I – AIR PERMEABILITY OF ELASTOMERS
(Permeability coefficient Q x 10^8 cm^2/sec/bar)

Elastomer	Temperature °C							
	40	60	80	100	120	140	160	180
Acrylic	5	13	30	38	45	60	75	93
Butyl	0.5	1.8	4.5	10	19	28	40	60
Neoprene (Cr)	2	4.5	10	17	26	38	52	71
Nitrile	2	4.5	8	15	22	31	46	63
SBR	3.5	9	18	30	45	70	100	115
Natural rubber	9.5	20	4.3	56	70	100	110	120
Silicone	290	380	475	580	660	755	900	1000
Fluorocarbon	1	3	9	20	36	60	90	150
Polyurethane	2	4.5	10	19	31	40	52	70

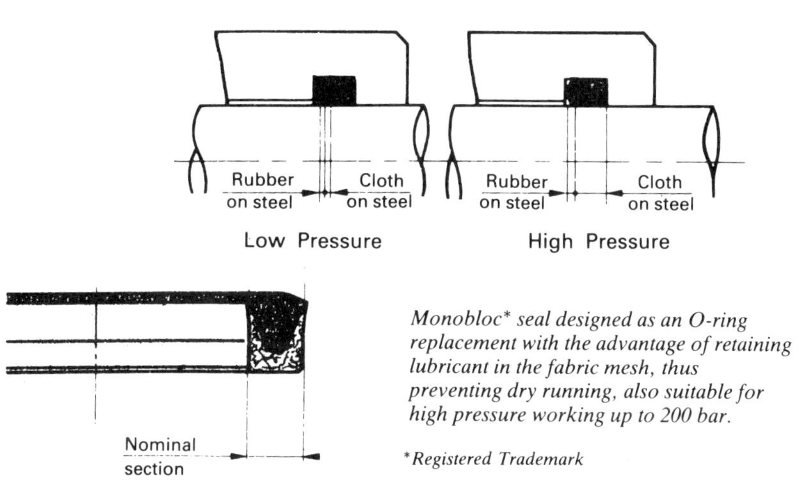

Monobloc seal designed as an O-ring replacement with the advantage of retaining lubricant in the fabric mesh, thus preventing dry running, also suitable for high pressure working up to 200 bar.*

**Registered Trademark*

proportion of the lubricant will be absorbed by the seal and promote swelling. Thus a common fault with all types of seals which operate under oil-mist or oil-fog lubrication is failure to provide for the amount of swell the seal is likely to undergo during its 'conditioning' period. This period may, in some cases, extend to several weeks of operation.

A period of immersion in fluid may also leach out a proportion of the plasticizer in the seal. If the seal is subsequently allowed to dry out it will shrink. For a dry seal, therefore, the material may have to be chosen specifically for a small amount of permanent set so that interference is not reduced too drastically in service. Seal sets and packings which can be adjusted can thus offer advantages in certain applications, although adjustment of packing pressure is not necessarily a satisfactory answer where consistency of performance is required. It may provide a means of adjusting sealing, but the friction will be modified both by the varying seal pressure and a variable coefficient of friction.

With suitable seal selection, however, frictional losses in cylinders, *etc,* should not exceed about 5%. While this is higher than the losses applicable to hydraulic cylinders, this is largely inevitable due to the greater sealing problem involved, *eg* lower viscosity fluid, lower pressure to energize flexible seals, and the less favourable operating conditions of the seal.

Static seals

Static seals used in pneumatic systems include cover gaskets for valves, filters, reservoirs, *etc*, end cover seals for cylinders and *gasket* seals for line accessories. All requirements can normally be met by simple seals or gaskets.

Sheet gasket materials are still widely used where a suitable flange area is available. Plain elastomeric sheet, however, has distinct limitations as a gasket material due to its tendency to spread under compression and also take on a high compression set. To perform satisfactorily it needs to be restrained in a groove. An O-ring will usually provide a better seal in such circumstances, at little or no extra cost. The O-ring, in fact, normally offers the simplest and most effective method of static sealing where a machined groove can be incorporated in the assembly, with the additional advantage that it can provide sealing in a minimum size flange or overlap. Another advantage is that it is effective with a much rougher, and therefore cheaper, surface finish than is required for gaskets. Standard groove sections will give positive sealing up to 105 bar (1500 lb/in^2). Very much higher pressures can be sealed if the joint is tightened right up, provided the surface finish is smooth enough to give tight closure to eliminate 'breathing' and possible extrusion of the O-ring. O-rings can also be used as crush seals, locating the ring against a flange or shoulder and applying pressure and squeeze by a chamfered edge on the mating member.

In certain types of static seals, such as valve seat seals, special precautions may have to be taken to prevent any possibility of the ring being blown out of its groove. Under certain circumstances this can occur at relatively low pressures, *eg* when the pressure force is directed radially outwards on the ring. Other treatments in this case include venting the O-ring groove to eliminate diametral pressure build-up, and mechanically restraining the ring with a specially-shaped groove or retaining ring.

All static seals of this type, where closure pressure distorts the O-ring section to produce a seal, must provide for positive retention of the ring, at the same time limiting the degree of squeeze applied to less than that which would permanently damage the ring. It will also be appreciated that due to the resilience of the ring material this type of static seal provides shock absorption as well as sealing.

Pneumatic fittings and assemblies

Sealing requirements are relatively moderate in the case of pneumatic fittings and couplings and with many low pressure systems face-to-face contact provides adequate sealing. Plastic couplings are used for pressures up to 3 bar (45 lb/in^2) while taper thread couplings are favoured for systems with maximum pressures up to 20 bar (300 lb/in^2). For higher pressure working the SAE joint is commonly specified, embodying a trapped O-ring (Figure 1).

Nitrile rubber with a hardness of 60 to 70 IRHD is suitable for O-rings within the temperature range −20 to +70°C and is fully compatible with oil, grease and dry or wet compressed air. Nitrile rubber becomes hard and brittle at higher temperatures when fluorocarbon rubber must be used. The majority of European users, in fact, tend to adopt fluorocarbon rubber O-rings for most applications, despite the higher cost.

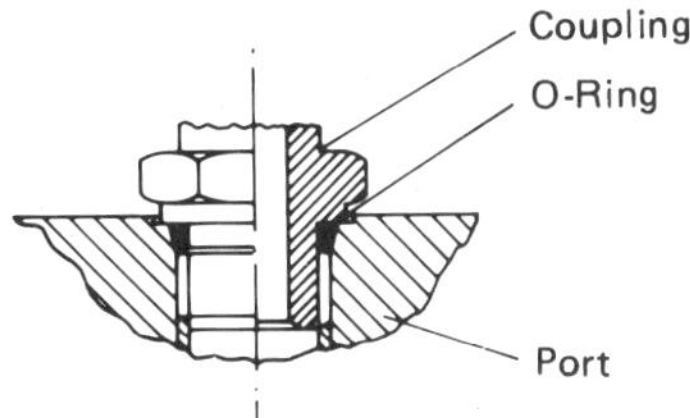

Figure 1
SAE joint.

Rather harder rubbers are required for O-rings used for dynamic seals – 80 IRHD being a typical choice for reciprocating and rotary seals. In modern pneumatic systems linear speeds of up to 1.4 m/sec are realized, and rotary speeds up to 0.05 radians/sec.

Dynamic seals

Dynamic seals embrace piston and rod seals for cylinders, seals for valve spools and valve stems, and rotary seals for rotating assemblies, *eg* shaft seals on air motors and seals for distribution chests on rotating cylinders.

The modern trend is towards simplification of seal design and assembly, replacing traditional type packings almost completely with fabricated seal rings either in homogeneous rubber or moulded rubber-fabric composites. A variety of sections may be used, although most are based on, or derived from, basic cross sections with proven performance. The exception is the O-ring, which is quite unique as a multi-purpose seal. This, too, has given rise to variants, but the O-ring remains a favourite choice where the simplest possible seal assembly is required for all types of duties, although it has its limitations.

O-rings

As a general rule O-rings are best suited for smaller sizes and lighter duties as reciprocating seals, although this is by no means an overall recommendation. O-rings have been used

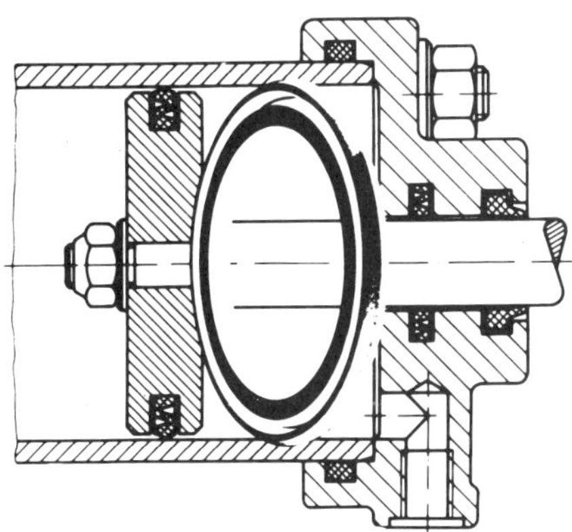

Special low friction piston head or rod seal for pneumatic cylinder.

successfully both as rod seals and piston seals on relatively large cylinder sizes. The main requirement is that the ring be used within its rating or capabilities, and mounted in a properly matched groove.

The conditions most favourable to O-ring reciprocating seals are a short stroke, adequate lubrication and moderate rubbing speeds. Conditions unfavourable to O-ring reciprocating seals are side and bending loads causing eccentricity of the moving system and high unit loads, very slow reciprocating speeds and very low pressures accompanied by slight leakage. Both the two latter conditions tend to develop very high friction which can be damaging to the seal.

Properly assembled and adequately lubricated, an O-ring should not normally tend to roll or twist under reciprocating motions because the contact area with the groove is far greater than the contact area with the sliding surface, with the torsional stiffness of the ring itself resisting twisting. The fact that the groove friction is static friction, and thus higher than the rubbing friction in any case, tends to hold the ring even better in its groove. Anything which modifies these conditions, however, can allow some parts of the ring to slide and other parts to roll simultaneously, resulting in an excessive amount of twist being developed in the ring. This can ultimately lead to ring failure, the break appearing as a tight spiral fracture around the ring, known as spiral failure.

Certain other sections developed from the O-ring, present a stiffer section torsionally and are less prone to spiral failure. They can be used as alternatives to O-rings in standard O-ring grooves, when a simple ring seal is required and the possibility of spiral failure is high. These sections can also be used for static seals. In general, however, the use of O-rings far exceeds the application of alternative simple ring sections for application as simple double-acting reciprocating seals. Where the duty involved is such that O-rings prove unsatisfactory (or are considered unsatisfactory), a heavier ring section would normally be selected.

U-rings

The U-ring, suitably proportioned and in a relatively soft rubber is capable of providing particularly good sealing with low friction at low pressures. Desirable features are thin lips for response to pressure energization, generous length for good flexibility, with the lips themselves designed to have low lip loads for low friction and at the same time enough contact area and pre-load for good sealing. A square heel section is also preferable to minimize heel wear and nibbling, (Figure 2).

U-ring seals and various modified forms developed specifically for pneumatic services, *eg* rod and piston seals, therefore tend to differ in cross-section from similar seals intended for hydraulic duties.

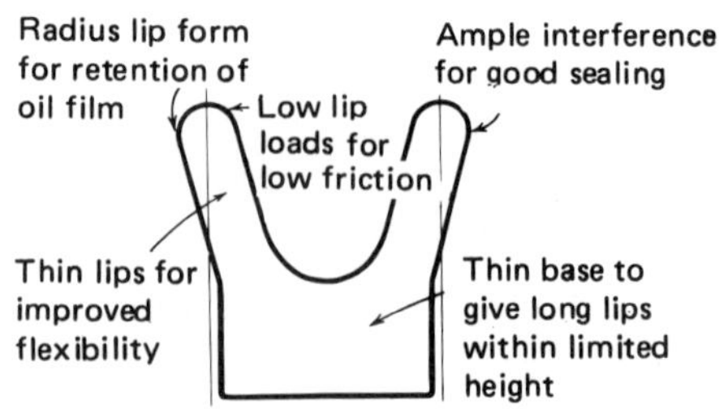

Figure 2
Basic requirements for a U-ring pneumatic seal.

Special sections

Modified U-ring forms, C-rings and modified C-rings and other special sections may also be used for piston and rod seals on pneumatic cylinders. A number of these have been specifically developed for pneumatic services. An example described as a *caliper* seal is shown in Figure 3. This is intended as a direct alternative to an O-ring, fitting in a similar groove and having similar dynamic friction to an O-ring but much lower break-out friction because of the lower cross-section nip and greater flexibility. Small grooves on the i.d. of the seal allow air to enter the seal and actuate the opposite lip against the side of the groove. The seal is thus fully pressure-energized under both static and dynamic conditions and will seal in either direction.

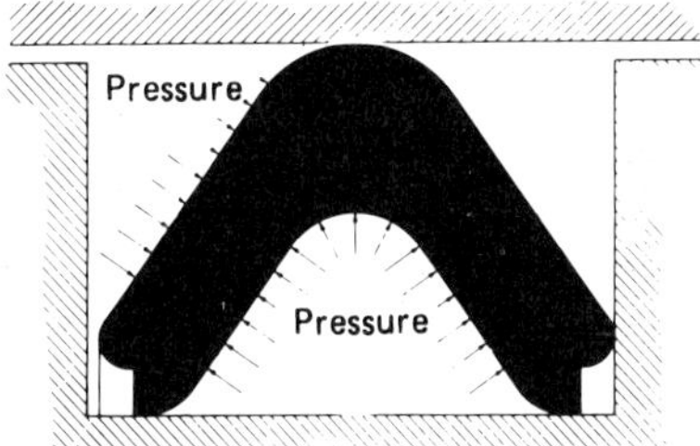

Figure 3
Caliper seal.

Composite seals

Composite 'hard' rings may also be used for specific applications, or where favoured by a particular designer. These take the form of a rigid ring section, which provides the rubbing and sealing surface, combined with a pliable cushion ring. The 'hard' section may be of metal or some other low friction material (*eg* PTFE). They are essentially compression seals, with resilience imparted by the cushion section. A possible advantage with rings of this type is that they are capable of acting as bearings as well as seals, with an ability to accommodate side loads and provide an automatic self-centring action.

A particular example here is the combination of an O-ring with a rigid PTFE ring, the latter providing a low friction sealing surface irrespective of whether or not lubricant is present. This type of seal can be either an internal or external type (Figure 4) and

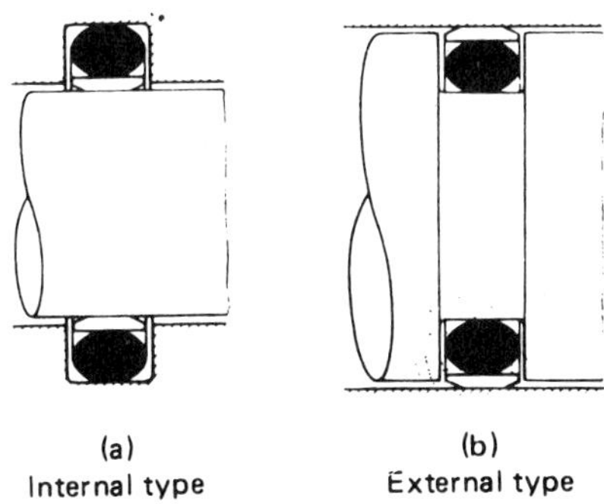

Figure 4

is produced by various manufacturers under different names (differing in the form of the PTFE ring). Best known in the UK is the *Angus Slip O-Ring*.* A comparison between the performance of this type of composite ring and a conventional O-ring is in Figures 5 and 6. Here it is interesting to note that the static friction is similar to that of an O-ring, although its performance under dynamic conditions is superior. Also noteworthy is that at very low pressures, lip type packings show the lowest friction (Figure 5). In this case, however, increasing pressure increases the lip area in contact and substantially increases the friction.

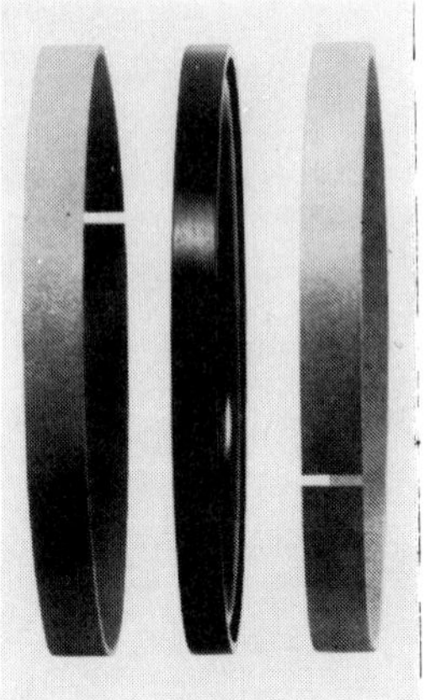

Low friction co-axial seal with loaded PTFE ring used between two load bearing rings (eg PTFE loaded with bronze). Operating pressures up to 400 bar are now possible with this type of seal.

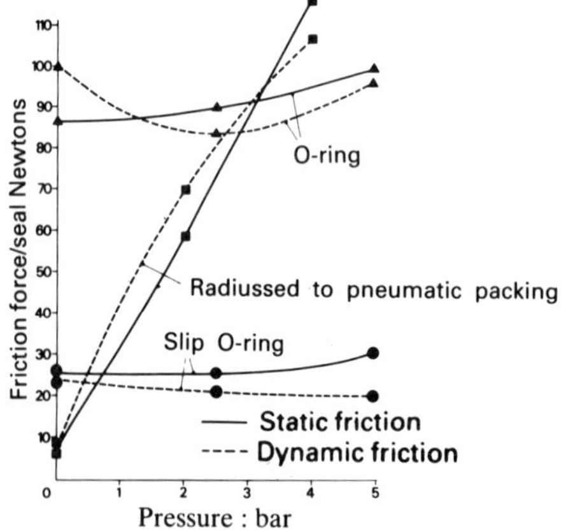

Friction force *versus* pressure for typical seals under dry conditions.

Figure 5

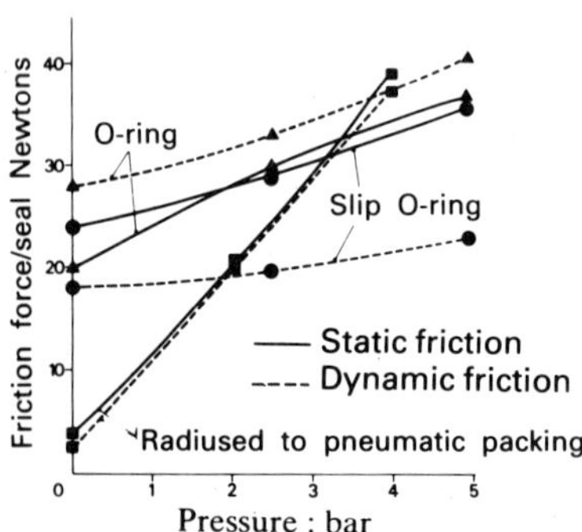

Friction force *versus* presure for typical seals under lubricated conditions.

Figure 6

**Registered Trademark*

Cylinder seals

Largely because of the considerable difference between rod and cylinder diameters, different types of seals may be used on the rod and piston in pneumatic cylinders. Some manufacturers also offer different types of seals with a given cylinder design. Even the use of packed glands or similar compression packings for rod seals is not entirely excluded from modern practice, despite their much higher friction than pressure-energized seals. There are applications where relatively high friction on the rod can be an advantage, *eg* to add stiffness to the movement in the case of a vertical air cylinder used to raise a weight.

An additional seal on the outside of the rod seal is also commonly used. This is a wiper seal, which scrapes over the entering rod surface and removes dirt, *etc*, thus preventing it from being drawn into the main rod seal where it could cause wear. Wiper seals can take the form of simple single lip seals, metallic scraper rings bonded to an elastomeric section, thin elastomeric sections backed by metal to provide support and rigidity, or even simple compression seals (*eg* felt rings). Polyurethane rubber wiper rings are being used more because of their good wear and abrasion-resistant properties. A gaiter may also be added for positive protection where the cylinder is used in particularly dirty conditions, or used in place of a wiper.

See also chapters on *Hydraulic Seals* and *Seal Friction,* (Section 1).

SECTION 7

Special Seal Types

BUSHING SEALS
POSITIVE ACTION SEALS
DIAPHRAGM SEALS
PISTON RINGS
CARBON SEALS
LIQUID RING SEALS
LIQUID BARRIER SEALS
INFLATABLE SEALS
FERROFLUIDIC SEALS
SELF-ADHESIVE COMPRESSION SEALS
HIGH TEMPERATURE SEALS
LARGE DIAMETER SEALS
LABYRINTH SEALS

Bushing Seals

BUSHING SEALS work on the principle of the throttling action provided by a long annular gap with a relatively small clearance. Because their effectiveness is directly related to fluid viscosity, they are suitable only for sealing liquids and, in particular, applications where some leakage can be tolerated. Their advantages are that they are simple and inexpensive, easy to install and require no maintenance.

With *fixed bushing* seals the sleeve producing the annular gap is fixed to the housing (Figure 1). It needs to be relatively long to keep leakage within reasonable limits while still maintaining acceptable clearance between the sleeve and shaft. This can accentuate alignment problems, further aggravated by any eccentricity in the rotating shaft system. The design normally allows for the fact that some rubbing contact is likely to occur, which favours the use of a soft low-friction metal for the sleeve, or a carbon, PTFE or molybdenum disulphide-loaded composite. *Babbitt* metals are commonly used for low temperature seals; aluminium alloys, bronze or carbon for higher temperature services.

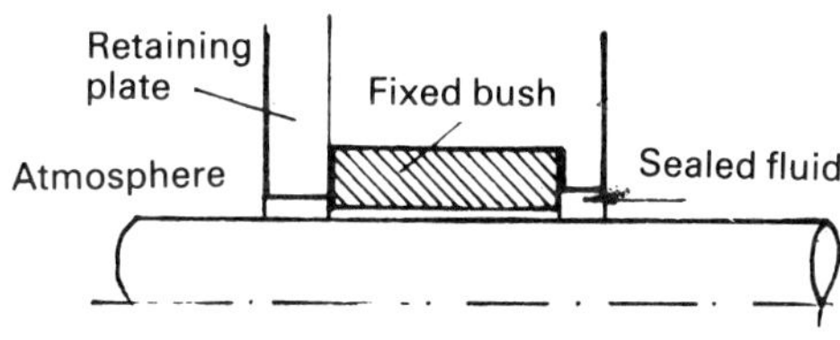

Figure 1

Theoretically the flow through a concentric annular orifice is given by:

$$Q = \frac{\pi D b^3 \Delta P g}{\nu \rho L}$$

where
- D = mean diameter of orifice
- b = radial clearance space
- ΔP = differential pressure
- ν = fluid viscosity
- ρ = fluid density
- L = length of origice.

From this it will be appreciated that the clearance space is far more significant than orifice length, *ie* leakage flow is proportional to b^3, but only directly proportional to length. This basic relationship applies only to flow through a concentric orifice. In a practical bushing seal a certain degree of eccentricity will almost inevitably be present, with unfavourable effects on the leakage rates. Thus it is virtually impossible to predict the performance of a fixed bushing seal, except on empirical data, although its likely performance can be calculated if the amount of eccentricity present is known.

In the case of an eccentric orifice, the corresponding flow formula is:

$$\text{flow through orifice} = \frac{\pi D b^3 \Delta P g}{\nu \rho L}\left(1 + \frac{3e^2}{2}\right)$$

where e = eccentricity of shaft relative to the bore.

The *floating bushing* seal overcomes this basic limitation in allowing the sleeve to follow radial shaft movements. The seal itself is designed so that the bushing can move radially, while being restrained in a normal static position by a spring (Figure 2). This means that the clearance can be reduced. At the same time the bushing must be restrained against rotation (*eg* by a dowel pin), and be smooth faced on the outward end to move smoothly up and down the contact face of its housing. This housing face must also be smooth finished.

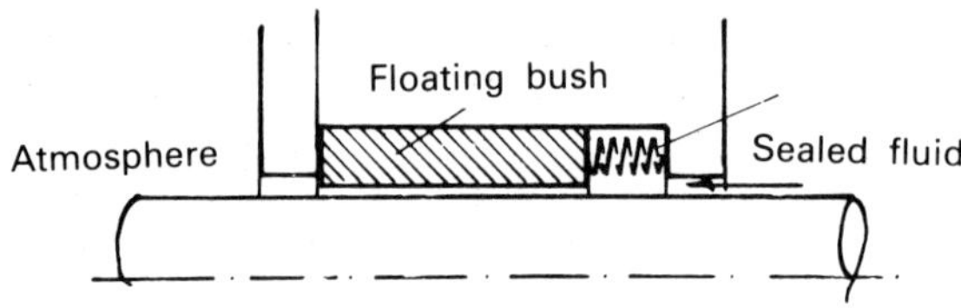

Figure 2

For sealing under low pressure, the spring holds the bushing in contact with the housing face. At moderate pressures, fluid pressure can provide the necessary force, eliminating the need for a spring (although one may still be included in the design). At still higher pressures it may be necessary to step the opposite end face of the seal to relieve the axial sealing force generated by fluid pressure. This is necessary because excessive axial force will make the bushing tend to stick to the housing face rather than float freely enough to follow shaft motions.

Some types of floating bushing seals are of balanced design, achieved by stepping the seal (or shaft) to provide 'back pressure' over a predetermined face area (Figure 3). Any required degree of hydraulic balance can be achieved in this way to make the bushing float freely.

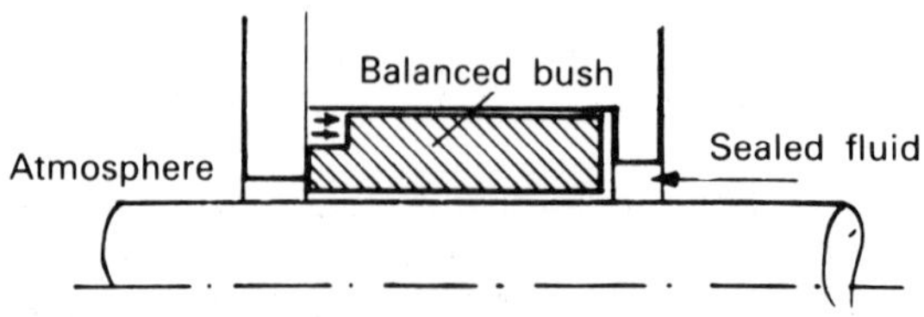

Figure 3

The *floating ring* seal overcomes this particular problem by splitting the length of the bushing into several rings, each capable of independent movement (Figure 4). Each ring can thus adjust automatically to shaft motions over its own length and because each ring only carries a proportion of the total pressure differential, it can move much more freely.

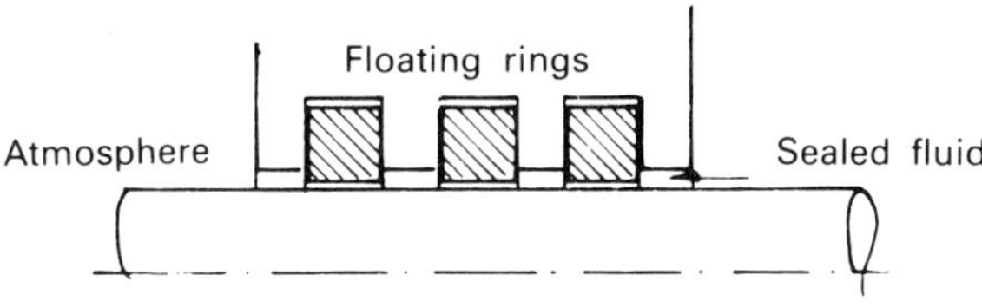

Figure 4

egmented-ring floating bushing seals differ again in employing a (single) floating ring associated with a retainer and a garter spring housed in a groove (Figure 5). The garter spring provides sealing contact with the shaft surface, while the angled interface between the ring and its retainer automatically provides radial sealing against the groove wall as the ring floats with shaft movement in a radial direction. Rings of this type can seal gases as well as liquids.

Because this type is a contact seal, the ring is usually made of Babbitt metal or bronze, or low-friction composites for sealing non-lubricants (*eg* water) and gases.

There are numerous variants on floating ring seal designs, including the incorporation of wear compensation on contacting rings (segmented rings), pressure balancing and buffering. The number of rings used in a complete seal may also vary with different applications.

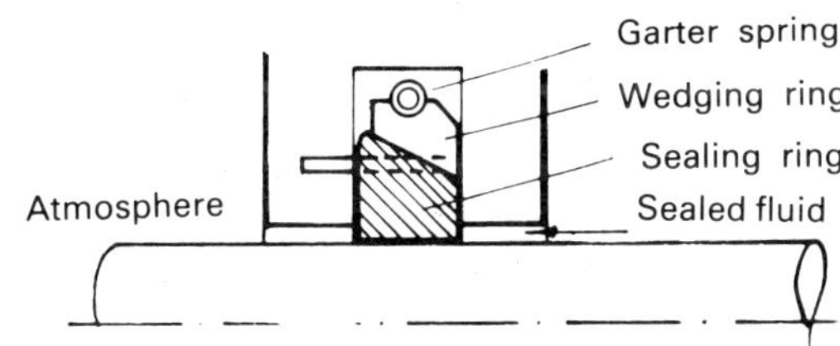

Figure 5

PTFE bearing tape

An inexpensive way of producing a simple bushing seal is from reinforced PTFE tape (*eg* glass-filled PTFE material). This is assembled in a shallow machined recess, as shown in Figure 6.

When used in this way, all the bushes shown are essentially the outboard seal of a double seal configuration. Their purpose is to contain a barrier fluid between the pumped product and atmosphere, the barrier fluid being of only low pressure. In addition, since the bushes are of PTFE, glass-filled PTFE or bronze, they can, as the pump bearings

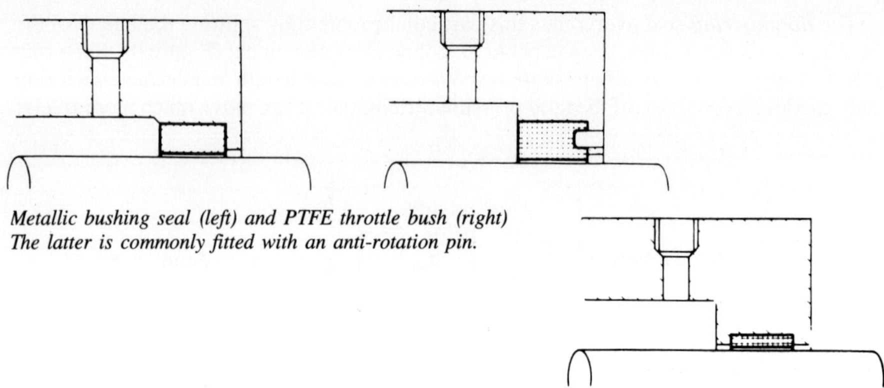

Metallic bushing seal (left) and PTFE throttle bush (right) The latter is commonly fitted with an anti-rotation pin.

Figure 6

become worn themselves, act as bearings, and without the risk of sparking. This is especially important in hazardous environments such as on oil refineries.

See also chapters on *Labyrinth Seals* and *Ferrofluidic Seals.*

Positive Action Seals

THIS DESCRIPTION is applied to radial seal designs which feature some form of positive back-pumping or back-pressure generation to inhibit flow through clearance spaces. They may also be referred to as *visco* seals, *screw* seals, *windback* seals, *etc.* They may be designed as clearance seals and contact seals. In the latter case the positive-action characteristic may be an additional feature incorporated on an otherwise conventional type of flexible lip seal. Nearly all individual forms of positive seals as such, however, are of a rigid type operating with positive clearance.

The simplest – as well as being the earliest – type is the screw form seal where a square screw form thread of fine pitch is cut on the shaft or housing (Figure 1). The sealing effect is provided by the viscosity of the fluid being pumped back through the clearance space as the shaft rotates. Such a seal can prove very effective for sealing liquids, with back pressure proportional to shaft speed (as well as liquid viscosity). It only seals when the shaft speed reaches a particular speed, however, and so for filled systems would need to be associated with a static seal. It is also a uni-directional seal. However, two such seals can be formed back-to-back to provide sealing in either direction.

While such seals can be used directly for sealing liquids they are relatively ineffective for sealing gases. However a liquid-filled screw seal can be used for gases with the seal itself maintaining a pressurized liquid barrier in the clearance space. A suitable fluid feed, recovery and recirculating system must then be incorporated to contain the fluid.

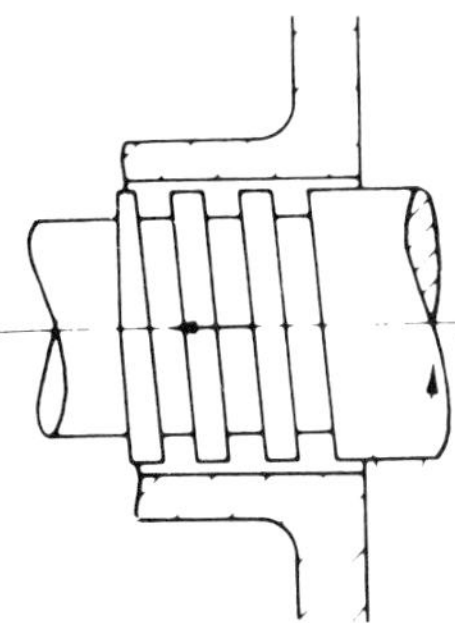

Figure 1
Windback type seal.

Here it can be noted that stopping the thread short of the full length of the seal within the bore can help retain fluid within the length of the bore.

Other design variations aim at reducing the clearance space to an absolute minimum, usually by incorporating the screw form in an elastically-suspended floating bushing or similar housing. This is then similar to the floating bushing seal, but with an internal windback feature. Other *floating spring-loaded screw thread* seals operate with light rubbing contact.

The most successful of screw-type seals has been the incorporation of windback or pumping action on normal lip-type seals. Such seals are easy to mould (instead of having to be machined) and are far less critical on dimensions than threaded bushing seals or similar configurations, and normally show improved performance. Also modern seals of this type do not necessarily function as 'pumping' seals. The primary sealing action is that provided by the meniscus of fluid under the seal lip on the external or low pressure side. The positive action effect incorporated in the shape of the lip section may be largely concerned with ensuring that this area of the seal is adequately fed with fluid to maintain satisfactory sealing under conditions of fully hydrodynamic lubrication. Most seals of this type are uni-directional. Double-acting seals have also been developed capable of giving the required pumping action regardless of the direction of rotation, *eg* see Figure 2.

An alternative approach is the employment of a conventional lip seal with the pumping effect provided by a fine thread form lightly machined into the shaft itself, *eg* see Figure 3. Although this would appear to promote wear on the seal lip, only very light machining is needed on the shaft to promote such an action. See also chapters on *Bushing Seals, Oil Seal Types* and *Labyrinth Seals.*

Figure 2
Windback seal with bi-directional action feature.

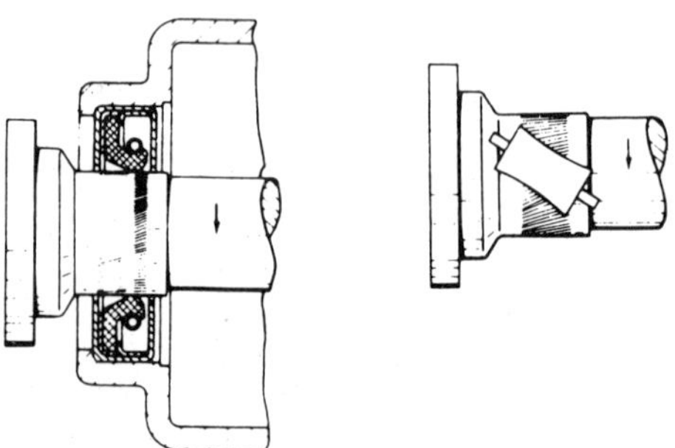

Figure 3
Helical grooves machined on shaft used in conjunction with a conventional lip seal.

Diaphragm Seals

A DIAPHRAGM is a flexible or elastomeric member constrained around its periphery and capable of movement substantially at right angles to this plane of restraint. The use of diaphragms as seals is, therefore, confined to reciprocating motions. While actual applications are somewhat limited and specialized, diaphragms are zero-leakage seals.

Diaphragm seals are normally made from elastomers or fabric-reinforced elastomers, although some use may be made of metallic diaphragms. Various configurations are possible, modifying both the travel or stroke of the diaphragm and the manner in which the diaphragm material is stressed.

Flat diaphragms are the simplest and cheapest, but have a limited stroke. The material is also directly stressed by stretching. Stroke normally needs to be limited by a mechanical stop to prevent over-stretching and relatively early fatigue failure is likely around the line of restraint.

Diaphragm seals.

Convoluted or *dished* diaphragms are capable of larger travel (more than twice the length of the convolutions in the former case and approximately twice the length of dishing in the second case). Both forms have the advantage that the material is stressed by flexing rather than stretching and so should have an appreciably longer service life.

Slack diaphragms have much greater stroke potential although the actual travel available is limited by the amount of slack which can be accommodated without the material wrinkling. The most usual type for long stroke application is the *rolling* diaphragm.

Diaphragm materials

The majority of applications of flat and convoluted diaphragms call for a reinforced rubber material to resist excessive deformation or 'ballooning'. Cotton or rayon is the usual textile fabric employed, impregnated with nitrile rubber. Properties of typical diaphragm materials are given in Table 1. Such materials are used to make individually moulded convoluted diaphragms and are also produced in the form of flat sheets (either moulded or calendered) from which both flat and convoluted diaphragms can be cut.

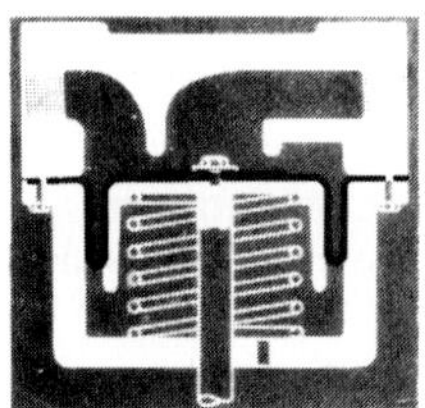

Demand valve

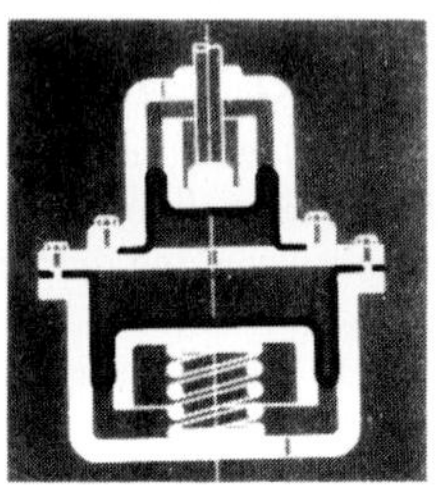

Dash pot

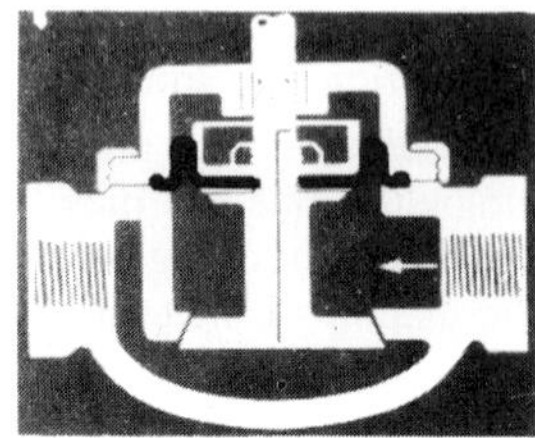

Balanced valve

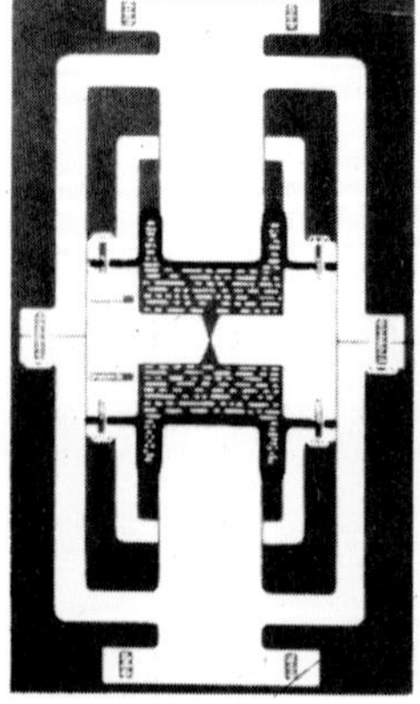

Damping device

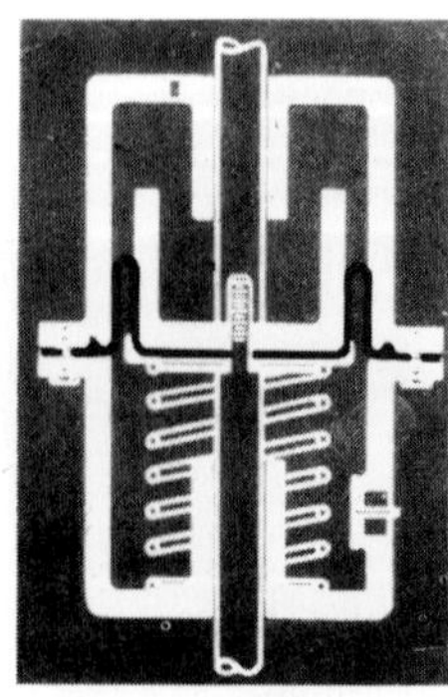

Snubber

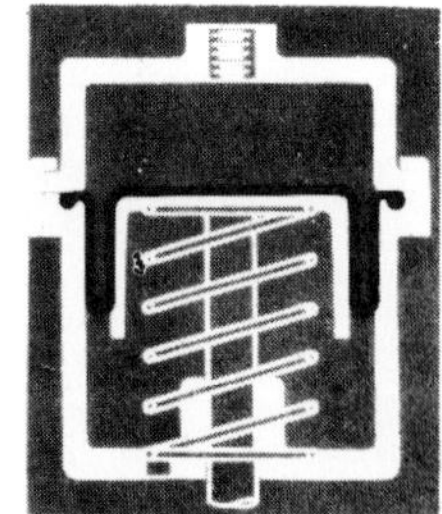

Expansion chamber

Application examples of rolling diaphragms.

TABLE 1 – PROPERTIES OF TYPICAL DIAPHRAGM MATERIALS

Reinforcement		Egyptian cotton	High-tenacity rayon	High-tenacity rayon	Egyptian cotton	High-tenacity rayon	High-tenacity rayon
Weight g/m²		75	133	240	75	133	240
Rubber coating		Nitrile/Viton	Nitrile/Viton	Nitrile/Viton	Nitrile/Viton	Nitrile/Viton	Nitrile/Viton
Finished thickness	mm	0.275±0.02	$0.375^{+0.05}_{-0.02}$	0.05±0.05	0.275±0.02	$0.375^{+0.05}_{-0.02}$	0.5±0.05
	in	0.011±0.001	$0.015^{+0.002}_{-0.001}$	0.020±0.002	0.011±0.001	$0.015^{+0.002}_{-0.001}$	0.020±0.002
Weight g/m²		320±20	450±30	630±40	320±20	450±30	630±40
Burst strength (mullen) bar		7	14	23	7	14	23
Porosity at 1.4 bar		Nil	Nil	Nil	Nil	Nil	Nil
Surface finish		Grey, matt	Grey, matt	Grey, matt	Black, smooth	Black, smooth	Black, smooth
Oil resistance (DED 2472/B/O) 24 hours 150 °C, weight %		–6	–4	–5	–6	–8	–7
Fuel resistance (35% ATF) 48 hours 40 °C, weight %		+6	+8	+7	+8	+7	+6
Dried out, weight %		–9	–10	–10	–8	–9	–8
Maximum working temperature °C approx.		100	100	100	100	100	100
Special characteristics		Oil and fuel resistant materials with medium strength					

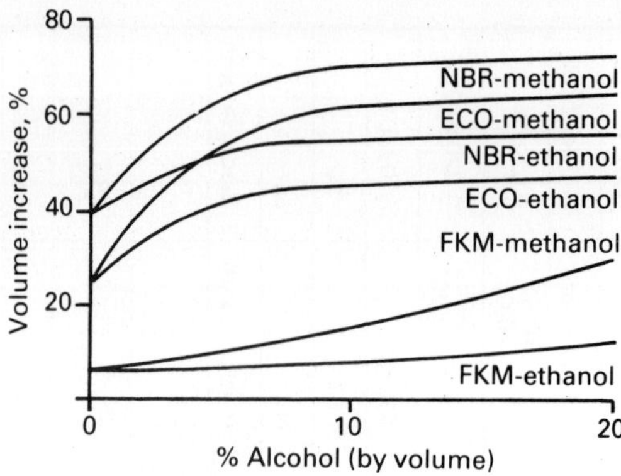

Piston Rings (setting on picture)

Effect on elastomers of gasoline/alcohol blends.

Rolling diaphragms

A *rolling* diaphragm is essentially a very slack diaphragm contained within a cylinder and capable of piston-type movement. The movement is normally energized in one direction only, the return motion being by spring action. Rolling diaphragms are normally employed as sealing devices, rather than pressure sensitive elements. A variant, the rolling sleeve, is employed as a pneumatic spring.

A proprietary type of rolling diaphragm is shown in Figure 1. The slack diaphragm is moulded in a suitable elastomer, usually reinforced with high tenacity rayon, PTFE or glass-fibre. Reinforcement is necessary to permit free circumferential elongation of the diaphragm with no axial elongation and so give a true rolling action. A plain elastomeric bag will not 'roll true' and will also develop friction.

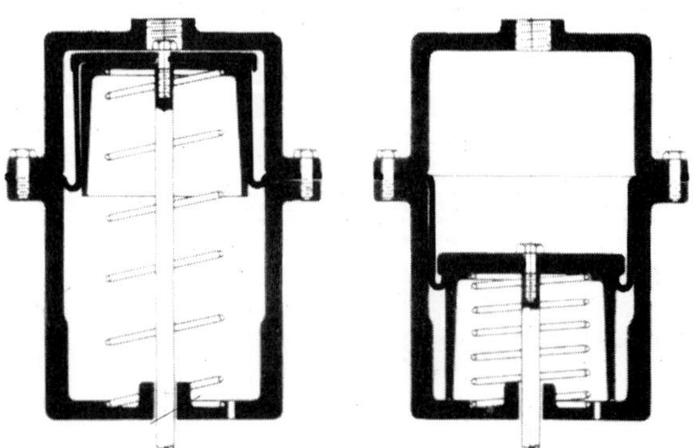

Figure 1

The rolling diaphragm is effective only when pressurized in one direction. The bulk of the pressure load is carried by a piston over which the diaphragm drapes, the slack in the diaphragm then forming a convolution of U-shaped section in the coaxial clearance space (Figure 2). Pressure on the diaphragm maintains it in intimate contact with the piston and cylinder bore. Piston movement merely causes the convolution to roll up or down the axial clearance space, travel being limited only by the amount of slack available. Thus a diaphragm of this type can have a very large travel, determined by the depth of the original bag shape. Diaphragm motion is smooth and continuous and free from friction (because the motion is rolling) and hysteresis effect (because the diaphragm is not stretched).

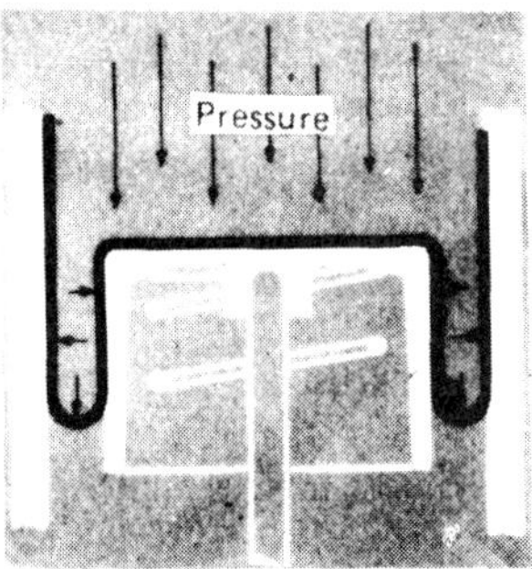

Figure 2

Rolling diaphragms have applications as frictionless and non-leaking seals, for sealing pressurized fluids (*eg* certain types of pneumatic actuators). Working pressures up to 35 bar (500 lb/in^2) can be accommodated, using readily available fabric materials, because the fabric stress is relatively low. This stress can be calculated directly as:

$$S = \frac{P \times C}{2} \quad \text{(in consistent units)}$$

where P is the applied pressure
and C is the radial clearance or convolution width.

The rolling diaphragm is a special case and is normally considered as part of a whole unit design. About the only mechanical requirements are the avoidance of sharp edges on the piston and at the clamping point of the diaphragm, and excess slack to ensure that the convolution is never pulled out straight.

Single diaphragms are usually located between flange areas closed by clamping, bolting, crimping or, less usually, riveting. The sealing areas may be further shaped or treated to provide an adequate gripping surface.

Bolted flanges are usually to be preferred as they are cheaper than clamps and readily capable of providing an adequate and consistent seal, provided the bolts are torqued evenly to ensure equal closed pressure over the whole of the sealing area. Bolted flanges are also easy to disassemble to replace a diaphragm.

Crimping is a low cost production method, particularly suitable for lighter assemblies and involves rolling and turning one flange up over the others. The actual crimping pressure applied can, however, be critical. Excess crimping pressure can wrinkle, cut or otherwise damage the diaphragm material. Insufficient crimping pressure can result in leakage at the joint. In either case the fault cannot be rectified since a crimped assembly is a permanent unit which cannot be disassembled and remade.

Rolling diaphragms are generally suitable for sealing pressures from about 0.15 to 25 bar (2 lb/in^2); and for working temperatures over the range 55°C to +175°C. Heat degradation may, however, limit the upper temperature limit for continuous duty.

Piston Rings

FOR SEALING cylinders with working temperatures higher than those which can be accommodated by elastomeric, fabric or polymeric materials, metallic piston rings similar to those used in i/c engines become a logical choice. There are two categories of such seals: conventional (outspringing) piston rings or *compression rings* of integral construction; and *metal ring* seals which may be outspringing or inspringing and may consist of split ring or two or more separate rings in nesting configuration. The latter are described in a separate chapter. In addition there are modified ring forms, more properly described as individual designs of seals, but included here because they work as compression rings.

Automotive type compression rings have been developed as well established forms, *eg* see Figure 1. Other types of piston rings have been developed for i/c engines, *eg* oil control rings and pressure-energized rings (*eg* Dykes rings).

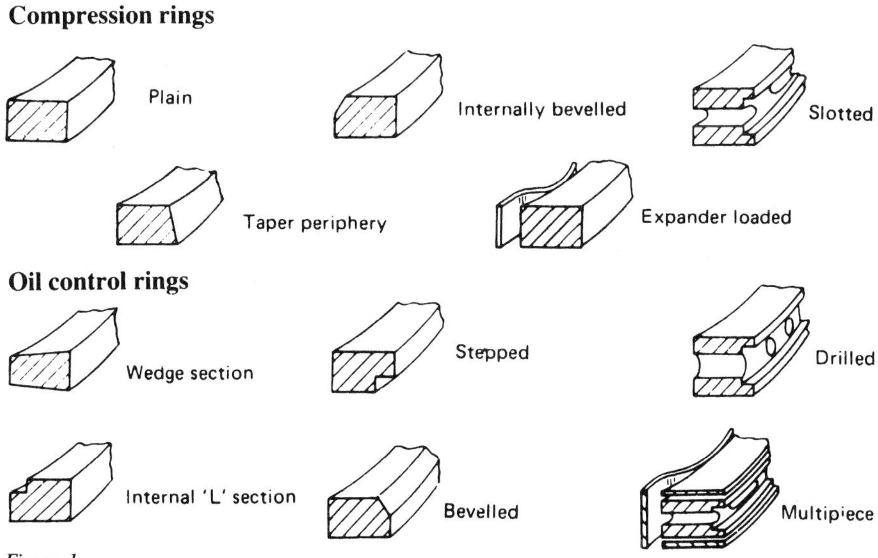

Compression rings

Oil control rings

Figure 1
Types of piston rings (BS 3627).

With high speed engines two compression rings are generally adequate because the use of a greater number invariably increases friction and the length of piston required. The conventional form is a plain rectangular section ring, although for rapid bedding down and better oil control, the lower rings may be tapered on the outer periphery or stepped on the inner periphery. A greater variety of sections are employed as oil control rings, including stepped, grooved, or stepped and bevelled sections. Ring grooves may also be inclined upwards slightly to inhibit twisting.

The majority of piston rings for i/c engines are designed to conform to BS 3627 which specifies normal diameters, maximum ideal thicknesses, widths, closed gap dimensions and minimum cylinder wall pressures for compression and oil control rings. Nomenclature for piston rings is specified additionally in BS 3556.

With piston rings, satisfactory performance is largely dependent on a high surface finish on both the bore and rubbing surface of the rings, and adequate support of the rings by their grooves. For assembly, the rings must be split and although this split is nominally closed this offers an additional leakage path. Leakage can be minimized by employing a stepped, scarfed or overlapping gap rather than a straight cut ring.

Good momentary sealing can be provided by suitable automotive type ring sets with relatively low rubbing friction in the presence of adequate lubricant; but the ability to hold pressure is limited. Performance also falls off rapidly with deterioration of the bore surface, or lack of concentricity. Better sealing on worn bores can be achieved by special forms of piston rings.

Typical forms of one-piece piston rings and their applications are summarized in Table 1 and also shown in Figure 2. *Balanced* piston rings are preferred for applications in which ring wear may be excessive. The use of such rings is desirable in many services involving excessive ring wear due to high loading, *eg* high pressures. These balanced piston rings greatly reduce ring wear (Figure 3).

TABLE 1 – TYPES OF ONE-PIECE PISTON RINGS
(see also Figures 2 and 3)

Description	Remarks
Uncut	Needs to be made in a stretchable material, *eg* PTFE or filled PTFE.
Straight cut	Most widely used type.
Angled cut	Similar leakage to straight cut; also widely used.
Step cut	Eliminates straight leakage path with reduced leakage.
Seal joint	More difficult to produce but can be designed to give virtually positive sealing with automatic compensation for wear.
Angle cut with expander	Similar leakage path to straight cut or angled cut ring, but expander can improve sealing performance.
Step cut with expander	Similar leakage path to step cut ring, but expander can improve sealing performance.
Balanced (single-acting)	Pressure balanced in one direction *via* circumferential groove and axial slots.
Balanced (double-acting)	Minimizes axial unbalance to eliminate side wear or pressure locking.

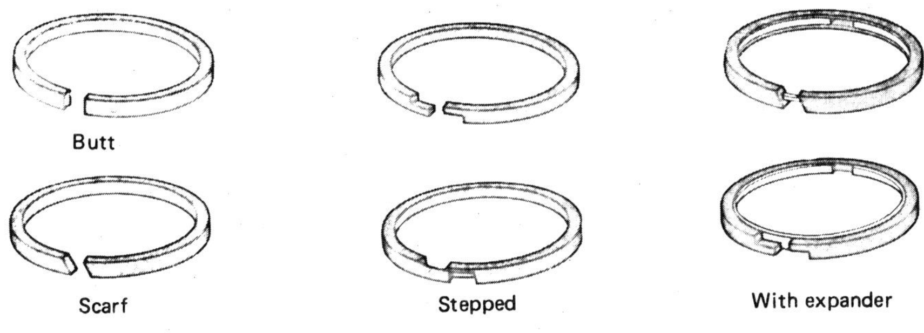

Figure 2

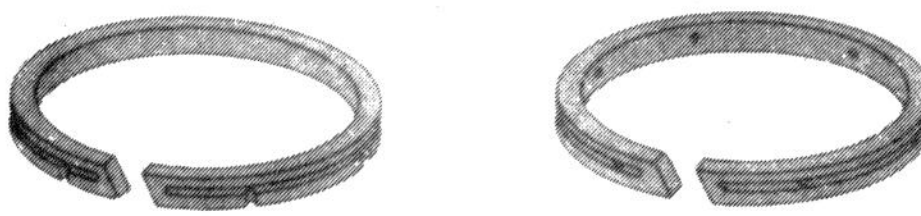

Figure 3
Balanced rings; single-acting (left) and double-acting (right).

Depending on the application, ring balancing may involve the ring face only or both the ring face and one side. Pressure balancing in one direction only (single acting) is accomplished by a groove and axial slots around the circumference of the ring face. For pressure balancing in both directions (double acting), radial holes are drilled into the groove. Where side wear or pressure locking may occur, piston ring performance can be improved by the use of dual balancing grooves which minimize axial unbalance to effectively reduce leakage and face wear.

Composite assemblies may also be used, *eg* twin or multiple rings (Figure 4) or segmental piston rings (Figure 5). Besides simplifying assembly, wear on piston rings having a small wall-to-diameter ratio can generally be more uniformly distributed when the ring is segmental.

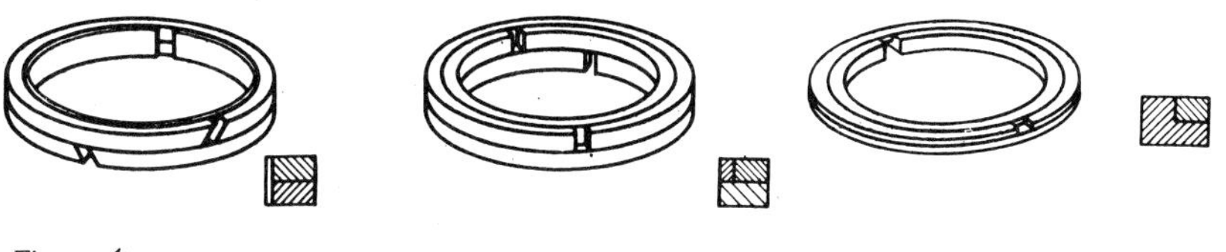

Figure 4
Multiple rings.

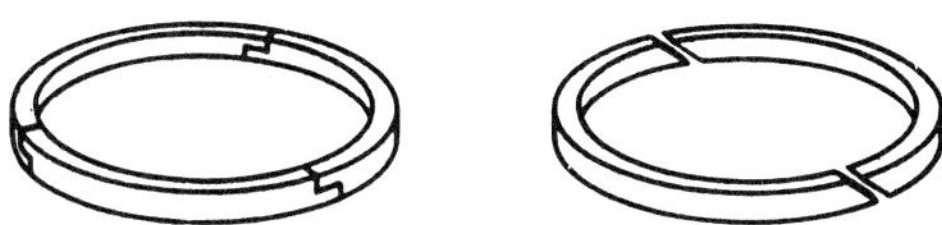

Figure 5
Segmental rings.

In complete assemblies, piston rings may be associated with rider rings or bull rings. These are used particularly in non-lubricated services to carry the weight of the piston and prevent piston-to-cylinder contact. Rider rings commonly have side relief or face relief grooves to prevent pressurizing. Typical forms are shown in Figure 6. See also Table 2 for typical dimensional data.

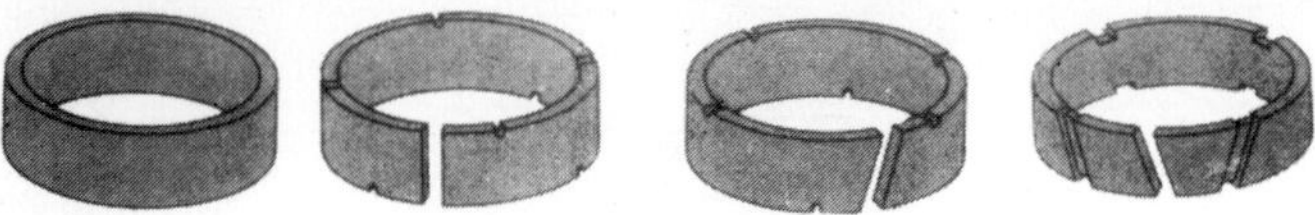

Figure 6
Examples of rider rings.

TABLE 2 – PISTON AND RIDER RING DATA

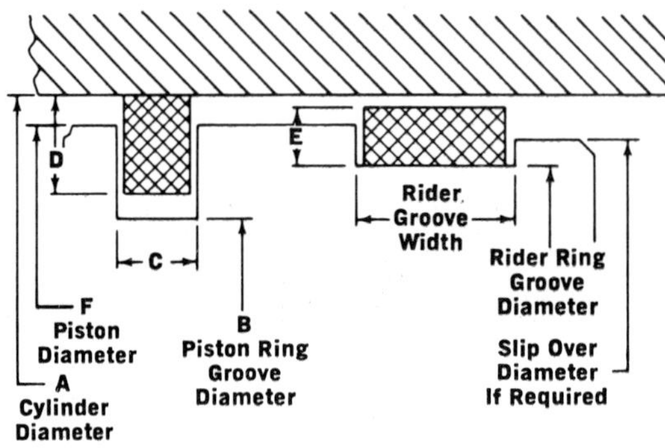

A	B	C	D	E		F
	Piston Rings			Rider Rings		Piston
Cylinder Diameter	Max Groove Diameter	Groove Width	Ring Wall	Solid Wall	Split Wall	Diameter*
¾ – 1¼	A–0.312	0.125	0.156	0.080	0.125	A–0.078
1¼ – 2½	A–0.500	0.188	0.218	0.094	0.188	A–0.094
2½ – 5	A–0.625	0.250	0.281	0.125	0.250	A–0.125
5 – 8	A–0.688	0.250	0.312	0.188	0.312	A–0.156
8–12	A–0.938	0.375	0.437	0.250	0.375	A–0.188
12–16	A–1.312	0.500	0.625	0.312	0.438	A–0.219
16–21	A–1.625	0.625	0.750	0.375	0.500	A–0.250
21–26	A–2.062	0.750	0.937	0.438	0.562	A–0.266
26–31	A–2.062	0.750	0.937	0.500	0.625	A–0.281
31–37	A–2.688	1.000	1.250	0.562	0.625	A–0.297
37 – up	A–2.688	1.000	1.250	0.625	0.625	A–0.312

*The piston-to-cylinder clearance shown is for non-lubricated service. In lubricated or minimal-lubricated service using PTFE rider rings, this clearance may be halved.
All dimensions are in inches

A modified form of automotive-type piston ring which has particular application to hydraulic cylinders is shown in Figure 7. In this example pairs of wedge shaped rings are employed, mounted in and supported by elastic cushion rings. These cushion rings provide initial diametral compression and also provide the metal rings with 'back up' pressure when distorted under the action of fluid pressure, extrusion of the cushion rings being prevented by the shape of the metal rings. Thus the seal is partially pressure-energized, with the advantage that contact pressure can be kept low at low pressures for minimum friction, increasing automatically with increasing pressure.

Such elastomeric-backed piston rings are normally employed in sets, as shown in Figure 8. The same principle may be employed for rod seals with inverted construction. In such types, however, the metal ring sets are normally separated by one or more 'spacer' rings in elastomeric material (Figure 9). This is sometimes referred to as a *cartridge type rod* packing. An advantage offered by this type of packing is that it can act as a bearing as well as a seal.

The same type of construction – such as stiff split rings of wedge form mounted in elastomeric cushion rings – may also utilize laminated plastic or other hard materials for the rubbing rings instead of metal, depending on the service conditions, operating temperature and pressure, and compatibility requirements.

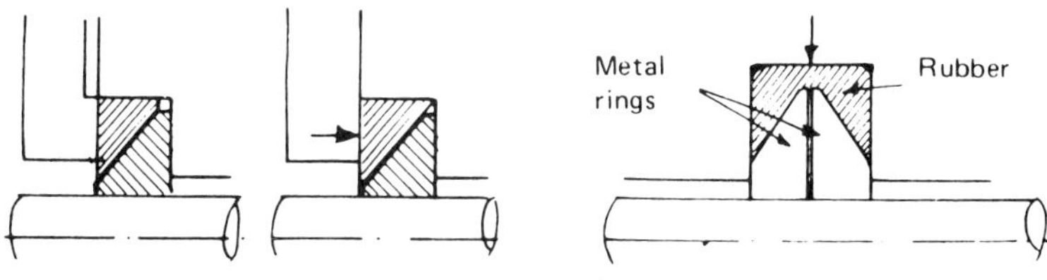

Figure 7

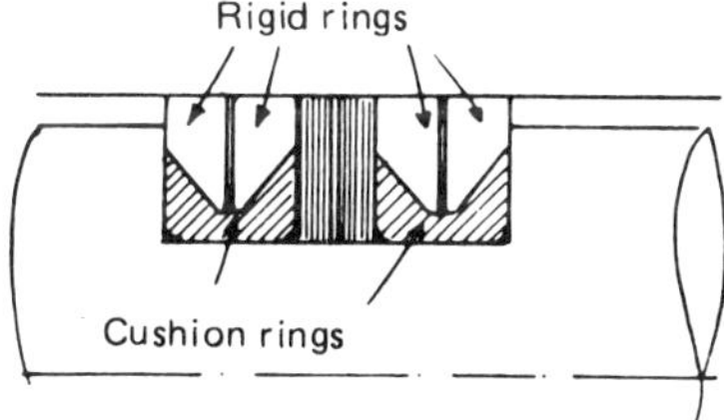

Figure 8

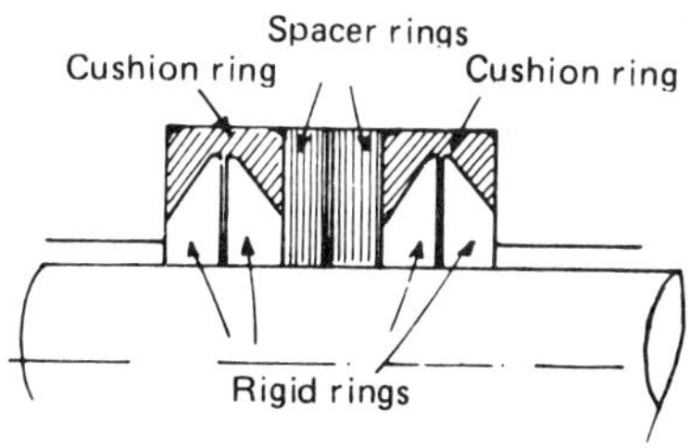

Figure 9

Materials

Metals commonly used for piston rings include: cast iron, alloy iron, steel and alloy steels, nickel alloys and cobalt alloys. Choice depends largely on the chemical and temperature resistance required. The former may also be improved by plating or coating the rings.

Cast iron piston rings provide good sealing over a wide temperature range in applications where adequate lubrication can be provided. Any tendency towards 'scuffing' can be inhibited by the application of a scuff-proof coating to the rings.

Bronze rings are to be preferred for high pressures and temperatures where lubrication may be marginal. They also have improved wear resistance over cast iron piston rings. Various grades of bronze are used, ranging from high lead bronzes to tin bronzes. Choice depends largely on the corrosion resistance required and working temperature.

Non-metallic rings produced in carbon are also suitable for high temperature applications. *Carbon-Bakelite* piston rings are also particularly suited to hydrocarbon gas and vapour services, being unaffected by sour or wet gases. Carbon-Bakelite rings are suitable for working temperatures up to 135°C (275°F); and asbestos-carbon-Bakelite up to 177°C (350°F). Plastic rings have a lower service temperature rating, but can offer specific advantages where they can be accommodated. For example, they are easier to produce and PTFE rings, in particular, have very favourable anti-friction properties. This makes them particularly suitable for piston seals operating with minimal lubrication (*eg* in oil-free compressors).

PTFE alone lacks the mechanical strength and wear resistance required from a piston ring material. These properties can be enhanced by filling. Initially PTFE filled with carbon/graphite or glass provided the necessary performance and largely replaced conventional oil-lubricated metallic *piston* rings and *bull* rings in oil-free compressors (and similar machines). Even better mechanical performance is now achieved by specialized grades of filled PTFE materials which have subsequently appeared.

Cord rings

In the cord segmental ring (Figure 10) the seal comprises a series of thin annular laminations or segments, one or more of them being cupped or dished. When assembled in the formation illustrated, lateral tension is exerted by the cupped segment sufficient to maintain a light contact with any axial abutments such as the sides of a groove turned in a shaft or the end plate of a bearing housing.

The segments are designed in such a way that the radial pressure exceeds the lateral pressure exerted by the cupping of the middle segment, which is just sufficient to ensure that the seal makes contact with the sides of the groove. Thus the seal will not revolve

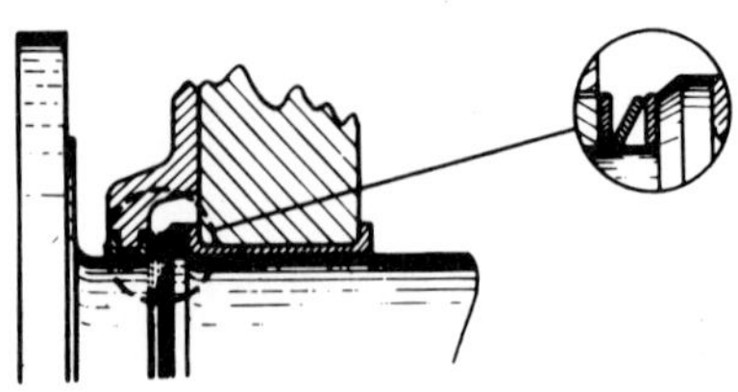

Figure 10
Cord ring used as a shaft seal.

Piston rings in reinforced PTFE.

in the crankshaft housing and very light pressure is exerted at the point where wear might occur, *ie* between groove and seal.

When the segments are assembled in a typical formation with the end gaps alternately at 180° to each other, a labyrinth is formed, thus reducing gap leakage to a minimum. Further, the fact that the several segments (in this case five) can move independently of one another enables a satisfactory seal to be made in the housing without demanding a better finish than would normally be required.

This application has a considerable potential where housings are of the split type, which make it necessary for a seal to be made in two sections. The segments in the seal are sufficiently flexible to enable them to be assembled in the groove on the shaft by winding them over the crank web or main bearing without fear of breakage or distortion; there are no difficulties in assembling the housing over the seal.

A feature of the seal is that, due to its design and the materials used, there is extremely little wear of the sealing ring itself or the shaft or housing.

For particular applications special materials can be used, such as stainless steel, or special finishes employed, such as hard chrome deposition, parkerizing, *etc*, or plating with a soft metal.

See also chapters on *Metal Ring Seals* (Section 4) and *Carbon Seals.*

Carbon Seals

SYNTHETIC CARBON has wide application as a sealing element for rotating and reciprocating movement, particularly where corrosive media are involved or where thermal stress does not permit lubrication. It is a self-lapping material which means that it can be 'run-in' to conform closely to its mating surface; but it is also a hard material, extremely resistant to wear after initial run-in.

Carbon rings are used as simple seals in their own right, and also as the sealing element in mechanical face seals. The former embrace both contact and non-contact seals.

Complex carbon ring seal for aircract jet engines.

Contact seal rings may be plain (*eg* simple sliding ring seals such as stem head glands) or segmented, in the case of radial seals. The former can act as a bearing as well as a seal both for reciprocating and rotary motions, with assembly tolerances as for carbon bearings. Splitting the seal ring into segments together with an external garter spring produces a true contact seal (Figure 1). The radial movement provided by the segment

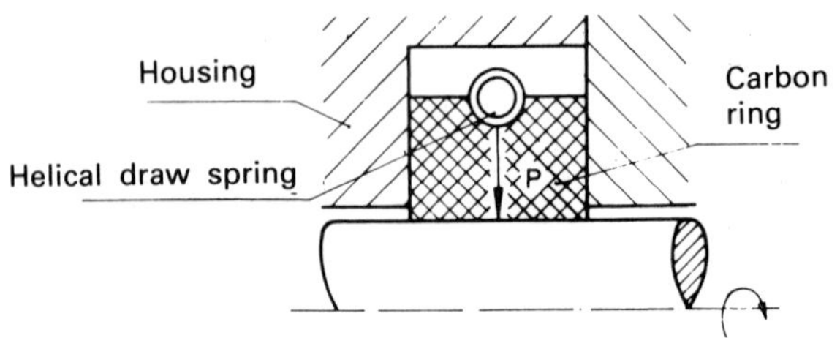

Figure 1

geometry and the helical tension spring is arranged so that the specific contact pressure (P) exerted on the shaft does not exceed 0.10 to 0.15 bar (1.5 to 2.2 lb/in ^{2}), depending on the carbon grade used. Such a construction is also usually self-adjusting to accommodate any wear on the i.d. of the seal.

Axial contact seals are plain moulded rings and may have the sealing surface lapped to very fine tolerances. They work as face contact seals, contact pressure being maintained by a compression spring (Figure 2). Split (segmental) rings may also be used as combined axial-radial seals, in which case the section is normally so shaped that a single helical draw spring provides both radial and axial contact pressure (Figure 2a).

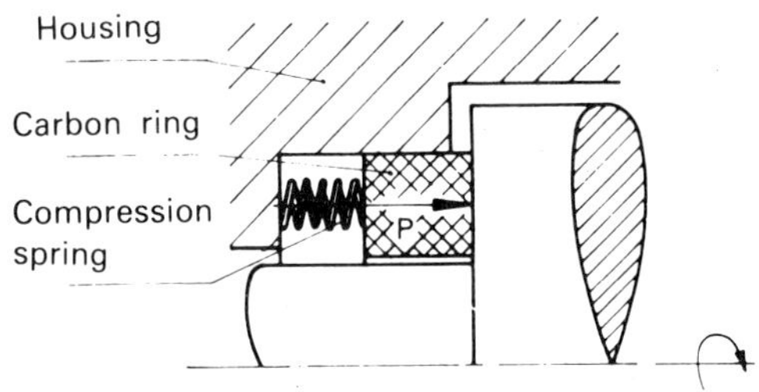

Figure 2

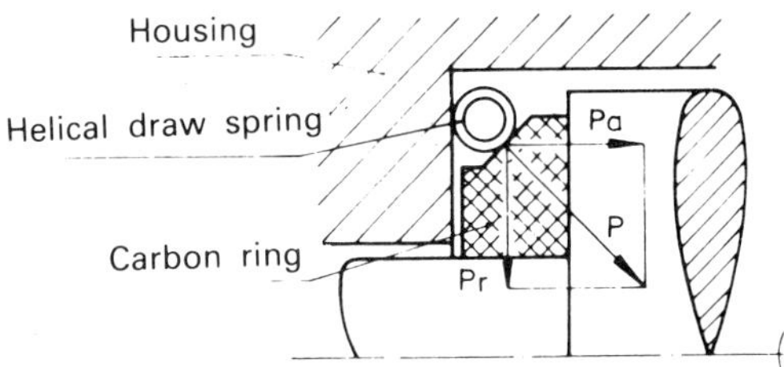

Figure 2a

As a general rule, carbon contact seals run best against a hard mating surface, such as cast iron, hardened steel, chrome-plated steel or ceramics or carbon for sealing highly corrosive media. Various carbon grades are also available to adjust the seal material to suit particular applications. These range from relatively porous carbon grades, for low to medium pressure oil sealing applications (the retention of oil by the carbon aiding seal face lubrication), to special grades for non-lubricated services, extreme corrosion resistance, or high resistance to oxidation or high temperatures.

Carbon bushings and gland rings are examples of non-contacting carbon seals (Figure 3). Close tolerance sizing and accurate assembly is necessary to obtain sufficiently close clearance(s) for the seal to be effective. A relatively long length of seal is also desirable, for the same reason. The fit must also take into account the difference in thermal expansion of the carbon and metal housing, at the working temperature of the seal.

Carbon is also an effective material to use for labyrinth seals (Figure 4). Here it has an advantage over metals that the seal is not necessarily damaged by contact between shaft and seal, so the difference in thermal expansion of the seal and shaft materials is less significant. This also means that a *carbon labyrinth* seal can be designed with

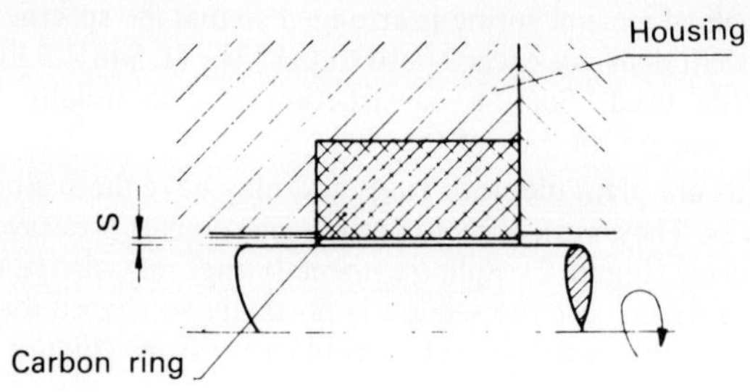

Figure 3

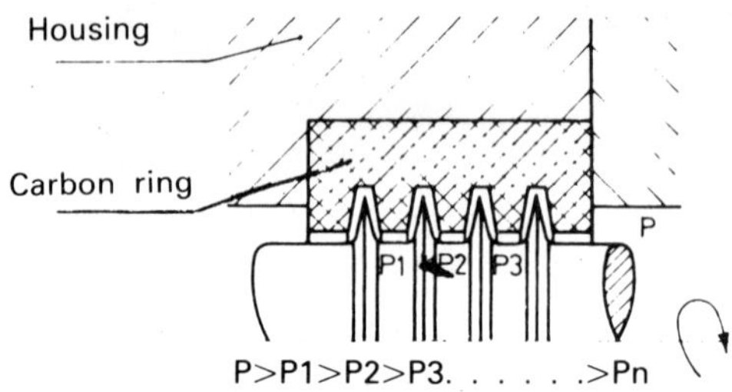

Figure 4

closer clearances than a *metal labyrinth,* so the length of the seal can be shorter for the same performance.

See the section on *Mechanical Face Seals,* and the chapters, *Compression Packings* (Section 4) and *Cemented Carbides* (Section 2).

Liquid Ring Seals

LIQUID RING seals are largely peculiar to pumps because they are in themselves pumps and are often incorporated in high speed pump designs as an alternative to mechanical or other shaft seals. The principle of operation is that an impeller rotating in a narrow chamber creates a ring of liquid around the periphery of the chamber with a pressure equal to that generated at the suction eye of the pump. The system is thus in balance, providing a 'rigid' *liquid* seal, and remains so as long as the pump is operating.

Typical components for a *liquid ring* seal applied to a pump are shown in Figure 1. Figure 2 illustrates the principle of operation. Starting with (a) the pump is stationary and full of liquid. At (b) the pump is rotating and a ring of liquid has been formed in the chamber housing the sealing impeller. Diagram (c) is appended to show that with the pump dry and rotating the sealing impeller runs with full clearance and so can run dry indefinitely without harm, unlike most other seal types. The fact that it produces no ring of sealing liquid under this condition is immaterial.There is no liquid present in the pump which requires sealing under this condition. Equally, however, once the pump stops there is no sealing action present even with a flooded pump. For this reason liquid ring seals applied to pumps operating with flooded suction must be associated with a secondary seal to provide sealing under static conditions.

Ideally, the static seal used should be simple and disengageable once the pump has started up and the liquid ring seal has been formed. One solution adopted is the use of a special section rubber seal ring which lifts clear of the shaft under centrifugal force once the pump is running. Alternatively, simple lip seals may be used. Yet another solution is to isolate this side of the pump automatically from the fluid with a shut-off valve

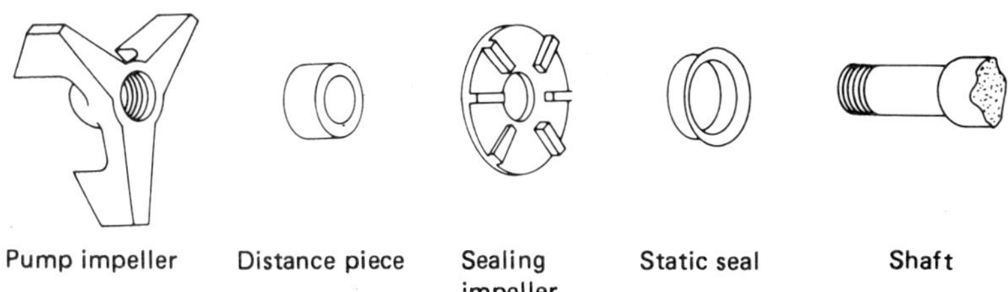

Figure 1
Typical liquid ring seal components.

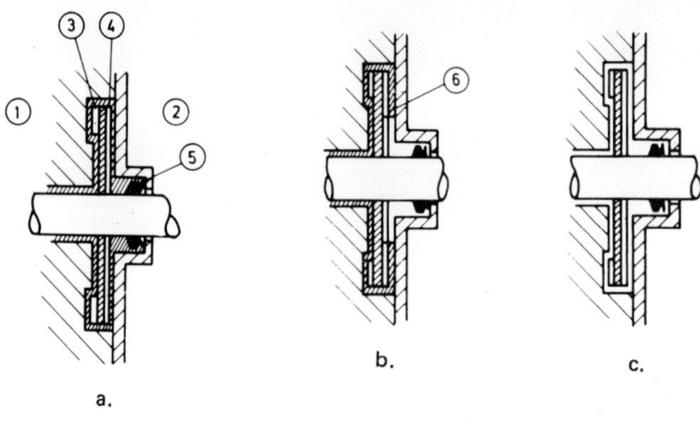

1. Pump side.
2. Atmospheric side.
3. Short vanes.
4. Long vanes.
5. Static seal.
6. Liquid level when rotating.

Figure 2
(a) Stationary pump full of liquid.
(b) Rotating pump full of liquid.
(c) Rotating pump dry.

immediately the pump stops. There is also the possibility of using an inflatable seal as a static seal, although this requires a separate pressurized air or fluid supply. Such problems with static sealing are, of course, only present with flooded suction.

A more general problem is the pressure limit of liquid ring seals. The higher the suction pressure the bigger the diameter of the sealing impeller required, absorbing more power and generating more fluid heating. Turbulence can also be a problem, interfering with the effectiveness of the seal; but a bigger problem can be air drawn into the pump when a pump is started dry and the static seal disengages. Additional problems can arise when the liquid being pumped is highly volatile.

The liquid ring seal is designed and built as part of the pump so such problems merely become part of the overall design of a pump for particular applications. Once solved for a specific application they may well provide solutions for other applications. Fields in which liquid ring seals have proved particularly effective are pumps handling chemical and pharmaceutical products, being free from corrosion and crystallization problems which can limit the life of conventional gland seals and even mechanical seals. They have also proved excellent on smaller size high speed pumps for water duties. Their main advantages appear to be as seals for high speed liquid pumps.

Liquid Barrier Seals

VARIOUS FORMS of *liquid barrier* seals have been developed for sealing gases. A basic example is the use of a lantern ring in an otherwise conventional stuffing box (Figure 1). Water or any other suitable fluid is fed under pressure to the centre of the stuffing box where it is distributed by the lantern ring. This produces a fluid sealed gland with zero leakage characteristics as far as the gas contained is concerned. There will, however, be leakage of the sealing fluid both inwards and outwards past the gland, and provision may have to be made to collect and drain this leakage on the outward side.

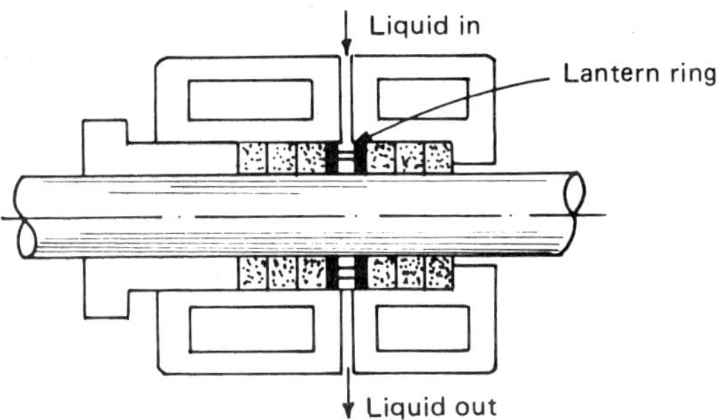

Figure 1

Both compressed air (or gas) and oil fluids can be used in this manner, although the precise control necessary normally limits the application of such seals to highly specialized machines, such as large machines handling gases or filled with gas, *eg* hydrogen cooled generators. In such cases face or shaft type seals may be employed, fed with pressurized oil so that a complete annulus of oil is maintained at the seal interface at a pressure higher than that of the gas to be sealed. The resulting seal has zero leakage characteristics as far as the gas is concerned, although there will normally be oil leaks in both directions past the seal. This can be restricted by back-up seals on either side of the main seal, with provision on each side to drain the oil away. In particular cases, however, it may be necessary to vacuum treat the oil in order to avoid reducing the purity of the gas being sealed, although a number of special seals of this type have such a low oil flow to the gas side that vacuum treatment is not required.

Compressed gas or *compressed air* seals operating on a similar principle are rather less common, although they offer certain inherent advantages. The main disadvantage is usually the complexity of the compressor and control gear requirements, although these may be simplified to a degree for simpler requirements.

Another form of *oil-fed* seal is shown in Figure 2. The sealing element is a double ring, free to float in a groove in a housing. Oil is then fed to the back of the ring at a slightly greater pressure than that of the gas being sealed, forcing the ring into contact with the shaft but also maintaining an oil film at the interface with flow both inwards and outwards. The flow rates in these two directions are proportional to the respective differential pressures. Using an elastomeric or resilient ring material, pressurization of the back of the ring provides automatic adjustment to shaft vibration or axial float. *Single ring oil-fed* seals of this type have also been produced, both with single and double flow. Single rings tend to be more effective with increasing pressures because of the reduction possible in ring distortion.

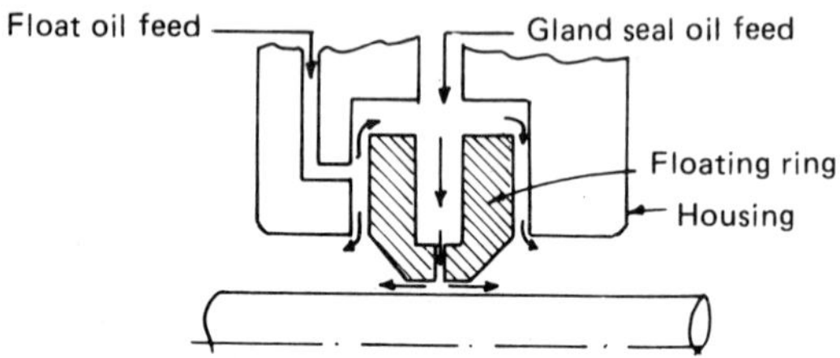

Figure 2

Centrifugal barrier seals

A more simple type of oil-fed shaft is the *centrifugal liquid barrier* seal (Figure 3). This comprises a plain rotor mounted on the shaft, rotating with positive clearance in a U-shaped chamber. Sealing fluid present between the two surfaces is thrown outwards under centrifugal force when the shaft is rotating providing a sealing pressure differential dependent on the centrifugal force, shaft speed and diameter, and actual length of the chamber. This form of seal has very low friction and is suitable for high shaft speeds and large shaft diameters. Sealing is only present when the shaft is rotating so an additional static seal (or seals) may be necessary to contain the fluid under static conditions. An inflatable seal is a possibility here.

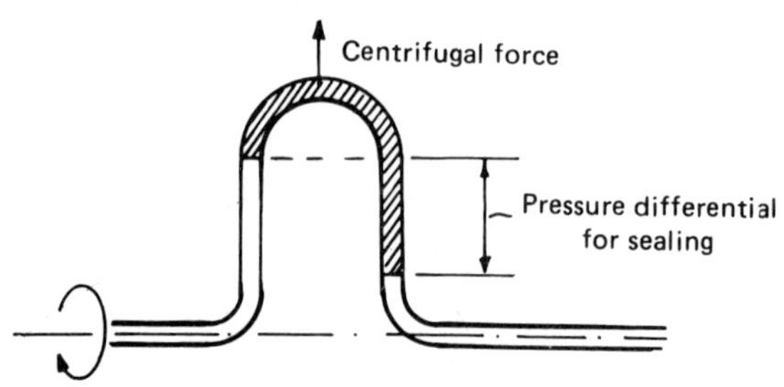

Figure 3

This same principle can be incorporated in the design of a mechanical seal (Figure 4). Here the mechanical face seal provides sealing when stationary or rotating at low speeds. At high speeds as centrifugal force is built up in the fluid, fluid pressure lifts the mechanical face seal off the rotor and sealing is then maintained by the differential pressure available in the liquid barrier.

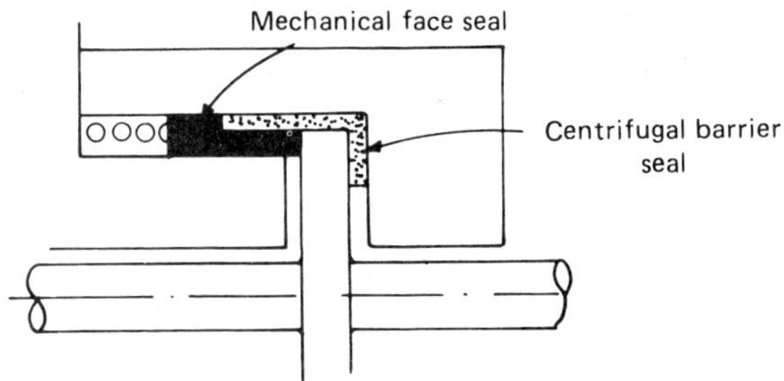

Figure 4

A further type operating on centrifugal force is the centrifugal impeller seal (Figure 5). Here an impeller with vanes on one side and a smooth face on the other rotates within a chamber with small radial and axial clearances. The liquid to be sealed is applied to the plain side of the impeller and the other side of the chamber is fed with pressurized gas. The liquid will flood the smooth side of the impeller and then tend to flow along the vane tip clearance space where it will be balanced by the gas pressure on that side. The gas pressure can be adjusted to establish the liquid/gas interface at a suitable position to provide a true zero leakage seal (although there will be gas leakage from the secondary shaft seal). Designs of this type may employ single or multiple impellers, or notched-bladed single impellers with a matching housing section. The latter type is generally more efficient in providing a labyrinth path as well as centrifugal sealing.

This particular form of seal is for sealing liquids, utilizing pressurized gas as the source of sealing. In this respect it is different from the liquid ring seal (described elsewhere) which is another type of centrifugal liquid barrier seal utilizing an impeller to generate the barrier (liquid ring), but working in the fluid being sealed.

See also chapter on *Liquid Ring Seals*.

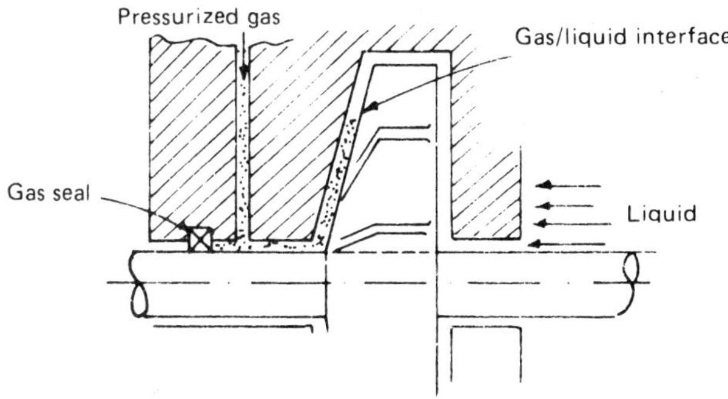

Figure 5

Inflatable Seals

INFLATABLE SEALS are highly specialized designs of gaskets providing closure of a gap by expansion of a hollow folded or 'concertinaed' seal section, through inflation of the seal *via* applied air or fluid pressure. Introduction of air (or fluid) causes the seal to expand to a partially rounded section filling the gap and applying positive pressure against the surface to be sealed. The seal section can also be shaped to provide high conformity and thus good sealing on uneven surfaces. The section can also be tailored to close quite large gaps, *eg* up to 75 mm (3 in), although large gap seals usually require high inflation pressures which may reduce the gasket life. Inflation pressures used may range from about 0.35 bar (5 lb/in^2) to 7 bar (100 lb/in^2).

Typical applications are seals for clean room doors, aircraft canopies and hatches, airlock doors, marine doors and hatches, pressure vessels, vacuum chambers, processing equipment, special doors and nuclear equipment. Such seals are normally manufactured from fabric reinforced rubber tube, moulded with the sealing face in the retracted or collapsed position, *eg* see Figure 1. Low pressure inflation gives rapid expansion to close the gap, an optimum gap size being about 80% of the free height of the fully inflated seal. When pressure is released the seal returns rapidly to its retracted shape.

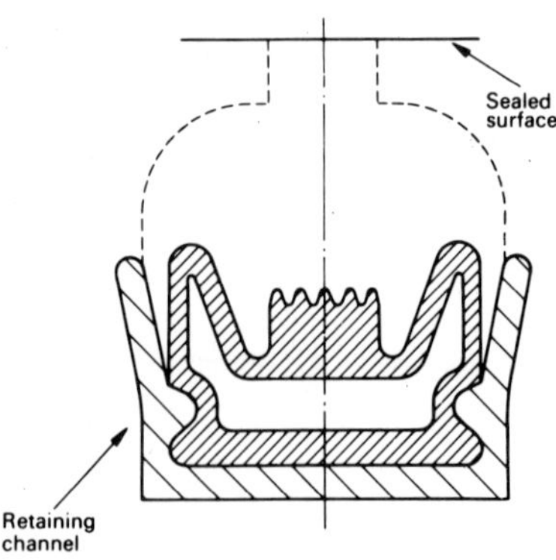

Figure 1
Typical inflatable seal.

Seals of this type are normally designed for specific industrial applications, taking into account the following considerations: gap to be sealed; periphery, or seal length; environment; pressures – both across the seal and internally; temperature; type of machined groove or retainer; inflation medium – gas or liquid; striking surface; inflation medium – gas or liquid; inlet type and configuration; minimum corner radius; plane of installation; and cycle frequency.

Assuming a groove can be machined in the frame, a seal is designed to either snap in or be cemented in this retention type recess. If not, then the seal is designed to be fitted in a suitable retaining channel section capable of positive location on assembly. The seal itself is then moulded in a continuous loop with high strength fabric reinforced elastomer to fit the periphery of the machined groove, or retainer. The seal is installed by inserting the valve stem and then inserting the balance of the seal by snapping into place or cementing to the groove configuration.

Operation of the seal is accomplished by inflating to the design pressure, permitting the seal to make positive contact with the surface to be sealed. Seals are designed to roll smoothly from the deflated to the inflated configuration and back again upon release of the inflating medium. This feature compensates for variations within the range of the seal. While the inflation medium may be any gas or liquid, compressed air is the most readily available, and thus is normally used. The use of liquids is desirable for applications such as on furnaces and degassing units, because of their cooling effect. When using liquids or when very rapid retraction is required an aspirator should be used. Also, a vacuum pump is most effective.

Inflatable seals may be operated at a very low pressure when used as a barrier to exclude light, heat, cold, noise, airborne contaminates, water *etc.* Higher pressures are encountered in pressure vessels, water gates and vacuum chambers.

A further type of (proprietary) inflatable seal is shown in Figure 2. This is designed for fitting outboard of conventional stern glands on marine propeller shafts. Its purpose is to provide a static seal with the shaft stationary, enabling the main gland to be serviced or checked without leakage, and without the necessity of trimming the ship bow-down to achieve this end. It can also be fitted to pump shafts to permit (main) seal maintenance without draining the system.

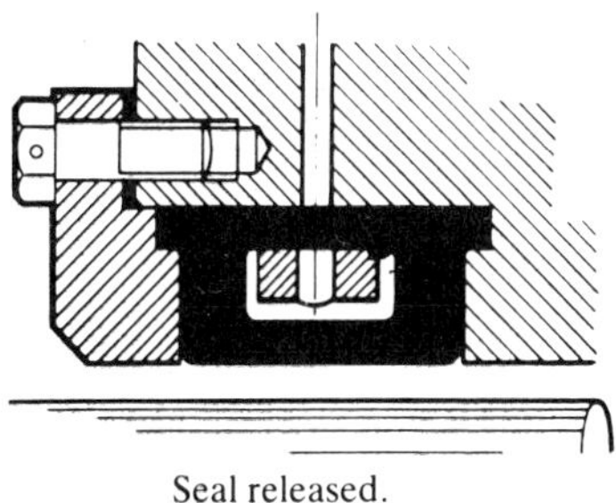

Seal released.

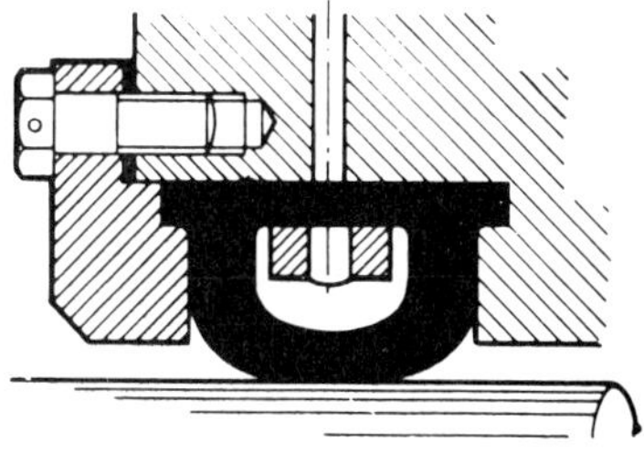

Seal inflated and gripping shaft.

Figure 2
Inflatable seal for marine propeller shafts.

Such seals are available as complete rings. They can also be split to fit over unbroken shafts, and bonded into complete rings on assembly. It should be noted that they are designed specifically as static seals in the inflated condition. When it is required that the shaft rotate the seal must be released.

Material selection is a vital factor in determining the longevity of the inflatable seal. Considering temperature and atmospheric conditions, flex requirements and motive operation, the use of nylon, *Dacron* or *Nomex* fabric combined with a range of elastomers enables each individual requirement to be satisfied.

Inflatable seals will conform to many contours. Moulded in a loop, they can be mounted in three planes: inflating radially inward, or outwards, or axially. Many applications have bends in more than one plane and, occasionally, the seals must twist.

Inflatable seals can also be used to apply forces that would not be possible using conventional rigid methods, *eg* in the following applications: actuators; force applicators; gripping devices; special plugs; and weight lifting devices. They are then referred to as flexible pressure vessels.

Ferrofluidic Seals

MAGNETIC FLUIDS, or *ferrofluidic* seals as they are known, consist of a colloidal suspension of sub-micron size magnetic iron oxide particles in various carrier fluids. While behaving like any true homogeneous fluid, they are influenced by a magnetic field, a property utilized in the design of ferrofluidic seals.

In its simplest form a ferrofluidic seal consists of two pole pieces separated by an axially polarized magnet and a ferrofluid (Figure 1). The flux lines formed in the closed magnetic circuit focus the ferrofluid in the gaps, producing a series of liquid O-ring seals with virtually immeasurable leakage, only low viscous drag and no contact wear. Viscous drag is independent of the pressure being sealed, produces no stiction and the low rotary drag means that the seal can operate successfully on high shaft speeds. At very high speeds, however, cooling may be necessary to prevent degradation of the fluid through the heat generated by viscous shear of the fluid.

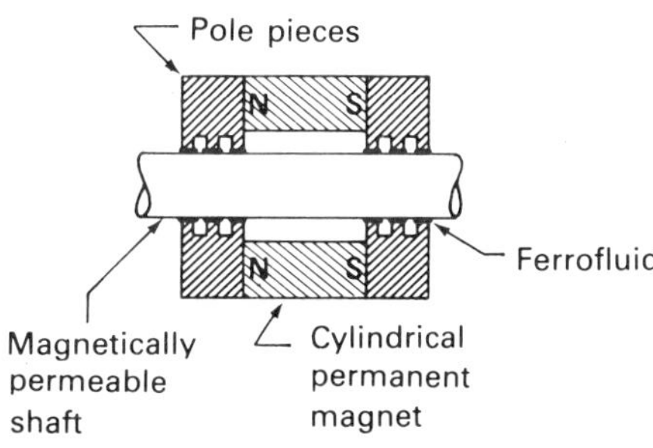

Figure 1

Design of a ferrofluidic seal can be tailored to meet specific differential pressure requirements. Control parameters are (i) the size and energy product of the magnet, (ii) the size of the gap, (iii) the magnetic saturation of the ferrofluid used, and (iv) the magnetic saturation of the shaft and pole pieces.

Magnets used range from low-energy ceramic magnets to high-energy Sm-Co or Nd-Fe-B magnets. Choice here is largely a compromise among the factor such as performance, volumetric efficiency and cost.

Choice of gap depends on several factors. The pressure capability of a ferrofluidic seal is to the first order inversely proportional to the size of the gap, favouring the use of a minimal gap. However, the ferrofluid itself is not load-bearing, so the shaft has to be carried by conventional bearings. Sufficient clearance must be present to prevent rubbing contact occurring between the pole pieces and shaft under all conditions. On the other hand, too large a gap will not only decrease the pressure capability of the seal, but also limit the maximum magnetic field intensity attainable because of fringe field effect.

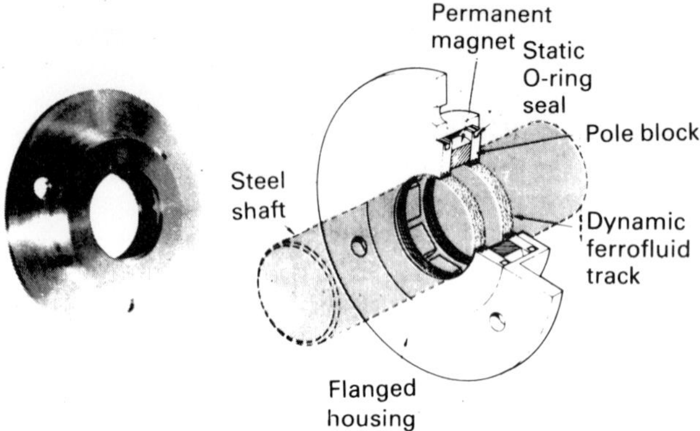

Typical construction of a simple ferrofluidic seal.

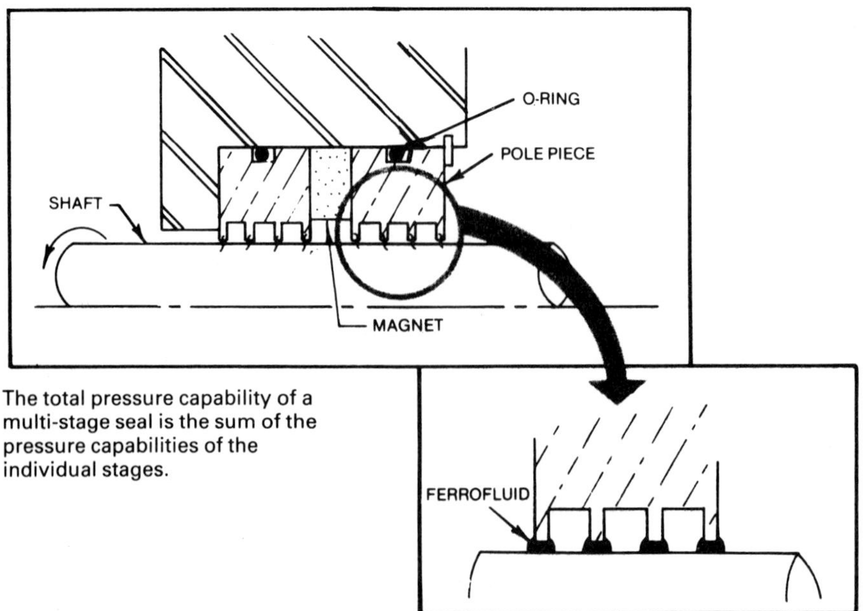

Multi-stage ferrofluidic seal.

In practice a typical gap figure is 0.05 to 0.125 mm (0.002 to 0.005 in), although this may be increased to 0.25 mm (0.010 in) to accommodate eccentric shaft motions.

A multi-stage seal may be used instead of an unstaged seal to improve the pressure capability. This can take the form of tooth-like projections on the shaft under the pole pieces, or on the pole pieces themselves (Figure 2). The effect is to focus ferrofluid on each tooth or segment, the total pressure capacity then being the sum of that of the individual stages.

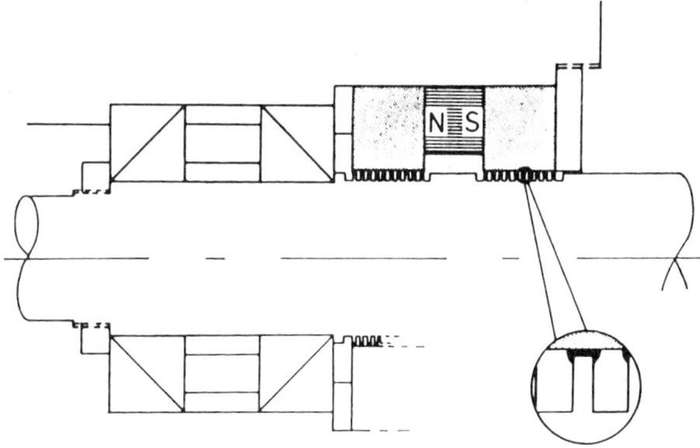

Figure 2

Ferrofluids themselves exhibit different levels of magnetic saturation, depending on the actual volume of magnetic solid colloidally suspended in the carrier liquid. Below saturation the fluid magnetization is more or less directly proportional to the applied field. Most commercial ferrofluids reach their saturation level at field strengths of the order of 2500 oersted. Above saturation the magnetization of the fluid remains constant, regardless of the strength of the applied field. Choice of carrier fluid is normally based on service requirements – see Table 1.

The fourth way of increasing pressure capacity of a ferrofluidic seal is to use materials having a higher magnetic permeability for the pole pieces and shaft. Here it should be noted that at high values of magnetization a magnetically 'soft' material approaches saturation while its permeability decreases relative to that of air. It is thus necessary to use soft magnetic materials well below their saturation level.

If the shaft is of non-magnetic material it must be fitted with a sleeve of magnetic material to complete the magnetic circuit. Equally, if the seal housing is of magnetic material, a non-magnetic sleeve must be fitted between the housing and seal to prevent a magnetic short circuit.

Currently ferrofluidic seals can be designed to seal gas pressures up to 35 bar (500 psi) or more by increasing the seal length and/or multi-staging.

Rotary drag

Magnetization does not affect the starting friction or stiction present in a ferrofluidic seal. At speed the only rotary drag force present is that due to viscous shear. This is controllable by selection of the fluid viscosity.

TABLE 1 — FERROFLUIDS FOR SEALING APPLICATIONS

ESTER BASED

- Used in all Standarad Feedthrus
- Compatible with water and humidity
- Low start-up torque

Hydrocarbon Based

- Recommended for heavy duty feedthrus
- Compatible with water and humidity
- High start-up torque
- Running power consumption same as ester based

Fluorocarbon Based

- Recommended for environments:
    - Reactive
    - Radiation
    - High Temperatures (no cooling)
    - Water & Humidity
- High start-up torque and power consumption

Low Viscosity Hydrocarbon Based

- Low temperature (aerospace) applications
- Compatible with water and humidity
- Low drag torque
- Limited life at high temperatures

POLYPHENYL ETHER

- Very low vapour pressure
- Oxidation resistant
- Compatible with high radiation levels
- High temp — no cooling necessary
- High start-up torque and power consumption

A ferrofluidic rotary seal

Losses due to viscous shear will show up as an increase in temperature of the fluid. This in turn will reduce fluid viscosity, so rotary drag remains low, even at very high speeds. Successful operation of ferrofluidic seals at speeds in excess of 100 000 rpm is not unusual. However, an excessive temperature rise must be avoided to prevent degradation of the fluid.

A further requirement for maintaining a reasonable low fluid temperature is to minimize evaporation and thus fluid loss from the seal.

Methods of reducing the working temperature of the fluid are:- using a fluid with a lower initial viscosity, increasing the gap (at the expense of pressure capability); cooling the seal. An example of a cooled ferrofluidic seal is shown in Figure 3.

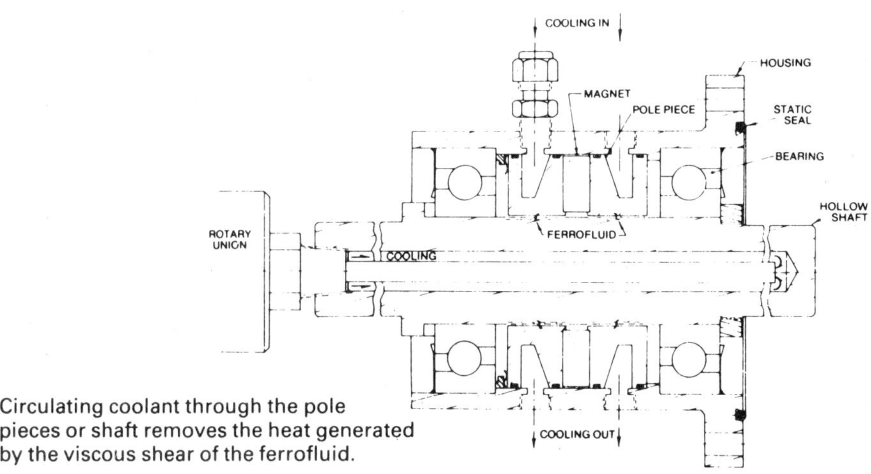

Figure 3

Choice of fluids

Carrier fluids used range from Esters through Polyphenyl Ethers. Choice is necessarily based on compatibility with the gaseous medium being sealed. Ester based fluids are probably the most commonly used carrier fluids with their very low vapour pressure and good heat resistance. Fluorocarbon fluids are resistant to reactive gases. For sealing radiation environments polyphenyl fluids are normally selected. Properties of commercial ferrofluids are listed in Table 2.

Applications

Ferrofluidic seals provide a simple, effective, virtually zero leakage shaft seal for gases. They are also frictionless, and can be used with any shaft size. They can provide sealing at pressures up to 35 bar (500 psi) or greater, and are also suitable for sealing under vacuum conditions.

Suitable designs will also work as exclusion seals under abrasive conditions. Abrasives can literally be held and floated to the surface of the ferrofluid without impairing its sealing action. In addition to solid particles, ferrofluidic seals can also exclude water, water vapour, mist and similar possible contaminants.

TABLE 2 — RANGE OF PHYSICAL PROPERTIES OF FERROFLUIDS

PROPERTIES OF COMMERCIAL FERROFLUIDS

Evaporation rate: $10^{-5}-10^{-10}\ \frac{gm}{cm-sec}$ at 80°C

Saturation magnetization: 5-1000 GAUSS

Viscosity: 1-100,000 cp at 27°C

Electrical resistant: $10^7-10^9\Omega$-cm at 100na, at 27°·C

Partical size: 40A°—100A°

Temal conductivity: 90-600 mw/m/·K, at 27°·C

Initial susceptibility (Xi): 0.60-0.4 GAUSS/oe at M = 100G

Permeability (pi): 1.06—1.4

Density: 0.9—2.5 gm/ml

Self-Adhesive Compression Seals

PLASTIC FOAM strip materials coated with a long-life high tack adhesive are widely used as light duty general purpose seals – mainly as draught or dust seals. Their particular advantages are that they are low cost, easy to install and can be suitable for permanent, semi-permanent and make-and-break joints. Because of the resilient nature of the material used, they can also be effective in reducing vibration and noise transmission.

Most seals of this type are manufactured from plasticized PVC-coated on one or sometimes both sides with a pressure-sensitive adhesive. They are primarily intended as dust seals or draught seals, using a soft foam to provide good conformability. Typical properties are shown in Table 1. PVC foams have excellent resistance to UV light, weather oxidation, salt water, weak acids and weak alkalis. They have poor resistance to some solvents.

The same method may also be available in sheet form or cut shapes for gaskets. Adequate sealing is normally provided by 25 to 50% compression (depending on the particular product). In the case of a *make-and-break* seal the clamping or locking system used should be designed to provide this compression. A refinement with products intended

TABLE 1 – PROPERTIES OF TYPICAL SELF-ADHESIVE PVC FOAM COMPRESSION DUST SEALS

Property	Value
Density (g/cm^3)	0.11 to 0.18
Colour	Usually white or black.
Compression required for seal	25 to 50% (usually 50%)
Force required to seal (g/cm^2)	90 to 250
Shore Hardness 'OO' scale	5 to 25
Service temperature (°C)	–40 to +60
Recovery from 24 hours compression	95 to 96%
Adhesion to steel	Greater than 500 g/25 mm

for make-and-break seals is the incorporation of a polyester backing to protect the foam from surface damage and also act as a barrier to plastomer migration, ensuring a clean make-and-break joint.

Water seals of self-adhesive foam may be made in Neoprene foam or PVC foam formulated to have similar characteristics. Primary requirements are a high closed cell content, good recovery after compression and a low sealing force. Water seals may be produced in a range of different densities – typically low/medium or medium densities for optimum conformability with awkward shapes and contours to high density, low compressibility foams with load-bearing properties for heavy duty applications. At the extreme of this end high tack rubber/resin formulations may be used instead of foam material. Examples of some proprietary self-adhesive PVC foam compression water seals are given in Table 2.

TABLE 2 – PROPERTIES OF TYPICAL SELF-ADHESIVE PVC FOAM COMPRESSION WATER SEALS

Properties	5360	5900	5375	5370
Colour	Black	Grey	Grey	White
Compression required for waterproof seal	30%	50%	25%	35%
Force required to achieve seal (g/cm^2)	400	1050	700	900
Density (g/cm^3)	0.18 to 0.24	0.20 to 0.26	0.22 to 0.28	0.24 to 0.30
Shore Hardness 'OO' sca scale	20 tc 40	40 to 50	50 to 60 40 to 50†	55 to 65
Application temperature	5 to 45 °C, but ideally room temperature			
Service temperature (°C)	–40 to +60	–40 to +70	–20 to +70	–20 to +70
Adhesion to steel	> 500 g/25 mm			
Recovery after 24 hours compression	97%	92.5%	97%	98%
Thickness (tolerance ± 10%) (mm)	1.5, 3, 5, 6, 7.5, 10, 12	1, 1.5, 2.5, 3.5	1.5, 2.5, 3.5, 6, 10, 12	4 only
Widths (tolerance ±1 mm)	6 to 900 mm in 1 mm increments			
Resistance to weather, oxidation, salt water, weak acids, weak alkalis, UV light	Excellent			

†(10, 12 mm)

Note: Type 5370 is a pre-retardant fire retardant PVC foam which does not support combustion and is classified as self-extinguishing when tested according to BS 2782, Part 5 1965, Method 508A.

High Temperature Seals

SEALS EMPLOYING elastomeric elements are inherently limited to low to moderate service temperatures of the order 125°C with some exceptions (*eg* silicone rubbers and fluoroelastomers up to 250°C). In particular applications the performance range of standard seal designs may, however, be increased by either cooling the fluid being sealed to maintain it at a suitable working temperature, or cooling the gland or seal to maintain that section at an acceptable working temperature. This can apply particularly to mechanical shaft seals where the construction often permits the use of alternative materials further to increase the working temperature range.

TABLE 1 – TEMPERATURE/MATERIAL SELECTION CHART

Seal Material	Temperature°C							
	50	75	100	150	200	250	300	Above 350
Perfluoroelastomer	S	S	S	S	S	S	S	LS
Natural rubber	S	S	LS					
Butyl	S	S	S	VLS				
Nitrile	S	S	S	LS				
Neoprene	S	S	S	LS				
Polyacrylate	S	S	S	S	VLS			
Viton	S	S	S	S	S	LS		
Polyurethane	S	S	LS	VLS				
Silicone	S	S	S	S	S	S	LS	
Asbestos	S	S	S	S	S	S	S	S
Metallic						S	S	S

Note: Where high temperature services are concerned, the possibility of cooling the fluid is often worth considering in order to maintain a fluid working temperature at which a conventional or low cost elastomer may be employed.

Key: S = Suitable.
LS = Limited suitability (*eg* short periods).
VLS = Very limited suitability (*eg* very short periods).

Whether such solutions are practical, or suitable, depends largely on the applications. Thus with a hydraulic system, cooling the fluid may be the most attractive solution because it simplifies system design and component selection. The only additional requirement is then the provision of an adequate heat exchanger to maintain the fluid temperature at the required level. With a pumping system, on the other hand, it may be strictly necessary to handle the fluid at a particular temperature, when gland cooling may be quite the simplest and most satisfactory solution. In yet another case the fluid to be handled may be both hot and toxic or hazardous and must not be contaminated in any way, so that only a special type of seal will be suitable for the job. This severely limits selection and may even call for special development. There are also specific high temperature sealing requirements where only non-metallic seals can be expected to work effectively without degradation.

Thermal expansion may also be a significant factor with high temperature seals, particularly as it may affect the clearance space. In particular applications the seal may be called upon to seal over a wide range of temperature (*eg* through 'warm-ups') and thus with varying clearance due to differential expansion of components.

TABLE 2 – WORKING TEMPERATURE VALUES FOR SEAL SPRING MATERIALS

Material	Typical Maximum Working Temperature Recommended °C
Spring brass	66
Phosphor bronze	107
Beryllium copper	120
Hard drawn steel	160
Carbon spring steel	190
Alloy spring steel	205
Stainless 18-8	290
Monel	220
K-Monel	230
Inconel	400
Inconel-X	450
Chrome-molybdenum-vanadium	510

As far as high temperature seals are concerned, failure is more likely to result from a combination of circumstances rather than the temperature factor alone, provided the seal chosen initially was of suitable type and material. This applies particularly where the seal may have to operate in corrosive surroundings as well as a high ambient temperature, where generalized data may be of little significance. On the other hand, the modern types of mechanical shaft seals have been developed to such a high state of performance that they can often maintain satisfactory working for extended periods under the most arduous conditions, provided a suitable type is chosen initially. Apart from design, this is largely because high temperature materials can be employed in such seals without degrading the mechanical performance of the seal.

All-metal seals, which are an obvious choice for elevated temperatures, have a distinctly limited performance as dynamic seals compared with flexible seals, although they may be used, and indeed be a standard choice, for certain applications. Seals which operate with positive clearance, such as labyrinth seals, are a typical example: while seal rings in pumps represent standard practice, the relatively high rate of wear realized is offset by making such rings readily replaceable. Metallic rings are also an obvious choice for high temperature rod and piston seals, with the advantage of offering positive bearing support. Metallic O-rings, on the other hand, while suitable for high temperature service, are not suitable for dynamic applications.

One of the major problems which has restricted the development of metallic lip seals for dynamic applications is that the wear characteristics tend to change abruptly at high temperatures. Thus, while conventional, and satisfactory, bearing characteristics may be maintained up to some 200 to 250°C (400 to 500°F), at higher temperatures severe galling and wear may occur with the same metals in contact. This also holds true to a large extent of piston ring seals once the temperature rises to a level at which the lubricating film becomes degraded.

A simple metallic *reed* seal is shown in Figure 1. This is a pressure-energized seal, the theoretical performance of which can be analyzed by 'beam' formulas, by which it should be possible to arrive at an optimum angle for the reed – the larger the angle the smaller the bending moment on the reed. Seals of this type would appear more suited to rotary than reciprocating motions, within the limitations imposed by galling or scoring, but are used with both types of motions.

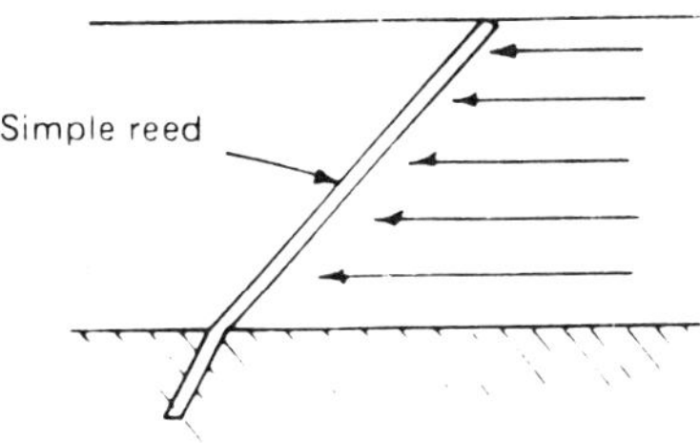

Figure 1

A rather more successful type is the truncated conical reed, shown in section in Figure 2. This is in effect a reed seal with lip allowing the actual sealing edge to be presented at a finer angle with somewhat better rubbing qualities. Much still remains to be done to develop seals of this and other types before greater use can be made of flexible metal seals of simple type, although this represents the most logical and direct approach to sealing fluids at very high temperatures.

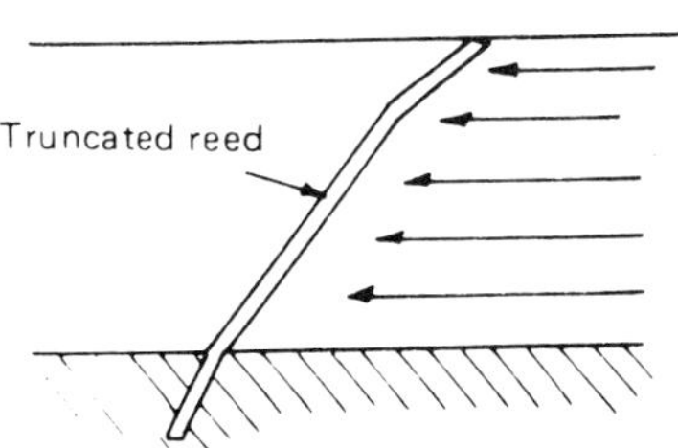

Figure 2

In applications where both static and dynamic sealing may be required a metallic seal with an adjustable sealing element may be considered. This would permit the seal unit load to be changed to meet static or dynamic demand, with leakage characteristics similar to that shown in Figure 3. Total average leakage could then be reduced to a very low level without overstressing the seal element or imposing an excessive breakout on the system, the main difficulty being the provision of the necessary control mechanism.

Figure 4 illustrates the difference in breakout force realized by such a form of control giving a static-dynamic seal.

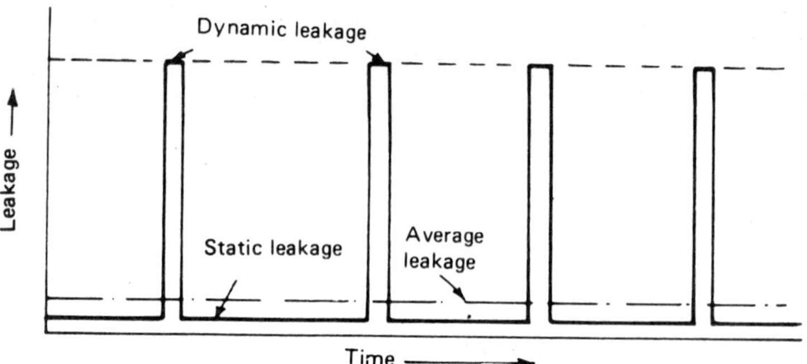

Figure 3

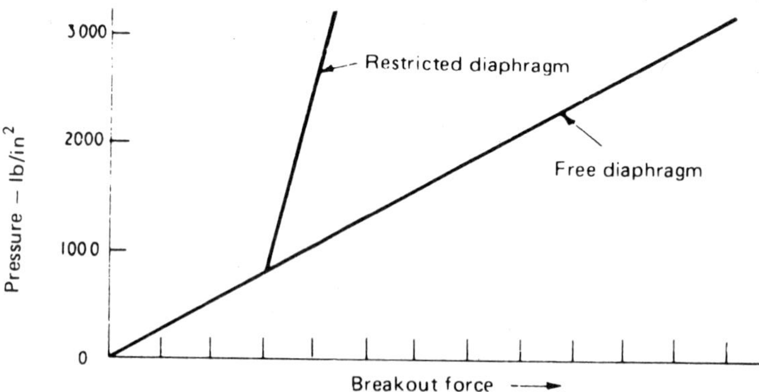

Figure 4

Major points to be considered in the design of metallic seals are:

(i) The breakout force is a function of the width of the sealing edge which favours the use of narrow, blunt knife-edged seals.

(ii) Narrow sealing edges can be formed to more exact shapes and give a better surface finish than wide edges, again favouring the use of thin section seal elements.

For purely static sealing applications a simple form of metallic reed seal is shown in Figure 5. The element is, in fact, a simple annulus with a curved cross-section which

TABLE 3 – TEMPERATURE RATING OF TYPICAL PACKINGS

Type	Construction	Max Pressure bar	Max Pressure lb/in²	Max Temperature °C	Max Temperature °F	Application
Asbestos	Plaited	70 to 140	1000 to 2000	345 to 540	650 to 1000	Reciprocating and rotary.
	Proofed	210	3000	120 to 260	250 to 500	Oil-resistant packings.
	Acid resistant	140 to 210	2000 to 3000	315	600	Rotary and reciprocating (acid duties).
	Braided sleeve	–	–	140	280	Rotary.
	Wire reinforced	50 to 140	750 to 2000	260	500	General.
	Rubber core	70	1000	315	600	General.
	PTFE impregnated	–	–	290	550	Chemical Services.
Cotton	Plaited	35	500	66	150	Pumps.
	Rubber impregnated	–	–	140	280	Water services.
	Metal reinforced	–	–	150	300	General low pressure.
Flax	Plaited	–	–	66 to 140	150 to 280	Fresh and salt water.
	Wire reinforced	–	–	140	280	Fresh and salt water.
Hemp	Lubricated	35	500	66 to 93	150 to 200	Water services
Glass	Braided	–	–	345	650	Chemical services.
Graphited fibre	–	–	–	+400	+750	Highly specialized.
Ceramic fibre	–	–	–	+2000	+3650	Highly specialized.
All-metal	Aluminium foil	–	–	540	1000	High temperature services.
Metal foil	Aluminium asbestos	–	–	540	1000	Rotary.
	Copper asbestos	–	–	650	1200	Rotary.
	Lead asbestos	–	–	230	450	Rotary.
	Babbit asbestos	–	–	120 to 200	250 to 400	Reciprocating and rotary.
	Babbit cotton	–	–	120 to 200	250 to 400	Reciprocating and rotary.
	Babbit flax	–	–	120 to 200	250 to 400	Reciprocating and rotary.
	Babbit hemp	–	–	120 to 200	250 to 400	Reciprocating and rotary.
Tubular metal	Babbit graphite	–	–	260	500	Reciprocating and rotary.
	Copper lubricated asbestos	–	–	650	1200	Reciprocating and rotary.
Tape	PTFE asbestos	–	–	290	550	Chemical services.
	PTFE glass fibre	–	–	315	600	Chemical services.
PTFE	Solid	–	–	290	550	Chemical services.
Kevlar/PTFE	Braided	200 to 1000	3000 to 15000	300	550	All services, including Chemical.

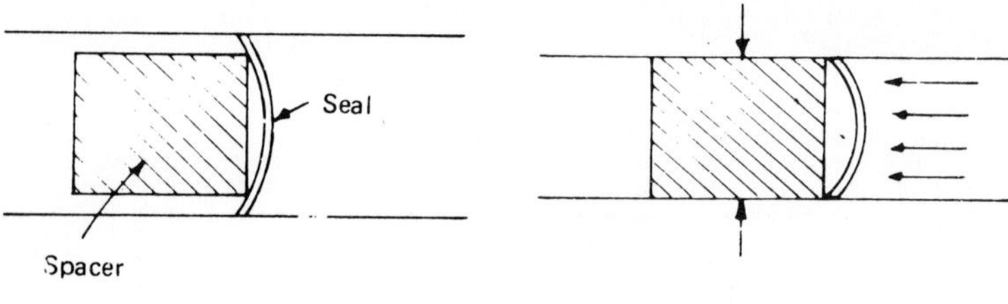

Figure 5

is further bowed under closure pressure. This degree of bowing can conveniently be controlled by a solid spacer, limiting the actual deflection of the sealing strip. The high unit load produced by deflection provides a positive, zero-leakage seal, and the pressure and sealing capacity is further enhanced by fluid pressure acting on the bowed element. It may not be necessary, in fact, for the faces to be closed completely against the spacer which can act primarily as a load-bearing element supporting the sealing strip against lateral displacement, and also serve to prevent over-compression and excessive bowing when the seal assembly is closed. With correct proportions ensuring that the sealing element is never over-deflected, the seal can be made a re-usable type.

For high temperature static sealing there is, in fact a wide range of gasket-type seals available, as well as metallic rings – see chapter on *Gaskets* (Section 3).

High temperature packings

Potential users more familiar with seal rings and mechanical face seals may tend to overlook the possibilities of packed glands as high temperature seals. Maximum service temperatures of asbestos packings can range up to 500°C or more; and that of pure expanded graphite packings up to 2500°C. The latter, being self-lubricating, also has a low coefficient of friction. Modern packings based on the aramid *Kevlar* may also be suitable for temperatures up to 300°C and are another low-friction packing.

Ceramic fibre packings, *eg* braided alumina-silicate yarn *Inconel* wire reinforcement have been developed for very high temperature services (*eg* up to 1260°C) and pressures up to 100 bar. Packings of this type have excellent chemical resistance (pH 0 to 9) and also high thermal insulation properties.

Consult the Manufacturers

Economically, selection of seals on a 'state of the art' basis is always to be preferred, if possible. Many potential users, however, overlook the fact that the bulk of the information for satisfactory seal selection lies with the individual seal manufacturers. In other words, the more severe the service conditions involved the more necessary it is to co-operate closely with the seal manufacturer both as regards seal selection and the manner in which the seal will be employed. This is then still working within the 'state of the art', but drawing on extensive practical experience of the manufacturer(s) approached.

See also chapters on *Compression Packings, O-rings (Metallic O-rings)* (Section 4), *Miscellaneous Static Seals* (Section 3) and *Mechanical Face Seals* (Section 5).

Large Diameter Seals

THEORETICALLY THERE is no limit to the diameter size of any type of dynamic seal. In practice there are a number of limiting factors which can influence the choice of seal type. These include:

(i) *Cost.* Certain types of seals (*eg* lip seals and compression packages) are relatively economic to consider in large sizes. Other types which have to be purpose made in the required size (*eg* mechanical face seals) will inevitably be more expensive. For the same reason, a compression packing can prove less costly than a ring seal in sizes which are not a standard production.

(ii) *Rubbing Speed.* This is more likely to be an important factor in rotary shaft seals than reciprocating seals, since rubbing speed increases in direct proportion to shaft diameter for the same rotational speed. All seal types with rubbing contact have maximum speed limits, depending on the seal form and material(s). Seals operating with clearance do not have this same limitation, but are not necessarily a satisfactory alternative for particular applications.

(iii) *Friction.* This is directly related to rubbing speed, seal material, lubrication and the finish of the surface against which the seal bears. It is the heating effect of friction, in fact, which determines acceptable rubbing speed limits. Friction is also proportional to contacting seal area, so a low friction seal is particularly desirable in large diameter seals to minimize power losses.

(iv) *Seal Stability.* With a large diameter lip seal or compression packing the seal section employed does not increase proportionally with seal diameter size. A particular problem which can arise with reciprocating (piston) seals as a result is the relatively large diametral growth of a large diameter cylinder under pressure increasing the clearance gap by a disproportionate amount and leading to extrusion of the seal.

Figure 1 is an example of a heavy-duty *V-ring* seal designed with this particular problem in mind. The extrusion gap that occurs when the cylinder bore 'grows' is bridged by the 'plastic' cylinder ring, which is energized by the rubber profile ring, outward in a radial direction.

It is not uncommon for large diameter single acting rams in special purpose presses to be installed with a self-adjusting type of packing. Figure 2 shows a typical large diameter

Figure 1
Large diameter heavy duty V-ring seal.

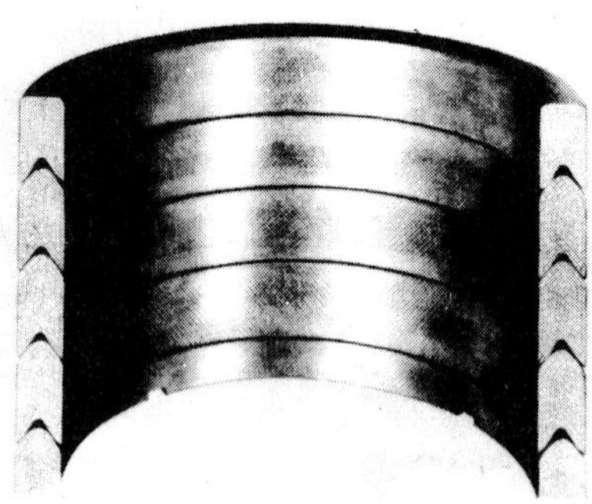

Heavy duty packing – suitable for assembly shown in Figure 2.

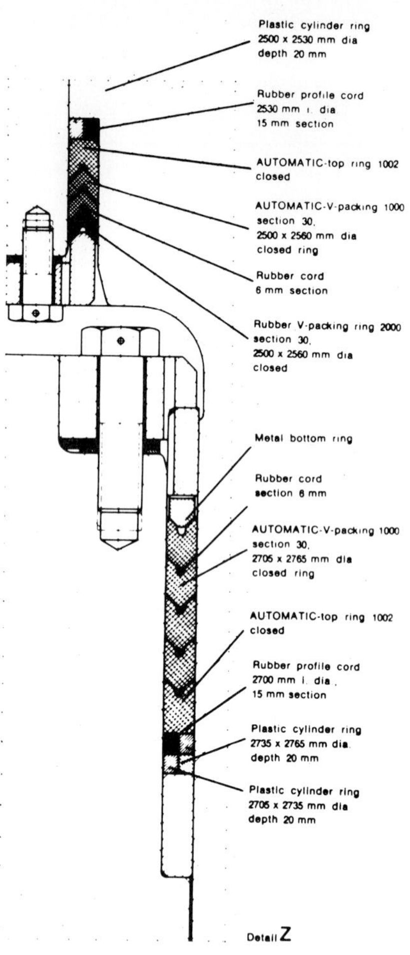

Figure 2
Detail of main ram seal.

seal set incorporating in-built rubber springs to give a constant axial loading to the seal and hence effect low pressure sealing. These types of seals are commonly made up to 1.8 m diameter and are of particular use where maintenance of multi-ram presses is extremely expensive and sometimes impossible. The rings, as can be seen from the photograph, are in split form so that when the inevitable seal change does occur they can be replaced without removing the ram.

Modern braided compression packings (*eg* Kevlar/PTFE) can also prove extremely effective as large diameter piston seals, working at pressures up to 1000 bar.

See also chapters on *Positive Action Seals* and *Metal Ring Seals* (Section 4).

Labyrinth Seals

LABYRINTH SEALS operate with positive clearance and provide sealing by virtue of a lengthy, tortuous gap path. Sealing then depends on the form of the labyrinth gap and the length of the leakage path. The basic labyrinth form can be provided by rings on the shaft and grooves in the housing (Figure 1), but more efficient forms are alternating teeth or alternating serrations (Figure 2). Tapered teeth are generally considered to be more effective in reducing leakage to a minimum with a given clearance and seal length, particularly when sealing gases.

The efficiency of a labryrinth seal increases in inverse proportion to the clearance gap, thus the smaller the gap the better, provided contact can be avoided within the seal under running conditions. The gap has to be larger than the axial or radial clearance on rolling bearings carrying the shaft, for example, or greater than the elastic deformation which may occur on anti-friction bearings. Typical assembly fits with anti-friction bearings are Housing: ISO K7, M7 and N7; Shaft: ISO h6, j6 or k6. The effect of thermal expansion closing the gap at operating temperatures must also be taken into account in determining a suitable gap.

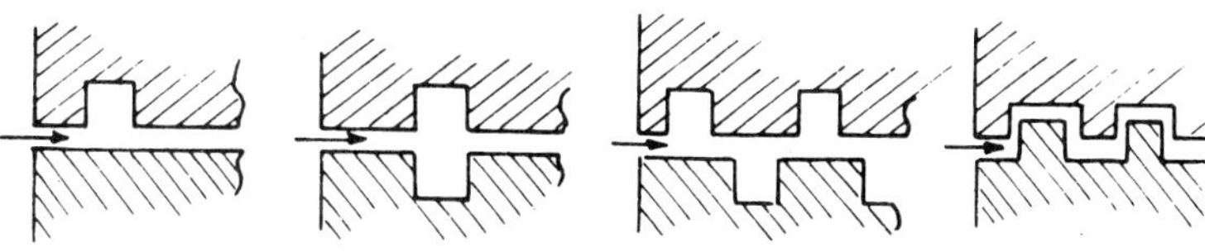

Figure 1

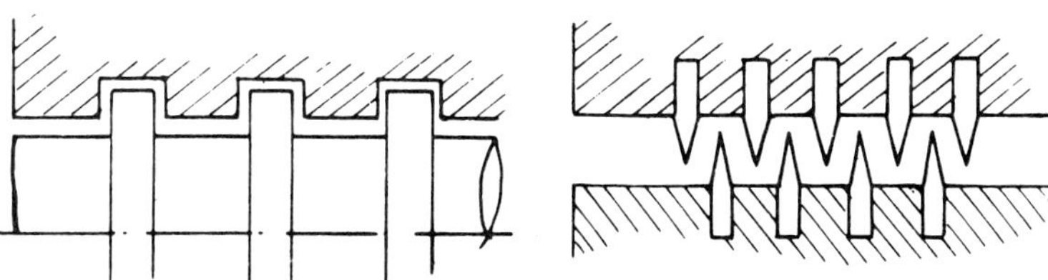

Figure 2

Alternatively, where there is a distinct possibility of contact occurring within the seal the teeth may be made of a material which sublimates rather than melts, so that, in the event of contact, galling or scoring of the mating member is avoided. Another solution is the use of carbon for one set of teeth where the effect of occasional rubbing contact will be mimimal. (More prolonged rubbing contact would run-in the carbon face(s) to a light 'bearing' fit).

Typical forms of proprietary labyrinth seals are shown in Figure 3. Both comprise an outer housing ring and an inner shaft ring formed with matching tooth profiles to operate with positive clearance. The second seal also has a number of peripheral slits through which any droplets of liquids which penetrate the seal are ejected under centrifugal force and then flow out through an annular groove with a drain hole in the housing. This is a preferred design where heavy splashing can occur.

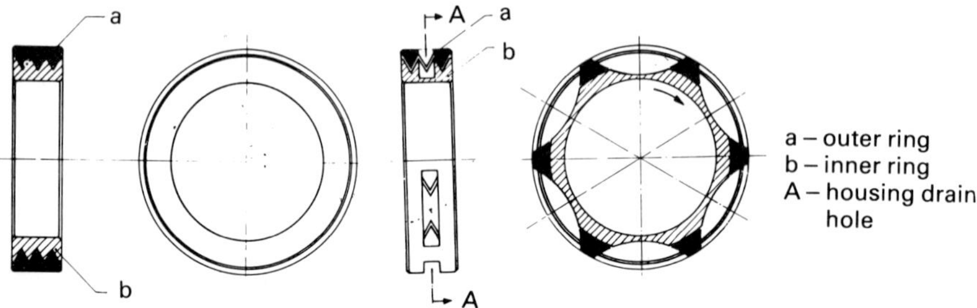

Figure 3
Proprietary labyrinth seals.

A further form of labyrinth seal is shown in Figure 4. Here the gap takes the form of a zig-zag, the seal being aligned so that the larger labyrinth diameter faces the fluid being sealed. Once fluid has entered the gap centrifugal force, assisted by the shape of the teeth, promotes a blocking action in the convergent leakage path, promoting better sealing. This is particularly effective when the fluid is a gas. Where the fluid being sealed is a liquid, performance can be improved by incorporating a discharge port in the inner or outer ring to reduce pressure build-up through the clearance path (Figure 5). This port must always be located at the bottom of the ring.

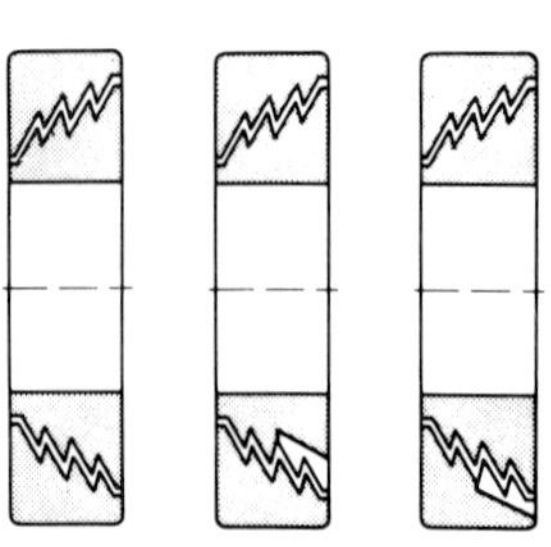

Figure 4

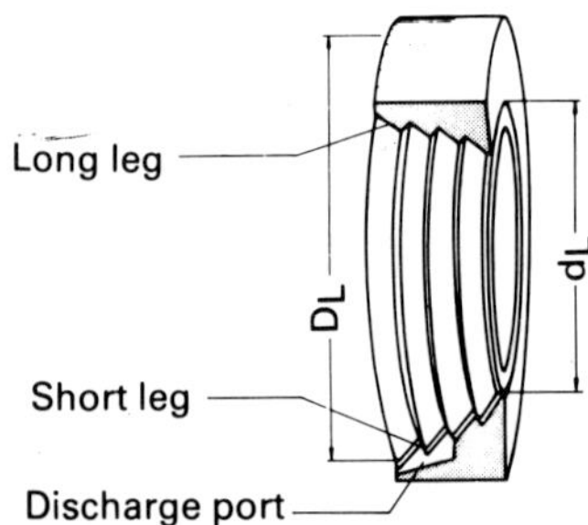

Figure 5

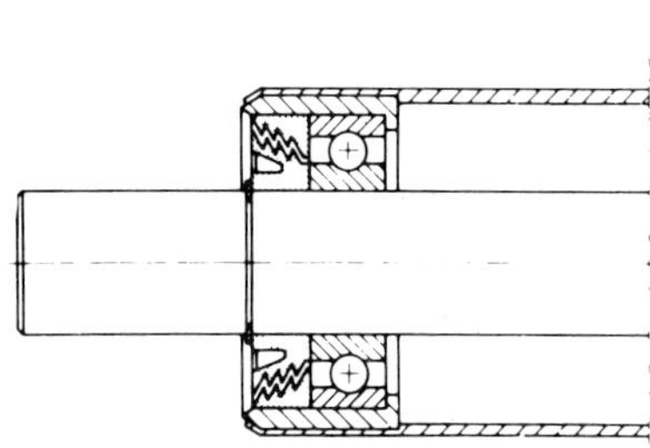

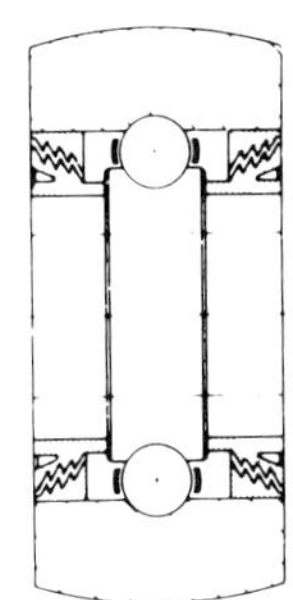

Single- and double-labyrinth seals for grease-lubricated ball-bearings.

Other variations on the simple labyrinth seal include the use of teeth set at an angle or with different pitch angles on rotating and stationary members so that a back-pumping effect is produced (*ie* the seal is given wind-back action and works as a *positive action hydrodynamic* seal).

Labyrinth seals are used quite widely as seals for rolling bearings, as well as for sealing shafts and spindles on heavy industrial machinery and other applications where relatively large leakage rates can be tolerated and a simple seal is required. Being *non-contact* seals they do not generate friction or heat and, with properly selected clearances, are suitable for high temperature services.

A further type of seal operating on pure labyrinth principles is the metal honeycomb ring working with light rubbing contact. In this case the labyrinth path is provided by the honeycomb structure. Seals of this type using a stainless steel or similar honeycomb are strong enough to resist erosion and degradation under pressure in such applications as *tip* seals for rotors.

See also chapter on *Carbon Seals*.

SECTION 8

Seal Selection

SEAL SELECTION GUIDES
CHEMICAL RESISTANCE AND COMPATIBILITY
SEAL COSTS

Seal Selection Guides

Gases or toxic fluids which need double mechanical seals or buffered glands

Acetic acid vapour
Ammonia gas
Carbon dioxide gas
Carbon monoxide gas
Chlorine gas
Coal gas
Furfural
Helium
Hydrocycanic acid
Hydrogen gas
Hydrogen fluoride gas
Krypton gas
Natural gas
Neon gas
Nitrogen
Oxygen
Steam
Sulphur dioxide gas
Xenon

Products liable to solidfy at the faces of mechanical seals

Asphalt
Bisphenol
Chocolate
Creosote
Ethylene oxide
Food products
Formaldehyde
Lard
Latex
Melamine resins
Maleic anhydride
Molasses
Paint
Phenol
Phenol formaldehyde mixture
Plasticizers
Phthalic anhydride
PVA emulsions
Resins
Saltcake
Sewage
Soap solutions
Sulphur
Tallow
Tar
Urea resins
Wax

Products for which PTFE fluoro- and perfluoroelastomers are suitable flexible seal materials

Acetic acid, hot
Acrylonitrile
Amines, mixed
Amyl ether 161 or 181
Brom 113
Bromine trifluoride
Bromine pentafluoride
Butyl acrylate
Butyl amine or N-Butyl amine
N Butyl ether
Chlorine trifluoride
1 Chloro 1 Nitroethane
Chlorosulphonic acid
Convelex 10
Dibutylamine
Dibutyl ether
Dichloro isopropyl ether
Dicyclohexylamine
Diethyl ether
Dinitro toluene
Ethers
Ethyl ether
Furan (Furfuran)
Hydrofluoric acid, hot
Lacquers
Lacquer solvents
Methyl ethyl ketone peroxide
Methyl isobutyl ketone
Methyl methacrylate
Nitromethane
Phenyl ethyl ether

*with performance limitations in some cases.

Vacuum seals

Polymeric materials used as seals in vacuum systems may contain residual volatile molecular components. In particular, hydrocarbons such as polyethylene with an average molecular mass of greater than 20000 amv can freely emit methane and similar gases at normal temperatures and dissociate liberated hydrogen and molecular fragments if heated. This can limit the useful application of such materials for vacuum seals.

High molecular weight greases and waxes can also be used for sealing purposes, specifically residual distillation products with vapour pressures in the uhv-range. Again, however, these materials can emit vapour molecules and also degradation products. A more significant characteristic, however, is that greases and waxes can trap gas which is subsequently released.

Freudenberg + Angus

= The Best of Both Worlds

Freudenberg -

A multi-national seal and precision moulding manufacturer, with all the resources and latest technological advances at its disposal throughout the world

Radial shaft seals

Hydraulic seals

Pneumatic seals

Diaphragms

O-rings

Rubber mouldings

Rubber/metal bonded components

PTFE parts

High quality plastic mouldings

Angus -

A British manufacturer, formerly George Angus & Co, producing only the highest quality precision parts for the world market

The Distribution Centre holds a vast stock of product lines, including those sourced from Freudenberg worldwide

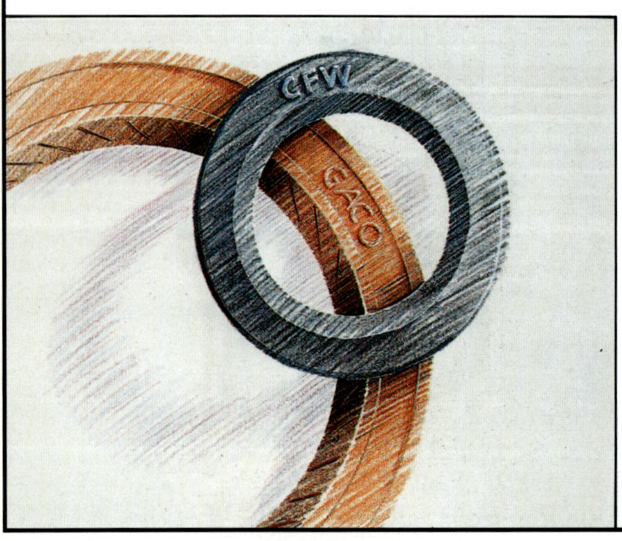

Freudenberg Angus LP
Coast Road
Wallsend
Tyne + Wear
NE28 9NR
Telephone: (091) 262 4551
Telefax: (091) 263 4301
Telex: 53628 GACO G

Distribution Centre:
Gilmorton Road
Lutterworth
Leicestershire
LE17 4DU
Telephone: (0455) 553081
Telefax: (0455) 554565
Telex: 347373 SIMRIT G

RUBBER SELECTION – SWELL, TEMPERATURE AND COST

Service temperature up to		Oil swell	Rubber	Cost ratio
°C	°F			
110	230	Low	Standard nitrile	1
130	266	Medium	Special nitriles	1.5
175	347	Low	Polyacrylic	3
200	400	Medium	Silicone	7
200	400	Very low	Fluoro-elastomer	18
300	570	Very low	Perfluoroelastomer	(Parts)

MATERIAL CHOICE FOR LUBRICANT SEALS

Fluid	Maximum operating temperature		Suitable material
	°C	°F	
Engine and machine oils	120	250	Nitrile Perfluoroelastomer
	150	300	Polyacetal
	162	325	Silicone
Transmission fluids	120	250	Nitrile Perfluoroelastomer
	150	300	Polyacetal
	162	325	Silicone
EP lubricants	100	212	Nitrile Perfluoroelastomer
	112	235	Polyacrylate
Greases	120	250	Nitrile
	200	400	Fluoro-elastomer
Synthetic lubricants: Diester	120	250	Nitrile
	160	325	Silicone Perfluoroelastomer
	200	400	Fluoro-elastomer
Phosphate esters	105	224	Butyl Perfluoroelastomer
	120	250	Ethylene propylene
	150	300	Butyl (resin cured)

ROTARY SHAFT SEALS

Type	Construction	Maximum Pressure lb/in²	bar	Maximum Speed ft/min	m/sec	Remarks
Oil seals	Homogeneous rubber or leather – spring loaded	5–10	0.5	500	2.5	Low pressures only
Oil seals 'reinforced)	Homogeneous rubber or leather with steel reinforcement spring loaded	10–30	0.75–2	1000–2000	5.0–10	Also special types *(eg* water pump seals).
Sealing rings	(i) All metal (ii) Laminated thermoset plastic (iii) Homogeneous non-metal	Depends on length and clearance		High		Primarily used on pumps
O-rings	Homogeneous rubber	–	–	–		Limited in performance and applications
Composite rings	Homogeneous rubber with moulded-on reinforcement ring	–	–	–	–	Various types but generally limited application †
Seal sets	Laminated fabric with or withour rubber impregnation	–	–	–	–	Various types but generally limited application
Compress-ion packings	(i) Plain	High		High		Performance adjustable by varying gland pressure
	(ii) Wire reinforced	Very high		Very high		Various types ranging from pure compression packings to gland rings
Mechanical seals (face seals)	Various designs	150 (typical)	10	2000 (typical)	10	Wide variety of types
Stern tube glands	Various designs	–	–	–	–	Special types of packed gland seals developed for marine propeller shaft sealing

Labyrinth* (inter-locking)	Rings, grooves or teeth	Moderate		Very high		Operate with positive clearance and thus have very low friction
Labyrinth* (non inter-locking	Rings, grooves or teeth	Low		Very high		Particularly suitable for large diameter gas seals
Honeycomb	Rings, grooves or teeth Stainless steel honeycomb rubbing element (also in other materials)	High		Very high		Works on labyrinth principle. Lightweight and particularly suitable for rotor blade seals
Lantern rings*	Fluid sealed gland	High		High		Some leakage both ways
Pressurized rings*	Gas sealed gland	High		Very high		Precise control difficult. May be used with mechanical back-up seals
Pressurized rings*	Liquid sealed gland	High		Very high		Various types and designs possible
Centrifugal liquid barrier*	Looped fluid passage	High, but depends on speed		Very high		Very low friction. Suitable for very high shaft speeds. May be used in conjunction with mechanical seals
Centrifugal Impeller*	Rotor form and pressurized gas feed	High		Very high		Seals at gas/liquid interface established when running
Frozen seals	'Cooled' section	—	—	—	—	Applicable to sealing hot liquids (*eg* molten metals) which freeze or solidify at low to moderate temperatures
Ferro-fluidic seals	Ferromagnetic fluid ring	500	35	Unlimited		Sealing gases only
Proprietary types	Various	—	—	—	—	—

*Particularly suitable for large diameter seals

† Some types of reciprocating seals may be suitable for rotary applications with specific performance limits

BASIC ELASTOMER CHOICE

Property required	Basic elastomer
Oil resistance	Butadiene, acrylic, nitrile
Maximum oil resistance	High nitrile
Phosphate ester fluids	Ethylene propylene
High sealing with low friction and torque	Nitrile
Low temperature flexibility	Natural rubber, silicone rubber, polyurethane
High temperature performance	Fluorinated polymers, silicone rubbers or perfluoroelastomers
Maximum temperature range	Silicone rubbers or perfluoroelastomers
Maximum strength	Natural rubber or polyurethane
Maximum chemical resistance	Fluorinated polymers, PTFE or perfluoroelastomers
Maximum abrasion resistance	Polyurethane
Maximum ozone resistance	Polyacrylic

MINIMUM SERVICE TEMPERATURE OF TYPICAL ELASTOMERS

Elastomer	Temperature	
	°C	°F
High nitrile	−15	5
Fluoro- and perfluoroelastomers	−15	5
Chlorosulphonated polyethylene	−40	−40
Butyl	−30	−22
Ethylene propylene	−40	−40
Chloroprene	−40	−40
Low nitrile	−45	−50
SBR	−50	−58
Fluorosilicone	−60	−76
Natural rubber	−60	−76
Silicone	−90	−130
Polyester	−40	−40
Halogenated polyolefin	−40	−40

VOLUMETRIC COEFFICIENT OF EXPANSION OF ELASTOMERS

Elastomer	per deg C x 10^{-4}
Natural rubber	5.25 to 6.6
Butyl	4.41
Chloroprene (neoprene)	4.11 to 5.46
Nitrile	3.33 to 5.25
Ethylene propylene	4.86
Fluoroelastomer	3.90 to 5.58
Silicone	5.31 to 9.3
Perfluoroelastomer	6.9

GENERAL RECOMMENDATIONS – LIMITS FOR SURFACE FINISH μin Ra

Type of Seal	Rubbing Surface Finish Normal Life	Rubbing Surface Finish Limited Life or Light-Duty	Housing or Gland Finish
O-rings	16*	16	32
Other elastomeric rings	16*	16	32
Cups – Rubber	16*	16	32
Cups – Semi-reinforced	16*	16	32
Cups – Rubber-impregnated fabric	16	32	32–48
Cups – Leather	32	48	48
U, V and C-rings – Rubber	16	16	32
Semi-reinforced	16	16	32
Rubber-impregnated fabric	16	32	32
Leather	32	48	48–60
Gland rings	16	16–32	32–48
Compression packings	16–32	32	32–48

* Or better

DYNAMIC SEAL TYPES FOR SPECIFIC SERVICES

Zero Leakage	**High Speed**	**Minimum Drag**
Flushed gland	Bushing seals	Bushing seals
Double seals	Labyrinth seals	Labyrinth seals
Windback seals	Centrifugal barrier	Centrifugal barrier
Liquid barrier seals	Ferrofluidic seals*	Ferrofluidic seals*
Ferrofluidic seals*		
Centrifugal barrier*		

* *Gases only*

Low Temperature	**Cryogenic**	**Vacuum**	**High Temperature**
Bellows seals	Bellows seals	Ferrofluidic seals	Bellows seals
Mechanical face seals	Mechanical face seals	Waxes and greases	Mechanical face seals
	Spring-energized sealing rings		Asbestos packings
			Frozen seals

SEAL TYPES – GENERAL GUIDE TO MAXIMUM TEMPERATURE RATINGS

Application	Seal type	Maximum Service Temperature °C	°F
Static	O-rings	Depends on elastomer	
	Asbestos jointings	400–540	750–1000
	Cellulose fibre	120–150	250– 300
	Cork compositions	120–150	250– 300
	High temperature cork-rubbers	200–230	400– 450
	Metal rings	400–540	750–1000
Rotary	Homogeneous rubber	Depends on elastomer and rubbing speed	
	Semi-reinforced	Depends on elastomer and rubbing speed	
	Laminated fabric	120–150	250– 300
	Mechanical	120 up	250 up depending on construction
Reciprocating (low pressure)	Homogeneous rubber	Depends on elastomer	
	Semi-reinforced	Depends on elastomer	
Reciprocating (high pressure)	Rubber-impregnated fabric	Depends on elastomer	
	Laminated fabric	120–150	250– 300
	Leather	52–150	125– 300
	Composite	Depends on materials	
	Metallic rings	400–540	750–1000

Note: When selecting a particular type of seal or material the manufacturer's recommendations for maximum working temperature should be followed. Actual performance at a particular working temperature will also be affected by compatibility with the fluid concerned.

TYPICAL PRESSURE RATING OF V-CUP SEAL SETS

Number of Rings	Rubber lb/in²	Rubber bar	Rubber and Impregnated Fabric lb/in²	Rubber and Impregnated Fabric bar	Leather lb/in²	Leather bar
2	500	35	1000	70		
3	1000	70	1500	105	500	35
4	2000	140	3000	210	1500	105
5	3000	210	5000	350	6000	420
6	5000	350	10000	700	20 000	1400

GENERAL RECOMMENDATIONS BY DUTY
(High Pressure Seals)

Seal type	Reciprocating: Light Duty Low Speeds	Reciprocating: Medium Duty	Reciprocating: Heavy Duty	Rotary: Light Duty	Rotary: Heavy Duty
O-rings	X	X		X	
Rectangular rings	X	X			
Quad rings	X	X		X	
Composite rings	X	X	X		
U-rings — Rubber	X	X		X	
Rubber impregnated fabric		X	X		
Leather		X	X		
V-rings — Rubber		X			
Rubber impregnated fabric		X	X		
Leather		X	X		
C-rings — Rubber	X	X		X	
Rubber impregnated fabric		X	X		
Gland rings — Rubber impregnated fabric	X	X		X*	
Asbestos fabric	X	X	X	X*	
Automatic — Rubber	X	X		X*	
Rubber impregnated fabric		X	X		
Cushioned rings		X	X	X*	X*
Metal piston rings	X	X	X		
Hat or flange — Rubber	X			X	
Rubber impregnated fabric	X	X		X	
Leather		X	X	X	

*Depending on type and construction

PRESSURE RATING RELATED TO MATERIALS
(pressure in lb/in^2 and bar)

Seal type	Homogeneous Rubber lb/in^2 (bar)	Semi-Reinforced lb/in^2 (bar)	Rubber-Impregnated Fabric lb/in^2 (bar)	Laminated Fabric lb/in^2 (bar)
O-rings, quad rings, etc	1000–1500 (70–100)	3000 (210*)		
U-rings	500–1000 (35–70)	1000 (70)	1000–3000 (70–210)	up to 5000 (up to 350)
V-rings	500–1000 (35–70)	1000–2000 (70–140)	2000–5000 (140–350)	5000–10000 (350–700)
C-rings	500–1000 (35–70)	1000–2000 (70–140)	2000–5000 (140–350)	5000 (350)
Gland rings	50–150 (3.5–10)	250–500 (18–35)	1000–5000 (70–350)	2000–5000 (140–350)
Composites		up to 3000 (up to 210)	up to 10000 (up to 700)	

*With anti-extrusion ring(s)

PISTON SEALS – HIGH PRESSURE WORKING

Seal Type	Construction	Maximum Pressure		Maximum Speed		Maximum Surface Roughness	Remarks
		lb/in^2	bar	ft/min	m/sec	μ in Ra	
No seal		5000–10 000	350–700	200	1	Fine as possible	Generally suitable for very small bores only with lapped piston fit
Piston rings	Solid metal	10000	700	100–150	0.5–0.75	16	Cast iron preferred if compatible
Segmental rings	Composite metal	10000	700	100	0.5	32	Maintain suitable sealing with worn bores
Cushioned rings	metal or non-metal rigid rings with elastomeric cushion	5000	350	100	0.5	30–40	For larger bores may be used in sets with intermediate plain rings
U-rings	Laminated rubber or leather	10000	700	100	0.5	32	Various proprietary types
U-rings	Rubber-impregnated fabric	5000	350	100	0.5	16	Various proprietary types
V-rings	Laminated fabric or leather	up to 20000	up to 1400	100	0.5	32	Pressure rating depending on number of rings
V-rings	Rubber impregnated fabric	up to 10000	up to 700	100	0.5	16	Pressure rating depends on number of rings
C-rings	Rubber-impregnated fabric	up to 5000	up to 350	100	0.5	16	Various proprietary types and modified forms. May also be used in seal sets
Automatic	Rubber-impregnated fabric	up to 5000	up to 350	100	0.5	16	Various proprietary designs. May also be used in seal sets
Composite rings	PTFE/rubber composite rings and variations	up to 10000	up to 700	–	–	–	Various proprietary designs and types
Gland rings	Various	2000 up	140 up	–	–	32	Various proprietary types; particularly suitable for heavy duty or special applications

Note: Numerous proprietary designs of seals are available as alternatives – consult manufacturers' data for performance characteristics and application(s).

PISTON SEALS – LOW TO MODERATE PRESSURE WORKING

Seal Type	Construction	Maximum Pressure		Maximum Speed		Maximum Surface Roughness	Remarks
		lb/in^2	bar	ft/min	m/sec	μ in Ra	
O-Rings	Homogeneous rubber	1000–1500	70–100	40	0.2	16	Up to 3000 lb/in^2 with back-up rings
Rectangular Rings	Homogeneous rubber	1000–1500	70–100	40	0.2	16	Somewhat better resistance to spiral failure
Quad Rings	Homogeneous rubber	1000–1500	70–100	35–40	0.2	16	Good resistance to spiral failure
Cups	Homogeneous rubber	1000–1200	70–85	70	0.35	16	Single-acting or back-to-back for double-acting
Cups	Semi-reinforced	500	35	60	0.3	16	Single-acting or back-to-back for double-acting
U-Rings	Homogeneous rubber	500–1000	35–70	60	0.3	16	Single-acting or back-to-back for double-acting
U-Rings	Leather	1000–1500	70–100	50	0.25	32	Single-acting or back-to-back for double-acting
C-Rings	Homogeneous rubber	500–1000	35–70	50–60	0.15–0.3	16	Lower friction than U-rings or V-rings
V-Rings	Homogeneous rubber	1000–1500	70–100	50–60	0.25–0.3	16	2 or 3 rings
V-Rings	Leather	1000–1500	70–100	50–60	0.25–0.3	32–48	2 or 3 rings
Gland Rings	Semi-reinforced	500	35	60	0.3	16	Offer advantages for heavy duty applications and large cylinder bores
Compression Packings	Various	500–1000	35–70	40–60	0.2–0.3	16–48	Not normally used unless operating conditions demand special high temperature materials. More suitable as rod seals

Note: Higher pressure type seals may also be used for low pressure working, particularly in homogeneous rubber or semi-reinforced construction. In such cases, *ie* low pressure application of high pressure seal sections, a rougher surface finish may be tolerated than specified for high duty service, providing rubbing speeds are not excessive.

STATIC SEAL SELECTION GUIDE

Seal Type	Typical Application(s)	Remarks	Relative Cost
O-ring	Flange joints	Also used as a crush seal	1
Rectangular ring	Alternative to O-ring for flange joints	Groove fitting – can be cut from flat sheet to any size	1 or less
Simple flat gasket	Flange joints	Basic materials: (i) Paper (ii) Cork (iii) Rubber-bonded cork (iv) Rubber (v) Compressed asbestos fibre	 1 2–3 2–3 2–10 3–5
Metal-clad gaskets	Flange joints, cylinder head gaskets, heat exchangers, petrochemical pressure vessels	Particularly used where sealing faces are narrow	4–5
Corrugated metal gaskets	Flange joints impervious to steam, water, air, chemicals, oils		
Spiral wound gaskets	Gaskets for joints where high mechanical strength and high recovery is required	Semi-metallic gaskets	
Printed gaskets	Precision flat sealing	Types are: (i) Metal based (high pressure) (ii) Paper based (iii) Pure print (soft joint)	
Envelope gaskets	Gaskets for chemical services		
Solid metal	Flange joints with good finish and high sealing stresses	Normally limited to fairly narrow faces	
Wire rings (circular or rectangular)	Groove-fitting flange seals	High temperature seals with high closure pressure	
All-metal (lip seals)	Face sealing of oils and greases at elevated temperatures		
Metallic O-rings	High temperature flange seals	Hollow section: (i) Plain (ii) Vented (iii) Gas-filled	
Flexible metal rings	High temperature flange seals	Various 'open' cross sections deforming to provide sealing under closure pressure	
Chamfer rings	High temperature static seals (*eg* in aircraft systems)	Wedge-action static seals	
Packings	Large flange joints, pressure vessels, tanks, *etc*		
Sealants	Flange joints, threaded joints	Types include: (i) Elastomeric sealants (ii) Anaerobic adhesives (iii) Thermosetting sealants (iv) Mastics	
Sealing washers	Sealing under bolts or nuts on threaded fasteners	Elastomeric, composite and all-metal	1–3
Bonded seals	Sealing under bolt heads, *etc*	Elastomeric sealing washer bonded to bolt	

DYNAMIC SEAL SELECTION GUIDE

Seal Type	Geometry	Operating Condition	Application(s)	Relative Cost
O-ring	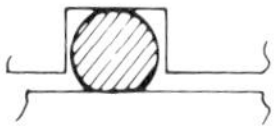	Reciprocating	Piston and rod seals on hydraulic and pneumatic cylinders, valve seals; numerous other applications.	1
		Rotary	'Waterproof' spindle seals; more limited application as rotating shaft seals.	1
Rectangular ring	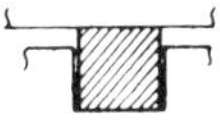	Reciprocating Rotary	Alternative to O-rings where these show limitations, *eg* under vibration or conditions tending to provide failure of O-rings. More flexible geometry than O-rings.	less than 1
Quad ring	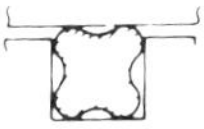	Reciprocating Rotary	More effective than O-rings at low speeds, with low backlash. Several different variations on actual section.	1–2 1–2
T-ring	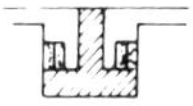	Reciprocating	High pressure applications, used with back-up rings.	3
Heart seal	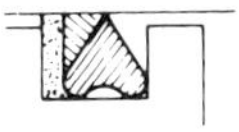	Rotary	Light duty oil seals only; normally used with back-up ring.	1–2
Delta rings, H-rings, D-rings, *etc*	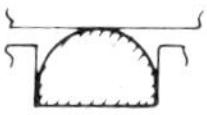	Reciprocating Rotary (some)	Alternative to O-rings offering advantages for specific applications, *eg* D-rings for rod seals.	1–3
Composite (co-axial)		Reciprocating	Proprietary designs of compact composite PTFE/rubber seal rings for piston and rod duties, many with high pressure ratings.	
		Rotary	Some types suitable.	
Composite (axial)	Various proprietary designs	Reciprocating	Compact piston and rod seals for hydraulic and pneumatic cylinders; single and double-acting types.	
		Rotary	Some types suitable.	
Lip seal	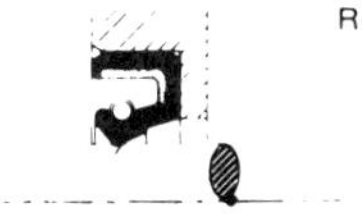	Rotary	Low pressure oil seals; many proprietary designs now incorporate 'wind back' action.	10–15
U-ring	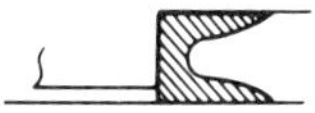	Reciprocating	Homogeneous rubber and reinforced for hydraulic cylinders; more flexible U-rings for pneumatic cylinders.	10
U-ring derivatives	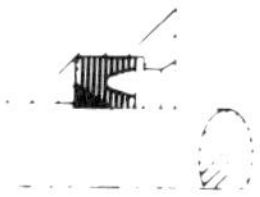	Reciprocating	Improved U-ring sections with or without reinforcement for similar applications as conventional U-rings, with uprated performance. Various proprietary designs.	10–15

cont...

DYNAMIC SEAL SELECTION GUIDE (contd.)

Seal Type	Geometry	Operating Conditions	Application(s)	Relative Cost
V-rings	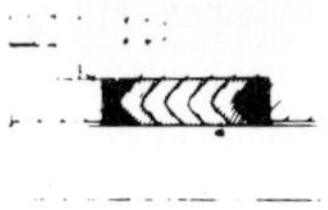	Reciprocating	Normally used in sets for high pressure hydraulic cylinder seals; pump glands; propeller shafts, *etc.*	50–100
C-rings	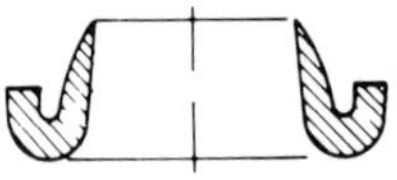	Reciprocating	Low-friction heavy duty seals for pneumatic cylinders; may be used with headers.	10–25
Automatic		Reciprocating	Normally used in sets for high pressure hydraulic cylinders, pump glands, *etc.*	50–100
Distributor	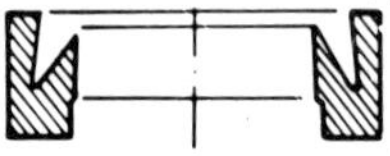	Reciprocating	Similar applications to automatic seals	50–100
Wedge-action U-ring		Reciprocating	Alternative to both O-rings and U-rings, combining the favourable properties of both. Various proprietary designs.	
Spring-energized U-ring		Reciprocating	U-ring alternative incorporating spring-loaded wedge in a composite construction. Various proprietary designs.	
Plastic U-ring		Reciprocating Rotary	Light duty flexible lip seal comprising a PTFE envelope and internal spring.	
Packed gland	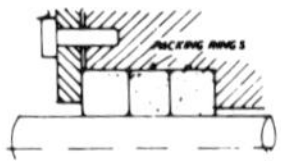	Reciprocating Rotary	High duty cylinder or rod seals; also primary choice for rotating shaft seals on pumps, *etc.*	
Bushing	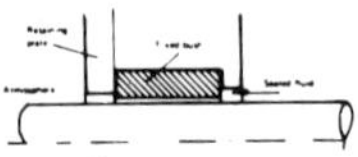	Rotary	Low cost shaft seal where leakage can be tolerated.	
Floating bush	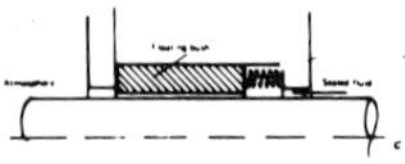	Rotary	Shaft seal with self-alignment properties to accommodate eccentricity, *etc.*	

cont...

DYNAMIC SEAL SELECTION GUIDE (contd.)

Seal Type	Geometry	Operating Condition	Application(s)	Relative Cost
Mechanical face seal		Rotary	High duty shaft seal for pumps, process equipment, *etc.*	50–5 000
Wiper		Rotary	Exclusion seal for rods or shafts	10
Scraper		Rotary	More robust form of wiper seal capable of removing and excluding abrasives	10–15
Boot (gaiter)	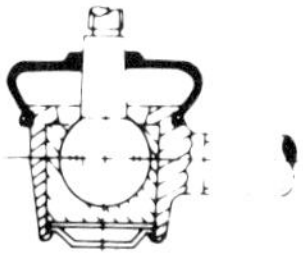	Reciprocating Semi-rotary	Exclusion seal for rods. Exclusion seal for swivel points, *etc.*	10
Metal bellows		Rotary	High duty shaft seal for high or low temperatures, handling corrosive fluids, *etc.*	100–1 000
Diaphragm		Short-stroke reciprocating	Fuel pumps	5
Convoluted diaphragm	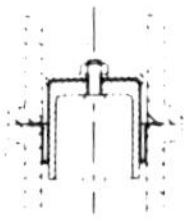	Long-stroke reciprocating	Metering pumps, servo drives, *etc.*	20–50
Labyrinth	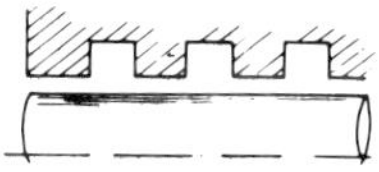	Rotary	Grease-lubricated bearings	

Chemical Resistance and Compatibility

COMPATIBILITY OF LEATHERS

(For fluids not mentioned either leather is not normally suitable, or data is lacking. Consult leather seal or packing manufacturers in such cases for specific information).

Fluid	COMPATIBILITY Good	Fair	Limited
Acetone	x		
Amyl acetate		x	
Amyl alcohol			x
Aniline			x
Asphalt			x
Beet sugar liquors			x
Benzene			x
Benzol			x
Butyl acetate		x	
Butyl alcohol			x
Cane sugar liquors			x
Castor oil	x		
Chlorinated solvents			x
Cod liver oil	x		
Corn oil	x		
Cotton seed oil	x		
Creosote			x
Diesel fuel	x		
Ethyl acetate		x	
Ethyl alcohol		x	
Ethylene glycol	x		
Fuel oil	x		
Gelatin	x		
Glue	x		
Glycerol	x		
Greases, all types	x		
Hydraulic oils, all types	x		
Kerosene		x	
Ketones	x		
Lacquers		x	
Lactic acid			x
Linseed oil	x		
Lubricating oils	x		
Mercury			x
Methyl alcohol		x	
Mineral oils	x		
Naphtha			x
Nitrobenzene			x
Paraffin		x	
Petrol			x
Petroleum oils, crude	x		
Sewage			x
Soya bean oil	x		
Steam			x
Tar			x
Toluene			x
Trichlorethylene			x
Tung oil	x		
Turpentine		x	

CHEMICAL RESISTANCE

Chemical	Hypalon®	Hytrel®	Neo-prene	Nordel®	Vamac®	Viton®
Acetaldehyde	C	–	C	A	B	C
Acetic acid, 20%	A	A	A	A	A	C
Acetic acid, 30%	A	A	A	A	A	C
Acetic acid, glacial	A-B	A	C	B	C	C
Acetic acid, glacial	–	B (100 °F)	–	–	–	–
Acetic anhydride	A	T	A	T	–	C
Acetone	B	B	B	A	C	C
Acetylene	B	A	B	A	–	A
Aluminium chloride solutions	A	T	A	A	A	A
Aluminium sulphate solutions	A (250 °F)	T	A (158 °F)	A	A	A
Ammonia, anhydrous	B	–	A	T	–	C
Ammonium chloride solutions	A	A	A	A	A	A
Ammonium hydroxide solutions	A (200 °F)	T	A (158 °F)	A	–	A
Ammonium sulphate solutions	A (200 °F)	A	A (158 °F)	A	A	A
Amyl acetate	C	B	C	A	C	C
Amyl alcohol	A (200 °F)	A	A (158 °F)	A	B-C	A (212 °F)
Aniline	B	C	C	A-B	X	A-B
Aniline	C (100 °F)	–	–	–	–	B (158 °F)
Aniline	–	–	–	–	–	C (300 °F)
ASTM oil No.1	A	A (300 °F)	A	C	(350 °F)	A (300 °F)
ASTM oil No.3	B (158 °F)	A (300 °F)	B (158 °F)	C	(300 °F)	A (350 °F)
ASTM reference fuel A	A	A (158 °F)	A	C	A	A
ASTM reference fuel B	C	A (158 °F)	C	C	B-C	A
ASTM reference fuel C	C	A	C	C	C	A
ASTM reference fuel C	–	B (158 °F)	–	–	–	–
Asphalt	B	T	B	A	T	A (400 °F)
Barium hydroxide solutions	A (200 °F)	T	A (158 °F)	A	T	A
Beer	A	A	A	A	T	A
Benzaldehyde	C	–	C	B	C	C
Benzene	C	B	C	C	C	B (158 °F)
Benzoyl chloride	C	–	C	C	X	B
Borax solutions	A (200 °F)	A	A (158 °F)	A	(212 °F)	A
Boric acid solutions	A (200 °F)	A	A (158 °F)	A	T	A
Bromine, anhydrous liquid	B	X	C	C	–	B (212 °F)

Rating Key:
A–Fluid has little or no effect.
B–Fluid has minor to moderate effect.
C–Fluid has severe effect.
T–No data – likely to be compatible.
X–No data – not likelv to be compatible.
Blanks indicate no evaluation has been attempted.

Unless otherwise noted, concentration of aqueous solutions are saturated.
All ratings are at room temperature unless specified.

*Reg U.S. Pat & Tm.Off.

cont...

CHEMICAL RESISTANCE (contd.)

Chemical	Hypalon*	Hytrel*	Neo-prene	Nordel*	Vamac*	Viton*
Butane	A	A	A	B	T	A
Butyl acetate	C	B	C	X	X	C
Butyraldehyde	B-C	–	B-C	B	C	C
Butyric acid	B-C	T	C	X	X	T
Calcium bisulfite solutions	A (200 °F)	–	A (158 °F)	T	T	A
Calcium chloride solutions	A	A	A	A	T	A
Calcium hydroxide solutions	A (200 °F)	T	A (158 °F)	A	–	A
Calcium hypochlorite, 5%	A	A	B	A	T	A
Calcium hypochlorite, 20%	A (200 °F)	–	B	A	T	B (158 °F)
Carbon bisulphate	C	–	C	T	X	A
Carbon dioxide	A (200 °F)	A	A	T	A (450 °F)	A
Carbon monoxide	A (200 °F)	A	A	T	–	T
Carbon tetrachloride	C	B-C	C	C	X	A (158 °F)
Castor oil	A (158 °F)	B	A (158 °F)	B	–	A
Chlorine gas, dry	B	X	B	X	–	A (212 °F)
Chlorine gas, wet	B	X	C	X	–	B
Chloroacetic acid	A	X	A	A	X	C
Chlorobenzene	X	X	X	X	C	A
Chloroform	C	C	C	C	X	A
Chlorosulphonic acid	C	C	C	C	X	C
Chromic acid, 10-50%	A (158 °F)	X	C	C	–	A
Citric acid solutions	A	A	A	A	–	A
Copper chloride solutions	A	A	A	A	A	A
Copper sulphate solutions	A	A	A	A	T	A
Cottonseed oil	A	A	A	A-B	A	A (300 °F)
Creosote oil	C	–	C	C	X	A (212 °F)
Cyclohexane	C	A	C	C	B	A
Dibutyl phthalate	C	A	C	A	C	B
Diethyl sebacate	B	A	C	B	X	B
Dioctyl phthalate	C	A	C	B	X	B
Dowtherm A	B	–	B	C	–	A (212 °F)
Dowtherm A	–	–	–	–	–	B (400 °F)
Epichlorohydrin	T	X	–	B	–	C (122 °F)
Ethyl acetate	C	B	C	A	C	C
Ethyl acetate	–	–	–	B (158 °F)	X (158 °F)	–

Rating Key:
A–Fluid has little or no effect.
B–Fluid has minor to moderate effect.
C–Fluid has severe effect.
T–No data – likely to be compatible.
X–No data – not likely to be compatible.
Blanks indicate no evaluation has been attempted.

Unless otherwise noted, concentration of aqueous solutions are saturated.
All ratings are at room temperature unless specified.

*Reg U.S. Pat & Tm.Off.

cont...

CHEMICAL RESISTANCE (contd.)

Chemical	Hypalon®	Hytrel®	Neo-prene	Nordel®	Vamac®	Viton®
Ethyl alcohol	A	A	A	A	B	A
	(200 °F)		(158 °F)			
Ethyl chloride	C	C	C	B	C	A
Ethyl ether	C	–	C	C	B	C
Ethylene dichloride	C	C	C	B	X	A-B
	(120 °F)		(120 °F)	(120 °F)		(120 °F)
Ethylene glycol	A	A	A	A	A	A
	(200 °F)		(158 °F)		(212 °F)	(250 °F)
Ethylene oxide	X	A	X	X	X	C
						(158 °F)
Exxon 2380 turbo oil (lubricant)	–	T	–	X	–	A
						(392 °F)
Ferric chloride solutions	A	T	A	A	T	A
	(200 °F)					
Fluosilicic acid	A	T	A	T	T	T
	(250 °F)		(158 °F)			
Formaldehyde, 40%	A	B	A	A	–	A
Formaldehyde, 40%	C	–	C	–	–	–
	(158 °F)		(158 °F)			
Formic acid	A	B	A	A	T	C
						(158 °F)
Freon-11†	A	A	A-B	C	B	A-B
Freon-11	T	–	B	–	–	T
	(130 °F)		(130 °F)			(130 °F)
Freon-12	A	A	A	B	A-B	A-B
Freon-12	A	–	A	–	–	B
	(130 °F)		(130 °F)			(130 °F)
Freon-22	A	–	A	C	C	C
Freon-22	A	–	A	–	–	X
	(130 °F)		(130 °F)			(130 °F)
Freon-113	A	A	A	C	B	A
Freon-113	A	A	A	–	–	T
	(130 °F)	(130 °F)	(130 °F)			(130 °F)
Freon-114	A	A	A	C	–	B
Freon-114	T	–	T	–	–	–
	(130 °F)		(130 °F)			
Furfural	B	–	B	B	C	C
						(158 °F)
Fyrquel 220 (hydraulic fluid)	–	T	–	–	–	
						(212 °F)
Gasoline	B	A	B	B-C	B-C	A
Glue	A	A	A	A	–	A
	(200 °F)		(158 °F)			
Glycerin	A	A	A	A	T	A
	(200 °F)		(158 °F)			(250 °F)
n-Hexane	A	A	A	C	–	A
Hydrazine	–	C	–	A	–	C
Hydrochloric acid, 20%	A	B	A	T	A	A
Hydrochloric acid, 20%	A	–	–	–	–	A
	(158 °F)					(230 °F)

Rating Key:
A–Fluid has little or no effect.
B–Fluid has minor to moderate effect.
C–Fluid has severe effect.
T–No data – likely to be compatible.
X–No data – not likely to be compatible.
Blanks indicate no evaluation has been attempted.

Unless otherwise noted, concentration of aqueous solutions are saturated.
All ratings are at room temperature unless specified.

*Reg U.S. Pat & Tm.Off.

cont...

CHEMICAL RESISTANCE (contd.)

Chemical	Hypalon®	Hytrel®	Neo-prene	Nordel®	Vamac®	Viton®
Hydrochloric acid, 37%	A (122 °F)	C	A	A-B	–	A (158 °F)
Hydrochloric acid, 37%	B (158 °F)	–	–	–	–	–
Hydrochloric acid, 37%	C (200 °F)	–	C (200 °F)	–	–	B (230 °F)
Hydrocyanic acid	A	T	A	A	–	A
Hydrofluoric acid, 48%	A (158 °F)	X	A	B	–	A
Hydrofluoric acid, 75%	A	X	B	C	–	B (158 °F)
Hydrofluoric acid, anhydrous	A	X	B	C	–	A
Hydrogen	A	A	A	A	T	A
Hydrogen peroxide, 90%	A	–	B	T	–	A
Hydrogen peroxide, 90%	–	–	–	–	–	C (270 °F)
Hydrogen sulphide	A	A	A	A	A (450 °F)	B (270 °F)
Isoctane	A	A	A	X	T	A
Isopropyl alcohol	A (200 °F)	A	A	T	T	A
Isopropyl ether	B	–	C	C	X	C
JP-4	C	A (100°F)	C	C	–	A (400 °F)
JP-5	C	-	C	C	–	A (400 °F)
JP-6	C	–	C	C	–	A (100 °F)
JP-6	–	–	–	–	–	C (550 °F)
Kerosene	B	A	C	C	B	A (158 °F)
Kerosene	–	–	–	–	–	B (400 °F)
Lacquer solvents	C	B	C	C	X	C
Lactic acid	A	T	A	A	–	A
Linseed oil	A	T	A	B	–	A
Lubricating oils	B (158 °F)	A	B (158 °F)	C	A (300 °F)	A (158 °F)
Magnesium chloride solutions	A (220 °F)	T	A (158 °F)	A	T	A
Magnesium hydroxide solutions	A (200 °F)	T	A (158 °F)	A	–	A
Mercuric chloride solutions	A	T	A	A	T	A
Mercury	A	A	A	A	T	A
Methyl alcohol	A	A	A (158 °F)	A	A	B
Methyl ethyl ketone	C	A	C	A	C	C
Methylene chloride	C	C	C (100 °F)	B	C	B (100 °F)
Mineral oil	A	A	A	C	T	A

Rating Key:
A–Fluid has little or no effect.
B–Fluid has minor to moderate effect.
C–Fluid has severe effect.
T–No data – likely to be compatible.
X–No data – not likely to be compatible.
Blanks indicate no evaluation has been attempted.

Unless otherwise noted, concentration of aqueous solutions are saturated.
All ratings are at room temperature unless specified.

*Reg U.S. Pat & Tm.Off.

cont...

CHEMICAL RESISTANCE (contd.)

Chemical	Hypalon*	Hytrel*	Neoprene	Nordel*	Vamac*	Viton*
Mobil XRM 206A (aircraft engine lube)	–	T	–	–	–	A (350 °F)
Naptha	C	A	C	C	X	A (158 °F)
Napthalene	C	B	C (176 °F)	C	–	A (176 °F)
Nitric acid, –10%	A	B-C	B	B	T	A
Nitric acid, 30%	A	C	C	B	A	A
Nitric acid, 30%	C (158 °F)	–	–	CV (158 °F)	–	–
Nitric acid, 60%	B	C	C	C	–	A
Nitric acid, 70%	C	C	C	C	X	A
Nitric acid, 70%	–	–	–	–	–	B (100 °F)
Nitric acid, red fuming	C	C	C	C	X	B
Nitric acid, red fuming	–	–	–	–	–	C (158 °F)
Nitrobenzene	C	C	C	A	X	B
Oleic acid	B	A	B	B	–	B
Oleum, 20-25%	B	C	C	C	X	A
Palmitic acid	B	A	B (158 °F)	B	–	A
Perchloroethylene	C	C	C	C	X	A (212 °F)
Phenol	C	C	C	B	X	A (212 °F)
Phenol	–	–	–	–	–	B (300 °F)
Phosphoric acid, 20%	A (200 °F)	–	A	A	T	A
Phosphoric acid, 60%	A (200 °F)	X	A	A	–	A (212 °F)
Phosphoric acid, 70%	A (200 °F)	X	A	A	–	A
Phosphoric acid, 85%	A (200 °F)	X	A	A	–	A
Pickling solution (20% nitric acid, 4% HF)	A	X	C	C	X	A
Pickling solution (17% nitric acid, 4% HF)	A (150 °F)	X	C	C	X	A
Pickling solution (17% nitric acid, 4% HF)	–	–	–	–	–	C (225 °F)
Picric acid	A	T	A	B	–	A
Potassium dichromate solutions	A (200 °F)	T	A	A	–	A
Potassium hydroxide, dilute solutions	A (200 °F)	A-B	A (158 °F)	A	–	A
Pydraul 312C	C	A	C	C	–	A
Pyridine	C	X	C	B	X	C
OFI-2023 (silicone brake fluid)	–	T	–	–	T	A (392 °F)

Rating Key:
A–Fluid has little or no effect.
B–Fluid has minor to moderate effect.
C–Fluid has severe effect.
T–No data – likely to be compatible.
X–No data – not likely to be compatible.
Blanks indicate no evaluation has been attempted.

Unless otherwise noted, concentration of aqueous solutions are saturated.
All ratings are at room temperature unless specified.

*Reg U.S. Pat & Tm.Off.

cont...

CHEMICAL RESISTANCE (contd.)

Chemical	Hypalon®	Hytrel®	Neo-prene	Nordel®	Vamac®	Viton®
SAE No.10 oil	C	A	C	C	A	A
Sea water	A	A	A	A	A	A
Shell turbine oil 307	T	T	T	X	–	B (392 °F)
Silicone grease	A	A	A	A	A (158 °F)	A
Skydrol 500	C	A	C	A (250 °F)	C (250 °F)	C
Skylube 450	–	–	–	–	–	C (392 °F)
Soap solutions	A (200 °F)	A	A (158 °F)	A (212 °F)	T	A
Sodium chloride solutions	A	A	A	A	A	A
Sodium dichromate, 20%	A (200 °F)	T	B	A	–	A
Sodium hydroxide, 20%	A (200 °F)	A-B	A	A	A	A
Sodium hydroxide, 46½%	A	B	A	A	–	A
Sodium hydroxide, 46½%	–	–	A (158 °F)	–	–	C (100°F)
Sodium hydroxide, 50%	A (285 °F)	–	A	A	–	C
Sodium hydroxide, 73%	A (280 °F)	X	A	A	C	C
Sodium hypochlorite, 5%	A	A	A	A	A	A
Sodium hypochlorite, 20%	A (158 °F)	T	B	A	A	B (158 °F)
Sodium peroxide solutions	A (200 °F)	–	A	A	–	A
Soybean oil	A	T	A	C	T	A (250 °F)
Stannic chloride	B	–	B	–	T	A
Stannous chloride, 15%	A (200 °F)	T	A (158 °F)	B	T	A
Steam (see water)	A	B (212 °F)	A	A (350 °F)	A (212 °F)	B (300 °F)
Steam	–	C (230 °F)	–	–	–	–
Stearic acid	B (158 °F)	T	B (158 °F)	B	–	T
Styrene	C	X	C	C	C	A
Sulphur, molten	A	T	A	A	T	A (250 °F)
Sulphur dioxide, gas	A	T	A	A	T	T
Sulphur dioxide, liquid	A	T	A	A	T	T
Sulphur trioxide	C	X	C	B	–	T
Sulphuric acid, up to 5%	A	A	A	A	A	A
Sulphuric acid, 5-10%	A	B	A	A	A	A
Sulphuric acid, 10-50%	A (250 °F)	C	A (158 °F)	B	A-B	A
Sulphuric acid, 50-80%	A (158 °F)	C	B-C	C	X	A

Rating Key:
A–Fluid has little or no effect.
B–Fluid has minor to moderate effect.
C–Fluid has severe effect.
T–No data – likely to be compatible.
X–No data – not likely to be compatible.
Blanks indicate no evaluation has been attempted.

Unless otherwise noted, concentration of aqueous solutions are saturated.
All ratings are at room temperature unless specified.

*Reg U.S. Pat & Tm.Off.

cont...

CHEMICAL RESISTANCE

Chemical	Hypalon®	Hytrel®	Neo-prene	Nordel®	Vamac®	Viton®
Sulphuric acid, 60%	A	C	B	C	X	A (250 °F)
Sulphuric acid, 90%	A	C	C	C	C	A (158 °F)
Sulphuric acid, 95%	A-B	C	C	C	X	A
Sulphuric acid, 95%	B (122 °F)	–	–	–	–	A (158 °F)
Sulphuric acid, fuming (20% oleum)	B-C	C	C	C	X	A
Sulphurous acid	A (158 °F)	B	C	C	–	C
Sunoco XS-820 (EP lubricant)	–	T	–	X	T	A (300 °F)
Tannic acid, 10%	A	A	A	A	T	A
Tartaric acid	A (200 °F)	T	A (158 °F)	B	T	A
Tetrahydrofuran	C	–	C	C	C	C
Toluene	C	B	C	C	C	B (100 °F)
Tributyl phosphate	C	–	C	C	X	C (212 °F)
Trichloroethylene	C	C	C	C	C	A
Trichloroethylene	–	–	–	–	C	B (158 °F)
Tricresyl phosphate	C	–	C	A (212 °F)	B	A (300 °F)
Triethanoalmine	A (158 °F)	C	A (158 °F)	A	A (158 °F)	C
Trisodium phosphate solutions	A	A	A	A	A	A
Tung oil	A	T	A	C	T	A
Turpentine	C	–	C	C	B-C	A (158 °F)
Water	A (158 °F)	A (158 °F)	A (158 °F)	A (158 °F)	A (212 °F)	A (158 °F)
Water	A (212 °F	A 212 °F)	A (212 °F)	A (212 °F)	–	A (212 °F)
Xylene	C	A-B	C	C	C	A
Xylene	–	–	–	–	–	B (158 °F)
Zinc chloride solutions	A (200 °F)	A	A	A	T	A

Rating Key:
A–Fluid has little or no effect.
B–Fluid has minor to moderate effect.
C–Fluid has severe effect.
T–No data – likely to be compatible.
X–No data – not likely to be compatible.
Blanks indicate no evaluation has been attempted

Unless otherwise noted, concentration of aqueous solutions are saturated.
All ratings are at room temperature unless specified.

*Reg U.S. Pat & Tm.Off.

PROPERTIES OF

	Properties	Natural rubber	Hypalon®	Hytrel®	
Mechanical properties	Hardness range (durometer A&D)	30–90A	40–95A	400	550
	Tensile strength, MPa (lb/in²) Pure gum	Over 20.7 (3,000)	Over 17.2 (2,500)	25.5 (3,700)	44.1 (6,400)
	Black loaded stocks	Over 20.7 (3,000)	Over 20.7 (3,000)		
	Specific gravity (base material)	0.93	1.12–1.28	1.17	1.20
	Vulcanizing characteristics	Excellent	Excellent	Unnecessary to vulcanize	
	Adhesion to metals	Excellent	Excellent	Good	Good
	Adhesion to fabrics	Excellent	Good	Good	Good
	Tear resistance	Good	Fair	Excellent	Outstanding
	Abrasion resistance	Excellent	Excellent	Outstanding	Outstanding
	Compression set	Good	Fair	Fair	Fair
	Rebound Cold	Excellent	Good	Very Good	Good
	Hot	Excellent	Good	Excellent	Very Good
Electrical properties	Dielectric strength	Excellent	Excellent	Fair/Good	Fair/Good
	Electrical insulation	Good/ Excellent	Good	Fair/Good	Fair/Good
Fluid resistance properties	Permeability to gases	Fair	Low– Very Low	Low	Low
	Acid resistance Dilute	Fair/Good	Excellent	Good	Good
	Concentrated	Fair/Good	Very Good	Poor	Poor

ELASTOMERS

Hytrel®		Neoprene	Nordel®	Vamac®	Viton®
630	720	40–95A	40–90A	40–95A	55–95A
40.0 (5,800)	40.0 (5,800)	Over 20.7 (3,000)			Over 12.4 (1,800)
		Over 20.7 (3,000)	Over 20.7 (3,000)	Over 17.2 (2,500)	Over 13.8 (2,000)
1.22	1.25	1.23	0.86	1.08–1.12	1.85
Unnecessary to vulcanize		Excellent	Excellent	Good	Good/Excellent
Good	Good	Excellent	Good/Excell.	Good	Good/Excellent
Good	Good	Excellent	Good	Good	Good/Excellent
Outstanding	Outstanding	Good	Good	Good	Fair/Good
Outstanding	Outstanding	Excellent	Excellent	Good	Good
Poor	Poor	Fair/Good	Good	Good	Good/Excellent
Fair	Fair	Very Good	Very Good	Poor	Good
Good	Good	Very Good	Very Good	Fair	Excellent
Fair/Good	Good	Good	Excellent	Good	Good
Fair/Good	Good	Fair/Good	Excellent	Fair/Good	Fair/Good
Low	Low	Low	Fair	Very Low	Very Low
Good	Good	Excellent	Excellent	Good	Excellent
Poor	Poor	Good	Excellent	Poor	Excellent

PROPERTIES OF

	Properties	Natural rubber	Hypalon ®	Hytrel ®	
Fluid resistance properties	Solvent resistance				
	Aliphatic hydrocarbons	Poor	Good	Excellent	Excellent
	Aromatic hydrocarbons	Poor	Fair	Good	Good
	Oxygenated (ketones, *etc*)	Fair/Good	Poor	Fair	Good
	Lacquer solvents	Poor	Poor	Fair	Fair/Good
	Resistance to:				
	Swelling in lubricating oil	Poor	Good/ Excellent	Good	Excellent
	Oil and gasoline	Poor	Good	Very Good	Excellent
	Animal and vegetable oils	Poor/Good	Good	Very Good	Excellent
	Water absorption (hydrolitic stability)	Very Good	Very Good	Very Good up to 100 °C (212 °F)	Very Good up to 100 °C (212 °F)
Resistance to environmental factors	Oxidation	Good	Excellent	Excellent	Excellent
	Ozone	Fair	Outstanding	Excellent	Excellent
	Sunlight ageing	Poor	Outstanding	Very Good†	Very Good†
	Heat ageing (upper limit continuous service)	85 °C (185 °F)	135 °C (275 °F)	100 °C (212 °F)	110 °C (230 °F)
	Flame*	Poor	Good	Good† (will melt)	Good† (will melt)
	Heat	Good	Excellent	Excellent	Excellent
	Cold	Excellent	Good	Excellent	Excellent

*These evaluations are qualitative and comparative only. They should not be construed as recommendations. Specific compounding is required to optimize performance. Elastomer choice should be based upon a practical consideration of the potential fire hazards involved in each individual case and, if applicable, the results of appropriate flame tests.

†With additives.

ELASTOMERS

Hytrel®		Neoprene	Nordel®	Vamac®	Viton®
Excellent	Excellent	Good	Poor	Good	Excellent
Good	Good	Fair	Poor	Fair	Excellent
Good	Good	Poor	Good	Poor	Poor
Good	Good	Poor	Poor	Poor	Poor
Excellent	Excellent	Good	Poor	Good	Excellent
Exclient	Excellent	Good	Poor	Poor	Excellent
Excellent	Excellent	Good	Good	Good	Excellent
Very Good up to 100 °C (212 °F)	Good up to 100 °C (212 °F)	Good	Very Good	Very Good up to 100 °C (212 °F)	Very Good
Excellent	Excellent	Excellent	Excellent	Excellent	Outstanding
Excellent	Excellent	Excellent	Outstanding	Outstanding	Outstanding
Very Good†	Very Good†	Very Good	Outstanding	Outstanding	Outstanding
110 °C (230 °F)	110 °C (230 °F)	95 °C (203 °F)	145 °C (293 °F)	170 °C (340 °F)	205 °C (401 °F)
Good† (will melt)	Good† (will melt)	Very Good	Poor	Poor	Excellent
Excellent	Excellent	Very Good	Excellent	Excellent	Outstanding
Excellent	Excellent	Very Good	Excellent	Good	Fair/Good

COMPATIBILITY OF RUBBERS USED FOR SEALS

		Nitrile (medium)	Nitrile (high)	Neoprene	Polyacrylic	Hypalon	SBR	Silicone	Viton	Ethylene propylene	Kalrez
		NBR	NBR	CR	ACM		SBR	MQ	FKM	EP	FFKM
ORGANIC COMPOUNDS											
Acids	*Acetic*	S	S	D	S	D	D	S	S	R	R
	Oleic	D	R	D	D	D	D	D	R	D	R
Anhydrides	*Acetic anhydride*	S	S	D	S	S	D	S	S	D	R
Acid chlorides	*Acetyl chloride*	S	S	S	S	S	S	S	S	D	R
Alcohols	*Ethanol*	R	R	R	D	R	R	R	R	R	R
Aldehydes	*Acetaldehyde*	S	S	S	S	S	S	S	S	R	R
Amines	*Ethanolamine*	D	D	D	S	D	D	D	S	D	R
	Aniline	S	S	D	S	D	S	R	D	D	R
Chlorinated	*Trichloroethylene*	S	S	S	S	S	S	S	R	S	D
	Chlorobenzene	S	S	S	S	S	S	S	D	S	D
Esters	*Tricresylphosphate*	S	S	S	S	S	R	R	R	R	R
	Dibutylphthalate	S	S	S	S	S	S	R	S	D	R
Glycols	*Ethylene*	R	R	R	R	R	R	R	R	R	R
Glycerine		R	R	R	S	R	R	R	R	R	R
Hydrocarbons	*Octane*	R	R	D	D	D	S	S	R	S	R
	Benzene	S	S	S	S	S	S	S	D	S	R
Ketones	*Acetone*	S	S	D	S	D	R	D	S	R	R
Nitro compounds	*Nitrobenzene*	S	S	S	S	S	S	S	S	S	R
Phenols	*Phenol*	S	S	D	S	R	S	R	R	D	R
GASES											
Chlorine (wet)		C	C	C	S	D	C	S	D	D	D
Calor gas		D	R	S	R	S	S	S	R	D	R
Carbon dioxide		R	R	R	R	R	R	R	R	D	R
Coal gas		D	R	D	R	D	D	D	R	D	R
Hydrogen		R	R	R	R	R	R	R	R	R	R
Natural gas		R	R	R	D	R	D	R	R	S	R
Nitrogen		R	R	R	R	R	R	R	R	R	R
Oxygen		R	R	R	R	R	R	R	R	R	R

INORGANIC											
Acids	*Chromic*	C	C	C	S	R	C	D	R*	C	R
	Hydrochloric	R	R	R	S	R	D	R	R*	C	R
	Nitric	C	C	C	C	D	C	C	D*	C	R
	Phosphoric	D	D	R	D	R	D	D	R*	D	R
	Sulphuric	C	C	R	C	R	R	C	R*	S	R
Alkali	*Caustic soda*	R	R	R	S	R	R	R	D*	R	R
	Ammonia	R	R	R	D	R	R	R	C	R	R
Carbon disulphide		S	D	S	S	S	S	S	R	S	R
Hydrogen peroxide		S	S	D	S	R	D	D	R	C	R
Steam up to 3.5 bar (50 lb/in^2)		D	D	D	C	S	D	D	D	R	R
Water		R	R	R	S	R	R	R	D	R	R
REFRIGERANTS											
Ammonia		R	R	R	S	R	R	R	C	R	D
Ethylene glycol		R	R	R	S	R	R	R	R	R	R
Freon 11 (Arcton 11)		S	R	S	S	S	R	S	D	S	D
Freon 12 (Arcton 12)		R	R	R	D	D	S	S	D	D	D
Freon 22 (Arcton 22)		S	S	D	S	D	S	S	S	D	D
Sulphur dioxide		S	S	D	S	R	D	R	R	R	R
LUBRICANTS AND GREASES											
Brake fluids	*castor based*	S	S	D	S	D	R	R	D	R	R
Cellosolve		S	S	D	S	D	D	S	S	D	R
Greases	*mineral*	R	R	D	R	D	S	D	R	S	R
	ester based										
	DTD 825	D	R	S	D	S	S	D	R	S	R
Hydraulic fluids	*aqueous EEL6*	R	R	R	S	R	R	R	R	D	R
	mineral										
	DTD 585	R	R	D	R	S	S	S	R	S	R
	ester based										
	(Skydrol 500)	S	S	S	S	S	S	S	R	S	R
Lubricating oils	*mineral DED 2472*	R	R	R	R	D	S	R	R	S	R
	hypoid	R	R	D	R	S	S	R	R	S	R
	ester DED 2487	D	R	S	D	S	S	D	R	S	R
Paraffin		D	R	D	R	S	S	S	R	S	R
Silicone oils		R	R	R	R	R	R	S	R	R	R
Transformer oil		R	R	D	R	S	S	D	R	S	R
Vegetable oil		R	R	R	R	D	D	R	R	R	R
FUELS											
Diesel oil		D	R	S	R	S	S	D	R	S	R
Gas oil		D	R	S	R	S	S	S	R	S	R
Fuel oil		D	R	S	R	S	S	S	R	S	R
Petrol		D	R	S	R	S	S	S	R	S	R
Solvent naphtha		S	S	S	S	S	S	S	R	S	R

Key: R — Very resistant to swelling
D — Some swelling, the amount depending on conditions
S — Excessive swelling — not recommended
C — Surface attack — not recommended

GENERAL GUIDE TO
(see also SEAL

	Materials	Natural Rubber and S.B. Rubber	Neoprene	Nitrile Rubbers	Thiokol	Chlorosulphonated polyethylene
GASES	Air or oxygen	A	SH(80°C)	SH	V(100°C)	SH(80°C)
	Halogens	E	E	E	E	LS
	Other gases and dry steam	A	SH(80°C)	SH(80°C)	V(100°C)	SH(80°C)
INORGANIC LIQUIDS	Water and neutral aqueous solutions	C	C	C	C(70°C)	C(80°C)
	Alkaline solutions	C	C	C	C(70°C)	C to LS
	Dilute acid solutions	C	C	C	C(70°C)	C(80°C)
	Strong hydrochloric acid	A	A	A	A	A
	Strong sulphuric acid	VA	A	A	A	A
	Conc. sulphuric acid	VA	VA	VA	SA	A
	Nitric acid	VA	VA	VA	SA	A
	0.880 ammonia	A	LS	C	C	SA
	Liquid ammonia	A	LS	C	C	C
	Liquid chlorine	VA	VA	VA	SA	LS
	Glacial acetic acid	LS	A	C	C	A
SOLVENTS	Aliphatics	MS	SS	C	C	MS
	Aromatics	MS	MS	SS	SS	MS
	Ketones	SS	MS	SS	SS	MS
	Chlorinated	MS	MS	MS	MS	MS
	Alcoholic	C	C	C	C	C
	Esters	MS	MS	SS	SS	MS
FUELS	Ordinary petrols	MS	SS	C	C	MS
	High octane petrols	MS	MS	SS	SS	MS
	Kerosene	MS	MS	C	C	MS
	Gas oil	MS	MS	C	C	MS
HYDRAULIC FLUIDS	Mineral base	MS	SS	C	C	SS
	Ester base (non-flam.)	MS	MS	SS	SS	MS
	Water base	SS	C	C	C	C
	Chlorinated type	MS	MS	MS	MS	MS
	Silicone base	LS	LS	LS	LS	LS
OILS	Mineral oils	MS	SS	C	C	SS
	Synthetic lubricants	MS	MS	SS	SS	MS

Note: Temperature figures represent upper limits where given in conjunction with key letters, *eg* SH (80°C) = Slow Hardening above 80°C

COMPATIBILITY OF ELASTOMER
SELECTION GUIDES)

Butyl Rubbers	Viton (Fluorocarbon elastomer)	Silicone Rubbers	Polyurethane Rubbers	Fluoro-silicones	PTFE (Polytetrafluoro-ethylene)	Kalrez Perfluoro-elastomer
C(100°C)	C(200°C	-70°C to 275°C	SH(80°C)	C(-60 C to 260°C)	C(-150 C to 250 C)	C(-200°C to +300°C
A	C	LS			C	C
LS(100°C)	C(200°C)	C(150°C)	A(Steam)	C(150 C)	C	C (300°C)
C	C	C	A (lot)	C	C	C
C	C	SA	A (lot)	SA	C	C
C	C	A	A (lot)	SA	C	C
C	C	A	VA	VA	C	C
A	C	VA	VA	VA	C	C
LS	C	VA	VA	VA	C	C
LS	LS	VA	VA	VA	C	C
C	A	C	VA	SA	C	C
C	E	SA	–	SA	C	C
E	LS	LS	–	–	C	C
LS	A	SA	VA	–	C	C
MS	C	MS	SS	SS	C	C
MS	C	MS	MS	SS	C	C
SS	MS	MS	MS	MS	C	C
MS	SS	MS	MS	LS	C	SS
C	C	SS	SS	C	C	C
C	MS	MS	MS	SS	C	C
MS	C	MS	C	SS	C	C
MS	C	MS	SS	SS	C	C
MS	C	MS	C	SS	C	C
MS	C	MS	C	SS	C	C
MS	C	SS	C	SS	C	C
C	MS	SS	SS	SS	C	C
C	C	–	SS	–	C	C
MS	C	MS	MS	MS	C	SS
LS	C	SS	LS	SS	C	C
MS	C	LS	SS	SS	C	C
SS	LS	LS	SS	SS	C	C

Key:

VA	–	Vigorously Attacked	C	–	Fully Compatible
A	–	Attacked	E	–	Embrittled
SA	–	Slight Attack	LS	–	Limited Serviceability
SH	–	Slow Hardening	SS	–	Slight to Moderate Swelling
V	–	Volatile	MS	–	Marked Swelling

Seal Costs

The true cost of seals

Evaluation of the *true cost effectiveness* of seals must take into account several parameters in addition to that of the initial cost of the seal; a fact that is obvious but so often overlooked when comparing possible alternative seal prices. More specifically, not all of the parameters involved may be taken into account. For example, the fact that the relative cost of a U-ring may be ten times that of an O-ring, does *not* mean that these two types of seals will be equally cost effective in a particular application if the U-ring lasts ten times as long as the O-ring. Yet this is the basis on which many users may be tempted to compare alternative seal types.

The parameters involved in true cost effectiveness are:

(i) Initial cost of the seal.

(ii) Effectiveness of the seal.

(iii) Time between necessary seal replacement.

(iv) True cost of seal replacement (seal cost plus labour time, plus any 'hidden' costs).

(v) Cost of equipment down time during seal replacement.

(vi) Reliability factor as affecting maintenance costs.

Items (i), (iii), (iv) and (v) are inter-related and cannot be used as truly comparative parameters on their own. Item (ii), however, can be considered on its own as a cost effectiveness parameter, in terms of reduction of product loss (and energy losses) with a more effective seal. A continuous drip from a pump gland can amount to 4 to 5 litres/hour product loss, or over 40000 litres loss in a full year. Equally, seal leakage can materially reduce the efficiency (and increase energy losses) of hydraulic and pneumatic cylinders. Seal efficiency may also be significant as an environmental parameter (which could affect choice of seal type).

Examples

As a simple example, the parameters relating to a typical packed gland fitted with a traditional packing (asbestos, cotton, hemp) and a modern synthetic fibre packing are outlined:

(i) Initial cost : traditional packing 1, modern packing 4. (Purchase price ratio of 1:4).

(ii) Effectiveness – ignored, as it is assumed that both types will give similar results, but in practice the modern synthetic fibre packing can be expected to be more effective, *eg* show a lower leakage rate.

(iv) Cost of replacement – labour time considered the same. However, there may be 'hidden' costs involved. For example, with a traditional packing there may be the need to re-sleeve the shaft at regular intervals. This may not be necessary at all with a modern synthetic packing. There may be other 'hidden' costs involved, *eg* the time taken to re-adjust a traditional gland packing on a day-to-day or week-to-week basis, whereas this may be minimal with a synthetic packing.

(v) Cost of equipment down time – assumed the same down time and production loss. However, this occurs *four times* a year with the traditional packing and only *once* a year with the modern synthetic packing.

(vi) Reliabililty factor – favourable to the modern synthetic packing in the same ratio as the life factor.

Summarizing, in this particular example, the traditional packing is showing a lower total cost only at the first gland fill. The modern synthetic fibre packing can show a straight total cost saving after three months and considerable savings thereafter. The *least* significant parameter in this study is thus (i), the initial cost of packing.

These parameters are re-stated with typical figures in the following Table, starting with a clean gland.

COMPARATIVE PACKING COSTS

	Traditional Packing		Modern Synthetic Fibre Packing	
	Stage Cost £	Accumulative Cost £	Stage Cost £	Accumulative Cost £
Packing cost	2.50		10.00	
Labour and o/h	8.00		8.00	
		10.50		**18.00**
Month 3 – repack				
Packing cost	2.50			
Labour and o/h	12.00		nil	
Down time cost				
(production loss)	A	**25.00 + A**		**18.00**
Month 6 – repack				
Packing cost	2.50			
Labour and o/h	12.00		nil	
Down time cost	A	**39.50 + 2A**		**18.00**
Month 9 – repack				
Packing cost	2.50			
Labour and o/h	12.00		nil	
Down time cost	A	**54.00 + 3A**		**18.00**
Month 12 – repack				
Packing cost	2.50		8.00	
Labour and o/h	12.00		12.00	
Re-sleeve	50.00		not necessary	
Down time cost	A	**118.50 + 4A**	A	**40.00 + A**
Saving after one year with modern synthetic fibre packing **£78.50 + 3A**				

SECTION 9

Engineering Data

Standards

Standards

UNTIL THE 1960's comparatively few standards existed for seals and sealing products. The most notable of the early standards was BS 1806 : 1962 covering Imperial sizes of O-rings (and based largely on American sizes). Metric sizes for O-rings were standardized in BS 4518 : 1969, this being based on the Swedish standard SMS 1588 widely used throughout Continental Europe at the time. These were at a later stage approved by the International Standards Organization (ISO) and specified as ISO R1077, R1078 and R1800.

Since 1970 many countries have co-operated in the preparation of further international standards, providing individual delegates to the ISO Seals Committee ISO TC131/SC7 (TC=Technical Committtee and SC=Sub Committee). With the exception of O-rings which are dimensionally specified, ISO standards for other seal designs have only their housing dimensions shown together with, where appropriate, quality acceptance criteria, storage, handling and fitting, method of performance rating and terminology.

eg For rotary shaft seals the following documents are approved by ISO or are in draft awaiting approval:

6194 Part 1. Rotary Shaft Seals. Nominal dimensions and tolerances.

6194 Part 2. Rotary Shaft Seals. Terminology (for Part 1).

6194 Part 3. Rotary Shaft Seals. Storage handling and fitting.

6194 Part 4. Rotary Shaft Seals. Method of performance rating.

Terminology documents are forthcoming for Parts 3 and 4.

In the UK new draft standards are generally compiled by trade associations and for sealing devices active bodies are the Association of Hydraulic Equipment Manufacturers (Committee B), British Rubber Manufacturers Association (Fluid Seal Technical Committee) and others which include the British Compressed Air Society. These drafts are usually passed to a BSI Mechanical Engineering and Equipment Committee, MEE 109, which, after deliberations with both manufacturer and user, passes it to the International Committee.

Similar organizations exist in the major industrialized countries and amongst the most active internationally are:

France – AFNOR – NF

Germany – Deutscher Normenausschuss – DIN

Japan – Japanese Industrial Standards Committee – JIS
UK – British Standards – BS
USA – Various Committees producing standards – ANSI, ASTM, SAE, NFPA, ASME
Sweden – Svenges Standardiserings Kommission – SMS
USSR – Komitet Standartov, Moscow Gosvdarstuennyi

France, Austria, Poland, Switzerland, *etc*, are usually represented at the sub committee meetings.

There are also aerospace/defence committees within ISO, one being ISO TC20/SC10 and the standards being produced are received from the various national bodies. In the UK they are the Society of British Aerospace Constructors and the Ministry of Defence while in the USA, US Standards are used for the Department of Defence for Military Procurement (ARP. AS. MS.).

Sometimes the work of the commercial ISO Committee TC131/SC7 overlaps the work of the aerospace committee TC20/SC10 which can result in two standards being issued which are very similar. This situation has arisen in recommendations for a range of metric O-rings in that two recommendations exist, *ie* ISO 3601 Part 1 (Commercial) and ISO 4613 Part 1 (Aerospace).

It should be emphasized that the ISO sub committees usually have working groups of elected national experts to carry out the necessary tasks. For sealing applications the following have been recently approved or are under discussion (DP = draft proposal).

ISO Standards

ISO 5597 Part 1 — Cylinder rod and piston seals for reciprocating applications.
ISO DP 6547 — Piston seals for reciprocating applications.
ISO DP 5597 Part 2. — Seals for single rod cylinders to ISO 6020 Part 2.
ISO DP 6195 Part 1. — Wiper ring housings, rod reciprocating applications.
ISO DP 6195 Part 2. — Wiper ring housings for compact cylinders.
ISO 3601 Part 1. — Metric O-rings, dimensions, tolerances, *etc*.
ISO DP 3601 Part 2. — Metric O-rings, design criteria for standard applications.
ISO 3601 Part 3. — Metric O-rings, quality assurance criteria.

(Other ISO Standards

R 1941:1970 Flat seal for hydraulic coupling.
R 383:1964 Interchangeable conical ground glass joints.
R 641:1968 Interchangeable spherical ground glass joints.
1749:1973 Aircraft-elastomeric sealing rings – packaging and identification.

2035:1974	Unplasticized polyvinyl chloride (PVC) moulded fittings for elastic sealing ring type joints for use under? pressure.
2045:1973	Single sockets for unplasticized polyvinyl chloride (PVC) pressure pipes with elastic sealing ring type joints – minimum depths of engagement.
2048:1973	Double socket fittings for unplasticized polyvinyl chloride (PVC) pressure pipes with elastic sealing ring type joints – minimum depths of engagement.
2084:1974	Pipeline flanges for general use – metric series – mating dimensions.
2441:1975	Pipeline flanges for general use – shapes and dimensions of pressure-tight surfaces.
3069:1974	End suction centrifugal pumps – dimensions of cavities for mechanical seals and for soft packings.
3603:1977	Fittings for unplasticized polyvinyl chloride (PVC) pressure pipes with elastic sealing ring type joints – pressure test for leakage.
3604:1976	Fittings for unplasticized polyvinyl chloride (PVC) pressure pipes with elastic sealing ring type joints – pressure test for leakage under conditions of external hydraulic pressure.

American Standard

American standards are issued by the following authorities:

MS:	Military Standard (for official use only).
MIL:	Military specification.
AN:	Army-Navy-Aeronautical Standards.
AS:	Aerospace Standards.
ANSI:	American National Standards Institute.
ASME:	American Society of Mechanical Engineers.
ASTM:	American Society for Testing and Materials.
SAE:	Society of Automotive Engineers.
NFPA:	National Fluid Power Association Standards.

(ASTM Standards (partial list)

ASTM C 834-76	Standard specification for latex sealing compounds.
ASTM D 1330-66	Sheet rubber gaskets.
ASTM D 1414-782	Rubber O-rings.
ASTM D 1418-81	Rubber and rubber latices.
ASTM F 37-68	Sealability of gasket materials.
B16.21-1962	Non-metallic gaskets for pipe flanges.
B2.1-1968	Pipe threads (except dryseal).
B2.2-1968	Dryseal pipe threads.
Z 261-2-73	Standard recommended practice for design and construction of non-metallic enveloped gaskets for corrosive service.

SAE Standards (partial list)

J110 b	Qualification and quality control tests for radial lip type seals for rotating shafts.
J111 b	Radial seal nomenclature and glossary. (Recommended Practice).
J476	Dryseal pipe threads. (Standard).
J515 a	Hydraulic O-rings. (Standard),
J654	Lathe cut seals. (Recommended Practice).
J780	Water pump seals. (Standard).
J929 a	Piston rings and pistons. (Standard).
J946 c	Guide to the application and use of radial lip type oil seals. (Recommended Practice).
J1601	Rubber cups for hydraulic actuating cylinders. (Standard).

Aerospace Standards (partial list)

Note: AS:Aerospace Standard. ARP:Aerospace Recommended Practice. Air:Aerospace Information Report. AMS:Aerospace Material Specification).

J110 b	Qualification and quality control tests for radial lip type seals for rotating shafts.
J111 b	Radial seal nomenclature and glossary. (Recommended Practice).
J476	Dryseal pipe threads. (Standard).
AS 568A	Aerospace size standard for O-rings.
ARP 674	Metal ring seal groove design.
AIR 851A	O-ring tension testing calculations.
AS 871A	Manufacturing and inspection standards for pre-formed packings.
ARP 979	Performance qualification testing for back-up rings and special sealing devices.
AIR 1077	Metallic seal rings for high temperature reciprocating hydraulic service.
AIR 1231	Gland design, elastomeric O-ring seals, general considerations.
ARP 1232A	Gland design, elastomeric O-ring seals, dynamic radial 1500 lb/in^2 maximum.
ARP 1233	Gland design, elastomeric O-ring seals, dynamic radial 1500 lb/in^2 maximum.
AMS 7260B	Rings, packing synthetic rubber, fuel and low temperature resistant 70 to 80.
AMS 7267C	Rings, sealing silicone rubber, heat resistant, low compression set, 70 to 80.
AMS 7273A	Rings sealing fluorosilicone rubber, high temperature fuel and oil resistant, 70 to 80.
AMS 7277B	Rings, sealing synthetic rubber, phosphate ester hydraulic fluid resistant, Butyl type, 70 to 85.

AMS 7310F Rings, piston cast iron.

AMS 7322B Rings, sealing cast bronze 80 Cu 19SN.

AMS 7325D Rings, sealing tubular metal, corrosion and heat resistant steel (AMS 5576).

British Standards (partial list)

903:Part A34: 1977 Determination of stress relaxation of rubber rings in compression.

1399 Rotary shaft lip seals.

1658:1970 Housings for hydraulic seals for reciprocating applications.

1806:1962 Dimensions of toroidal sealing rings (O-seals and their housings).

1832:1972 Oil resistant compressed asbestos fibre.

2494:1976 Materials for elastomeric joint rings for pipework and pipelines.

2815:1973 Compressed asbestos fibre jointing.

3063:1965 Dimensions of gaskets for pipe flanges.

3381:1973 Metallic spiral wound gaskets for use with flanges to BS 1560 : Parts 1 and 2.

3627:1963 Dimensions of compression and oil control rings and piston ring grooves for i/c engines.

3712 Methods of test for building sealants.

4243:1967 Cork/paper jointing. Amendment AMD 824, November 1971.

4255:Part 2: 1975 Cellular gaskets.

4332:1968 Composition cork jointing.

4371:1968 Fibrous gland packings.

4375:1968 Unsintered PTFE tape for thread sealing applications.

4518:1974 Metric dimensions of toroidal sealing rings (O-rings) and their housings.

4865 Dimensions of gaskets for pipe flanges to BS 4504.

4865:Part 1: Dimensions of non-metallic gaskets for pressure up to 64 bar.

4865:Part 2: 1973 Dimensions of metallic spiral wound gaskets for pressures 10 bar to 250 bar.

5106:1974 Dimensions of spiral anti-extrusion back-up rings and their housings.

5341: Specification for piston rings up to 200 mm diameter for reciprocating internal combustion engines.

5351:Part 1: Designs, dimensions, materials and designations for single piece rings.

5341:Part 2: Designs, dimensions, materials and designations for multipiece oil control rings.

5341:Part 5: 1976	Ring grooves.
AU 120:1966	Cork rubber gasket materials.
AU 121:1966	Composition cork gasket materials.
F66:1965	Rubber bonded cork sheets for aeronautical purposes (replacing Ministry of Aviation specification DTD 762). PD 5800, March 1966.
F125:1973	Rubber bonded compressed asbestos fibre jointing.

French Standards (AFNOR)

NF T 48-001 Juin 1967	HOM	Feuilles en amiante et élastomères comprimés pour joints, caractéristiques.
NF T 48-101 Juin 1967	HOM	Feuilles en amiante et élastomères comprimés pour joints, mesures des dimensions.
NF T 48-102 Juin 1967	HOM	Feuilles en amiante et élastomères comprimés pour joints, détermination de la densité relative.
NF T 48-103 Juin 1967	HOM	Feuilles en amiante et élastomères comprimés pour joints, détermination de la résistance à là rupture par traction.
NF T 48-104 Juin 1967	HOM	Feuilles en amiante et élastomères comprimés pour joints, détermination de la perte au feu.
NF T 48-105 Juin 1967	ENR	Feuilles en amiante et élastomères comprimés pour joints, essai d'immersion dans les liquides.
NF T 48-106 Juin 1967	HOM	Feuilles en amiante et élastomères comprimés pour joints, résistance à la rupture par traction après immersion dans les liquides.
NF T 48-107 Juin 1967	ENR	Feuilles en amiante et élastomères comprimés pour joints, essais de tenue à la pression hydraulique à température ambiante.
NF T 48-108 Juin 1967	HOM	Feuilles en amiante et élastomères comprimés pour joints, essais de tenue au cintrage.
NF T 48-109 Juin 1967	HOM	Feuilles en amiante et élastomères comprimés pour joints, essais de vieillissement artificiel à l'étuve.

NF M 87-621 Mai 1979	ENR	Joints spirales pour brides à face de joint surélevée. (Avec NF M 87-622, NF M 87-631, NF M 87-641 et NF M 87-642, Mai 1979, remplace NF M 87-615, Juin 1973).
NF M 87-622 Mai 1979	ENR	Joints spirales pour brides à emboîtements. (Avec NF M 87-621, NF M 87-631, NF M 87-641 et NF M 87-642, Mai 1979, remplace NF M 87-615, Juin 1973).
NF M 87-631	ENR	Joints métalloplastiques pour brides à face de

Mai 1979		joint surélevée. (Avec NF M 87-621, NF M 87-622, NF M 87-641 et NF M 87-642 Mai 1979, remplace NF M 87-615, Juin 1973).
NF M 87-641 Mai 1979	ENR	Joints en amiante et élastomères comprimés pour brides à emboîtements. (Avec NF M 87-621, NF M 87-622, NF , 87-631, et NF M 87-641, Mai 1979, remplace NF M 87-615, Juin 1973).
NF M 87-642 Mai 1979	ENR	Joints en amiante et élastomères pour brides à emboîtements.(Avec NF M 87-621, NF M 87-622, NF M 87-631, et NF M 87-641, Mai 1979, remplace NF M 87-615, Juin 1973).

Bride

E 48-054 ?85? 056	carrées et rectangulaires pour collets à souder.
A 48-501 à 515	de canalisation en fonte.
A 48-808 et 809 A 72-120	de tuyauterie pour fluide gazeux.
C 68-381	de fixation de carter (Matériel antidé na grant).
E 21-501 à 506	de serrage.
E 22-604	pré-alésées pour transmissions à cardan industrielles.
E 29-004	Emboîtement
E 29-021, 023 024 et C 51-105	
Série E 29-2	de tuyauterie industrielle.
E 29-720 à 723 et E 29-728	pour vide.

Bride (joints pour)

E 29-911 à 956	et collets (tuyauterie industrielle).
E 51-010	circulaires (ventilateurs).
E 62-	09?de fixation de têtes multibroches.
F 01-048	de ressorts à lames.
F 37-027	d'obturation (wagons-citernes).
F 55-025	pour boîtes de jonction.
F 55-028	pour boîtes de dérivation.
F 56631 et 332	pour potelets et traverses (lignés aériennes).
F 81-001 et 002	de fixation de tuyauterie.
S/classe J 22- et série J 82-2	Construction navale.
L 34-112	à 4 trous pour arbres carrés.
L 43-601 à 611	Raccordements
L 43-711	à section en V pour tuyauteries.
L 86-310	de fixation de magnétos.
M 82-45	de tuyauteries à rotule.

M 87-500 à 576	Industrie du pétrole.
R 11-403	de fixation de démarreur.
R 16-301 et R 16-401	de carburateur ovale à 2 trous

(Amiante

G 28-001	Essais des fils.
Série T48	Essais des feuilles
0 51-001	Carton.

Élastomère

E 29-911	Joints.
L 46-030 à 033	
G 18-001 à 006	Fils élastiques (Élastodîene).
Série G 37-1	Supports textiles revêtus. utilisés en construction aérospatiale (Essais, qualification, contrôle, caractéristiques, marquage, emballage, stockage).
S/classe L17-	pour calfeutrement étanche.
Série Pà5	pour calfeutrement étanche.
T 40-001	Terminologie générale.
T 40-101	Atmosphère de référence.
T 40-102, 103	Propriétés générales.
S/classe T43-	Élastomères bruts.
Série T 45-	Matières premières.
Série T 46-	Vulcanisé (Essais).
S/classe T 47-	Objets manufacturés.
Série T 48-	Comprimés pour joints.
S/classe T 56-	Produits alvéolaires à base d'élastomères ou de matières plastiques.
T47-001	Produits moulés et extrudés.

Étanchéité

A 48-722, 756, 801 eté	Bague d' pour canalisations en fonte.
C 20-620 à 629	Essais des composants des matériels électriques.
E 29-122	des robinets de gaz.
E 29-991	Garnitures mécaniques.
H 14-001	des récipients paraffinés.
H 44-020	des mécanismes de valve (récipients aérosols).
J 84-201 à 205	des panneaux de claires-voies avec verres.
L 40-114	radiale des raccordements.
L 46-312	axiale des bouchons d'obturation.

M 60-301	des emballages pour matières radio-actives.
M 61-002	des sources scellées.
M 62-200	des enceintes de confinement (Classification).
M 88-088	des véhicules – citernes routiers.
P 10-203 (D.T.U.20.12)	toiture.
P 78-101	miroiterie et viterie.
P 78-202 (D.T.U. 39.4)	
P 78-501 à 504	
P 30-303 et 304	de la construction.
S/classe Pà	des joints mastics.
P 84-203 (D.T.U.43)	code des conditions d'éxécution.
P 84-305	Produits asphaltiques.
P 84-307, 308	Feutres de verre bitumés.
P 84-313	
P 84-352	Rêvetements (essai)
Q 15-004	Papier pour récipient étanche paraffiné.
R 93-920	Joints circulaires.
S 81-515	des montres.
S 81-517	des montres-bracelets.
S 86-605	des boîtes de montres.
T 54-038 à 042	canalisations.
U 57-012	des serres.

German (DIN) Standards (partial list)

470	Sealing washers.
1367	Part 1. Seal wires of galvanized steel wire.
2690	Gaskets for flanges with flat sealing faces.
2691	Gaskets for tongued and grooved flanges.
2692	Gaskets for recessed flanges.
2693	O-rings for profiled flanges with grooves.
2695	Diaphragm seal rings and diaphragm welded seal for flanged connections.
2696	Sealing discs and disc packings for flanged connections.
3485	Seals for compressed air fittings.
3750	Seals: nomenclature.
3754	Sheet sealing material.
3760	Rotary shaft lip seals.
3770	Elastomeric O-rings.
3780	Seals and packings; stuffing box diameters and related packing widths.
4000	Part 7. Special performance characteristics for seals.

20006 Flat gaskets for pipes with flanges and collars.

24909 Part 1. Piston rings; list of types and general requirements.

24960 Part 1. Mechanical seals; single mechanical seals; principal mounting dimensions, cavities.
Part 2. Mechanical seals; single mechanical seals; designation and material codes.

28040 Flat gaskets for flanged joints on equipment.

53505 Testing of elastomers; Shore A and D hardness testing.

70907 Piston rings; test of quality characteristics; terms, definitions, measuring procedure.

Japanese Standards (partial list)

B 0116:1966 Glossary of terms for packing and gaskets.

B 2401:1977 O-rings.

B 2402:1976 Oil seals.

B 2403:1977 V-packings.

B 2404:1977 Spiral wound gaskets for pipe flanges.

B 2405:1977 General rules for mechanical seals.

B 2406:1977 Installation and gland design for O-rings.

B 2407:1977 Back-up rings for O-rings.

B 8032:1970 Piston rings for reciprocating internal combustion engines.

K 6820:1977 Fluid sealants.

K 6885:1976 Unsintered polytetrafluoroethylene tapes for thread sealing.

R 3452:1959 Asbestos braided packing.

W 1516:1962 Packings and gaskets, pre-formed, petroleum hydraulic fluid resistant.

W1529:1973 Oil resistant O-ring packing and gasket for aircraft.

W 2006:1963 Packing; installation and gland design, hydraulic general, specification for.

Swedish (SMA) Standards (partial list)

705 American pipe threads; type dryseal. Internal cylindrical pipe threads NPSF.

706 American pipe threads; type dryseal. External tapered pipe threads NPTF.

707 American pipe threads; type dryseal. Internal tapered. 1961.

708 American pipe threads; type dryseal. External tapered PTF-SPL short. 1961.

1586 O-rings. Dimensions. 1974.

1587 O-rings. Materials. 1974.

1588 O-rings. Housings. 1974.

2290 Sealing rings for rotating shafts.

2291 Sealing rings for rotating shafts. Mountings.

2292 O-rings – flat back-up rings.

2293 O-rings – helical back-up rings.

2294 O-rings – installation with back-up rings.

3126 Gland packings for rotating shafts, square cross-section. Lubricated braided or plaited white asbestos packings. 1975.

3132 Gland packings for rotating shafts, square cross-section. PTFE-impregnated braided or plaited graphite yarn packings. 1975.

3133 Gland packings for rotating shafts, square cross-section. Braided or plaited PTFE-packings – with or without PTFE impregnations. 1975.

See also chapter on *Gaskets* (Section 2).

Glossary

Abrasion Resistance – The resistance of a rubber composition to wearing away by contact with a moving abrasive surface.

Aeration – Air (or gas) bubbles entrained or accumulated in a liquid.

Ageing – Change in characteristics of rubbers with time specifically influenced by environmental factors (*eg* light, heat, *etc*).

Air Side – The side of a seal which faces out wards or towards atmosphere, as opposed to the fluid being sealed.

Aniline Point – General indication of the aromatic content of an oil fluid determined as the lowest temperature at which the oil is miscible with an equal volume of aniline.

Anti-Extrusion Ring (Device) – A ring or similar device assembled with a seal to prevent the seal extruding into the clearance space.

Antioxidant – Additive in a rubber mix to resist oxidation.

Antiozonant – Additive in a rubber to resist degradation caused by ozone.

Automatic Seal – General term applied to describe seal designs that are pressure energized. More specifically it is used to classify certain types of flexible lip seals.

Axial Clearance – Clearance between a sealing element and the inside face of the cover.

Axial Interference – Clearance or dimensional difference between the id or od of a seal and the assembled rod (or shaft) or housing diameter respectively.

Back (of seal) – The side of a seal facing outwards or opposite to that facing the fluid being sealed.

Back-up Ring – Alternative description for an anti-extrusion ring.

Back-up Washer – Alternative name for anti-extrusion ring.

Bellows – Pseudo-static 'boot' type exclusion seal, also known as a garter.

Bellows Seal – Face-type seal with contact pressure generated by a bellows.

Braid – Hollow or solid structure of round, square or polygonal section constructed from interlocking filaments or yarn strand laid obliquely to the axis of the braid.

Braid-over-Braid – A braid produced by more than one pass through a multiple-carrier braiding machine.

Break-out Friction – Frictional force to be overcome to initiate movement. Specifically, *static friction.*

Brittler Point – Temperature at which an elastomer becomes brittle.

Bull Ring – A type of piston ring. A rigid or semi-rigid ring employed at one end or both ends of a packing to exclude extrusion of the packing into the clearance space.

Case – Metal component of a seal to which the sealing element is bounded, clamped or otherwise contained.

Centring Ring (Centering Ring) – An extension of a gasket designed to locate the gasket centrally on a flange.

Checking – Cracking or crazing of the surface of an elastomer due to the action of sunlight.

Chevron Seal – Seal ring (or ring set) of V-shaped cross-section.

Chrysotile – Fibrous magnesium silicate, or 'white' asbestos mineral.

Clearance – Dimensional difference between sealing element and related component.

Co-axial Seal – A composite seal in the form of two (or more) co-axial ring members.

Collar – Characteristic type of flexible lip rod seal, also known as a hat seal or hat ring.

Composite Seal – Seal comprising two (or more) separate materials, usually bonded together.

Compression Modulus – Ratio of compression stress to resulting compression strain expressed as percentage of the original dimension.

Compression Packing – Resilient sealing material, usually in plaited form, for fitting in a gland or stuffing box – and compressed to expand radially by tightening of the gland cover to produce a seal.

Compression Seal – Seal working on the principle of being compressed to fill the clearance space.

Compression Set – Permanent deformation of rubber after subscription to compression for a period of time. Specifically determined as the ratio of dimensional change to compression strain.

Conductivity – Elastomers are considered conductive when they possess a direct current resistivity of less than 10^5 ohm/cm.

Contact Force (Contact Load) – Total interface pressure between a seal and the adjacent surface.

Controlled Gap Seal – A seal designed to maintain constant clearance with shaft.

Copolymer – A polymeric material comprising molecules of two or more different kinds.

Cover – Member or casing protecting or strengthening a seal element.

Creep – Movement or deformation of a substance under the effect of prolonged load stress.

Crocidolite – Fibrous iron silicate, or 'blue' asbestos.

Crown Height – The height of a (gasket) sealing element above the surface of the retainer.

Cup – Specific type of piston seal defined by its geometry (*ie* cup-shaped); but also utilized as a description of a (seal) case.

Cure – Vulcanization process applied to rubbers.

Diametral Clearance – Difference between id of a seal and the shaft or rod diameter; or the id of a seal and its housing.

Dielectric Strength – The voltage required to puncture a sample of known thickness and is expressed as volts per mm of thickness.

Die Formed Ring – A packing ring mechanically compacted into an (apparently) homogeneous form.

Double Acting Seal – Seal for reciprocating movements capable of sealing with both directions of movement.

Dry Running – Rubbing contact without any liquid being present at the interface.

Durometer Hardness – Arbitrary measurement of hardness related to the resistance to penetration of an indentor point on a durometer.

Dynamic Friction – Friction generated when relative movement takes place between two contacting surfaces.

Dynamic Seal – A seal capable of working with relative movement (either reciprocating or rotary) between components being sealed.

Elasticity – Property inherent in elastomeric materials of readily returning to its original form when released from a deforming load.

Elastomer – Rubbers or rubber-like materials possessing elasticity.

Elongation – Strain defined as the extension between bench marks produced by a tensile force applied to specimen and is expressed as a percentage of the original distance between the marks. Ultimate elongation is the elongation at the moment of rupture.

Face – Front surface of a seal (where appropriate).

Face Seal – Seal embodying two faces in rubbing contact in a plane at right angles to the axis of the seal.

Filler Ring – Elastic ring assembled with a U-ring or V-ring to consolidate the section.

Finger Ring – A spring form with flexible fingers.

Flash Line – Exaggerated degree of flash due to clearance or gap between mould parts.

Flex Cracking – Surface cracks resulting from repeated flexual cycling.

Flex Fatigue Resistance – The ability to withstand fatigue resulting from repeated distortion by bending, extension or compression.

Flexual Modulus – Stress at a certain strain – not a ratio and not a constant, but merely the co-ordinate of a point on the stress-strain curve.

Flinger (Ring) – Washer-form ring mounted next to a gland or gland follower for directing any leakage away from the shaft.

Flinger Ring – Secondary seal element in the form of a ring generating 'windback' action.

Flow Crack – Imperfection in a moulding due to imperfect flow of material during moulding.

Flow Mark – Imperfection in a moulding due to incomplete flow of material in the mould.

Follower – See **Gland Follower.**

Front – The side of a seal facing the fluid to be sealed. Specifically applied to rotary shaft in seal descriptions.

Garter – Pseudo-static exclusion seal, usually in the form of elastomeric bellows.

Garter Spring – Helical wire spring of circular geometry fitted to a lip seal (specifically an oil seal) to enhance lip contact pressure.

Gasket – Static seal made from deformable sheet material sandwiched and compressed between two mating plane surfaces.

Gland – General description of housing or cavity for accommodating compression packings or sealing rings.

Gland Cover – Fixed gland member fitted on the non-pressure side of a gland to retain the seal against the action of pressure.

Gland Follower – Adjustable gland member which can be tightened to compress and expand radially the packing in a gland.

Hard Face – A facing of high hardness applied to softer materials.

Hardness – Resistance to indentation of a rubber, normally measured and specified in IRHD (International Rubber Hardness Degrees).

Hat Ring – Alternative name for a collar seal.

Head – Portion of a seal carrying the sealing edge.

Header – Ring of hard material used in conjuntion with seal ring(s) to locate the seal(s) and eliminate axial movement.

Heart Seal – Solid elastomeric ring seal with a heart shaped cross-section.

Heat Build-up – The temperature rise in a rubber body resulting from hysteresis.

Heel – The part of the seal cross-section adjacent to the clearance gap on the non-pressurized side; or adjacent to the shaft on the back of an oil seal.

Housing – Annular recess into which a shaft or rod seal is assembled.

H-Ring – Solid elastomeric seal ring of H-shaped cross-section.

Hydraulic Packing – A packing specifically designed for the sealing of hydraulic fluids in cylinders, *etc.*

Hysteresis – The percent energy lost per cycle of deformation, or 100% minus the resilience percentage. Results from internal friction and is manifest by the conversion of mechanical energy into heat.

Interference – Negative dimensional difference between a seal id or od and the final seal assembly diameter.

Junk Ring – Alternative name for an anti-extrusion ring.

Lantern Ring – Ring with radial ports located at an intermediate position in a gland to allow coolant or lubricant to be introduced.

Lead-in (Chamfer) – Chamber introduced in component(s) to facilitate assembly of seal on to a rod or shaft, or into a cylinder or housing.

Lip – The part or edge of a seal which forms the sealing surface.

Lip Opening Pressure – Air pressure required to lift a lip seal off its shaft and allow air leakage at the rate of 10 l/min.

Lip Seal – A seal where the sealing surface is in the form of a flexible lip.

Load – Actual pressure at sealing face of a seal, normally the sum of the interference load and fluid pressure acting on the seal.

Migration – Degradation products removed from an elastomeric sealing element and escaping to other parts of the system.

Mix – General description of a rubber compound formulation.

Modulus – Specifically the shape of the stress/strain curve for a material at a given elongation.

Modulus of elasticity – The ratio of stress to the strain produced by that stress when stress is proportional to strain. But in rubber, modulus measurements are made in comparison or shear, rather than in tension, and they are only valid for strains up to about 15%.

Moulding Shrinkage – Loss of dimension of a moulded product after removal from the mould and subsequent cooling.

Mould Mark – Imperfection in a mould duplicating a surface defect on the mould itself.

Multiple Seal – A seal set comprising two or more seal rings of sealing elements.

Neck Bush – Throttle bush fitted at the bottom of a stuffing box or gland.

Offset – Step or break in the surface of a moulded product due to faulty register of the mould.

Oil-side – The side of a seal facing the fluid being sealed (specifically applied in the case of an oil seal).

O-ring – Solid elastomer ring seal of circular cross-section.

Ozone Cracking – Surface cracking of rubber due to the degrading effect of ozone.

Packing – General name for a compression type dynamic seal housed in a gland. Also applicable to the materials used in this type of seal.

Panting – Movement between sealed surfaces of a static seal due to pressure fluctuations and insufficient clamping or tightening.

Pedestal Ring – Support for a U-ring seal.

Permanent Set – The amount of residual displacement in a rubber part after a distorting load has been removed.

Permeability – A measure of the ease with which a liquid or gas can pass through a rubber film.

Pitting – Surface voids produced by mechanical erosion (wear) or chemical action.

Plasticizer – Constituent of a rubber mix controlling the hardness and plasticity of the final product.

Polymer – Materials with long-chain molecules, *eg* natural rubber and synthetic rubbers.

Pseudo-static Seal – An exclusion seal (*eg* bellows or garter) for excluding dust, dirt, *etc*, but also capable of accommodating relative movement between the components to which it is attached.

Pusher-Type Seal – A mechanical seal in which the secondary seal is automatically pushed along the shaft or sleeve to compensate for wear.

Quad Ring – Solid elastomeric ring seal of modified circular cross-section giving four sealing ridges.

Quench – A neutral fluid introduced into a seal cavity to dilute fluid which may have leaked past the seal.

Radial Clearance – Clearance between the shaft and internal diameter of an oil seal cover.

Radial Interference – Negative dimensional difference between the radial dimension of a seal and its housing or space into which it is fitted.

Radial Load – Total load carried by the lip of a rotary shaft seal or rod seal.

Relaxation – Decrease in stress occurring with time under constant load or deformation.

Rectangular Seal – Solid elastomeric ring seal of rectangular cross-section.

Resilience – Energy recovery property of an elastomer under deformation cycles. Specifically the ratio of energy returned to energy input, per cycle of rapid deformation.

Rider Ring – Wear or load-carrying ring associated with some form of ring seal (usually a metallic ring or piston ring).

Ring – Any circular seal or seal element.

Ring Gasket – A flange gasket which lies wholly within the ring of bolts.

Rotary Seal – Seal type specifically suitable for sealing rotating shafts of rotary motions.

Rotary Seal Ring – Driven or rotating face of a mechanical seal.

Rubber – Elastomeric substance, either natural or synthetic.

Rubber Face – Rubber coating applied to a seal case to provide a sealing surface against the seal housing.

Run-in – Period of initial operation and wear during which a seal becomes properly bedded down.

Running Friction – Friction generated by a seal under dynamic operating conditions.

Scraper (Scraper Ring) – Heavy duty wiper seal to exclude grit and heavier contaminants with reciprocating rod movements.

Scorch – Premature curing of vulcanized rubber due to excessive heat.

Scuffing – Surface roughness produced by mechanical wear.

Sealing edge – The extreme section of an oil seal which provides the actual seal.

Sealing Element – Portion of the seal section or seal element covering the sealing edge.

Sealing Land – Flat portion of sealing edge of an oil seal after prolonged contact with shaft.

Seal Plate – Alternative description for a gland cover.

Seal Width – The overall axial dimensions of a seal.

Secondary Seal – O-ring, bellows or similar device which accommodates leakage from the primary seal of a mechanical face seal.

Separator – An intermediate ring of thin, stiff material which allows individual rings in a seal assembly to slide over one another.

Shaft Run-out – A dimension equivalent to twice the displacement of the axis of a shaft from normal at any particular point on the shaft under running conditions.

Shell – Case of an oil seal.

Single Acting Seal – Dynamic seal capable of sealing in one direction of movement only.

Slinger Ring – Alternative name for a flinger.

Slipper Seal – Co-axial seal comprising an O-ring and a hard low-friction PTFE bearing ring (slipper ring).

Soft Packing – Gland packing of soft resilient material.

Spew – Excess material forced from a mould during moulding process.

Split-ring Seal – Split rigid section ring (usually metallic) similar in form and principle to a piston ring.

Split Seal – Elastomeric seal ring split to facilitate assembly.

Squeeze – Deformation of a seal produced when assembled with an interference fit.

Static Friction – Instantaneous or "holding" friction of a seal under static conditions.

Stationary Seal Ring – Static face of a mechanical seal.

Stick-Slip – Jerky or irregular motion when a seal is operating under varying static-dynamic friction.

Stiction – Initial friction or break-out friction when motion is started.

Strain Relaxation or Creep – Is that characteristic of all elastomers to show gradual increase in deformation under constant load with passage of time. It is usually expressed as percent relative creep, which equals total deformation minus intial deformation divided by intitial deformation times 100.

Stress relaxation – The loss in stress when an elastomer is held at a constant strain over a period of time.

Stuffing Box – Alternative name for a gland for containing packings or seal rings.

Swell – Increase in volume of a seal or elastomeric material when in contact with a fluid.

Tear Resistance – The force per unit of thickness required to propagate a nick or cut in a direction normal to the direction of the applied force, or to initiate tearing in a direction normal to the direction of the stress.

Tensile strength – The force per unit of the original cross-sectional area which is applied at the time of the rupture of a specimen.

Tensile Stress – More commonly called "modulus", is the stress required to produce a certain elongation.

Tension Set – Increase in normal (unstressed) length of an elastomeric specimen after initial stretching and release.

Thread Stretch – Loosening of bolted assemblies or unions due to bedding down of threads.

Throttle Bush – Restrictive bush fitted at the bottom of a stuffing box or gland; also descriptive of a bushing seal.

Toroidal Seal – Alternative name for an O-ring.

Track – Mark made on a shaft by a rotary seal.

Trapped O-Ring – Static seal using an O-ring in a special groove.

Trim Diameter – Of an oil seal, without spring.

Trim face – Front surface of a rotary oil seal trimmed to an angle to firm the sealing edge.

Trimming Angle – Angle between the trimmed face of a seal lip and the seal axis.

Unidirectional Seal – A seal which provides fluid sealing from one side only.

Unirotational Seal – A rotary shaft seal designed for application with one direction of shaft rotation only.

Unit Seal – A seal ring consisting of a single ring, and not normally subject to axial compression.

U-Ring – Flexible lip seal of U-shaped cross-section.

V-Ring – Flexible lip seal of V-shaped cross-section – also known as a chevron seal.

Vulcanization – Heat process treatment for rubber to stabilize and harden it.

Wear Bond – Mark made by a rotary seal on a shaft.

Wear Ring – A ring of hard material associated with a seal or seal assembly and intended to take rubbing wear.

Weepage – Small amount of leakage from a seal, arbitrarily defined as leakage rate of less than one drop per minute.

SECTION 10

Buyer's Guide

Sub-section (a)

TRADE NAMES INDEX

ANGUS - Oil seals - Freudenberg Angus Ltd
APAX - Automotive aftermarket - Freudenberg Angus Ltd
AQUAFLOW - Packing - James Walker & Co Ltd
AQUAGRAF - Packing - James Walker & Co Ltd
AQ-SEAL - Seal designs - Shamban Europa Ltd
ARGOSELE - Hydraulic Seal - James Walker & Co Ltd
AROTHERM - Special valve seal (combination of pure graphite and PTFE) - Feodor Burgmann Dichtungswerke Gmbh & Co
BCX. - Asbestos-free fibre - Beldam Crossley Ltd
BELDAMITE - Anti-seize compounds - Beldam Crossley Ltd
BURASIL - Gasketing sheet, non-asbestos - Feodor Burgmann Dichtungswerke Gmbh & Co
CEFIGRAF - Expanded graphite - Le Carbone-Lorraine
CEFILAIR - Inflatable seals - Le Carbone-Lorraine
CEFIPRINT - Printed gaskets - Le Carbone-Lorraine
CFW - Seals, moulded parts - Freudenberg Angus Ltd
CHESTERTON - - A.W. Chesterton Co
CHEVRON - Hydraulic seal - James Walker & Co Ltd
COMBI-SLYDRING - Combined seal & bearing design - Shamban Europa (UK) Ltd
CORTECO - Seals, moulded parts - Freudenberg Angus Ltd
CROSSFLON - PTFE-based moulded materials - Beldam Crossley Ltd
DITHERSEAL - High speed/frequency rod seal - Tetrafluor Inc
DOUBLE DELTA - Seal designs - Shamban Europa (UK) Ltd
DOUBLE-PLY - Patented 2-ply metal bellows - EG&G Sealol
DOWPRINT - - Dowty Seals Ltd
DOWSHIELD - - Dowty Seals Ltd
DOWTEC - - Dowty Seals Ltd
DP SEALS - - Dowty Seals
DURAMID - Gland Packing - James Walker & Co Ltd
ENERRING (TM) - All metal static seals in O-ring, C-ring, V-ring configurations - Advanced Products (Seals & Gaskets) Ltd
ENERSEAL (TM) - PTFE spring energised seal for dynamic or static applications - Advanced Products (Seals & Gaskets) Ltd
EXCLUDER - - Shamban Europa UK Ltd
EXPANDING PACKING - Packing - James Walker Mfg Co
EXTENDED PERFORMANCE - Longer/lower leakage mechanical seals for critical services - EG&G Sealol
EZE-LON - - A.W. Chesterton Co

EZ-1 - Metal bellows cartridge seal - EG&G Sealol
FARGRAF - Stuffing box packing - Le Carbone-Lorraine
FILATRON - Glass fibre stuffing box packing, PTFE impregnated - Feodor Burgmann Dichtungswerke Gmbh & Co
FLEXIBOX - Mechanical seals - Flexibox International
FLEXICARB - Expanded graphite materials for use in gaskets and seals - Flexitallic Ltd
FLEXISERVICE - Seal reconditioning - Flexibox International
FLEXITALLIC - Metallic and semi-metallic industrial gaskets including spiral wound API ring type and clamp joints - Flexitallic Ltd
FLEXITE - Non metallic, non-asbestos spiral wound gasket filler material - Flexitallic Ltd
FLEXMET - Gland packing - James Walker & Co Ltd
FLUOBOND - Bondable PTFE - James Walker & Co Ltd
FLUOCORD - Valve packing - James Walker & Co LtdFluograf - Gland packing - James Walker & Co Ltd
FLUOLION - PTFE components - James Walker & Co Ltd
FORTRESS - Gland packing - James Walker & Co Ltd
FORTUNA - Gland packing - James Walker & Co Ltd
FREUDENBERG - Seals, moulded parts - Freudenberg Angus Ltd
G.P.A. - Mechanical seals - Le Carbone-Lorraine
GACO - Oil seals - Freudenberg Angus Ltd
GALON - PTFE seals - Freudenberg Angus Ltd
GASKOID - Jointing - James Walker & Co Ltd
GLYDRING - Seal designs - Shamban Europa UK Ltd
GOLDEND - - A.W. Chesterton Co
GRAYDAYE - Jointing - James Walker & Co Ltd
GULLIVER - Mechanical seals - Le Carbone-Lorraine
HACHSELE - Marine hatch seal - James Walker & Co Ltd
HALLITE HYSLIP - Thermoplastic elastomers - Hallite Seals International Ltd
HALLITE - Hydraulic and pneumatic seals - Hallite Seals International Ltd
HALLPRENE - Elastomers - Hallite Seals International Ltd
HATSEAL - Seal designs - Shamban Europa UK Ltd
HELICOFLEX - Metallic sealing rings with elastic core - Le Carbone-Lorraine
HEPHAISTOS - High temp seals & insulation - Latty International
HORNET - Gland packing - James Walker & Co Ltd
HYPAK - Hydraulic seal - James Walker & Co Ltd
HYTHANE - Polyurethane elastomers - Hallite Seals International Ltd
IMPAX - Mechanical seal - Pioneer Weston Ltd
INCOVAL - Valve packing - James Walker & Co Ltd
IPC - Rotary shaft seals - Freudenberg Angus Ltd
JOUDOL - Universal separating agent and lubricant - Feodor Burgmann Ltd
KALREZ - Perfluorelastomer - Du Pont de Nemours (Belgium)
KALREZ - Perfluoroelastomer seals - Du Pont de Nemours International SA
KENOL - Solid metal seals - Le Carbone-Lorraine
KLINGER SEALEX - PTFE joint sealant - R Klinger Ltd
KLINGERFLOW - PTFE materials (gaskets, sheaths, etc.) - R Klinger Ltd
KLINGERIT - Compressed asbestos jointing - R Klinger Ltd
KLINGERSIL - Compressed synthetic jointing - R Klinger Ltd
LATTYFLON - Packing - Latty International
LATTYGOLD - Gaskets & jointings - Latty International
LATTYGRAF - Packing - Latty International
LATTYGRAF REFLEX - Gaskets & jointings - Latty International
LATTYGRAF - Gaskets & jointings - Latty International
LATTYRING - Packing - Latty International
LATTYRIT - Gaskets & jointings - Latty International
LATTYSEAL - Mechanical seal - Latty International
LATTYSERVICE - Service products - Latty International
LATTYTEX - Packing - Latty International
LIDPACK - Chemical service hatch seals - Beldam Crossley Ltd
LION - Moulded rubber components - James Walker & Co Ltd
LIONCELLE - Jointing - James Walker & Co Ltd
LYCORA - Gland packing - James Walker & Co Ltd
MARATHON - PTFE seal - Pioneer Weston Ltd
MAYBURY - Joints - James Walker & Co Ltd
METAFLEX - Spiral-wound gaskets - James Walker & Co Ltd
METASTREAM - Flexible PT couplings - Flexibox International
MONOSEAL - - A.W. Chesterton Co

Sub-section (b)

CLASSIFIED INDEX OF MANUFACTURERS BY PRODUCT CATEGORY

Aircraft Seals
Tetrafluor Inc

Aerospace Seals
Bal Seal Engineering
Beldam Crossley Ltd
Dowty Seals Ltd
Du Pont de Nemours (Belgium)
Du Pont de Nemours (Switzerland)
Feodor Burgmann
Flexibox International
Freudenberg Angus Ltd
Furon Co
James Walker & Co Ltd
John Crane UK Ltd
Le Carbone-Lorraine
Seal Master Corporation
Tetrafluor Inc
Wills Engineered Polymers Ltd
Wynns' Precision UK Ltd
Wynn's Precision Canada Ltd

Automotive Seals
Asberit Ltda
Bal Seal Engineering
Dowty Seals Ltd
Du Pont de Nemours (Switzerland)
Feodor Burgmann
Freudenberg Angus
Hillman Newby Ltd
John Crane UK Ltd
John Crane Inc
Le Carbone-Lorraine
Pac-Seal Inc International
Pioneer Weston Ltd
Seal Master Corporation
Tetrafluor Inc
Truseal Ltd
Wynn's Precision Canada Ltd

Aluminium Smelter Seals
Tetrafluor Inc

Back-up Rings (PTFE)
Tetrafluor Inc

Back-up Rings
Wills Engineered Polymers Ltd

Barrier Fluid Systems
AES Engineering
Feodor Burgmann
Flexibox International
James Walker & Co Ltd
John Crane UK Ltd
John Crane Inc
Pioneer Weston Ltd
Seal Master Corporation
Tetrafluor Inc

Bellows Seals
A W Chesterton Co Ltd
Claron Hydraulic Seals Ltd
Du Pont de Nemours Ltd
Feodor Burgmann
Flexibox International
James Walker & Co Ltd
John Crane UK Ltd
John Crane Inc
Latty International
Le Carbone-Lorraine
Seal Master Corporation
Wynn's Precision UK Ltd

Pac-Seal Inc International
Pioneer Weston Ltd

Mechanical Face Seals (Special Design)
AES Engineering
British Seals & Rubber Mouldings Ltd
BT Tenute
Feodor Burgmann
Flexibox International
Freudenberg Angus Ltd
James Walker & Co Ltd
John Crane UK Ltd
John Crane Inc
Latty International
Le Carbone-Lorraine
Martin Merkel Gmbh
Pac-Seal Inc International
Pioneer Weston Ltd
Tetrafluor Inc

Mining Seals
AES Engineering
A W Chesterton Co Ltd
BT Tenute
Dowty Seals Ltd
Flexibox International
Hallite Seals International
Hillman Newby Ltd
James Walker & Co Ltd
John Crane UK Ltd
Mantek Manufacturing Ltd
Martin Merkel Gmbh
Pioneer Weston Ltd
Tetrafluor Inc
Wynn's Precision UK Ltd

Non-metallic Bearings
Hallite Seals International

Oil Seals
Beldam Crossley Ltd
British Seals & Rubber Mouldings Ltd
Dowty Seals Ltd
Du Pont de Nemours (Belgium)
Freudenberg Angus Ltd
Furon Co
Hillman Newby Ltd
James Walker Mfg
James Walker & Co Ltd
John Crane UK Ltd
John Crane Inc
Mantek Manufacturing Ltd
Martin Merkel Gmbh
Pioneer Weston Ltd
Tetrafluor Inc

O-Rings (Metallic)
Beldam Crossley Ltd
British Seals & Rubber Mouldings Ltd
Furon Co
James Walker & Co Ltd
John Crane Inc
Le Carbone-Lorraine
Mantek Manufacturing Ltd
Tetrafluor Inc
Wills Engineered Polymers Ltd
Wynn's Precision Canada Ltd

O-Rings (Rubber)
AES Engineering
Beldam Crossley Ltd
British Seals & Rubber Mouldings Ltd
Claron Hydraulic Seals Ltd
Dowty Seals Ltd
Du Pont de Nemours (Belgium)
Du Pont de Nemours (Switzerland)
Freudenberg Angus Ltd
Furon Co
Hallite Seals International
Hillman Newby Ltd
James Walker Mfg
James Walker & Co Ltd
John Crane Inc
Le Carbone-Lorraine
Mantek Manufacturing Ltd
Martin Merkel Gmbh
Seal Master Corporation
Wynn's Precision UK Ltd
Wynn's Precision Canada Ltd

O-Rings (Composite)
AES Engineering
A W Chesterton Co Ltd
Beldam Crossley Ltd
British Seals Ltd
Claron Hydraulic Seals Ltd
Dowty Seals Ltd
James Walker & Co Ltd
Le Carbone-Lorraine
Mantek Manufacturing Ltd
Martin Merkel Gmbh

O-Rings (Teflon)
Tetrafluor Inc

Paper & Pulp Mill Seals
Tetrafluor Inc

Packings (Automatic)
Asberit Ltda
A W Chesterton Co
Beldam Crossley Ltd
Feodor Burgmann
Freudenberg Angus Ltd
James Walker Mfg
James Walker & Co Ltd
Latty International
Martin Merkel Gmbh
Teadit Vertriebsgesellschaft mbH

Packings (Composite)
Asberit Ltda
A W Chesterton Co
Beldam Crossley Ltd
Feodor Burgmann
Freudenberg Angus Ltd
James Walker Mfg
James Walker & Co Ltd
John Crane Inc
Latty International

Sub-section (c)

ALPHABETICAL LIST OF MANUFACTURERS WITH ADDRESSES, TELEPHONE , FAX AND TELEX NUMBERS OF HEAD OFFICE

ADVANCED PRODUCTS (SEALS & GASKETS) LTD, Unit 25a, Number One Industrial Estate, Consett, Co Durham DH8 6SR
Telephone: 0207 500317 Facsimile: 0207 501210

A.E.S. ENGINEERING LTD, Mangham Road, Barbot Hall Industrial Estate, Rotherham S61 4RJ
Telephone: 0709 369966 Telex: 547834 AESS G
Facsimile: 0709 374 919

ASBERIT LTDA, Avenida Automovel Clube 8939, Colégio, Cep 21530, Rio De Janeiro, Brazil
Telephone: 21 372 7755 Telex: (21) 23720 ASBE
Facsimile: 21 371 2170

A W CHESTERTON COMPANY, 225 Fallon Road, Stoneham, MA 02180, USA
Telephone: 617 438 7000 Telex: 94-9417 Facsimile: 617 438 2930

BAL SEAL ENGINEERING, 620 W Warner Ave, Santa Ana, CA 92707, USA
Telephone: 714 557 5192 Facsimile: 714 241 0185

BELDAM CROSSLEY LTD, P O Box 7, Hill Mill, Temple Road, Astley Bridge, Bolton, Lancs BL1 6PB
Telephone: 0204 494711 Telex: 63149 Facsimile: 0204 40550

BRITISH SEALS & RUBBER MOULDINGS LTD, 4 Bridge Close, Oldchurch Road, Romford, Essex RM7 0AP
Telephone: 0708 725311 Telex: 897044 Facsimile: 0705 762603

BT TENUTE MECCANICHE ROTANTI S.P.A., Via Leonardo da Vinci 3/6, I-36057, Arcugnano, Italy
Telephone: 0444 566399 Telex: 481046 BT SEAL
Facsimile: 0444 543289

CLARON HYDRAULIC SEALS LTD, Station Road, Cradley Heath, Warley, West Midlands B64 6PN
Telephone: 021 559 9711

DOWTY SEALS LTD, Aschurch, Tewkesbury, Glos GL20 8JS
Telephone: 0684 299111 Telex: 43163 Facsimile: 0684 852210

DU PONT DE NEMOURS (BELGIUM), Polymer Products Department, Antoon Spinoystraat 6, B-2800 Mechelen, Belgium
Telephone: (15) 401411 Telex: 22554 Facsimile: (15) 411408

DU PONT DE NEMOURS INTERNATIONAL S.A., 2 Chemin du Pavillon, P O Box 50, CH-1218 Le Grand Saconnex, Geneva, Switzerland
Telephone: (022) 717 53 69 Telex: 415 777 DUP CH
Facsimile: (022) 717 51 09

EG&G SEALOL, Coronation Road, High Wycombe, Bucks HP12 3TP
Telephone: 0494 451661 Telex: 837007 Facsimile: 0494 452425

FEODOR BURGMANN DICHTUNGSWERKE GMBH & CO, Äussere Sauerlacher Str 6-10, D-8190, Wolfratshausen 1, P O Box 1260, West Germany
Telephone: 08171 23-0 Telex: 527801 Facsimile: 08171 23214

FLEXIBOX INTERNATIONAL, Nash Road, Trafford Park, Manchester M17 1SS
Telephone: 061 872 2484 Telex: 667281 Facsimile: 061 872 1654

FLEXITALLIC LTD, P O Box 3, Marsh Works, Dewsbury Road, Cleckheaton, West Yorkshire BD19 5BT
Telephone: 0274 851273 Telex: 518313 Facsimile: 0274 851386

FREUDENBERG ANGUS LTD, Coast Road, Wallsend, Tyne & Wear NE28 9NR
Telephone: 091 262 4551 Telex: 53628 GACO G
Facsimile: 091 263 4301

FURON CO., Mechanical Seal Div, 4412 Corporate Center Dr, Los Alamitos, CA 90720, USA
Telephone: 714 995 1818 Facsimile: 714 761 1270

HALLITE SEALS INTERNATIONAL LTD, 130 Oldfield Road, Hampton, Middlesex TW12 2HT
Telephone: 081 941 2244 Telex: 916028 Facsimile: 081 783 1669

HILLMAN NEWBY LTD, Thornleigh Trading Estate, Clee Road, Dudley, West Midlands DY2 8UE
Telephone: 0384 257351 Facsimile: 0384 236211

INMARCO INDUSTRIAL MAINTENANCE (P) LTD, 1117 Raheja Chambers, 213 Nariman Point, Bombay - 400 021, India
Telephone: 242802/242885 Telex: 11-4360 RMD IN
Facsimile: 91 22 287 2759

JAMES WALKER MFG CO, 511 W 195 Street, P O Box 467, Glenwood, IL 60466, USA
Telephone: 708 754 4020 Telex: 25-2460 Facsimile: 708 754 4058

JAMES WALKER & CO LTD, Lion Works, Woking, Surrey GU22 8AP
Telephone: 0483 757575 Telex: 859221 LIONWK G
Facsimile: 0483 755711

JOHN CRANE UK LTD, Crossbow House, 40 Liverpool Road, Slough SL1 4QX
Telephone: 0753 31122 Telex: 848176 Facsimile: 0753 73677

JOHN CRANE INC, 6400 W Oakton Street, Morton Grove, IL 60053, USA
Telephone: 708 967 2400 Facsimile: 708 967 2863

JOHN TALENT & CO LTD, 79 Clumber Road, Poynton, Stockport, Cheshire SK12 1NW
Telephone: 0625 873686

LATTY INTERNATIONAL, Rue Xavier Latty, 28160 Brou, France
Telephone: (33) 37.47.40.30 Telex: 760966 LATTY F
Facsimile: (33) 37.47.82.96

LE CARBONE-LORRAINE, 41 Rue Jean Jaures, F-92231, Gennevilliers, France
Telephone: (33) 1 47 92 43 00 Telex: F-620847 LCLGV
Facsimile: (33) 1 47 92 43 11

MANTEK (MANUFACTURING) LTD, Unit G, Holder Road, Aldershot, Hants GU12 4RH
Telephone: 0252 343335 Telex: 858418 Facsimile: 0252 343570

MARTIN MERKEL GMBH & CO KG, Sanitasstr. 17-21, Postfach 93 02 80, D-2102, Hamburg 93, West Germany
Telephone: 040 7511-0 Telex: 2163522 mer d.
Facsimile: 040 7511 440

PAC-SEAL INC INTERNATIONAL, 211 Frontage Road, Burr Ridge, IL 60521, USA
Telephone: 708 986 0430 Facsimile: 708 986 1033

PIONEER WESTON LTD, Douglas Green, Pendleton, Salford M6 6FT
Telephone: 061 736 5811 Telex: 667255 Facsimile: 061 736 5107

R KLINGER LTD, Sidcup, Kent DA14 5AG
Telephone: 081 300 7777 Telex: 28709 KGTLDN
Facsimile: 081 302 8145

SEAL MASTER CORPORATION, 368 Martinel Drive, Kent OH 44240, USA
Telephone: 216 673 8410 Facsimile: 216 673 8242

SHAMBAN EUROPA (UK) LTD, Ewart House, 5 St James's Terrace, Nottingham NG1 6FW
Telephone: 0602 411866 Telex: 377280 Facsimile: 0602 411278

TEADIT VERTRIEBSGESELLSHAFT MBH, Salzburger Strasse 17, A-6382, Kirchdorf, Tyrol, Austria
Telephone: 5352 3616 Telex: 51286 TEAD A
Facsimile: 5352 2951-83

TETRAFLUOR, INC, 2051 E Maple Avenue, El Segundo, CA 90245, USA
Telephone: 213 322 8030 Telex: 910-348-6697
Facsimile: 213 322 9148

TRUSEAL LTD, Molly Millars Bridge, Wokingham, Berskhire RG11 2RQ
Telephone: 0734 775454 Facsimile: 0734 771513

WILLS ENGINEERED POLYMERS LTD, Dunball Park, Dunball, Bridgewater, Somerset TA6 4TP
Telephone: 0278 684888 Telex: 46207 Facsimile: 0278 685051

WYNNS PRECISION UK LTD, Hoskins Place, Watchetts Road, Camberley, Surrey GU15 2PB
Telephone: 0276 24676 Facsimile: 0276 26411

WYNN'S PRECISION CANADA LTD, P O Box 190, Precision Drive, Orillia, Ontario, Canada L3V 6J3
Telephone: 705 325 2391 Telex: 06-875584 Facsimile: 705 325 5721

Editorial Index

Index to Advertisers